LONDON MATHEMATICAL SOCIETY
MONOGRAPHS NEW SERIES

Series Editors
H. G. Dales   Peter M. Neumann

LONDON MATHEMATICAL SOCIETY MONOGRAPHS NEW SERIES

Previous volumes of the LMS Monographs were published by Academic Press, to whom all enquiries should be addressed. Volumes in the New Series will be published by Oxford University Press throughout the world.

NEW SERIES

1. *Diophantine inequalities* R. C. Baker
2. *The Schur multiplier* Gregory Karpilovsky
3. *Existentially closed groups* Graham Higman and Elizabeth Scott
4. *The asymptotic solution of linear differential systems* M. S. P. Eastham
5. *The restricted Burnside problem* Michael Vaughan-Lee
6. *Pluripotential theory* Maciej Klimek
7. *Free Lie algebras* Christophe Reutenauer
8. *The restricted Burnside problem* (2nd edition) Michael Vaughan-Lee
9. *The geometry of topological stability* Andrew du Plessis and Terry Wall
10. *Spectral decompositions and analytic sheaves* J. Eschmeier and M. Putinar
11. *An atlas of Brauer characters* C. Jansen, K. Lux, R. Parker, and R. Wilson
12. *Fundamentals of semigroup theory* John M. Howie
13. *Area, lattice points, and exponential sums* M. N. Huxley
14. *Super-real fields* H. G. Dales and W. H. Woodin
15. *Integrability, self-duality, and twistor theory* L. Mason and N. M. J. Woodhouse
16. *Categories of symmetries and infinite-dimensional groups* Yu. A. Neretin
17. *Interpolation, identification, and sampling* Jonathan R. Partington
18. *Metric number theory* Glyn Harman
19. *Profinite groups* John S. Wilson
20. *An introduction to local spectral theory* K. B. Laursen and M. M. Neumann
21. *Characters of finite Coxeter groups and Iwahori-Hecke algebras* M. Geck and G. Pfeiffer
22. *Classical harmonic analysis and locally compact groups* Hans Reiter and Jan D. Stegeman
23. *Operator spaces* E. G. Effros and Z.-J. Ruan
24. *Banach Algebras and automatic continuity* H. G. Dales
25. *The mysteries of the real prime* M. J. Shai Haran

The Mysteries of the Real Prime

M. J. Shai Haran
Department of Mathematics
Technion–Israel Institute of Technology

CLARENDON PRESS • OXFORD
2001

OXFORD
UNIVERSITY PRESS
Great Clarendon Street, Oxford OX2 6DP

Oxford University Press is a department of the University of Oxford.
It furthers the University's objective of excellence in research, scholarship,
and education by publishing worldwide in

Oxford New York

Athens Auckland Bangkok Bogotá Buenos Aires
Cape Town Dar es Salaam Delhi Florence Hong Kong Istanbul
Karachi Kolkata Kuala Lumpur Madrid Melbourne Mexico City Mumbai
Nairobi Paris São Paulo Shanghai Singapore Taipei Tokyo Toronto Warsaw
and associated companies in Berlin Ibadan

Oxford is a registered trade mark of Oxford University Press
in the UK and in certain other countries

Published in the United States
by Oxford University Press Inc., New York

First published 2001

A catalogue record for this title is available from the British Library

Library of Congress Cataloging in Publication Data
(Data available)

ISBN 0 19 850868 9

10 9 8 7 6 5 4 3 2 1

Typeset by Integra Software Services Pvt. Ltd., Pondicherry, India
www.integra-india.com
Printed in Great Britain
on acid-free paper by
T.J. International Ltd,
Padstow, Cornwall

Preface

The continuum of real numbers $\mathbb{R}$ is the completion of the rational numbers $\mathbb{Q}$. As such, it is one of many completions: for each prime number p, we have the p-adic completion $\mathbb{Q}_p$, the p-adic numbers. The real numbers were discovered long before the discovery of the p-adic numbers. Ever since, the tendency has been to try and carry theories that were initially developed over the reals to the p-adic setting. Yet the p-adic numbers are simpler than the reals (e.g. $\Sigma_n a_n$ converges in $\mathbb{Q}_p$ if and only if a_n tends to 0), and in many respects are more natural—they are the majority of the completions of $\mathbb{Q}$. In this work, we reverse the direction and develop a 'p-adic attitude' towards the reals.

The problem with the real numbers is the singularity of the operation of addition: we carry the reminder in the wrong direction, from the smaller to the larger scale. Thus the interval $[-1, 1]$ is not closed under addition. We do not have a (discrete valuation) ring. Algebraic geometry, which is based on rings, fails to see the real prime, and hence fails to produce global theorems. Worse, we do not even have an abelian group under addition. Thus the tools of homological algebra, which are based on abelian categories, are not applicable to the truly global arithmetic situation. To some extant the real prime can be added to the finite primes, *deus ex machina*, as is done in Arakelov's theory. This gives a powerful tool, as demonstrated by the proof of the Mordell conjecture by Faltings, but this is just the tip of the iceberg. Many of the most basic concepts of geometry have their own mysterious and singular manifestations when carried over to the real setting.

For the p-adic numbers we have the subring of p-adic integers $\mathbb{Z}_p \subseteq \mathbb{Q}_p$, and we have reduction modulo powers of p, giving rise to inverse limit expressions for the $\mathbb{Z}_p$-points of various functors, in particular the projective line $\mathbb{P}^1$,

$$\mathbb{Q}_p \cup \{\infty\} = \mathbb{P}^1(\mathbb{Q}_p) = \mathbb{P}^1(\mathbb{Z}_p) = \varprojlim_N \mathbb{P}^1(\mathbb{Z}/p^N\mathbb{Z}); \tag{0.1}$$

the additive group $\mathbb{G}_a$,

$$\mathbb{Z}_p = \mathbb{G}_a(\mathbb{Z}_p) = \varprojlim_N \mathbb{G}_a(\mathbb{Z}/p^N\mathbb{Z}) = \varprojlim_N \mathbb{Z}/p^N\mathbb{Z}; \tag{0.2}$$

and the multiplicative group $\mathbb{G}_m$,

$$\mathbb{Z}_p^* = \mathbb{G}_m(\mathbb{Z}_p) = \varprojlim_N \mathbb{G}_m(\mathbb{Z}/p^N\mathbb{Z}) = \varprojlim_N (\mathbb{Z}/p^N\mathbb{Z})^* . \tag{0.3}$$

Our main concern will be to understand the real analogue of these. We cannot reduce points in $\mathbb{P}^1(\mathbb{R})$ modulo powers of the real prime, but we can 'reduce' complex-valued functions on $\mathbb{P}^1(\mathbb{R})$. We will see that the answer is given by the theory of the classical orthogonal polynomials: Jacobi for $\mathbb{P}^1$, Laguerre for $\mathbb{G}_a$, and Hermite for $\mathbb{G}_m$. As a by-product we develop a new approach to the classical orthogonal polynomials based on the Heisenberg relations 'up the ladder' (applicable to the most general polynomials of Askey and Wilson). For the p-adic numbers the inverse limit (0.1) means that we can view $\mathbb{P}^1(\mathbb{Q}_p)$ as the boundary of a tree. For the reals we shall view $\mathbb{P}^1(\mathbb{R})$ as the boundary of a graph which is not a tree, and the Heisenberg relations 'up the ladder' replace the lack of tree structure for the reals. Reversing the directions again, we shall obtain the p-adic analogue of the classical orthogonal polynomials. We would like to view these observations as prolegomena to a truly global arithmetic that will appear in the future as geometry—we can envision a reformulation of arithmetic algebraic geometry using complex-valued functions on varieties.

Physicists describing the real world using real numbers have found the Heisenberg relations and quantum mechanics to be the basic phenomena of the microscopic world. Yet these phenomena are mathematical ones, not physical ones, and have to do with the very nature of the real numbers. The situation resembles a painter depicting a landscape, who discovers instead the nature of his paints and brushes. Quantum mechanics and the uncertainty principle are valid over the p-adic numbers as well. Moreover, when one considers the global theory of the Heisenberg group, one sees that it is equivalent to the theory of elliptic curves and theta functions. This, in turn, gives rise to the 'q-world' (of basic hypergeometric functions, quantum groups, and q-commutative algebra) which can be used to interpolate between the p-adic world ($q = 1/p^n \longrightarrow 0$ as $n \to \infty$) and the real world ($q \to 1$). This gives an excellent source of insight into the analogies between the p-adic numbers and the reals.

The primes by themselves are just a set; there is nothing we can say about them as such. Interesting problems about the primes arise only when we throw them into other sets. We can throw them into $\mathbb{Z}/N\mathbb{Z}$, where all but finitely many of them will fall into $(\mathbb{Z}/N\mathbb{Z})^*$, and Dirichlet's theorem tells us that they are equidistributed within $(\mathbb{Z}/N\mathbb{Z})^*$. Note, however, that the finer formulation of Dirichlet's theorem, Čebotarev's density theorem, shows that

$$\frac{\#\{p \in [1,x], p \equiv a \bmod N\}}{\#\{p \in [1,x]\}} \longrightarrow \frac{1}{\varphi(N)} \quad \text{as } x \to \infty \quad (a \in (\mathbb{Z}/N\mathbb{Z})^*), \tag{0.4}$$

and this involves also, through $p \in [1, x]$, the distribution of the primes within the reals, and is equivalent to the analytic statement

$$L(1,\chi) = \sum_{n \geq 1} \frac{\chi(n)}{n} \neq 0 .$$

Here $(\mathbb{Z}/N\mathbb{Z})^*$ is the Galois group of the cyclotomic field, and Dirichlet's theorem can be generalized by throwing the primes into (conjugacy classes of) non-abelian

Galois groups. We can also let $N = p^m$, and take the limit as $m \to \infty$, obtaining the distribution of the primes within $\mathbb{Z}_p^* \subseteq \mathbb{Q}_p^*$. However, we remain almost ignorant about the most basic problem of the distribution of the primes within the continuum of real numbers $\mathbb{R}^+$. The prime number theorem tells us that

$$\pi(x) = \#\{p \in [1, x]\} \sim Li(x) = \int_0^x \frac{dy}{\log y} \quad \text{as } x \to \infty, \tag{0.5}$$

and this is equivalent to the analytic statement that $\zeta(1 + it) = `L(1, |\cdot|^{it})' \neq 0$. However, we have no control over the error term $\pi(x) - Li(x)$; the fact that this is of size $O(x^{1/2+\varepsilon})$, for any $\varepsilon > 0$ (or even of size $O(x^{1/2}\log x)$), is equivalent to the Riemann hypothesis: $\zeta_{\mathbb{A}}(s) = 0 \Rightarrow \mathrm{Re}(s) = 1/2$. The Riemann hypothesis has numerous formulations, some of which are recalled in the text, for example the equidistribution of the rationals $\mathbb{Q}^+$ within $\mathbb{R}^+$ (cf. §5.3). The problem is thus not with the primes, but with their interaction with the real numbers, and, as we shall see, this is intimately connected with the mysteries of the real prime.

In Chapter 1, we shall recall the well-known analogy between arithmetic and geometry, motivating the point of view that the real numbers $\mathbb{R} = \mathbb{Q}_\eta$ correspond to just another point η in spec $\mathbb{Z}$. We shall also recall briefly Arakelov's philosophy: for the 'real integers $\mathbb{Z}_\eta$', and a real vector space V, a $\mathbb{Z}_\eta$-structure on V corresponds to a metric on V, or equivalently, $\mathrm{GL}_N(\mathbb{Z}_\eta) = \mathrm{O}_N$, the orthogonal group.

In Chapter 2 we shall look at Tate's representation of the local components of the zeta function, motivating the point of view that the 'characteristic function of $\mathbb{Z}_\eta$' is the Gaussian $\phi_{\mathbb{Z}_\eta}(x) = e^{-\pi x^2}$. We shall study the gamma distribution both locally and globally.

In Chapter 3, in order to avoid the complicated Gaussian, we shall take a product of two gamma distributions on the plane, and project them onto the projective line, obtaining the beta distribution. The beta distribution is expressible in terms of the canonical distance function on $\mathbb{P}^1$, and we shall explain how the resulting beta integral is a generalization of Tate's gamma integral (obtained by taking one of the parameters in the beta integral to infinity). This gives our first insight into $\mathbb{Z}_\eta$ as the infinitesimal neighbourhoods of zero within the reals.

In Chapter 4 we shall begin by explaining the relations between (oriented) graphs and (Markov) chains. We shall describe some technical constructions with these (products and quotients) that will be used later on. We shall explain how, when the underlying graph is a tree, there is a bijection between chains and probability measures on the boundary. We then apply this principle to obtain the p-adic gamma chain on the tree corresponding to equation (0.2), and the p-adic beta chain on the tree corresponding to equation (0.1). We shall study quotients of these chains, noting that our chains are induced from their $\mathbb{Z}_p^*$-quotient, and that the underlying tree of this quotient is independent of the prime p.

In Chapter 5 we shall give some well-known real analogues of the p-adic beta chain: the usual hyperbolic plane; the $(N+1)$-regular tree corresponding to N-adic expansion of a real number; and the Farey tree corresponding to continued fraction expansion. We rule these out as the true analogue, demanding a discrete, canonical, and local chain with harmonic measure equal to the real beta distribution.

In Chapter 6 we shall enter the q-world, or Ramanujan's garden of q-series. We define the q-zeta function as the inverse of the q-shifted factorial

$$\zeta_{(q)}(\beta) = \prod_{n\geq 0}(1 - q^{\beta\,|\,n})^{-1},$$

and show how it interpolates between the p-adic and real components of the zeta function. For us, as from the days of Jacobi, q in the unit disc represents the elliptic curve $E_q = \mathbb{C}^*/q^{\mathbb{Z}}$ (with a fixed period on it), and β represents the point on the curve $q^\beta \bmod q^{\mathbb{Z}}$ (lifted to the universal covering). We shall explain the periodicities of $\zeta_{(q)}(\beta)$ and its compatibilities with isogenies. In the p-adic limit, $\zeta_p(\beta) = (1 - p^{-\beta})^{-1}$ is $(2\pi\mathrm{i}/\log p)$-periodic, and in the real limit, $\zeta_\eta(\beta) = \pi^{-\beta/2}\Gamma(\beta/2)$ is essentially $2\mathbb{Z}$-periodic. The p-adic and real limits are obtained by taking an infinite series of isogenies, and thus they appear on 'equal footing'. We shall then give a crash course on q-series, showing the dominant role of the q-beta sum as the common root of all q-identities that we shall need.

In Chapter 7 we shall give the basic q-chain, modelled on $\zeta_{(q)}(\beta)$, from which all our chains will be constructed by the processes of taking products, quotients, and specializing the parameters q and β. In particular, we shall construct the q-gamma chain and the q-beta chain. The underlying graph of these chains is not a tree, which leads to another uncertainty principle: the history of a particle walking in our chain cannot be retrieved from its position at a given time, and the representation of a point on the boundary as a path in our chain is far from being unique. As a result, the Martin kernel does not give an embedding or a projection between the Hilbert spaces corresponding to the harmonic measure on the boundary and its finite-dimensional approximations. To overcome this difficulty, we introduce the finite difference operator, and its adjoint, with a shift in the parameter β to $\beta + 1$, showing that they satisfy the Heisenberg relations 'up the ladder', hence giving a compatible orthogonal basis on the boundary (the little q-Jacobi polynomials) and on the finite approximation level (the q-Hahn polynomials) diagonalizing the Martin kernel. We show how these converge in the p-adic limit to the orthogonal basis associated with (a quotient of) the inverse limit of (0.1).

In Chapter 8 we shall specialize the q-beta chain to its real limit, obtaining the real beta chain: the space–time weighted random walk on $\mathbb{N} \times \mathbb{N}$. In the real limit we obtain the Heisenberg up the ladder operators, giving rise to the Jacobi polynomials, and their finite approximations—the Hahn polynomials. This is our real analogue of (the $\mathbb{Z}_p^*$-quotient of) the inverse limit (0.1). We shall also show how to add the $\{\pm 1\}$ real units into the game, obtaining the induced chain on $\mathbb{Z} \times \mathbb{Z}/\{\pm 1\}$ with boundary $\mathbb{P}^1(\mathbb{R})$ and harmonic measure the beta distribution. Thus, for us a real number is represented (non-uniquely!) as a path in $\mathbb{Z} \times \mathbb{Z}/\{\pm 1\}$, and $\mathbb{P}^1(\mathbb{Z}/\eta^N\mathbb{Z})$, the real analogue of $\mathbb{P}^1(\mathbb{Z}/p^N\mathbb{Z})$, turns out to be the $2N$-gon. Taking one of the parameters to infinity, we obtain the Laguerre basis of the Hilbert space associated with the gamma distribution (the real analogue of (0.2)). Similarly, taking both parameters to infinity, we obtain the (unweighted) space–time random walk on $\mathbb{N} \times \mathbb{N}$, the (ordinary) Heisenberg operators of creation–annihilation of quantum mechanics, giving rise to the Hermite polynomials and their finite approximations—the elementary symmetric functions. We shall interpret these as the missing real units: the real analogue of (0.3) corresponding to the principal p-adic units $\mathrm{e}^{2p\mathbb{Z}_p} = 1 + 2p\mathbb{Z}_p \subseteq \mathbb{Z}_p^*$.

In Chapter 9 we shall indicate how the local beta chains glue together to give the adelic beta 'chain'. The point is that the adelic 'chain' is not a Markov chain in the usual sense because its 'time' is the direct sum of the local times; we do not have one Markovian operator, but a commutative algebra generated by the local operators. We shall also indicate how our approach generalizes to higher dimensions resulting in the 'quantization' of $\mathbb{P}^r(\mathbb{R})$, $r > 1$.

In Chapter 10 we shall study the action of the multiplicative group on the Hilbert space corresponding to the Tate distribution, viewing $\phi_{\mathbb{Z}_p}$ as a cyclic vector, giving rise to the beta function at imaginary argument as the associated positive measure on the dual group. We do this on the q-level, obtaining the real and p-adic limits as a result. We shall then study the Fourier transform as applied to $\mathbb{Z}_p^*$-invariant functions, and its generalization to the context of the Tate distribution—the Fourier–Bessel transform. We analyse what is left from addition after the division by the units $\mathbb{Z}_p^*$—the symmetric convolution, including the q-analogue interpolating between the reals and the p-adic numbers. We show that the q-Fourier–Bessel operator takes the q-Laguerre basis to a very simple basis—the basic basis, which is independent of the parameter β, and which converges in the p-adic limit to the basis $\{\phi_{\mathbb{Z}_p}, \phi_{p^{-n}\mathbb{Z}_p^*}, n \geq 1\}$, and in the real limit to the basis $\{(\pi x^2)^n \cdot \mathrm{e}^{-\pi x^2}, n \geq 0\}$. We shall introduce the q-Heisenberg algebra with its natural action on the basic basis, and the translated action to the q-Laguerre basis, explaining how the Heisenberg relations up the ladder result from this. We shall describe briefly how the same approach works in the context of the beta measure, giving rise to the Fourier–Jacobi transform. When we break our spaces into even–odd functions with respect to the Fourier–Bessel(–Jacobi) operator, we obtain the pure gamma (beta) basis, whose Mellin transforms have all their zeros on the critical line. The pure basis gives a finite-dimensional approximation to our Hilbert space that can be characterized multiplicatively or analytically. Over the p-adic numbers we also have an additive characterization of the pure gamma spaces as band- and time-limited functions, and the real analogue of this is the space formed with the prolate spheriodal wave functions, the 'cut-off' basis (which converges to the pure gamma basis). We describe the q-analogue of these, obtaining in the real limit the generalization of the prolate spheriodal wave functions for all $\beta > 0$. Finally we shall show how the Askey–Wilson polynomials arise from the Heisenberg relations up the ladder, and how they give rise to the pure basis alternating in sign.

In Chapter 11 we shall describe the action of the quantum enveloping algebra associated with $\mathcal{S}\ell(2)$ on the spaces of the q-gamma measure. Since we want to specialize q to a real parameter, we are forced to work with the 'algebraic' (rather than 'formal') formulation, and with the $*$-structure corresponding to the quantum $\mathrm{SU}(1,1)$ (rather than $\mathrm{SL}_2(\mathbb{R})$). Moreover, since we are interested in the p-adic limit, we are forced to work with an algebra U_q which resembles, but is not isomorphic to, $U_q(SU(1,1))$; the latter algebra is invariant under $q \leftrightarrow q^{-1}$, and is modelled on the symmetric q-numbers

$$\langle\beta\rangle_{q^{1/2}} = \frac{q^{\beta/2} - q^{-\beta/2}}{q^{1/2} - q^{-1/2}},$$

while U_q is modelled on the non-symmetric q-numbers

$$[\beta]_q = \frac{1 - q^\beta}{1 - q}.$$

We show that U_q has all the nice properties of $U_q(SU(1,1))$ and more; for example, U_q is semi-simple and the finite-dimensional irreducible U_q representations correspond bijectively to those of its real limit $U(SU(1,1))$ (while for $U_q(SU(1,1))$ there is an inherent ± 1 ambiguity). The highest weight U_q-module is the space associated with the q-gamma measure. There are two natural subalgebras U_q^+ and U_q^- inside U_q: the algebra U_q^+ converges in the p-adic limit, and its action on the p-adic gamma measure space generates the Toeplitz algebra; the algebra U_q^- acts on the highest weight module through 'differential' operators induced from the action of the q-Heisenberg algebra. We shall then describe the Hopf algebra structure on U_q: taking the tensor product of two highest weight U_q-modules we show that the Clebsch–Gordan coefficients are given by the q-Hahn polynomials. Finally, we shall describe the quasi-triangular structure, showing how the universal R-matrix repairs the lost symmetry in the parameters of the q-Hahn basis.

In Chapter 12 we shall study the Heisenberg group and its fundamental representation, beginning with the Schrödinger model, showing the parallel theories over the p-adic numbers and the reals. We shall study the matrix coefficients and explain the uncertainty principle. Specializing to the matrix coefficients of $\phi_{\mathbb{Z}_p}$, we derive the local lattice model over $\mathbb{Q}_p$, and its real analogue, the Bargmann–Fock model, motivating the understanding of 'Heis$(\mathbb{Z}_\eta)$' as the 'holomorphic direction'. For the reals, $SL_2(\mathbb{Z}_\eta) = SO_2$ being commutative, we obtain a decomposition of the fundamental representation into one-dimensional subspaces, and the associated Hermite basis. After giving the decomposition of the p-adic fundamental representation with respect to the non-commutative $SL_2(\mathbb{Z}_p)$, we give some p-adic Hermite basis: these can be obtained either by taking commutative subgroups of $SL_2(\mathbb{Z}_p)$ associated with quadratic extensions of $\mathbb{Q}_p$, or better, by enlarging the commutant of $SL_2(\mathbb{Z}_p)$ respecting the $\mathbb{Z}_p^*$-decomposition. We shall then study the global lattice model and show how it induces local models, namely the Weil–Berezin model over the reals and its p-adic analogue. Finally we shall consider the automorphic forms on the Heisenberg group, $L_2(\mathrm{Heis}(\mathbb{Q})\backslash\mathrm{Heis}(\mathbb{A})/\mathrm{Heis}(\widehat{\mathbb{Z}})\mathrm{Heis}(\mathbb{Z}_\eta))$ with our understanding of Heis$(\mathbb{Z}_\eta)$, and show that it is equivalent to the study of theta functions and elliptic curves.

In Chapter 13 we shall study the global zeta function. We begin by giving the functional equation from the point of view of the Heisenberg group. We then study the 'explicit sums' which give the precise dictionary between the distribution of prime powers within $\mathbb{R}^+$ and the zeros of the zeta function. To put the contribution of the real prime to the explicit sums on an equal footing with the contribution of the finite primes, we shall follow our general approach of using the q-world. We define the q-Weil distribution on both the gamma and beta levels, and show how they give the limiting real and p-adic distributions. We shall then study the Eisenstein series, and recall its various connections with the zeta function. We give a crash course on the unramified principal series of representations of PGL_2 over the p-adic numbers and the reals, including the intertwining operator, the zonal spherical functions, and the Plancherel measure. We shall see how the constant term and its adjoint—the local Eisenstein integral—transform the hyperbolic Laplacian (or Hecke operator) into the infinitesimal generator of the commuting action of the multiplicative group, and give the q-analogue interpolating between the p-adic and the real theories. We then give a brief description of the global Eisenstein series, showing how its analytic behaviour is controlled by the intertwining operator, and in return how it controls the spectral theory of automorphic forms. We translate the explicit sums into formulae involving the normalization of the trace of the action of the multiplicative group $\mathbb{R}^+ = \mathbb{A}^*/\mathbb{A}^{(1)}$ (where $\mathbb{A}^{(1)} = \mathbb{Q}^* \cdot \widehat{\mathbb{Z}}^* = \ker\{|\cdot| : \mathbb{A}^* \longrightarrow \mathbb{R}^+\}$) on $\mathbb{A}/\mathbb{A}^{(1)}$ and on $\mathbb{P}^1(\mathbb{A})/\mathbb{A}^{(1)}$; the normalizations are given by the heat kernel of the number operator, the intertwining operator of PGL_2, or the cut-off. We end by giving a heuristic proof of the Riemann hypothesis. This proof works in the function field case and shows that the proof of the Riemann hypothesis is distinctly related to the mysteries of the real prime.

In this book we have fused some original observations with some well-known tenets, and we have not drawn a line between the two (an impossible task anyway). In fact the original idea of reducing functions modulo powers of the real prime was so strange at first that we have tried to connect this point of view with known tenets (special functions, quantum groups, eliptic curves, the Heisenberg group, etc.) resulting in this book. Many topics that have direct connections with our subject matter (e.g. the Jacobi and metaplectic groups) are not mentioned in order to keep this book to a reasonable length. Many calculations are left out so that the reader will not lose sight of the forest among the many trees. As a partial remedy for this, we have supplemented the book with an extensive bibliography, and some remarks on appropriate references for specific topics.

This research began when the author visited the Isaac Newton Institute for Mathematical Sciences, during the Special Program on L-functions and Arithmetic in the spring of 1993. Special thanks are due to the organizers of that historic conference, and especially to John Coates for creating a sanctuary for contemplation. Thanks are also due to my students, Ofer Porat, Hadar Porat (§9.2 is due to them), and Uri Onn.

Cambridge
May 2001

Contents

1 The real prime 1

1.1 Algebra and geometry 1

1.2 The real prime 3

1.3 The mystery 4

2 The zeta function and gamma distribution 6

2.1 The local zeta function 6

2.2 The gamma distribution 7

2.3 Remarks on global theory 8

3 The beta distribution 10

3.1 Phase space 10

3.2 The beta distribution 11

3.3 Special values 13

3.4 Remarks on the global theory 15

4 The *p*-adic hyperbolic point of view 17

4.1 Chains and trees 17

4.2 The *p*-adic gamma chain 21

4.3 The *p*-adic symmetric beta chain 24

4.4 The *p*-adic beta chain 28

5 Some real hyperbolic chains 31

5.1 The hyperbolic plane 31

5.2 *N*-adic expansion 31

5.3 Continued fraction expansion 31

6 Ramanujan's garden 34

6.1 The *q*-zeta function 34

6.2 Elliptic curves 36

6.3 *q*-series 37

7 The *q*-gamma and *q*-beta chains 44

7.1 The *q*-gamma chain 44

7.2 The *q*-beta chains 47

7.3 The Heisenberg relation and special basis 52

8 The real beta chains 65
- 8.1 The beta chain 65
- 8.2 The Heisenberg relations and the special basis 67
- 8.3 The real units 76

9 Global 'chains' and higher dimensions 84
- 9.1 Restricted direct products of chains 84
- 9.2 Higher dimensional beta chains 87

10 The Fourier transform 98
- 10.1 The Tate distribution and the beta function at imaginary argument 98
- 10.2 The Fourier–Bessel transform 101
- 10.3 Symmetric convolution 103
- 10.4 The basic basis and the Laguerre basis 105
- 10.5 The beta measure 110
- 10.6 The pure gamma basis and the cut-off basis 112
- 10.7 The pure beta basis 122
- 10.8 The Askey–Wilson polynomials 124

11 The quantum group SU(1,1) 129
- 11.1 The quantum enveloping algebra U_q 129
- 11.2 Highest weight representation 131
- 11.3 The Hopf algebra structure 135
- 11.4 The universal R-matrix 139

12 The Heisenberg group 142
- 12.1 The Heisenberg group and its fundamental representation 142
- 12.2 Twisted convolution and multiplication 145
- 12.3 Matrix coefficients 147
- 12.4 The local lattice model 149
- 12.5 Special basis 154
- 12.6 The global lattice model 161
- 12.7 Automorphic forms on the Heisenberg group 167

13 The Riemann zeta function 173
- 13.1 The Riemann zeta function and the theta function 173
- 13.2 The explicit sums 176
- 13.3 The Eisenstein series connections 181
- 13.4 The Eisenstein series and the intertwining operator 188
- 13.5 The Riesz potential connection 207

Bibliography 226

Index 233

1
The real prime

1.1 ALGEBRA AND GEOMETRY

The arithmetic of *number fields*, which are finite extensions of the field of rational numbers $\mathbb{Q}$, resembles the geometric theory of *function fields*, the fields of meromorphic functions on the one-dimensional objects of geometry—the curves. Over the complex numbers $\mathbb{C}$, a curve Σ is a compact Riemann surface, and its field of meromorphic functions $\mathbb{C}(\Sigma)$ is a finite extension of the field of rational functions $\mathbb{C}(\mathbb{P}^1) = \mathbb{C}(z)$. The geometry of Σ can be retrieved from the arithmetic of the field $\mathbb{C}(\Sigma)$. There is a one-to-one correspondence between the points of Σ and valuations of the field $\mathbb{C}(\Sigma)$; fixing $q \in (0,1)$, we have

$$\begin{aligned} \Sigma &\leftrightarrow \{|\cdot| : \mathbb{C}(\Sigma)^*/\mathbb{C}^* \twoheadrightarrow q^{\mathbb{Z}} \mid |f_1 + f_2| \le \max\{|f_1|, |f_2|\}\}, \\ \alpha &\leftrightarrow |\cdot|_\alpha, |f|_\alpha = q^{v_\alpha(f)}, \end{aligned} \tag{1.1.1}$$

where $v_\alpha(f) =$ order of zero (or minus order of pole) of f at α. For the simplest Riemann surface, the Riemann sphere $\mathbb{P}^1_{\mathbb{C}}$, this correspondence is given explicitly by

$$\begin{aligned} \mathbb{C} \cup \{\infty\} = \mathbb{P}^1(\mathbb{C}) &\leftrightarrow \{|\cdot| : \mathbb{C}(z)^*/\mathbb{C}^* \twoheadrightarrow q^{\mathbb{Z}} \mid |f_1 + f_2| \le \max\{|f_1|, |f_2|\}\}, \\ \mathbb{C} \ni \alpha &\leftrightarrow \left|(z-\alpha)^k \cdot \frac{n(z)}{d(z)}\right|_\alpha = q^k, \quad k \in \mathbb{Z}, \quad n(\alpha)d(\alpha) \neq 0, \\ \infty &\leftrightarrow \left|\frac{n(z)}{d(z)}\right|_\infty = q^{\deg d(z) - \deg n(z)}. \end{aligned}$$

At each point $\alpha \in \Sigma$ we have the discrete valuation ring

$$\mathcal{O}_{\Sigma,\alpha} = \{f \in \mathbb{C}(\Sigma) \mid |f|_\alpha \le 1\},$$

with its unique maximal ideal

$$m_{\Sigma,\alpha} = \{f \in \mathbb{C}(\Sigma) \mid |f|_\alpha < 1\}.$$

The completion $\mathbb{C}(\Sigma)_\alpha$ of the field $\mathbb{C}(\Sigma)$ with respect to the valuation $|\cdot|_\alpha$ is the field of fractions of the completion $\widehat{\mathcal{O}}_{\Sigma,\alpha}$ of $\mathcal{O}_{\Sigma,\alpha}$, where $\widehat{\mathcal{O}}_{\Sigma,\alpha}$ is described algebraically as the inverse limit

$$\widehat{\mathcal{O}}_{\Sigma,\alpha} = \varprojlim_N \mathcal{O}_{\Sigma,\alpha}/m_{\Sigma,\alpha}^N.$$

The embedding $\mathcal{O}_{\Sigma,\alpha} \hookrightarrow \widehat{\mathcal{O}}_{\Sigma,\alpha}$ (respectively $\mathbb{C}(\Sigma) \hookrightarrow \mathbb{C}(\Sigma)_\alpha$) corresponds to expanding a holomorphic (respectively meromorphic) function in a power series (respectively Laurent series) around the point α. The projection $\mathcal{O}_{\Sigma,\alpha} \twoheadrightarrow \mathcal{O}_{\Sigma,\alpha}/m^N_{\Sigma,\alpha}$ corresponds to giving the 'degree N polynomial approximation' of a function holomorphic at α. For $\Sigma = \mathbb{P}^1_{\mathbb{C}}$, we have $\mathbb{C}(\mathbb{P}^1)_\alpha = \mathbb{C}((t_\alpha))$, the field of Laurent series, and $\widehat{\mathcal{O}}_{\mathbb{P}^1,\alpha} = \mathbb{C}[\![t_\alpha]\!]$ the ring of power series, with $t_\alpha = (z - \alpha)$ for $\alpha \in \mathbb{C}$ and $t_\infty = 1/z$.

The (non-constant) maps between Riemann surfaces $f\colon \Sigma_1 \to \Sigma_2$ correspond bijectively to the field embeddings $f^\#\colon \mathbb{C}(\Sigma_2) \hookrightarrow \mathbb{C}(\Sigma_1)$. For example, with $f \in \mathbb{C}(\Sigma) \setminus \mathbb{C}$, we have the embedding $f^\#\colon \mathbb{C}(\mathbb{P}^1) = \mathbb{C}(z) \hookrightarrow \mathbb{C}(\Sigma), z \mapsto f$, which corresponds to the map $f\colon \Sigma \to \mathbb{P}^1_{\mathbb{C}}$. The fact that a meromorphic function $f\colon \Sigma \to \mathbb{P}^1_{\mathbb{C}}$ has the same number of poles and zeros (counting multiplicities) can be written as the product formula

$$\prod_{x\in\Sigma} |f|_x = 1\,.$$

The above dictionary between the geometry of curves and the arithmetic of their function fields works for curves over any algebraically closed field, such as $\overline{\mathbb{F}}_q$, the algebraic closure of a finite field $\mathbb{F}_q$. When we have a (smooth, projective) curve Σ defined over the finite field $\mathbb{F}_q$, we have the field of rational functions defined over $\mathbb{F}_q$, $f\colon \Sigma \to \mathbb{P}^1$; we denote this field by $\mathbb{F}_q(\Sigma)$. Embeddings of fields $f^\#\colon \mathbb{F}_q(\Sigma_2) \hookrightarrow \mathbb{F}_q(\Sigma_1)$ now correspond to maps of curves $f\colon \Sigma_1 \to \Sigma_2$ defined over $\mathbb{F}_q$. The valuations of $\mathbb{F}_q(\Sigma)$ correspond to the $\mathrm{Gal}(\overline{\mathbb{F}}_q/\mathbb{F}_q)$-orbits of valuations of $\overline{\mathbb{F}}_q(\Sigma)$, and so we have a one-to-one correspondence

$$\Sigma(\overline{\mathbb{F}}_q)/\mathrm{Gal}(\overline{\mathbb{F}}_q/\mathbb{F}_q) \leftrightarrow \Big\{ |\cdot|\colon \mathbb{F}_q(\Sigma)^*/\mathbb{F}_q^* \twoheadrightarrow q^{\mathbb{Z}} \,\Big|\, |f_1 + f_2| \le \max\{|f_1|, f_2|\} \Big\}. \quad (1.1.2)$$

For example, the valuations on $\mathbb{F}_q(\mathbb{P}^1) = \mathbb{F}_q(z)$ correspond now to $\mathrm{Gal}(\overline{\mathbb{F}}_q/\mathbb{F}_q)$-orbits on $\mathbb{P}^1(\overline{\mathbb{F}}_q) = \overline{\mathbb{F}}_q \cup \{\infty\}$. If $\alpha \in \overline{\mathbb{F}}_q$, its orbit $\{\sigma_1(\alpha), \ldots, \sigma_d(\alpha)\}$ corresponds to the irreducible polynomial $p(z) = \prod_i (z - \sigma_i(\alpha))$, and thus the set of valuations of $\mathbb{F}_q(z)$ can alternatively be described as

$$\{\infty\} \cup \{\text{primes of } \mathbb{F}_q[z]\}\,.$$

Given a valuation $|\cdot|_\alpha$ of $\mathbb{F}_q(\Sigma)$, we have the associated discrete valuation ring $\mathcal{O}_{\Sigma,\alpha} = \{f \in \mathbb{F}_q(\Sigma) \,|\, |f|_\alpha \le 1\}$, with its maximal ideal $m_{\Sigma,\alpha}$, and now $\mathcal{O}_{\Sigma,\alpha}/m_{\Sigma,\alpha} = \mathbb{F}_{q^d}$, where d is $\deg(\alpha)$. The completion of $\mathcal{O}_{\Sigma,\alpha}$ can be described as a power series ring,

$$\widehat{\mathcal{O}}_{\Sigma,\alpha} = \varprojlim_N \mathcal{O}_{\Sigma,\alpha}/m^N_{\Sigma,\alpha} = \mathbb{F}_{q^d}[\![t_\alpha]\!]\,,$$

where t_α is a uniformizer at α. The completion of $\mathbb{F}_q(\Sigma)$ with respect to $|\cdot|_\alpha$ is the fraction field: $\mathbb{F}_q(\Sigma)_\alpha = \mathbb{F}_{q^d}((t_\alpha))$. The valuation $|\cdot|_\alpha$ corresponds to d points in $\Sigma(\overline{\mathbb{F}}_q)$, and the normalized valuation is defined by $\|f\|_\alpha = |f|^d_\alpha$. We now have the product formula

$$1 = \prod_{x\in\Sigma(\overline{\mathbb{F}}_q)} |f|_x = \prod_{\alpha\in\Sigma(\overline{\mathbb{F}}_q)/\mathrm{Gal}(\overline{\mathbb{F}}_q/\mathbb{F}_q)} \|f\|_\alpha \quad \text{for all } f \in \mathbb{F}_q(\Sigma)^*\,.$$

1.2 THE REAL PRIME

In arithmetic we replace the fields $\mathbb{F}_q(\Sigma)$ by number fields, and try to retrieve the geometry behind them. We shall concentrate on the simplest number field, the field of rational numbers $\mathbb{Q}$. Letting Prime $(\mathbb{Z}) = \{2, 3, 5, 7, \ldots\}$ denote the set of prime numbers, fixing $q \in (0, 1)$, and imitating the algebra–geometry dictionary (1.1.2), we obtain

$$\begin{aligned} \text{Prime}(\mathbb{Z}) &\leftrightarrow \{|\cdot|: \mathbb{Q}^* \twoheadrightarrow q^{\mathbb{Z}} \,|\, |f_1 + f_2| \leq \max\{|f_1|, |f_2|\}\} \\ p &\leftrightarrow |p^k \cdot n/d|_p = q^k, \quad k \in \mathbb{Z},\ p \nmid nd\,. \end{aligned} \tag{1.2.1}$$

Replacing q by $q' = q^\alpha$ has the effect of changing the metric $|\cdot|$ to the equivalent metric $|\cdot|^\alpha$, so the set in (1.2.1) can alternatively be described as the valuations of $\mathbb{Q}$ modulo equivalence. There is no natural q, and we shall always normalize by taking $q = 1/p$ for the metric $|\cdot|_p$ corresponding to a prime p. The set

$$\{|\cdot|: \mathbb{Q}^* \to \mathbb{R}^+ \,|\, |f_1 + f_2| \leq \max\{|f_1|, |f_2|\}\} \tag{1.2.2}$$

is the cone on the set Prime $(\mathbb{Z})$: for each prime p we have a line $\{|\cdot|_p^\alpha, \alpha \in \mathbb{R}^+\}$, and these lines are joined together at the 'generic point' g, where $|n/d|_g \equiv 1 \equiv |n/d|_p^0$ is the trivial valuation. Relaxing the conditions in (1.2.2) slightly, the set

$$\{|\cdot|: \mathbb{Q}^* \to \mathbb{R}^+ \,|\, |f_1 + f_2| \leq |f_1| + |f_2|\} \tag{1.2.3}$$

is the cone of (1.2.2) enriched with the half-line $\{|\cdot|_\eta^\alpha,\ 0 \leq \alpha \leq 1\}$, where η, *the real prime*, corresponds to the usual absolute value: $|\pm n/d|_\eta = n/d,\ n \cdot d > 0$. The completion of $\mathbb{Q}$ with respect to the valuations $|\cdot|_p$ is the field of p-adic numbers $\mathbb{Q}_p$, and the completion with respect to $|\cdot|_\eta$ is the field of real numbers $\mathbb{Q}_\eta = \mathbb{R}$. We shall write '$p \neq \eta$' for the finite primes, and '$p \geq \eta$' for all primes *including* the real prime. We now have the product formula

$$1 = \prod_{p \geq \eta} |f|_p \quad \text{for } f \in \mathbb{Q}^*\,.$$

Thus the mystery of the real prime η is that we have only the ℓ_1 triangle inequality (1.2.3), and not the ℓ_∞ inequality (1.2.2). Thus,

addition is a more singular operation at the real prime.

For a finite prime $p \neq \eta$, we have a discrete valuation ring

$$\mathcal{O}_p = \{f \in \mathbb{Q} \,|\, |f|_p \leq 1\}\,,$$

with its maximal ideal $p \cdot \mathcal{O}_p$, residue field $\mathcal{O}_p/p \cdot \mathcal{O}_p = \mathbb{Z}/(p) = \mathbb{F}_p$, and the associated completion, the ring of p-adic integers

$$\mathbb{Z}_p = \varprojlim_N \mathbb{Z}/(p^N)\,. \tag{1.2.4}$$

In simple terms, elements of $\mathbb{Z}_p$ are represented as 'power series'

$$a_0 + a_1 \cdot p + a_2 \cdot p^2 + a_3 \cdot p^3 + \cdots, \quad 0 \leq a_i < p\,.$$

The p-adic numbers $\mathbb{Q}_p$ are the fraction field of $\mathbb{Z}_p$, $\mathbb{Q}_p = \mathbb{Z}_p[1/p]$, so that elements of $\mathbb{Q}_p$ are represented as 'Laurent series'

$$a_{-N} \cdot p^{-N} + \cdots + a_{-1} \cdot p^{-1} + a_0 + a_1 \cdot p + \cdots, \quad 0 \le a_i < p.$$

We get from (1.2.4) an inverse limit expression for $\mathrm{G}(\mathbb{Z}_p)$, the $\mathbb{Z}_p$-points of various functors G. In particular for the projective line $\mathrm{G} = \mathbb{P}^1$, we obtain

$$\mathbb{Q}_p \cup \{\infty\} = \mathbb{P}^1(\mathbb{Q}_p) = \mathbb{P}^1(\mathbb{Z}_p) = \varprojlim_N \mathbb{P}^1(\mathbb{Z}/p^N), \tag{1.2.5}$$

and for the multiplicative group $\mathrm{G} = \mathbb{G}_m$ we obtain the p-adic units

$$\mathbb{Z}_p^* = \mathbb{G}_m(\mathbb{Z}_p) = \varprojlim_N (\mathbb{Z}/p^N)^*. \tag{1.2.6}$$

1.3 THE MYSTERY

The basic question emerges: what is $\mathbb{Z}_\eta$?

If we try to follow the definition $\mathbb{Z}_p = \{x \in \mathbb{Q}_p \mid |x|_p \le 1\}$, we obtain the interval $[-1,1] = \{x \in \mathbb{Q}_\eta \mid |x|_\eta \le 1\}$, but in the absence of the ℓ_∞-inequality this interval is not closed under addition. We do not have a (discrete valuation) ring, and, worse, we do not have an abelian group under addition. Thus Grothendieck's algebraic geometry, which is based on rings, fails to 'see' the real prime, and hence fails to produce truly global arithmetical theorems. Similarly we do not have an abelian category of sheaves over spec $\mathbb{Z} \cup \{\eta\}$, and the possibilities of applying the tools of homological algebra and K-theory collapse. We have a few hints, however.

Denote the 'standard' maximal symmetric metric on $\mathbb{Q}_p^{\oplus N}$ by

$$|x|_p = |x_1, \ldots, x_N|_p = \begin{cases} \max\{|x_1|_p, \ldots, |x_N|_p\} & p \neq \eta, \\ (x_1^2 + \cdots + x_N^2)^{1/2} & p = \eta. \end{cases} \tag{1.3.1}$$

We have for $p \neq \eta$

$$\mathrm{GL}_N(\mathbb{Z}_p) = \left\{ g \in \mathrm{GL}_N(\mathbb{Q}_p) \mid |gx|_p = |x|_p \right\}, \tag{1.3.2}$$

a maximal compact subgroup of $\mathrm{GL}_N(\mathbb{Q}_p)$. Taking (1.3.2) as the definition of $\mathrm{GL}_N(\mathbb{Z}_p)$ for the real prime $p = \eta$, we obtain

$$\mathrm{GL}_N(\mathbb{Z}_\eta) = \left\{ g \in \mathrm{GL}_N(\mathbb{R}) \mid |gx|_\eta = |x|_\eta \right\} = \mathrm{O}_N, \tag{1.3.3}$$

the orthogonal group, a maximal compact subgroup of $\mathrm{GL}_N(\mathbb{Q}_\eta)$.

We shall similarly use the notations

$$\mathrm{SL}_N(\mathbb{Z}_\eta) = \mathrm{SO}_N, \quad \mathrm{PGL}_N(\mathbb{Z}_\eta) = \mathrm{PO}_N = \mathrm{O}_N/\{\pm \mathrm{I}\}, \quad \text{etc}. \tag{1.3.4}$$

Thus we know what $\mathrm{G}(\mathbb{Z}_\eta)$ is for G as above, but we do not know what $\mathrm{G}(\mathbb{Z}_\eta)$ is for the additive group $\mathrm{G} = \mathbb{G}_a$.

The above hint leads to Arakelov's philosophy that for the real prime η, a '$\mathbb{Z}_\eta$-structure' on a $\mathbb{Q}_\eta$-vector space is given by a positive-definite quadratic form (for a *complex* place η of a number field, $\mathrm{GL}_N(\mathbb{Z}_\eta) = \mathrm{U}_N$, the unitary group, and a '$\mathbb{Z}_\eta$-structure' on a $\mathbb{C}$-vector space is given by a Hermitian form).

In this work we shall answer the following question: what is the real analogue of the inverse limits (1.2.4), (1.2.5), and (1.2.6)? That is, what is the meaning of reduction modulo powers of the real prime η^N?

Our answer will be presented in Chapter 8. We shall first obtain the answer for the projective line (1.2.5) via the beta chain and Jacobi basis; from it we shall deduce the additive theory (1.2.4) via the gamma measure and Laguerre basis, and finally, in §8.3, we shall obtain the multiplicative theory (1.2.6) via the Hermite basis.

2

The zeta function and gamma distribution

2.1 THE LOCAL ZETA FUNCTION

The mystery of the real prime is reflected in the difference between the local components of the (completed) zeta function

$$\zeta_{\mathbb{A}}(s) = \prod_{p \geq \eta} \zeta_p(s), \quad \mathrm{Re}(s) > 1\,. \tag{2.1.1}$$

We summarize these differences as follows:

$\underline{p \neq \eta}$	$\underline{p = \eta}$	
$\zeta_p(s) = (1 - p^{-s})^{-1}$	$\zeta_\eta(s) = \pi^{-s/2}\Gamma(s/2)$	(2.1.2)
periodicity: $\zeta_p\left(s + \dfrac{2\pi \mathrm{i}}{\log p}\right) = \zeta_p(s)$	$\zeta_\eta(s+2) = \zeta_\eta(s) \cdot (s/2\pi)$	(2.1.3)
poles: $\dfrac{2\pi \mathrm{i}}{\log p}\mathbb{Z}$	$-2\mathbb{N} = \{0, -2, -4, \ldots\}$	(2.1.4)
duplication:		(2.1.5)
$\zeta_p(Ns) = \prod_{0 \leq k < N} \zeta_p\left(s + \dfrac{2\pi \mathrm{i}}{\log p} \cdot \dfrac{k}{N}\right)$	$\zeta_\eta(Ns) = \left(\dfrac{N^s}{2}\right)^{N/2} \cdot \prod_{0 \leq k < N} \zeta_\eta\left(s + 2 \cdot \dfrac{k}{N}\right)$	

For example, for $N = 2$:

$\dfrac{\zeta_p(2s)}{\zeta_p(s)} = (1 + p^{-s})^{-1}$	$\dfrac{\zeta_\eta(2s)}{\zeta_\eta(s)} = 2^{s-1}\zeta_\eta(s+1)$	(2.1.6)

A major difference is the asymptotic behaviour for $\mathrm{Re}(s) \to \infty$: for $p \neq \eta$, $\zeta_p(s) \approx 1$, while for the real prime we have Stirling's formula

$$\zeta_\eta(s) \approx \left(\frac{s}{2\pi}\right)^{(s-1)/2} \mathrm{e}^{-s/2}\sqrt{2}\exp\left(\sum_{n \geq 1} \frac{\mathrm{B}_n}{2n(2n+1)} s^{1-2n}\right). \tag{2.1.7}$$

It follows that we have for β fixed

$$\frac{\zeta_\eta(s+\beta)}{\zeta_\eta(s)} \approx \left(\frac{s}{2\pi}\right)^{\beta/2}, \quad \mathrm{Re}(s) \to \infty . \tag{2.1.8}$$

2.2 THE GAMMA DISTRIBUTION

Another hint concerning $\mathbb{Z}_\eta$ comes from Tate's representation of $\zeta_p(s)$, for all $p \geq \eta$, as the Mellin transform of the *vacuum* $\phi_{\mathbb{Z}_p}$:

$$\text{gamma integral: } \zeta_p(s) = \int_{\mathbb{Q}_p^*} \phi_{\mathbb{Z}_p}(x)|x|_p^s \,\mathrm{d}^*x, \quad \mathrm{Re}(s) > 0\,,\; p \geq \eta . \tag{2.2.1}$$

The measure d^*x is the multiplicative Haar measure, $\mathrm{d}^*x = \zeta_p(1)|x|_p^{-1}\mathrm{d}x$, normalized by $\mathrm{d}^*x(\mathbb{Z}_p^*) = 1$ for $p \neq \eta$.

Here, for $p \neq \eta$, $\phi_{\mathbb{Z}_p}$ is the characteristic function of $\mathbb{Z}_p$, while for the real prime $p = \eta$, the 'characteristic function of $\mathbb{Z}_\eta$' turns out to be the Gaussian

$$\phi_{\mathbb{Z}_\eta}(x) = \mathrm{e}^{-\pi x^2} = \sum_{n\geq 0} \left(\operatorname*{Res}_{s=-2n} \zeta_\eta(s)\right) \cdot x^{2n} . \tag{2.2.2}$$

The Gaussian is not a projection in the algebra $\mathrm{C}_0(\mathbb{R})$; we have only

$$\phi_{\mathbb{Z}_\eta}(x)^2 = \phi_{\mathbb{Z}_\eta}\left(\sqrt{2}x\right), \tag{2.2.3}$$

and for a real number x 'being a member of $\mathbb{Z}_\eta$' has only a probabilistic meaning, $\phi_{\mathbb{Z}_\eta}(x) \in (0,1]$, not a $\{0,1\}$ answer.

Using the central limit theorem, we can write

$$\int_{\mathbb{R}} \mathrm{d}x\, \phi_{\mathbb{Z}_\eta}(x) f(x) = \lim_{N\to\infty} \frac{1}{2^N} \int_{-1}^{1} \mathrm{d}x_1 \cdots \int_{-1}^{1} \mathrm{d}x_N f\left(\frac{x_1 + \cdots + x_N}{\sqrt{\frac{2}{3}\pi N}}\right),$$

so we 'close' the interval $[-1,1]$ under addition by taking the Nth convolution powers of its characteristic function, and we scale by a factor $\sqrt{2\pi N/3}$.

For all $p \geq \eta$, we have the *Tate distribution*:

$$\tau_p^\beta = |x|_p^\beta \frac{\mathrm{d}^*x}{\zeta_p(\beta)}, \quad \mathrm{Re}(\beta) > 0 . \tag{2.2.4}$$

It is the unique (up to a constant multiple) β-homogeneous distribution on $\mathbb{Q}_p$; it is normalized by $\tau_p^\beta(\phi_{\mathbb{Z}_p}) = 1$ and has an analytic continuation to all β.

For real and positive $\beta \geq 0$, τ_p^β is a positive measure, and we have a probability measure, the *gamma distribution*:

$$\tau_{\mathbb{Z}_p}^\beta = \phi_{\mathbb{Z}_p} \cdot \tau_p^\beta = \phi_{\mathbb{Z}_p}(x)|x|_p^\beta \frac{\mathrm{d}^*x}{\zeta_p(\beta)} . \tag{2.2.5}$$

We denote the associated Hilbert spaces by

$$\widetilde{\mathscr{H}}_p^\beta = \mathrm{L}_2\Big(\mathbb{Q}_p\,, \tau_p^\beta\Big), \quad \widetilde{\mathscr{H}}_{\mathbb{Z}_p}^\beta = \mathrm{L}_2\Big(\mathbb{Z}_p\,, \tau_{\mathbb{Z}_p}^\beta\Big). \tag{2.2.6}$$

We note that τ_p^β and $\phi_{\mathbb{Z}_p}$ are $\mathbb{Z}_p^*$-invariant, where $\mathbb{Z}_\eta^* = \{\pm 1\}$. We shall be interested in the subspaces of $\widetilde{\mathscr{H}}_p^\beta$ and $\widetilde{\mathscr{H}}_{\mathbb{Z}_p}^\beta$ consisting of $\mathbb{Z}_p^*$-invariant functions; we denote them by

$$\mathscr{H}_p^\beta = \Big(\widetilde{\mathscr{H}}_p^\beta\Big)^{\mathbb{Z}_p^*}, \quad \mathscr{H}_{\mathbb{Z}_p}^\beta = \Big(\widetilde{\mathscr{H}}_{\mathbb{Z}_p}^\beta\Big)^{\mathbb{Z}_p^*}. \tag{2.2.7}$$

We have unitary representations of $\mathbb{Q}_p^*$ on $\widetilde{\mathscr{H}}_p^\beta$, and of $\mathbb{Q}_p^*/\mathbb{Z}_p^*$ on $\mathscr{H}_p^\beta$, given by

$$\pi_p^\beta(a)\varphi(x) = |a|_p^{-\beta/2}\varphi(a^{-1}x)\,. \tag{2.2.8}$$

On the spaces $\mathscr{H}_{\mathbb{Z}_p}^\beta$ with $p \neq \eta$, we do not have a unitary representation, but only the isometry (respectively co-isometry) for $|a|_p \leq 1$, (respectively for $|a|_p \geq 1$):

$$\pi_{\mathbb{Z}_p}^\beta(a)\varphi(x) = \phi_{\mathbb{Z}_p}(a^{-1}x)|a|_p^{-\beta/2}\varphi(a^{-1}x)\,, \tag{2.2.9}$$

and $\pi_{\mathbb{Z}_p}^\beta(a_1)\pi_{\mathbb{Z}_p}^\beta(a_2) \equiv \pi_{\mathbb{Z}_p}^\beta(a_1 \cdot a_2)$ only modulo finite rank operators on $\mathscr{H}_{\mathbb{Z}_p}^\beta$.

For $p = \eta$, since $\phi_{\mathbb{Z}_\eta}$ is supported on all of $\mathbb{R}$, we have a unitary representation also on $\mathscr{H}_{\mathbb{Z}_\eta}^\beta$,

$$\pi_{\mathbb{Z}_\eta}^\beta(a)\varphi(x) = \frac{\phi_{\mathbb{Z}_\eta}(a^{-1}x)^{1/2}}{\phi_{\mathbb{Z}_\eta}(x)^{1/2}}|a|_\eta^{-\beta/2}\varphi(a^{-1}x)\,. \tag{2.2.10}$$

2.3 REMARKS ON GLOBAL THEORY

In the global theory we work over the ring of adeles $\mathbb{A}$. The global Tate distribution $\tau_{\mathbb{A}}^\beta = \otimes_{p\geq\eta}\tau_p^\beta$ is the unique β-homogeneous distribution on $\mathbb{A}$, normalized by $\tau_{\mathbb{A}}(\phi_{\mathbb{Z}_{\mathbb{A}}}) = 1$, where $\phi_{\mathbb{Z}_{\mathbb{A}}}(\{x_p\}) = \prod_{p\geq\eta}\phi_{\mathbb{Z}_p}(x_p)$; for $\beta > 0$ it is a positive measure and we have the associated Hilbert spaces

$$\widetilde{\mathscr{H}}_{\mathbb{A}}^\beta = \mathrm{L}_2\Big(\mathbb{A}, \tau_{\mathbb{A}}^\beta\Big) = \bigotimes_{p\geq\eta}\widetilde{\mathscr{H}}_p^\beta \quad (\otimes \text{ with respect to } \phi_{\mathbb{Z}_p})\,.$$

The measure $\tau_{\mathbb{A}}^\beta$ is invariant under the units $\mathbb{Z}_{\mathbb{A}}^* = \prod_{p\geq\eta}\mathbb{Z}_p^*$. The subspace of $\widetilde{\mathscr{H}}_{\mathbb{A}}^\beta$ consisting of $\mathbb{Z}_{\mathbb{A}}^*$-invariant functions will be denoted

$$\mathscr{H}_{\mathbb{A}}^\beta = \mathrm{L}_2\Big(\mathbb{A}, \tau_{\mathbb{A}}^\beta\Big)^{\mathbb{Z}_{\mathbb{A}}^*} = \bigotimes_{p\geq\eta}\mathscr{H}_p^\beta \quad (\otimes \text{ with respect to } \phi_{\mathbb{Z}_p})\,. \tag{2.3.1}$$

We have unitary representations of the group of ideles $\mathbb{A}^*$ on $\widetilde{\mathscr{H}}_{\mathbb{A}}^\beta$, and of $\mathbb{A}^*/\mathbb{Z}_{\mathbb{A}}^* = \mathbb{Q}^+ \times \mathbb{R}^+$ on $\mathscr{H}_{\mathbb{A}}^\beta$,

$$\pi_{\mathbb{A}}^\beta(a)\varphi(x) = |a|_{\mathbb{A}}^{-\beta/2}\varphi(a^{-1}x)\,. \tag{2.3.2}$$

Similarly we have the global probability measure $\tau_{\mathbb{Z}_{\mathbb{A}}}^\beta = \phi_{\mathbb{Z}_{\mathbb{A}}} \cdot \tau_{\mathbb{A}}^\beta = \otimes_{p\geq\eta}\tau_{\mathbb{Z}_p}^\beta$, and the associated Hilbert spaces

$$\widetilde{\mathscr{H}}^{\beta}_{\mathbb{Z}_{\mathbb{A}}} = \mathrm{L}_2\left(\widehat{\mathbb{Z}} \times \mathbb{R}, \tau^{\beta}_{\mathbb{Z}_{\mathbb{A}}}\right) = \bigotimes_{p \geq \eta} \widetilde{\mathscr{H}}^{\beta}_{\mathbb{Z}_p} \quad (\otimes \text{ with respect to } \mathbb{1}, \text{ the constant function } 1),$$

$$\mathscr{H}^{\beta}_{\mathbb{Z}_{\mathbb{A}}} = \left(\widetilde{\mathscr{H}}^{\beta}_{\mathbb{Z}_{\mathbb{A}}}\right)^{\mathbb{Z}^*_{\mathbb{A}}} = \bigotimes_{p \geq \eta} \mathscr{H}^{\beta}_{\mathbb{Z}_p} \quad (\otimes \text{ with respect to } \mathbb{1}). \tag{2.3.3}$$

For $\beta > 1$ the measure $\tau^{\beta}_{\mathbb{A}}$ is supported on the ideles $\mathbb{A}^* \subseteq \mathbb{A}$, but for $\beta \in (0,1]$ the action of $\mathbb{A}^*$ on the measure space $\left(\mathbb{A}, \tau^{\beta}_{\mathbb{A}}\right)$ is ergodic. For $\beta \in (0,1], \mathbb{A}^* \ltimes \mathrm{L}_\infty\left(\mathbb{A}, \tau^{\beta}_{\mathbb{A}}\right)$ is the hyperfinite III_1 factor; it is induced from the II_∞ factor $\mathbb{Q}^+ \ltimes \mathrm{L}_\infty\left(\mathbb{A}, \tau^{\beta}_{\mathbb{A}}\right)^{\mathbb{Z}^*_{\mathbb{A}}}$. The orbit space $\mathbb{A}/\mathbb{Q}^*$, and its factor relevant for the Riemann zeta function, $\mathbb{A}/\mathbb{Q}^*\widehat{\mathbb{Z}}^* = \mathbb{A}/\mathbb{A}^{(1)}$, are 'non-commutative spaces'. When a group G acts nicely on a space X, the commutative algebra describing the orbit space $\mathrm{C(X/G)} = \mathrm{C(X)}^G$ is Morita equivalent to the (non-commutative) algebra semi-direct product $\mathrm{G} \ltimes \mathrm{C(X)}$. When G acts ergodically on X, so that $\mathrm{C(X)}^G$ are just the constant functions, we are left to work with the algebra $\mathrm{G} \ltimes \mathrm{C(X)}$. This philosophy leads to many non-commutative spaces in arithmetic. The most basic are the (related) spaces given by the obvious action of G on X with

	I	II	III	IV	V
G:	$\mathbb{Z}$	$\mathbb{Q}^*$	$\mathbb{Q}^* \ltimes \mathbb{Q}$	$\mathbb{Q}^*$	$\mathbb{Q}^*$
X:	$\widehat{\mathbb{Z}}$	$\mathbb{A}/\mathbb{Q}$	$\mathbb{A}$	$\mathbb{A}$	$\mathbb{A} \oplus \mathbb{A}/\mathbb{Q} \oplus \mathbb{Q}$

(2.3.4)

The spaces I and II are related since $\mathbb{A}/\mathbb{Q} = (\widehat{\mathbb{Z}} \times \mathbb{R})/\mathbb{Z}$. The spaces II, III, IV are obviously related. Finally the spaces IV and V are related since (cf. Chapter 12) the Hilbert space $\mathrm{L}_2(\mathbb{A})$ is isomorphic to the space of L_2-sections of the 'theta' line bundle over the 'torus' $\mathbb{A} \oplus \mathbb{A}/\mathbb{Q} \oplus \mathbb{Q}$, and the action of a rational number $q \in \mathbb{Q}^*$ on the adeles $\mathbb{A}$ translates into the action of

$$\begin{pmatrix} q & 0 \\ 0 & q^{-1} \end{pmatrix}$$

on the torus. These basic non-commutative spaces, or ergodic actions, will be shown to be related to the Riemann hypothesis in Chapter 13. There are other, well-known, and less basic ergodic actions related to the Riemann hypothesis 'coming from SL_2': the geodesic flow on the modular curve (for a hint on this see (§13.3.26)), the dynamics of the Gauss map, and continued fractions (for a hint on this see §5.3).

3

The beta distribution

3.1 PHASE SPACE

In order to see the similarities between the finite primes and the real prime, it will be necessary to pass from the gamma distribution to the beta distribution.

Let

$$V = \mathbb{A}^1 \oplus \mathbb{A}^1, \ \langle (x_1, y_1), (x_2, y_2) \rangle = x_1 y_2 - y_1 x_2 \,,$$

denote the two-dimensional plane with the symplectic form $\langle \cdot , \cdot \rangle$. Let $\mathbb{P}^1 = \mathbb{P}^1(V)$ denote the projective line with $0 = (1:0), \infty = (0:1) \in \mathbb{P}^1$ corresponding to the above polarization. For all $p \geq \eta$, we have the $\mathbb{Q}_p$-points $V(\mathbb{Q}_p) = \mathbb{Q}_p \oplus \mathbb{Q}_p$, on which we have the standard metric $|(x,y)|_p$.

For $p \neq \eta$, this metric corresponds to the $\mathbb{Z}_p$-lattice

$$V(\mathbb{Z}_p) = \mathbb{Z}_p \oplus \mathbb{Z}_p = \Big\{ (x,y) \in V(\mathbb{Q}_p) \, \big| \, |(x,y)|_p \leq 1 \Big\} . \tag{3.1.1}$$

It induces a $\mathbb{Z}_p$-structure on $\mathbb{P}^1$ given by

$$\mathbb{P}^1_{\mathbb{Z}_p} = \operatorname{Proj} \operatorname{Symm}_{\mathbb{Z}_p}(V(\mathbb{Z}_p)) = \operatorname{Proj} \mathbb{Z}_p[x,y] \,.$$

For all $p \geq \eta$, we obtain a metric ρ_p on $\mathbb{P}^1(\mathbb{Q}_p)$:

$$\rho_p((x_1 : y_1), (x_2 : y_2)) = \frac{|x_1 y_2 - y_1 x_2|_p}{|(x_1, y_1)|_p \cdot |(x_2, y_2)|_p} \,. \tag{3.1.2}$$

We have

$$\begin{aligned} &0 \leq \rho_p(z_1, z_2) \leq 1, \quad z_i \in \mathbb{P}^1(\mathbb{Q}_p) \\ &\rho_p(z_1, z_2) = 0 \Longleftrightarrow z_1 = z_2 \\ &\rho_p(z_1, z_2) = \rho_p(z_2, z_1) \\ &\rho_p(z_1, z_3) \leq \begin{cases} \max\{\rho_p(z_1, z_2), \rho_p(z_2, z_3)\} & p \neq \eta \,, \\ \rho_\eta(z_1, z_2) + \rho_\eta(z_2, z_3) & p = \eta \,. \end{cases} \end{aligned} \tag{3.1.3}$$

For $p \neq \eta$ this metric has the simple meaning

$$\rho_p(z_1, z_2) \leq p^{-N} \Longleftrightarrow z_1 \equiv z_2 (\operatorname{mod} p^N) \,. \tag{3.1.4}$$

We shall identify $x \in \mathbb{Q}_p^*$ with $(1:x) \in \mathbb{P}^1(\mathbb{Q}_p)$, and we denote its distances from 0 and ∞ by

$$\begin{aligned} \rho_{0_p}(x) &= \rho_p((1:0), (1:x)) = |(1, x^{-1})|_p^{-1} \\ \rho_{\infty_p}(x) &= \rho_p((0:1), (1:x)) = |(1, x)|_p^{-1} \,. \end{aligned} \tag{3.1.5}$$

We have

$$\begin{aligned} \max\{\rho_{0_p}(x), \rho_{\infty_p}(x)\} &= 1 \quad \text{for } p \neq \eta\,, \\ \rho_{0_\eta}(x)^2 + \rho_{\infty_\eta}(x)^2 &= 1 \quad \text{for } p = \eta\,. \end{aligned} \tag{3.1.6}$$

Note that, for all $p \geq \eta$, we have

$$|x|_p = \frac{\rho_{0_p}(x)}{\rho_{\infty_p}(x)}\,. \tag{3.1.7}$$

Thus a rational number $a \in \mathbb{Q}^*$ has at the real prime η both a zero and a pole, with 'degree' given by $\rho_{0_\eta}(a)$ and $\rho_{\infty_\eta}(a)$.

For $p \neq \eta$, these are related to the Toeplitz operator $\pi^{\beta}_{\mathbb{Z}_p}(a)$:

$$\begin{aligned} -\log_p \rho_{0_p}(a) &= \dim\ker \pi^{\beta}_{\mathbb{Z}_p}(a)\,, \\ -\log_p \rho_{\infty_p}(a) &= \dim\operatorname{coker} \pi^{\beta}_{\mathbb{Z}_p}(a)\,, \\ -\log_p |a|_p &= \operatorname{index}\left\{\pi^{\beta}_{\mathbb{Z}_p}(a) : \mathscr{H}^{\beta}_{\mathbb{Z}_p} \to \mathscr{H}^{\beta}_{\mathbb{Z}_p}\right\}. \end{aligned} \tag{3.1.8}$$

3.2 THE BETA DISTRIBUTION

Taking the product of two gamma distributions we have the probability measure $\tau^{\beta}_{\mathbb{Z}_p} \otimes \tau^{\alpha}_{\mathbb{Z}_p}$ on $V(\mathbb{Q}_p)$, and projecting this measure onto $\mathbb{P}^1(\mathbb{Q}_p)$ we obtain the beta distribution $\tau_p^{\alpha,\beta}$. A straightforward computation gives for all $p \geq \eta$, the equation

$$\tau_p^{\alpha,\beta}(\mathrm{d}x) = \rho_{\infty_p}(x)^{\alpha} \cdot \rho_{0_p}(x)^{\beta} \cdot \frac{\mathrm{d}^*x}{\zeta_p(\alpha,\beta)}\,, \tag{3.2.1}$$

where d^*x is the multiplicative Haar measure ($\mathrm{d}^*x(\mathbb{Z}_p^*) = 1$ for $p \neq \eta$), and the normalizing factor is the beta function

$$\zeta_p(\alpha,\beta) = \frac{\zeta_p(\alpha)\zeta_p(\beta)}{\zeta_p(\alpha+\beta)}\,. \tag{3.2.2}$$

That $\tau_p^{\alpha,\beta}(\mathrm{d}x)$ is a probability measure is expressed in the

$$\text{beta integral: } \zeta_p(\alpha,\beta) = \int_{\mathbb{Q}_p^*} \rho_{\infty_p}(x)^{\alpha} \cdot \rho_{0_p}(x)^{\beta}\, \mathrm{d}^*x, \quad \mathrm{Re}(\alpha), \mathrm{Re}(\beta) > 0,\ p \geq \eta\,. \tag{3.2.3}$$

The beta integral has a few advantages over the gamma integral (2.2.1). The distances ρ_{∞_p} and ρ_{0_p} appear on an equal footing for the finite and the real primes, and we avoid the complicated Gaussian $\phi_{\mathbb{Z}_\eta}$. Moreover, we obtain the gamma integral as the $\alpha \to \infty$ limit of the beta integral. For $p \neq \eta$, as $\alpha \to \infty$,

$$\zeta_p(\alpha,\beta) \to \zeta_p(\beta), \quad \rho_{\infty_p}(x)^{\alpha} \cdot \rho_{0_p}(x)^{\beta} \to \phi_{\mathbb{Z}_p}(x) \cdot |x|_p^{\beta}\,. \tag{3.2.4}$$

For the real prime the limit is more delicate:

$$\left(\frac{\alpha}{2\pi}\right)^{\beta/2} \cdot \zeta_\eta(\alpha,\beta) \to \zeta_\eta(\beta)\,, \tag{3.2.5}$$

so we multiply the beta integral by $(\alpha/2\pi)^{\beta/2}$ and then substitute $(\alpha/2\pi)^{-1/2}\cdot x$ for x in the integral. Then as $\alpha\to\infty$, we obtain

$$\rho_{\infty_\eta}\left(\left(\frac{\alpha}{2\pi}\right)^{-1/2}x\right)^{\alpha}\to\phi_{\mathbb{Z}_\eta}(x),\quad \left(\frac{\alpha}{2\pi}\right)^{\beta/2}\cdot\rho_{0_\eta}\left(\left(\frac{\alpha}{2\pi}\right)^{-1/2}x\right)^{\beta}\to|x|_\eta^\beta\,. \tag{3.2.6}$$

The beta integral thus has the merit of showing the common origin of $\phi_{\mathbb{Z}_p}(x)$, and of $|x|_p^\beta$, in the integrand of the gamma integral.

The beta distribution $\tau_p^{\alpha,\beta}$ can be considered for complex $\alpha,\beta\in\mathbb{C}$, where $\mathrm{Re}(\alpha)$, $\mathrm{Re}(\beta)>0$, and it has meromorphic continuation to all $\alpha,\beta\in\mathbb{C}$. For example, if $p\neq\eta$ and φ is a locally constant function on $\mathbb{P}^1(\mathbb{Q}_p)$, let $N\gg0$ be such that φ is constant on the sets $\{x\,|\,\rho_{0_p}(x)\le p^{-N}\},\{x\,|\,\rho_{\infty_p}(x)\le p^{-N}\}$; then for all α, β we have

$$\begin{aligned}\tau_p^{\alpha,\beta}(\varphi)&=\frac{\zeta_p(\alpha+\beta)}{\zeta_p(\alpha)}p^{-N\beta}\varphi(0)+\frac{\zeta_p(\alpha+\beta)}{\zeta_p(\beta)}p^{-N\alpha}\varphi(\infty)\\&\quad+\int_{p^{-N}<|x|_p<p^N}\varphi(x)\rho_{\infty_p}(x)^\alpha\cdot\rho_{0_p}(x)^\beta\frac{\mathrm{d}^*x}{\zeta_p(\alpha,\beta)}\,.\end{aligned}$$

The beta integral (3.2.3) follows from the gamma integral (2.2.1) using (3.1.7), and the following basic equation which defines $\rho_{0_p}(x)^\beta$ for all $p\ge\eta$, $\beta>0$:

$$\rho_{0_p}(x)^\beta=\left(\phi_{\mathbb{Z}_p},\phi_{\mathbb{Z}_p}^{(x)}\right)_{\mathscr{H}_p^\beta}=\int\frac{\mathrm{d}^*y}{\zeta_p(\beta)}|y|_p^\beta\phi_{\mathbb{Z}_p}(y)\phi_{\mathbb{Z}_p}(x^{-1}y)\,, \tag{3.2.7}$$

where $\phi_{\mathbb{Z}_p}^{(x)}(y)=\phi_{\mathbb{Z}_p}(x^{-1}y)$.

Alternatively, we can write this equation in the form

$$\tilde\rho_p(x)^\beta\overset{\text{def}}{=}\rho_{\infty_p}(x)^{\beta/2}\cdot\rho_{0_p}(x)^{\beta/2}=|x|_p^{-\beta/2}\rho_{0_p}(x)^\beta=\left(\phi_{\mathbb{Z}_p},\pi_p^\beta(x)\phi_{\mathbb{Z}_p}\right)_{\mathscr{H}_p^\beta} \tag{3.2.7$'$}$$

showing that the function $\tilde\rho_p(x)^\beta$ is a positive-definite function on $\mathbb{Q}_p^*/\mathbb{Z}_p^*$, for all $p\ge\eta$, and $\beta>0$. The beta integral can be written in a non-symmetric form as a Mellin transform, namely

$$\int\rho_{\infty_p}(x)^\beta|x|_p^s\,\mathrm{d}^*x=\zeta_p(s,\beta-s)\,. \tag{3.2.8}$$

Alternatively, it is the Mellin transform of the positive-definite function $\tilde\rho_p$:

$$\int\tilde\rho_p(x)^\beta|x|_p^s\,\mathrm{d}^*x=\zeta_p\left(\frac{\beta}{2}+s,\frac{\beta}{2}-s\right), \tag{3.2.8$'$}$$

which is a positive function for $s\in\mathrm{i}\mathbb{R}$.

When α, β are real and positive, $\tau_p^{\alpha,\beta}$ is a probability measure. We have the associated Hilbert space

$$\widetilde{\mathscr{H}}_p^{\alpha,\beta}=\mathrm{L}_2\left(\mathbb{P}^1(\mathbb{Q}_p),\tau_p^{\alpha,\beta}\right), \tag{3.2.9}$$

and its subspace of $\mathbb{Z}_p^*$-invariant functions

$$\mathscr{H}_p^{\alpha,\beta}=\left(\widetilde{\mathscr{H}}_p^{\alpha,\beta}\right)^{\mathbb{Z}_p^*}. \tag{3.2.10}$$

On this spaces $\mathbb{Q}_p^*$ acts unitarily via

$$\pi_p^{\alpha,\beta}(a)\varphi(x) = \frac{\rho_{\infty_p}(a^{-1}x)^{\alpha/2}}{\rho_{\infty_p}(x)^{\alpha/2}} \cdot \frac{\rho_{0_p}(a^{-1}x)^{\beta/2}}{\rho_{0_p}(x)^{\beta/2}} \cdot \varphi(a^{-1}x)\,.$$

We have an isomorphism of Hilbert spaces

$$\begin{gathered}\mathscr{H}_p^{\alpha,\beta} \otimes \mathscr{H}_p^{\alpha+\beta} \simeq \mathscr{H}_p^{\alpha} \otimes \mathscr{H}_p^{\beta} \quad \left(\text{and } \mathscr{H}_p^{\alpha,\beta} \otimes \mathscr{H}_{\mathbb{Z}_p}^{\alpha+\beta} \simeq \mathscr{H}_{\mathbb{Z}_p}^{\alpha} \otimes \mathscr{H}_{\mathbb{Z}_p}^{\beta}\right) \\ \varphi \otimes f \mapsto \varphi \otimes f(x,y) = \varphi(x:y) \cdot f(|x,y|_p)\,,\end{gathered} \tag{3.2.11}$$

intertwining the action of the multiplicative group

$$\pi_p^{\alpha,\beta}(a) \otimes \pi_p^{\alpha+\beta}(a) \simeq \pi_p^{\alpha}(a) \otimes \pi_p^{\beta}(a)\,. \tag{3.2.12}$$

3.3 SPECIAL VALUES

The parameters $(\alpha,\beta) \in \mathbb{R}^+ \times \mathbb{R}^+$ of the beta distribution form the following picture:

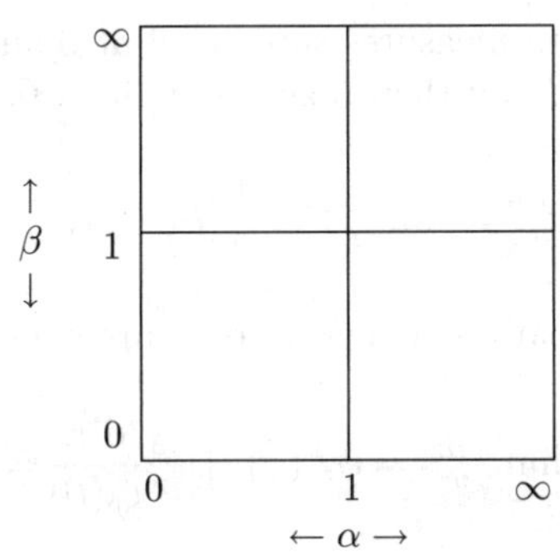

(3.3.1)

We shall next examine special values of (α,β), as well as its boundary limits as α, β approach 0 and ∞.

For $\alpha = \beta = 1$, and all $p \geq \eta$, the beta measure $\tau_p^{1,1}$ is the unique probability measure on $\mathbb{P}^1(\mathbb{Q}_p)$ which is invariant under the standard maximal compact subgroup $\mathrm{GL}_2(\mathbb{Z}_p) \subseteq \mathrm{GL}_2(\mathbb{Q}_p)$; that is, $\tau_p^{1,1}$ is the image of the Haar measure of $\mathrm{GL}_2(\mathbb{Z}_p)$ via the projection $\mathrm{GL}_2(\mathbb{Z}_p)/\mathrm{B}(\mathbb{Z}_p) \simeq \mathbb{P}^1(\mathbb{Q}_p)$, where $\mathrm{B}(\mathbb{Z}_p)$ is the Borel subgroup stabilizing ∞.

For $\alpha = \beta = n$ an integer, and for finite $p \neq \eta$, we can view $\tau_p^{n,n}$ as a probability measure on $\mathbb{Z}_p^*\backslash\mathbb{P}^1(\mathbb{Q}_p) = \{\infty\} \cup p^{\mathbb{Z}} \cup \{0\}$, and this measure is obtained from $\tau_p^{1,1}$ by replacing p by p^n everywhere. This can be interpreted as follows: let $\mathbb{Z}_{p,n} = \mathbb{Z}_p[\mu_{p^n-1}]$ denote the integers in the unique degree n unramified extension $\mathbb{Q}_{p,n}$ of $\mathbb{Q}_p$; then $\tau_p^{n,n}$ is the image of the Haar measure on $\mathrm{GL}_2(\mathbb{Z}_{p,n})$ via the projection onto

$$\mathbb{Z}_{p,n}^*\backslash\mathrm{GL}_2(\mathbb{Z}_{p,n})/\mathrm{B}(\mathbb{Z}_{p,n}) \simeq \mathbb{Z}_{p,n}^*\backslash\mathbb{P}^1(\mathbb{Z}_{p,n}) \simeq \mathbb{Z}_p^*\backslash\mathbb{P}^1(\mathbb{Z}_p)\,. \tag{3.3.2}$$

For the real prime η, we can similarly view $\tau_\eta^{n,n}$ as a probability measure on $\mathbb{Z}_\eta^*\backslash\mathbb{P}^1(\mathbb{Q}_\eta) = \{\pm 1\} \setminus \mathbb{P}^1(\mathbb{R})$, and we use the Cayley transform to identify this set with the interval $[-1,1]$:

$$\begin{aligned} \{\pm1\}\backslash\mathbb{P}^1(\mathbb{R}) &\tilde{\to}[-1,1] \\ x \mapsto \rho_{\infty_\eta}(x)^2 - \rho_{0_\eta}(x)^2 &= \frac{1-x^2}{1+x^2}. \end{aligned} \tag{3.3.3}$$

We can now view $\tau_\eta^{n,n}$ as a probability measure on $[-1,1]$, and it is the image of the Haar measure on $\mathrm{GL}_{n+1}(\mathbb{Z}_\eta) = \mathrm{O}_{n+1}$ via the projection

$$\begin{aligned} &\mathrm{O}_{n+1}/\mathrm{O}_n \simeq \mathrm{S}^n \twoheadrightarrow \mathrm{O}_n\backslash\mathrm{O}_{n+1}/\mathrm{O}_n = [-1,1] \\ &(x_0, x_1, \ldots, x_n) \mapsto x_0\,. \end{aligned} \tag{3.3.4}$$

Let us now turn to the boundary limits of the beta measure, and first to the easy limit as one of the parameters approaches 0. We have for all $p \geq \eta$

$$\tau_p^{\alpha,0} = \lim_{\beta\to 0} \tau_p^{\alpha,\beta} = \delta_{0_p}, \quad \tau_p^{0,\beta} = \lim_{\alpha\to 0} \tau^{\alpha,\beta} = \delta_{\infty_p}, \tag{3.3.5}$$

where $\delta_{0_p}, \delta_{\infty_p}$ are the delta measures supported at 0 and at ∞.

We can also set $\alpha = \beta$ and then take the limit at 0, for all $p \geq \eta$:

$$\tau_p^{0,0} = \lim_{\alpha\to 0} \tau_p^{\alpha,\alpha} = \frac{1}{2}\left(\delta_{0_p} + \delta_{\infty_p}\right). \tag{3.3.6}$$

The limits as one of the parameters goes to ∞ are simple for the finite $p \neq \eta$:

$$\tau_p^{\infty,\beta} = \lim_{\alpha\to\infty} \tau_p^{\alpha,\beta} = \phi_{\mathbb{Z}_p}(x)\cdot|x|_p^\beta \frac{\mathrm{d}^*x}{\zeta_p(\beta)} = \tau_{\mathbb{Z}_p}^\beta(x)\,, \tag{3.3.7}$$

the gamma measure. Similarly,

$$\tau_p^{\alpha,\infty} = \lim_{\beta\to\infty} \tau_p^{\alpha,\beta} = \phi_{\mathbb{Z}_p}(x^{-1})\cdot|x^{-1}|_p^\alpha \frac{\mathrm{d}^*x}{\zeta_p(\alpha)} = \tau_{\mathbb{Z}_p}^\alpha(x^{-1})\,. \tag{3.3.8}$$

We can also let both parameters go to ∞ obtaining

$$\tau_p^{\infty,\infty} = \lim_{\alpha,\beta\to\infty} \tau_p^{\alpha,\beta} = \phi_{\mathbb{Z}_p^*}(x)\mathrm{d}^*x\,, \tag{3.3.9}$$

the Haar measure on $\mathbb{Z}_p^*$. For the real prime the limits as α, β go to ∞ degenerate:

$$\tau_\eta^{\infty,\beta} = \lim_{\alpha\to\infty} \tau_\eta^{\alpha,\beta} = \delta_{0_\eta}, \quad \tau_\eta^{\alpha,\infty} = \lim_{\beta\to\infty} \tau_\eta^{\alpha,\beta} = \delta_{\infty_\eta}. \tag{3.3.10}$$

Similarly, the limit when both α and β go to ∞ simultaneously degenerate:

$$\tau_\eta^{\infty,\infty} = \lim_{\alpha\to\infty} \tau_\eta^{\alpha,\alpha} = \tfrac{1}{2}(\delta_1 + \delta_{-1})\,, \tag{3.3.11}$$

the Haar measure on $\mathbb{Z}_\eta^* = \{\pm 1\}$. Nevertheless, we can also obtain the gamma measure as a limit of the beta measure for the real prime, if we use the more delicate limit procedure of (3.2.5), (3.2.6):

$$\begin{aligned} \lim_{\alpha\to\infty} \tau_\eta^{\alpha,\beta}\left(\left(\frac{\alpha}{2\pi}\right)^{-1/2} \mathrm{d}x\right) &= \phi_{\mathbb{Z}_\eta}(x)|x|_\eta^\beta \frac{\mathrm{d}^* x}{\zeta_\eta(\beta)} = \tau_{\mathbb{Z}_\eta}^\beta(\mathrm{d}x) \\ \lim_{\beta\to\infty} \tau_\eta^{\alpha,\beta}\left(\left(\frac{\beta}{2\pi}\right)^{1/2} \mathrm{d}x\right) &= \phi_{\mathbb{Z}_\eta}(x^{-1})|x^{-1}|_\eta^\alpha \frac{\mathrm{d}^* x}{\zeta_\eta(\alpha)} = \tau_{\mathbb{Z}_\eta}^\alpha(\mathrm{d}x^{-1}) \,. \end{aligned} \tag{3.3.12}$$

Thus for $p = \eta$ in the $\alpha \to \infty$ limit we have to scale by a constant multiple of $\alpha^{1/2}$ in order to concentrate on smaller and smaller neighbourhoods of 0 in $\mathbb{R}$.

One face of the mystery of the real prime is that for $p \neq \eta$, $\mathbb{Q}_p^*/\mathbb{Z}_p^* = p^{\mathbb{Z}}$ is discrete, while $\mathbb{Q}_\eta^*/\mathbb{Z}_\eta^* = \mathbb{R}^*/\{\pm 1\} = \mathbb{R}^+$ is the continuum; this can be interpreted also as the degeneration of $\mathbb{Z}_\eta^*$ into $\{\pm 1\}$, and we are especially interested in the 'missing part' of $\mathbb{Z}_\eta^*$. One way to approach this is to take the limit as $\alpha = \beta \to \infty$, as in the p-adic analogue (3.3.9), but we use the Cayley transform (3.3.3) and a more delicate limit (as in (3.3.12)) to avoid the degeneration (3.3.11). The result is that for φ, a Schwartz function on $\mathbb{R}$, we have

$$\lim_{\alpha\to\infty} \int_{\mathbb{P}^1(\mathbb{R})} \varphi\left(\left(\frac{\alpha}{2\pi}\right)^{1/2} \frac{1-x^2}{1+x^2}\right) \tau_\eta^{\alpha,\alpha}(\mathrm{d}x) = \int_{\mathbb{R}} \varphi(x)\phi_{\mathbb{Z}_\eta}(x)\,\mathrm{d}x \,. \tag{3.3.13}$$

Thus just as the Gaussian appeared in (3.3.12) as the real analogue of $\mathbb{Z}_p$, or as describing the 'union of infinitesimal neighbourhoods of 0_η' (and similarly of ∞_η), here the Gaussian appears as the real analogue of $1 + 2p\mathbb{Z}_p = \mathrm{e}^{2p\mathbb{Z}_p}$, or as describing the 'union of infinitesimal neighbourhoods of 1_η'; for more details see §8.3 below.

3.4 REMARKS ON THE GLOBAL THEORY

Having a probability measure $\tau_p^{\alpha,\beta}$ for each $p \geq \eta$, we obtain a probability measure $\tau_{\mathbb{A}}^{\alpha,\beta} = \otimes_{p\geq\eta}\tau_p^{\alpha,\beta}$ on the product space $\mathbb{P}^1(\mathbb{A}) = \prod_{p\geq\eta}\mathbb{P}^1(\mathbb{Q}_p)$. The associated Hilbert spaces are

$$\widetilde{\mathscr{H}}_{\mathbb{A}}^{\alpha,\beta} = \mathrm{L}_2\left(\mathbb{P}^1(\mathbb{A}), \tau_{\mathbb{A}}^{\alpha,\beta}\right) = \bigotimes_{p\geq\eta} \widetilde{\mathscr{H}}_p^{\alpha,\beta} \quad (\otimes \text{ with respect to } \mathbb{1}) \,. \tag{3.4.1}$$

and the subspace of $\mathbb{Z}_{\mathbb{A}}^*$-invariant functions

$$\mathscr{H}_{\mathbb{A}}^{\alpha,\beta} = \left(\widetilde{\mathscr{H}}_{\mathbb{A}}^{\alpha,\beta}\right)^{\mathbb{Z}_{\mathbb{A}}^*} = \bigotimes_{p\geq\eta} \mathscr{H}_p^{\alpha,\beta} \quad (\otimes \text{ with respect to } \mathbb{1}) \,. \tag{3.4.2}$$

We have the unitary representation $\pi_{\mathbb{A}}^{\alpha,\beta} = \otimes_{p\geq\eta}\pi_p^{\alpha,\beta}$ of $\mathbb{A}^*$ (respectively $\mathbb{A}^*/\mathbb{Z}_{\mathbb{A}}^* = \mathbb{Q}^+ \times \mathbb{R}^+$) on $\widetilde{\mathscr{H}}_{\mathbb{A}}^{\alpha,\beta}$ $\left(\text{respectively } \mathscr{H}_{\mathbb{A}}^{\alpha,\beta}\right)$. We also have isomorphisms of Hilbert spaces, interwining with the $\mathbb{A}^*$-action,

$$\mathscr{H}_{\mathbb{A}}^{\alpha,\beta} \otimes \mathscr{H}_{\mathbb{A}}^{\alpha+\beta} \simeq \mathscr{H}_{\mathbb{A}}^{\alpha} \otimes \mathscr{H}_{\mathbb{A}}^{\beta} \quad \left(\text{and } \mathscr{H}_{\mathbb{A}}^{\alpha,\beta} \otimes \mathscr{H}_{\mathbb{Z}_{\mathbb{A}}}^{\alpha+\beta} \simeq \mathscr{H}_{\mathbb{Z}_{\mathbb{A}}}^{\alpha} \otimes \mathscr{H}_{\mathbb{Z}_{\mathbb{A}}}^{\beta}\right). \tag{3.4.3}$$

For $\alpha+\beta>1$, the product of two gamma measures $\tau^{\alpha}_{\mathbb{Z}_{\mathbb{A}}}\otimes\tau^{\beta}_{\mathbb{Z}_{\mathbb{A}}}$ on $V(\mathbb{A})=\mathbb{A}\oplus\mathbb{A}$ is supported on the subset

$$\begin{aligned} V^*(\mathbb{A}) &= \prod_{p\geq\eta} V^*(\mathbb{Q}_p):V^*(\mathbb{Z}_p) \quad \text{(restricted direct product)} \\ &= \{(x,y)\in V(\mathbb{A})\big|(x_p,y_p)\neq(0,0) \text{ for all } p\geq\eta, \\ &\qquad |(x_p,y_p)|_p=1 \text{ for almost all } p\}, \end{aligned} \tag{3.4.4}$$

and the beta measure $\tau^{\alpha,\beta}_{\mathbb{A}}$ is its projection via $V^*(\mathbb{A})/\mathbb{A}^*\simeq\mathbb{P}^1(\mathbb{A})$.

For $\alpha,\beta>1$, we have

$$\sum_{p\neq\eta}\left|1-\tau^{\alpha,\beta}_p(\mathbb{Z}^*_p)\right| = \sum_{p\neq\eta}\left|1-\frac{1}{\zeta_p(\alpha,\beta)}\right| = \sum_{p\neq\eta}\frac{p^{-\alpha}+p^{-\beta}-2p^{-(\alpha+\beta)}}{1-p^{-(\alpha+\beta)}}<\infty, \tag{3.4.5}$$

and so the measure $\tau^{\alpha,\beta}_{\mathbb{A}}$ is supported on the ideles $\mathbb{A}^*\subseteq\mathbb{P}^1(\mathbb{A})$.

For $\beta\leq 1<\alpha$, we have

$$\begin{aligned} \sum_{p\neq\eta}\left|1-\tau^{\alpha,\beta}_p(\mathbb{Z}_p)\right| &= \sum_{p\neq\eta}\left|1-\frac{\zeta_p(\alpha+\beta)}{\zeta_p(\alpha)}\right| \\ &= \sum_{p\neq\eta}p^{-\alpha}\cdot\frac{1-p^{-\beta}}{1-p^{-(\alpha+\beta)}}<\infty, \end{aligned} \tag{3.4.6}$$

and $\tau^{\alpha,\beta}_{\mathbb{A}}$ is supported on the adeles $\mathbb{A}=\{(1:x),x\in\mathbb{A}\}\subseteq\mathbb{P}^1(\mathbb{A})$.

We have the involution $\mathrm{r}\colon(x,y)\mapsto(y,x)$ on $V(\mathbb{A})$ inducing the involutions $\mathrm{r}\colon(x:y)\mapsto(y:x)$ on $\mathbb{P}^1(\mathbb{A})$, and $\mathrm{r}\colon x\mapsto x^{-1}$ on $\mathbb{A}^*$. It gives

$$\mathrm{r}_*\tau^{\alpha,\beta}=\tau^{\beta,\alpha}, \quad \mathrm{r}\colon\mathscr{H}^{\alpha,\beta}\xrightarrow{\sim}\mathscr{H}^{\beta,\alpha}, \tag{3.4.7}$$

and we can always restrict our attention to the case $\alpha\geq\beta$.

Thus for $\alpha\leq 1<\beta$, $\tau^{\alpha,\beta}_{\mathbb{A}}$ is supported on the set $\mathrm{r}\mathbb{A}=\{(x:1)\,|\,x\in\mathbb{A}\}\subseteq\mathbb{P}(\mathbb{A})$. For $\alpha,\beta\leq 1$, $\tau^{\alpha,\beta}_{\mathbb{A}}$ is supported on the full $\mathbb{P}^1(\mathbb{A})$.

We can summarize the support of $\tau^{\alpha,\beta}_{\mathbb{A}}$ with respect to our parameter picture (3.3.1):

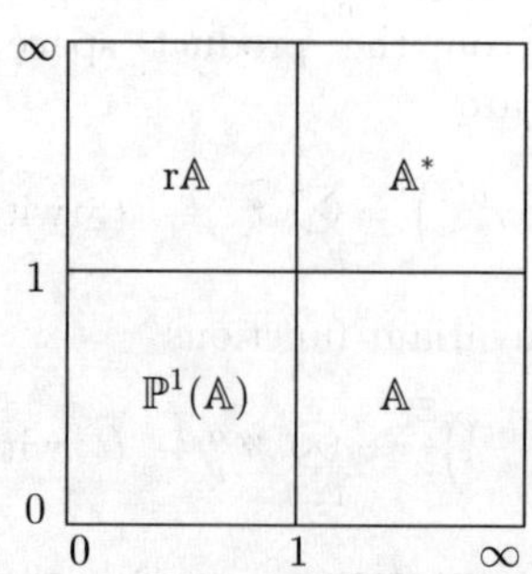

(3.4.8)

When $\alpha\leq 1$ or when $\beta\leq 1$, the two 'critical strips', the action of $\mathbb{A}^*$ on the measure space $\left(\mathbb{P}^1(\mathbb{A}),\tau^{\alpha,\beta}_{\mathbb{A}}\right)$ is ergodic, and we obtain the Araki–Woods hyperfinite III_1 factor $\mathbb{A}^*\ltimes\mathrm{L}_\infty\left(\mathbb{P}^1(\mathbb{A}),\tau^{\alpha,\beta}_{\mathbb{A}}\right)$.

4

The p-adic hyperbolic point of view

4.1 CHAINS AND TREES

A (Markov) *chain* is a (denumerable) set X, the *state space*, together with a function $P_X: X \times X \to [0, 1]$, the *transition probabilities*, satisfying

$$P_X(x, X) = 1 \quad \text{for all } x \in X, \tag{4.1.1}$$

where for any subset $Y \subseteq X$ we write $P_X(x, Y) = \sum_{y \in Y} P_X(x, y)$. We think of a particle walking between the states of X, where $P_X(x, y)$ is the probability of walking from x to y.

A map $f: X \to Y$ of chains is a map of the underlying state spaces satisfying

$$P_X(x, f^{-1}(y)) = P_Y(f(x), y) \quad \text{for all } x \in X,\ y \in Y. \tag{4.1.2}$$

By a *graph* $\mathcal{J} = (\mathcal{J}_0, \mathcal{J}_1)$ we shall mean a set $\mathcal{J}_0$, the *vertices*, and a subset $\mathcal{J}_1 \subseteq \mathcal{J}_0 \times \mathcal{J}_0$, the *edges*. Thus all our graphs are oriented. A map of graphs $f: \mathcal{J} \to \mathcal{J}'$ is a map of the vertices $f: \mathcal{J}_0 \to \mathcal{J}'_0$ such that $f \times f(\mathcal{J}_1) \subseteq \mathcal{J}'_1$. To a chain X, we can associate the graph $\mathcal{J}(X)$, with states as vertices, and with an edge going from x to y (i.e. $(x, y) \in \mathcal{J}_1$) whenever $P_X(x, y) > 0$. This defines a functor from the category of chains to the category of graphs.

A graph $\mathcal{J} = (\mathcal{J}_0, \mathcal{J}_1)$ is *locally finite* if

$$v_x = \#\{y \in \mathcal{J}_0 \mid (x, y) \in \mathcal{J}_1\} \quad \text{is finite for all } x \in \mathcal{J}_0.$$

For such a graph we can define a chain, the *random walk on* $\mathcal{J}$, with state space $\mathcal{J}_0$, and with transition probabilities

$$P(x, y) = \begin{cases} 1/v_x & \text{if } (x, y) \in \mathcal{J}_1, \\ 0 & \text{otherwise}. \end{cases} \tag{4.1.3}$$

More generally, given a chain X and a graph $\mathcal{J} = (\mathcal{J}_0, \mathcal{J}_1)$, a map of graphs $f: \mathcal{J} \to \mathcal{J}(X)$ is *locally finite* if

$$v_{x,y} = \#\{y' \in \mathcal{J}_0 \mid (y, y') \in \mathcal{J}_1, f(y') = x\} \quad \text{is finite for all } x \in X, y \in \mathcal{J}_0.$$

For such a map f we can define a chain $\mathcal{J}_X$, the *random walk on* $\mathcal{J}$ *over* X, with state space $\mathcal{J}_0$, and with transition probabilities

$$P(y, y') = \begin{cases} 1/v_{f(y'),y} \cdot P_X(f(y), f(y')) & \text{if } (y, y') \in \mathcal{J}_1, \\ 0 & \text{otherwise}. \end{cases} \tag{4.1.4}$$

The map $f: \mathcal{J}_X \to X$ is now a map of chains.

We say that a group G acts by automorphisms on a chain X if G acts on the state space by permutations satisfying

$$\mathrm{P_X}(gx, gy) = \mathrm{P_X}(x, y), \quad x, y \in \mathrm{X}, \ g \in \mathrm{G}\,. \tag{4.1.5}$$

In this case, we can form the quotient chain X/G, where states are the G-orbits in X, and with transition probabilities

$$\mathrm{P_{X/G}}(\mathrm{G}\cdot x, \mathrm{G}\cdot y) = \mathrm{P_X}(x, \mathrm{G}\cdot y)\,. \tag{4.1.6}$$

There is a canonical map of chains $\mathrm{X} \to \mathrm{X/G}$.

Given two chains X, Y we can form their *direct product* $\mathrm{X} \times \mathrm{Y}$, with transition probabilities

$$\mathrm{P_{X\times Y}}((x, y), (x', y')) = \mathrm{P_X}(x, x')\cdot \mathrm{P_Y}(y, y'), \quad x, x' \in \mathrm{X}, y, \ y' \in \mathrm{Y}\,. \tag{4.1.7}$$

The projections are chain maps from $\mathrm{X}\times\mathrm{Y}$ to X and Y, but $\mathrm{X}\times\mathrm{Y}$ is not the product in the category of chains. Given maps of chains $f\colon \mathrm{Z}\to\mathrm{X}$, $g\colon \mathrm{Z}\to\mathrm{Y}$, they determine a map of chains $f\times g\colon \mathrm{Z}\to\mathrm{X}\times\mathrm{Y}$ if and only if the maps f, g are *independent* in the sense that

$$\mathrm{P_Z}(z, f^{-1}(x)\cap g^{-1}(y)) = \mathrm{P_Z}(z, f^{-1}(x))\cdot \mathrm{P_Z}(z, g^{-1}(y)), \quad x \in \mathrm{X}, \ y \in \mathrm{Y}, \ z \in \mathrm{Z}\,.$$

When our particle walks in the product chain $\mathrm{X}\times\mathrm{Y}$ according to the rule (4.1.7), at each step we simultaneously throw two independent dice, one which dictates the change in the X-coordinate, the other for the Y-coordinate. It will be useful to have another product, the *semi-direct product* $\mathrm{X} \rtimes (\mathrm{Y})$, where we do one thing at a time. We first throw a die dictating the change in the X-coordinate, and only if there is no change in the X-coordinate do we throw the other die dictating the change in the Y-coordinate. Formally, the state space is still $\mathrm{X}\times\mathrm{Y}$, but the transition probabilities are now

$$\mathrm{P_{X\rtimes(Y)}}((x, y), (x', y')) = \begin{cases} \mathrm{P_X}(x, x') & x \neq x', y = y'\,, \\ \mathrm{P_X}(x, x')\cdot \mathrm{P_Y}(y, y') & x = x'\,, \\ 0 & x \neq x', y \neq y'\,. \end{cases} \tag{4.1.8}$$

Another useful construction will be to pass from a chain X to the *associated non-stop chain* $\mathrm{X}_!$, where the particle is not allowed to stop at any state of X. An *absorbing state* of a chain X is a state $x \in \mathrm{X}$ with $\mathrm{P_X}(x, x) = 1$. If the chain X has no absorbing states, the associated non-stop chain $\mathrm{X}_!$ has the same state space X, but with transition probabilities

$$\mathrm{P_{X_!}}(x, y) = \begin{cases} \dfrac{\mathrm{P_X}(x, y)}{1 - \mathrm{P_X}(x, x)} & x \neq y\,, \\ 0 & x = y\,. \end{cases} \tag{4.1.9}$$

Given a graph $\mathcal{J} = (\mathcal{J}_0, \mathcal{J}_1)$ and a point $x_0 \in \mathcal{J}_0$, we denote by $\Omega^n_{\mathcal{J},x_0}$ the collection of paths of length n beginning at x_0, so that

$$\Omega^n_{\mathcal{J},x_0} = \{(x_0, x_1, \ldots, x_n) \mid x_i \in \mathcal{J}_0, (x_i, x_{i+1}) \in \mathcal{J}_1\}\,, \tag{4.1.10}$$

and we let $\Omega_{\mathcal{J},x_0} = \coprod_{n\geq 0} \Omega^n_{\mathcal{J},x_0}$. Then $\Omega_{\mathcal{J},x_0}$ has the structure of a graph, with edges going from $(x_0, \ldots, x_{n-1})$ to $(x_0, \ldots, x_n)$ for each $(x_0, \ldots, x_n) \in \Omega_{\mathcal{J},x_0}$, and we have a canonical map of graphs $\Omega_{\mathcal{J},x_0} \to \mathcal{J}, (x_0, \ldots, x_n) \longmapsto x_n$. The graph $\Omega_{\mathcal{J},x_0}$ is in fact a *tree*, and we call it the *universal covering of* $\mathcal{J}$ *at* x_0.

Given a chain X, and a point $x_0 \in \mathrm{X}$, the universal covering $\Omega_{\mathcal{J},x_0}$ of the graph $\mathcal{J} = \mathcal{J}(\mathrm{X})$ has a natural chain structure with transition probabilities

$$\mathrm{P}((x_0, \ldots, x_n), (x'_0, \ldots, x'_m)) = \begin{cases} \mathrm{P_X}(x_n, x'_m) & \text{if } m = n+1, \text{ and } x_0 = x'_0, \ldots, x_n = x'_n\,, \\ 0 & \text{otherwise}\,. \end{cases} \tag{4.1.11}$$

The canonical map $\Omega_{\mathcal{J},x_0} \to \mathrm{X}$ is a map of chains. We call $\Omega_{\mathcal{J},x_0}$ the *universal covering chain of* X *at* x_0, and denote it by Ω_{X,x_0}.

All our chains X (and graphs) will have a fixed origin state, denoted $\mathrm{O_X}$, from which all other states may be reached:

$$\Omega_{\mathrm{X}} = \Omega_{\mathrm{X},\mathrm{O_X}} \to \mathrm{X} \quad \text{is onto}\,. \tag{4.1.12}$$

We shall deal mainly with transient chains, where (4.1.12) makes the potential theory of the chain X simpler (cf. Doob 1959). Maps of chains will always be assumed to map the origin to the origin.

We can view $\mathrm{P_X}$ as an operator on functions on the state space,

$$\mathrm{P_X} f(y) = \sum_x \mathrm{P_X}(y, x) f(x)\,, \tag{4.1.13}$$

satisfying

$$\begin{aligned} &\text{(i)} \quad f \geq 0 \Rightarrow \mathrm{P_X} f \geq 0\,, \\ &\text{(ii)} \quad \mathrm{P_X}\mathbb{1} = \mathbb{1}\,. \end{aligned} \tag{4.1.14}$$

Similarly we have the formal adjoint operator $\mathrm{P}^*_{\mathrm{X}}$

$$\mathrm{P}^*_{\mathrm{X}} f(y) = \sum_x \mathrm{P_X}(x, y) f(x)\,. \tag{4.1.15}$$

A positive function f on the state space $\mathrm{X}, f\colon \mathrm{X} \to [0, \infty)$, is *harmonic* (or *regular*) if

$$\mathrm{P_X} f = f, \quad \text{and} \quad f(\mathrm{O_X}) = 1\,. \tag{4.1.16}$$

The set of harmonic functions is a convex set whose extreme points correspond to the extreme points of the Martin boundary, which we describe next.

The *Green kernel* is given by

$$\mathrm{G_X}(x, y) = \sum_{n \geq 0} (\mathrm{P_X})^n (x, y)\,. \tag{4.1.17}$$

The *Martin kernel* is given by the normalization

$$\mathrm{K_X}(x, y) = \frac{\mathrm{G_X}(x, y)}{\mathrm{G_X}(\mathrm{O_X}, y)}\,. \tag{4.1.18}$$

A sequence of states $x_n \in \mathrm{X}$ is a Cauchy sequence with respect to the Martin metric if and only if

$$\mathrm{K_X}(x, x_n) \text{ is a Cauchy sequence (of real numbers) for all } x \in \mathrm{X}\,. \tag{4.1.19}$$

By completing X with respect to the Martin metric, we obtain a compact space $X \cup \partial X$; we say that ∂X is the *Martin boundary.*

The Martin Kernel $K_X(x, y)$ extends by continuity to the boundary

$$K_X(x, y) = \lim_{x_n \to y} K_X(x, x_n), \quad x \in X, \ y \in \partial X.$$

A sequence $x_n \in X \cup \partial X$ converges to $x_\infty \in X \cup \partial X$ if and only if for all $x \in X$ the sequence (of real numbers) $K_X(x, x_n)$ converges to $K_X(x, x_\infty)$.

For each $y \in \partial X$, $K_X(x, y)$ is harmonic as a function of $x \in X$. Integration against the kernel $K_X(x, y)$ induces a one-to-one correspondence between harmonic functions f on X, and probability measures μ supported on the (extreme points of the) boundary ∂X_{ext}:

$$f\colon X \to [0, \infty), \quad f(O_X) = 1, \quad P_X f = f \Longleftrightarrow f(x) = \int_{\partial X_{\text{ext}}} K_X(x, y)\mu(dy). \tag{4.1.20}$$

In particular, corresponding to the constant harmonic function $\mathbb{1}$ is the *harmonic measure* on ∂X_{ext}, denoted by τ_X, satisfying

$$\int_{\partial X} K_X(x, y)\tau_X(dy) = 1 \quad \text{for all } x \in X. \tag{4.1.21}$$

Let us denote by $\mathbb{N}$ the *unit shift* chain, with state space the natural numbers $\mathbb{N} = \{0, 1, 2, \dots, \}$ and transition probabilities

$$P_{\mathbb{N}}(i, j) = \begin{cases} 1 & j = i + 1, \\ 0 & j \neq i + 1. \end{cases} \tag{4.1.22}$$

A chain X together with a map of chains $X \to \mathbb{N}$ will be called a *timed chain*; equivalently, the state space X is a disjoint union $X = \coprod_{N \geq 0} X_N, X_0 = \{O_X\}$, and our particle walks from states in X_N only to states in X_{N+1}. Denoting by δ_{O_X} the delta function supported at O_X, we have a probability measure supported on X_N, namely

$$\tau_{X_N}(x) = (P_X^*)^N \delta_{O_X}(x) \tag{4.1.23}$$

is the probability of hitting $x \in X_N$ starting at O_X. We view τ_{X_N} as the Nth approximation of the harmonic measure τ_X. For a timed chain the computation of the Green (and hence Martin) Kernel is simplified:

$$G_X(x, x') = P_X^{N'-N}(x, x') \quad \text{for } x \in X_N, \ x' \in X_{N'}, \tag{4.1.24}$$

and $G_X(x, x')$ is the probability of hitting x' starting at x (after $N'-N$ steps).

The simplest kind of timed chain arises when the graph associated with our chain X is a tree (with origin, and with the orientation going out from the origin to the boundary); such a chain is timed with X_N, the sphere of radius N around the origin O_X. In this case the Martin boundary can be identified with (a subset of) the space of infinite paths beginning at O_X, or in other words, the inverse limit of the sets X_N with respect to the natural 'source' maps $X_{N+1} \to X_N$. We obtain canonical projections

$$\pi_N\colon \partial X \subseteq \varprojlim_N X_N \to X_N. \tag{4.1.25}$$

To each state $x \in \mathrm{X}$ we can associate the 'interval' $\mathrm{I}_x \subseteq \partial\mathrm{X}$ consisting of all paths passing through x, or in other words, the inverse image of $x \in \mathrm{X}_N$ under $\pi_N\colon \mathrm{I}_x = \pi_N^{-1}(\{x\})$. The transition probabilities for such a tree chain are completely determined by the harmonic measure τ_{X}:

$$\mathrm{P}_{\mathrm{X}}(x, x') = \frac{\tau_{\mathrm{X}}(\mathrm{I}_{x'})}{\tau_{\mathrm{X}}(\mathrm{I}_x)}, \quad x \in \mathrm{X}_N, x' \in \mathrm{X}_{N+1}. \tag{4.1.26}$$

The probability measure $\tau_{\mathrm{X}_N} = (\mathrm{P}_{\mathrm{X}}^*)^N \delta_{\mathrm{O}_{\mathrm{X}}}$ is just the image of the harmonic measure τ_{X} under the canonical projection: $\tau_{\mathrm{X}_N} = \pi_{N*}(\tau_{\mathrm{X}})$. For such a tree chain $\mathrm{X} = \coprod_{N\geq 0} \mathrm{X}_N$, integration against the Martin kernel induces the embeddings and projections

$$\ell_2(\mathrm{X}_N, \tau_{\mathrm{X}_N}) \underset{\mathrm{K}^*_{\mathrm{X}(N,M)}}{\overset{\mathrm{K}_{\mathrm{X}(N,M)}}{\rightleftarrows}} \ell_2(\mathrm{X}_M, \tau_{\mathrm{X}_M}), \quad N < M, \tag{4.1.27}$$

so we can identify $\ell_2(\mathrm{X}_N, \tau_{\mathrm{X}_N})$ with a subspace of $\ell_2(\mathrm{X}_M, \tau_{\mathrm{X}_M})$. Similarly in the limit $M \to \infty$, we obtain the embeddings and projections

$$\ell_2(\mathrm{X}_N, \tau_{\mathrm{X}_N}) \underset{\mathrm{K}^*_{\mathrm{X}(N)}}{\overset{\mathrm{K}_{\mathrm{X}(N)}}{\rightleftarrows}} \mathrm{L}_2(\partial\mathrm{X}, \tau_{\mathrm{X}}), \tag{4.1.28}$$

so we can identify $\ell_2(\mathrm{X}_N, \tau_{\mathrm{X}_N})$ with a subspace of $\mathrm{L}_2(\partial\mathrm{X}, \tau_{\mathrm{X}})$.

For a tree chain we define the *number operator* $\mathbb{N}_{\mathrm{X}}$ on $\mathrm{L}_2(\partial\mathrm{X}, \tau_{\mathrm{X}})$ by

$$\mathbb{N}_{\mathrm{X}} = \sum_{N\geq 0} \left(1 - \mathrm{K}^*_{\mathrm{X}(N)}\mathrm{K}_{\mathrm{X}(N)}\right). \tag{4.1.29}$$

Its spectral decomposition is given by

$$\mathrm{L}_2(\partial\mathrm{X}, \tau_{\mathrm{X}}) = \bigoplus_{n\geq 0} \mathscr{H}_{\mathrm{X}}(n), \quad \mathbb{N}_{\mathrm{X}} \equiv n \text{ on } \mathscr{H}_{\mathrm{X}}(n), \tag{4.1.30}$$

where $\mathscr{H}_{\mathrm{X}}(n) \cong \ker\{\mathrm{K}_{\mathrm{X}(n-1,n)}\}$ can be identified with the orthogonal complement of $\ell_2(\mathrm{X}_{n-1}, \tau_{\mathrm{X}_{n-1}})$ in $\ell_2(\mathrm{X}_n, \tau_{\mathrm{X}_n})$. We can identify the '$N$th approximation' with the subspace where the number operator has bounded spectrum:

$$\ell_2(\mathrm{X}_N, \tau_{\mathrm{X}_N}) = \bigoplus_{0\leq n\leq N} \mathscr{H}_{\mathrm{X}}(n). \tag{4.1.31}$$

4.2 THE p-ADIC GAMMA CHAIN

Fix a finite prime p, and consider the tree

$$\widetilde{\mathrm{X}}_p^0 = \coprod_{N\geq 0} \widetilde{\mathrm{X}}^0_{p(N)}, \quad \widetilde{\mathrm{X}}^0_{p(N)} = \mathbb{Z}/p^N\mathbb{Z}, \quad \widetilde{\mathrm{X}}^0_{p(0)} = \{\mathrm{O}_{\widetilde{\mathrm{X}}_p^0}\}, \tag{4.2.1}$$

with edges going from $x \in \widetilde{\mathrm{X}}^0_{p(N)}$ to its p pre-images in $\widetilde{\mathrm{X}}^0_{p(N+1)}$, and its boundary

$$\partial\widetilde{\mathrm{X}}_p^0 = \varprojlim_N \mathrm{X}_N\widetilde{\mathrm{X}}^0_{p(N)} = \mathbb{Z}_p. \tag{4.2.2}$$

The probability measure $\tau^{\beta}_{\mathbb{Z}_p}$ on the boundary $\partial\widetilde{\mathrm{X}}^0_p$ induces a unique chain $\mathrm{P}^{\beta}_{\widetilde{\mathrm{X}}^0_p}$ on the state space $\widetilde{\mathrm{X}}^0_p$, with harmonic measure $\tau^{\beta}_{\mathbb{Z}_p}$; we call it the *p-adic gamma chain*. For example, for $\beta = 1$, $\tau^1_{\mathbb{Z}_p}$ is the Haar measure on $\mathbb{Z}_p$, and the chain $\mathrm{P}^1_{\widetilde{\mathrm{X}}^0_p}$ is a random walk on the tree $\widetilde{\mathrm{X}}^0_p$ (with probability $1/p$ for each edge).

Since the measure $\tau^{\beta}_{\mathbb{Z}_p}$ is invariant under the $\mathbb{Z}_p^*$-action, the group $\mathbb{Z}_p^*$ acts by chain automorphisms on the p-adic gamma chain, and we have the associated quotient chain $\mathrm{P}^{\beta}_{\mathrm{X}^0_p}$; its state space is the quotient tree $\mathrm{X}^0_p = \widetilde{\mathrm{X}}^0_p/\mathbb{Z}_p^*$. We shall identify the state space of X^0_p with the set $\{(N,j) \in \mathbb{N}\times\mathbb{N} \,|\, 0 \le j \le N\}$ by using the parametrization

$$\mathrm{X}^0_p = \coprod_{N\ge 0} \mathrm{X}^0_{p(N)}, \quad \mathrm{X}^0_{p(N)} = (\mathbb{Z}/p^N\mathbb{Z})/(\mathbb{Z}/p^N\mathbb{Z})^* \cong \{(N,j), 0 \le j \le N\}$$

$$[p^{N-j}] \leftrightarrow (N,j)\,. \tag{4.2.3}$$

The boundary is now

$$\partial\mathrm{X}^0_p = \lim_{\overleftarrow{N}} \mathrm{X}^0_{p(N)} = \mathbb{Z}_p/\mathbb{Z}_p^* \cong p^{\mathbb{N}} \cup \{p^\infty = 0\}\,. \tag{4.2.4}$$

The transition probabilities for this simple chain are given in Fig. 4.1. The finite-dimensional 'mod p^N approximation' of the harmonic measure $\tau^{\beta}_{\mathbb{Z}_p}$ is given on $\mathrm{X}^0_{p(N)}$ by

$$\tau^{\beta}_{p(N)}(N,j) = \begin{cases} (1-p^{-\beta})\cdot p^{-(N-j)\beta} & N \ge j > 0\,,\\ p^{-N\beta} & N > j = 0\,. \end{cases} \tag{4.2.5}$$

The extension of the Martin kernel to the boundary is given by

$$\mathrm{K}^{\beta}_{\mathrm{X}^0_p}((N,j),p^n) = \begin{cases} \zeta_p(\beta)\cdot p^{(N-j)\beta} & n = N-j, j>0\,,\\ p^{N\beta} & n \ge N > j = 0\,,\\ 1 & n \ge N = j = 0\,,\\ 0 & \text{otherwise}\,. \end{cases} \tag{4.2.6}$$

Note that, unlike $\widetilde{\mathrm{X}}^0_p$, the shape of the underlying tree X^0_p is independent of the prime p. We have

$$\#\mathrm{X}^0_{p(N)} = N+1 \quad \text{and} \quad \#\widetilde{\mathrm{X}}^0_{p(N)} = p^N\,, \tag{4.2.7}$$

and so we obtain for the associated eigenspaces of the number operator

$$\dim \mathscr{H}^{\beta}_{\mathbb{Z}_p}(N) = 1 \quad \text{and} \quad \dim \widetilde{\mathscr{H}}^{\beta}_{\mathbb{Z}_p}(N) = \begin{cases} p^N(1-p^{-1}) & N \ge 1\,,\\ 1 & N = 0\,. \end{cases} \tag{4.2.8}$$

Equivalently, we obtain for the trace of the 'Heat kernel' $p^{-t\mathrm{N}}$,

$$\mathrm{tr}\left(p^{-t\mathrm{N}_{\mathrm{X}^0_p}} \,|\, \mathscr{H}^{\beta}_{\mathbb{Z}_p}\right) = \sum_{N\ge 0} p^{-tN} = \zeta_p(t), \quad \mathrm{Re}(t) > 0\,, \tag{4.2.9}$$

and

$$\mathrm{tr}\left(p^{-t\mathrm{N}_{\widetilde{\mathrm{X}}^0_p}} \,|\, \widetilde{\mathscr{H}}^{\beta}_{\mathbb{Z}_p}\right) = 1 + \sum_{N\ge 1}(1-p^{-1})p^{(1-t)N} = \frac{\zeta_p(t-1)}{\zeta_p(t)}, \quad \mathrm{Re}(t) > 1\,. \tag{4.2.10}$$

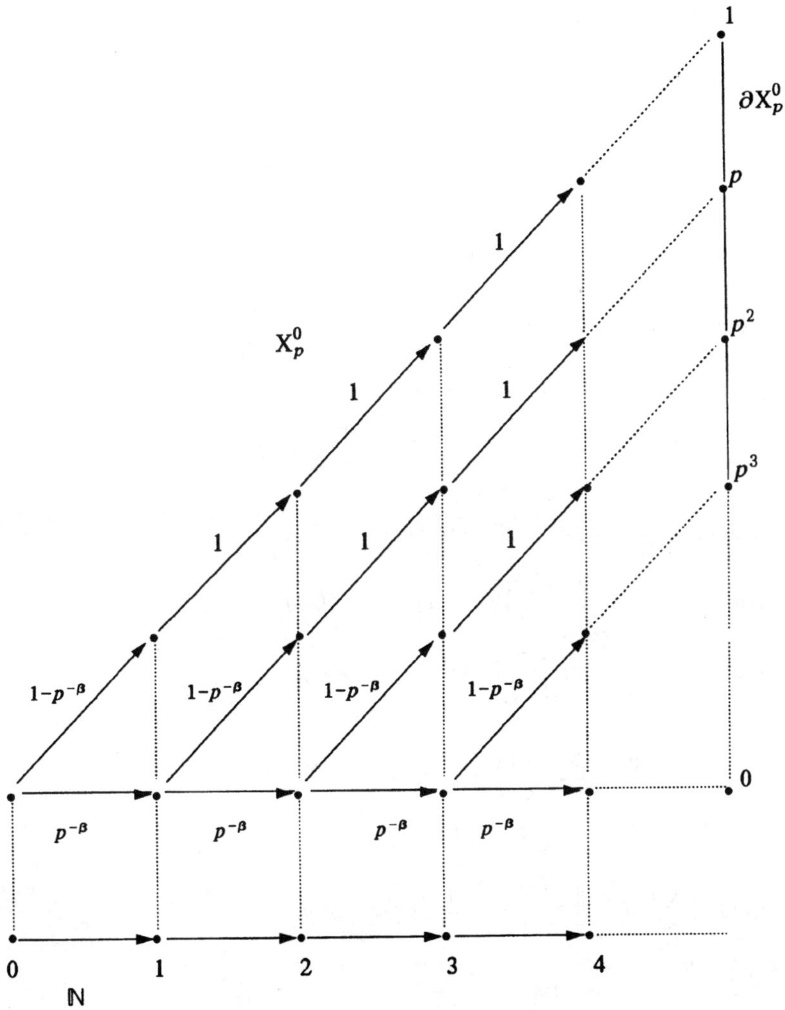

FIG. 4.1 The p-adic gamma chain, X_p^0.

Since $\dim \mathscr{H}^{\beta}_{\mathbb{Z}_p}(m) = 1$, we have a unique (up to constant multiple) eigenfunction $\varphi^{\beta}_{\mathbb{Z}_p,m} \in \mathscr{H}^{\beta}_{\mathbb{Z}_p}(m)$. We normalize by taking

$$\varphi^{\beta}_{\mathbb{Z}_p,0} = \phi_{\mathbb{Z}_p},$$
$$\varphi^{\beta}_{\mathbb{Z}_{p,m}} = (-1)^m p^{m\beta} \phi_{p^m\mathbb{Z}_p} + (-1)^{m-1} p^{(m-1)\beta} \phi_{p^{(m-1)}\mathbb{Z}_p}, \quad m \geq 1. \tag{4.2.11}$$

It is an orthogonal basis for $\mathscr{H}^{\beta}_{\mathbb{Z}_p}$, which we call the *p-adic Laguerre basis.* Note that $\varphi^{\beta}_{\mathbb{Z}_p,m}$ can also be viewed as a function defined mode p^N for $N \geq m$, and $\{\varphi^{\beta}_{\mathbb{Z}_p,m}\}_{0\leq m\leq N}$ give an orthogonal basis for the finite-dimensional 'mod p^N approximation'

$$\mathscr{H}^{\beta}_{p(N)} = \ell_2(X^0_{p(N)}, \tau^{\beta}_{p(N)}) = \bigoplus_{0\leq m\leq N} \mathscr{H}^{\beta}_{\mathbb{Z}_p}(m). \tag{4.2.12}$$

When viewed as functions on $\mathrm{X}^0_{p(N)}$ we shall write $\varphi^{\beta}_{p(N),m}$ instead of $\varphi^{\beta}_{\mathbb{Z}_p,m}$. Explicitly

$$\varphi^{\beta}_{p(N),0} \equiv \mathbb{1}$$

$$\varphi^{\beta}_{p(N),m}(N,j) = \begin{cases} 0 & N-j < m-1\,, \\ (-1)^{m-1}p^{(m-1)\beta} & N-j = m-1\,, \quad m>0 \\ (-1)^m p^{m\beta}(1-p^{-\beta}) & N-j \geq m\,. \end{cases} \tag{4.2.13}$$

The relation between $\varphi^{\beta}_{\mathbb{Z}_p,m}$ and the $\varphi^{\beta}_{p(N),m}$ is given by

$$\text{reduction mod}\, p^N\colon\ \mathrm{K}^{\beta}_{p(N)}\varphi^{\beta}_{\mathbb{Z}_p,m} = \begin{cases} \varphi^{\beta}_{p(N),m} & 0 \leq m \leq N\,, \\ 0 & N < m\,. \end{cases} \tag{4.2.14}$$

The basis $\{\varphi^{\beta}_{\mathbb{Z}_p,m}\}$ enjoys the following properties:

$$\text{norm:}\quad \left\|\varphi^{\beta}_{\mathbb{Z}_p,m}\right\|^2 = \left\{\begin{matrix} 1 & m=0 \\ p^{m\beta} - p^{(m-1)\beta} & m \geq 1 \end{matrix}\right\} \stackrel{\text{def}}{=} \mathrm{C}^{\beta}_p(m) \tag{4.2.15}$$

$$\text{special value:}\quad \varphi^{\beta}_{\mathbb{Z}_p,m}(0) = (-1)^m \cdot \mathrm{C}^{\beta}_p(m) \tag{4.2.16}$$

$$\text{Mellin transform:}\quad \tau^s_{\mathbb{Z}_p}\left(\varphi^{\beta}_{\mathbb{Z}_p,m}\right) = (-1)^m \cdot \mathrm{C}^{\beta-s}_p(m) \tag{4.2.17}$$

$$\beta'\!-\!\beta \text{ relation:}\quad \varphi^{\beta'}_{\mathbb{Z}_p,m} = \sum_{0\leq j\leq m} (-1)^j \mathrm{C}^{\beta'-\beta}_p(j)\varphi^{\beta}_{\mathbb{Z}_p,m-j} \tag{4.2.18}$$

$$\text{generating function:}\quad \sum_{m\geq 0}(-1)^m p^{-tm}\left\|\varphi^{\beta}_{\mathbb{Z}_p,m}\right\|^{-2}\varphi^{\beta}_{\mathbb{Z}_p,m}(x) = \zeta_p(t,\beta)\cdot\left(1-|px|^t_p\right)\phi_{\mathbb{Z}_p}(x) \tag{4.2.19}$$

$$\text{heat kernel:}\quad \sum_{m\geq 0} p^{-tm}\left\|\varphi^{\beta}_{\mathbb{Z}_p,m}\right\|^{-2}\varphi^{\beta}_{\mathbb{Z}_p,m}(x)\varphi^{\beta}_{\mathbb{Z}_p,m}(y) = \zeta_p(\beta)p^{-t}|x|^{t-\beta}_p\phi_{\mathbb{Z}_p^*}\left(\frac{x}{y}\right) + \frac{\zeta_p(t-\beta)}{\zeta_p(t)}\cdot\left(1-|px,py|^{t-\beta}_p\right). \tag{4.2.20}$$

4.3 THE p-ADIC SYMMETRIC BETA CHAIN

Let $\mathrm{G} = \mathrm{PGL}_2$, $\mathrm{B} = \mathbb{G}_m \ltimes \mathbb{G}_a$ its Borel subgroup stabilizing ∞, and $\mathbb{P}^1 = \mathrm{G}/\mathrm{B}$. Consider the *p-adic hyperbolic plane*, the $(p+1)$-regular tree

$$\widetilde{\mathrm{X}}_p = \mathrm{G}(\mathbb{Q}_p)/\mathrm{G}(\mathbb{Z}_p)\,. \tag{4.3.1}$$

We fix the origin $\mathrm{O}_{\widetilde{\mathrm{X}}_p}$ ($= \mathrm{G}(\mathbb{Z}_p)$), and give $\widetilde{\mathrm{X}}_p$ the orientation going out from $\mathrm{O}_{\widetilde{\mathrm{X}}_p}$ to the boundary. We have natural identifications

$$\widetilde{\mathrm{X}}_p = \coprod_{N\geq 0}\widetilde{\mathrm{X}}_{p(N)}, \quad \widetilde{\mathrm{X}}_{p(N)} \cong \mathbb{P}^1(\mathbb{Z}/p^N\mathbb{Z})\,,$$
$$\partial\widetilde{\mathrm{X}}_p = \varprojlim_N \widetilde{\mathrm{X}}_{p(N)} = \mathbb{P}^1(\mathbb{Z}_p)\,. \tag{4.3.2}$$

The probability measure $\tau_p^{\alpha,\beta}$ on the boundary $\partial\widetilde{\mathrm{X}}_p$ induces a unique chain $\mathrm{P}^{\alpha,\beta}_{\widetilde{\mathrm{X}}_p}$ on the state space $\widetilde{\mathrm{X}}_p$ with harmonic measure $\tau_p^{\alpha,\beta}$; we call it the *p-adic symmetric beta chain*. For example, for $\alpha = \beta = 1$, the chain $\mathrm{P}^{1,1}_{\widetilde{\mathrm{X}}_p}$ is a random walk on the tree $\widetilde{\mathrm{X}}_p$ (with probability $1/(p+1)$ for the first step, and probability $1/p$ for all other steps).

Since $\tau_p^{\alpha,\beta}$ is $\mathbb{Z}_p^*$-invariant, we have an action of $\mathbb{Z}_p^*$ by automorphisms on the beta chain, and we may pass to the quotient chain $\mathrm{P}^{\alpha,\beta}_{\mathrm{X}_p}$; its state space is the quotient tree

$$\mathrm{X}_p = \mathbb{Z}_p^*\backslash\widetilde{\mathrm{X}}_p = \coprod_{N\geq 0} \mathrm{X}_{p(N)}\,.$$

We identify X_p with $\mathbb{N}\times\mathbb{N}$ using the parametrization

$$\mathrm{X}_{p(N)} = (\mathbb{Z}/p^N\mathbb{Z})^*\backslash\mathbb{P}^1(\mathbb{Z}/p^N\mathbb{Z}) \cong \{(i,j) \mid \max\{i,j\} = N\} \tag{4.3.3}$$
$$(p^{N-i}{:}\,p^{N-j}) \leftrightarrow (i,j)\,.$$

The boundary is now

$$\partial\mathrm{X}_p = \varprojlim_N \mathrm{X}_{p(N)} = \mathbb{Z}_p^*\backslash\mathbb{P}^1(\mathbb{Z}_p) \cong \{\infty\}\cup p^{\mathbb{Z}}\cup\{0\}\,. \tag{4.3.4}$$

The transition probabilities for this simple chain are given in Fig. 4.2. The finite-dimensional 'mod p^N' approximations of the harmonic measure $\tau_p^{\alpha,\beta}$ are given on $\mathrm{X}_{p(N)}$ by

$$\tau_{p(N)}^{\alpha,\beta}(i,j) = \begin{cases} \dfrac{\zeta_p(\alpha+\beta)}{\zeta_p(\alpha)}p^{-N\beta} & j=0\,,\\ \zeta_p(\alpha,\beta)^{-1}\cdot|1,p^{i-j}|_p^{-\alpha}\cdot|1,p^{j-i}|_p^{-\beta} & i,j>0\,,\\ \dfrac{\zeta_p(\alpha+\beta)}{\zeta_p(\beta)}p^{-N\alpha} i=0\,. \end{cases} \tag{4.3.5}$$

The extension of the Martin kernel to the boundary is given by

$$\mathrm{K}^{\alpha,\beta}_{\mathrm{X}_p}((i,j),p^n) = \begin{cases} 1 & i=j=0\,,\\ \zeta_p(\alpha,\beta)\cdot|1,p^n|_p^{\alpha}\cdot|1,p^{-n}|_p^{\beta}, & n=i-j, i,j>0\,,\\ \dfrac{\zeta_p(\alpha)}{\zeta_p(\alpha+\beta)}\cdot p^{\beta\cdot i} & n\geq i>j=0\,,\\ \dfrac{\zeta_p(\beta)}{\zeta_p(\alpha+\beta)}\cdot p^{\alpha\cdot j} & n\leq -j<i=0\,,\\ 0 & \text{otherwise}\,. \end{cases} \tag{4.3.6}$$

Note again that the shape of the underlying tree X_p is independent of the prime p. We have a chain map $\widetilde{\mathrm{X}}_p \to \mathrm{X}_p$, and the beta chain on $\widetilde{\mathrm{X}}_p$ is a random walk over the beta chain on X_p.

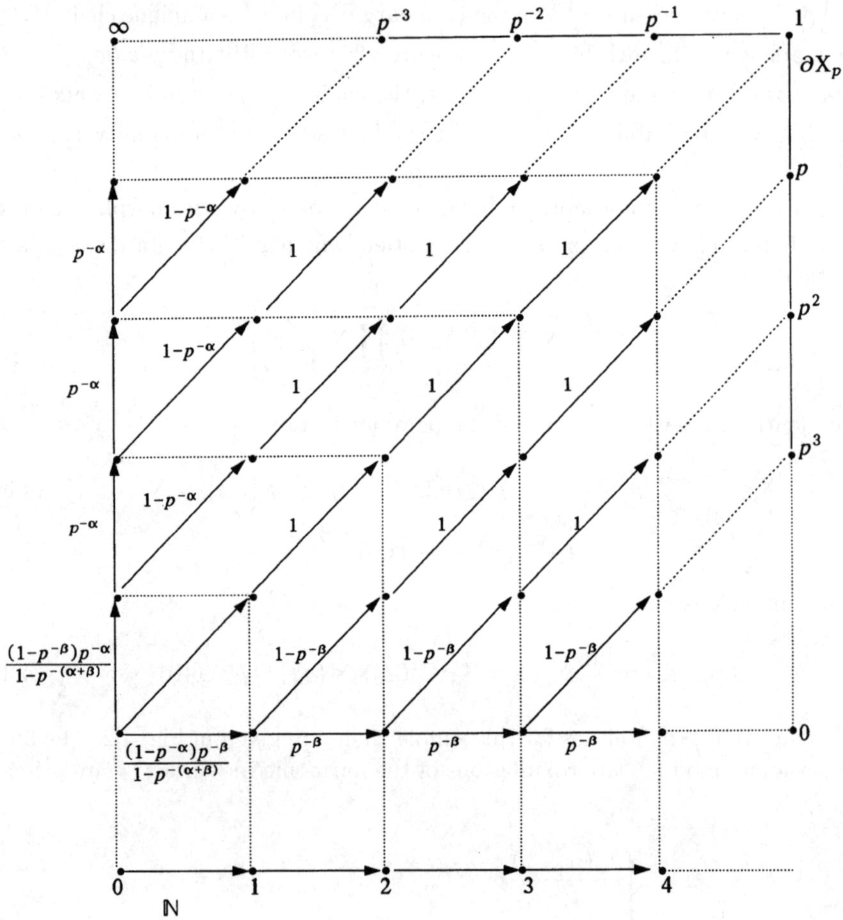

FIG. 4.2 The p-adic symmetric beta chain, X_p.

We have

$$\#X_{p(N)} = 2N+1 \quad \text{and} \quad \#\widetilde{X}_{p(N)} = \begin{cases} 1 & N=0\,, \\ p^N(1+p^{-1}) & N\geq 1\,, \end{cases} \tag{4.3.7}$$

and we obtain for the associated eigenspaces of the number operator

$$\dim \mathscr{H}_p^{\alpha,\beta}(N) = \begin{cases} 1 & N=0 \\ 2 & N\geq 1 \end{cases} \quad \text{and} \quad \dim \widetilde{\mathscr{H}}_p^{\alpha,\beta}(N) = \begin{cases} 1 & N=0\,, \\ p & N=1\,, \\ p^N\cdot(1-p^{-2}) & N\geq 2\,. \end{cases} \tag{4.3.8}$$

Equivalently, the trace of the heat kernels $p^{-t\mathbb{N}}$ is given by

$$\operatorname{tr}\left(p^{-t\mathbb{N}_{X_p}} \mid \mathscr{H}_p^{\alpha,\beta}\right) = 1 + \sum_{N\geq 1} 2p^{-tN} = \frac{\zeta_p(t)^2}{\zeta_p(2t)}, \quad \operatorname{Re}(t)>0\,, \tag{4.3.9}$$

and

$$\operatorname{tr}\left(p^{-t\mathbb{N}_{\widetilde{\mathrm{X}}_p}} \,\middle|\, \widetilde{\mathscr{H}}_p^{\alpha,\beta}\right) = 1 + p^{1-t} + \sum_{N\geq 2}(1-p^{-2})p^{N(1-t)} = \frac{\zeta_p(t-1)}{\zeta_p(2t)}, \quad \mathrm{Re}(t) > 1. \tag{4.3.10}$$

It is clear that in the limit $\alpha \to \infty$, the symmetric beta chain $\mathrm{P}^{\alpha,\beta}$ on X_p (or $\widetilde{\mathrm{X}}_p$) 'converges' to the gamma chain P^{β} on X_p^0 (or $\widetilde{\mathrm{X}}_p^0$); compare Figs 4.1 and 4.2. In the limit $\alpha, \beta \to \infty$, the symmetric beta chain $\mathrm{P}^{\alpha,\beta}_{\widetilde{\mathrm{X}}_p}$ on $\widetilde{\mathrm{X}}_p$ converges to the random walk on the tree $\widetilde{\mathrm{X}}_p^{(1)}$ associated with the multiplicative group of units $\mathbb{Z}_p^*$ (with probability $1/(p-1)$ for the first step, and probability $1/p$ for all other steps):

$$\widetilde{\mathrm{X}}_p^{(1)} = \coprod_{N\geq 0} \widetilde{\mathrm{X}}_{p(N)}^{(1)} = \coprod_{N\geq 0} (\mathbb{Z}/p^N\mathbb{Z})^*, \tag{4.3.11}$$

with boundary

$$\partial \widetilde{\mathrm{X}}_p^{(1)} = \varprojlim_N \widetilde{\mathrm{X}}_{p(N)}^{(1)} = \mathbb{Z}_p^*. \tag{4.3.12}$$

The harmonic measure is the normalized Haar measure on $\mathbb{Z}_p^*$, so that

$$\mathrm{d}^*a = \zeta_p(1)\frac{\mathrm{d}a}{|a|_p}.$$

We have

$$\#\widetilde{\mathrm{X}}_{p(N)}^{(1)} = \begin{cases} 1 & N = 0, \\ p^N(1-p^{-1}) & N \geq 1, \end{cases} \tag{4.3.13}$$

and we obtain for the trace of the heat kernel $p^{-t\mathbb{N}}$

$$\begin{aligned} \operatorname{tr}\left(p^{-t\mathbb{N}} \,\middle|\, \mathrm{L}_2(\mathbb{Z}_p^*)\right) &= 1 + (p-2)p^{-t} + \sum_{N\geq 2}(1-p^{-1})^2 p^{N(1-t)} \\ &= \frac{\zeta_p(t-1)}{\zeta_p(t)^2}, \quad \mathrm{Re}(t) > 1. \end{aligned} \tag{4.3.14}$$

We have descended from the p-adic hyperbolic plane $\widetilde{\mathrm{X}}_p$ to the tree X_p, dividing by the action of $\mathbb{Z}_p^*$. But the essential chain is the one living on X_p, since the chain on $\widetilde{\mathrm{X}}_p$ is just the random walk over X_p, and can be reconstructed from it by means of (4.1.4). We can ascend further in a similar way. For any commutative ring R, let us write

$$V^*(\mathrm{R}) = \{(x,y) \in \mathrm{R}\times\mathrm{R} \,|\, x\mathrm{R} + y\mathrm{R} = \mathrm{R}\} \cong \mathrm{GL}_2(\mathrm{R}) \Big/ \left\{\begin{pmatrix} * & * \\ 0 & 1 \end{pmatrix}\right\}. \tag{4.3.15}$$

The units R^* act naturally on $V^*(\mathrm{R})$, $u\cdot(x,y) = (ux,uy)$, $u \in \mathrm{R}^*$, and the quotient is just $\mathrm{R}^*\backslash V^*(\mathrm{R}) = \mathbb{P}^1(\mathrm{R})$. We now have a graph over $\widetilde{\mathrm{X}}_p$,

$$\widetilde{\widetilde{\mathrm{X}}}_p = V(\mathbb{Q}_p)/V(\mathbb{Z}_p) \cong \coprod_{N\geq 0} V^*(\mathbb{Z}/p^N\mathbb{Z}), \tag{4.3.16}$$

$$\left(\frac{x}{p^N}, \frac{y}{p^N}\right) \leftrightarrow (x,y) \bmod p^N,$$

with boundary

$$\partial\widetilde{\widetilde{\mathrm{X}}}_p = \varprojlim_N V^*(\mathbb{Z}/p^N\mathbb{Z}) = V^*(\mathbb{Z}_p) = \left\{(x,y) \in V(\mathbb{Q}_p) \,\middle|\, |(x,y)|_p = 1\right\}. \quad (4.3.17)$$

The symmetric beta chain on $\widetilde{\mathrm{X}}_p$ induces a chain on $\widetilde{\widetilde{\mathrm{X}}}_p$, the random walk over $\widetilde{\mathrm{X}}_p$; cf. (4.1.4). The group $\mathbb{Z}_p^*$ acts by automorphisms on the chain on $\widetilde{\widetilde{\mathrm{X}}}_p$, and the quotient is the chain on $\widetilde{\mathrm{X}}_p$. In summary, we have the tower of chains

$$\widetilde{\widetilde{\mathrm{X}}}_p \to \widetilde{\mathrm{X}}_p \to \mathrm{X}_p\,, \quad (4.3.18)$$

where each arrow is obtained by dividing by the action of $\mathbb{Z}_p^*$.

In the language of elliptic curves, $\widetilde{\widetilde{\mathrm{X}}}_p$ describes the p power torsion points, $\widetilde{\mathrm{X}}_p$ describes the p-power isogenies, and the arrow $\widetilde{\widetilde{\mathrm{X}}}_p \to \widetilde{\mathrm{X}}_p$ is the association of an isogeny to a torsion point by dividing the elliptic curve by the subgroup generated by the point.

4.4 THE p-ADIC BETA CHAIN

We can take smaller quotients of the p-adic symmetric beta chain. For example, we may divide by the full automorphism group of the underlying tree $\widetilde{\mathrm{X}}_p$ (preserving the origin), namely $\mathrm{G}(\mathbb{Z}_p)$, even though $\mathrm{G}(\mathbb{Z}_p)$ does not act by chain automorphisms! The associated state space is

$$\mathrm{G}(\mathbb{Z}_p)\backslash\widetilde{\mathrm{X}}_p = \mathrm{G}(\mathbb{Z}_p)\backslash\mathrm{G}(\mathbb{Q}_p)/\mathrm{G}(\mathbb{Z}_p) \cong \mathbb{N}\,, \quad (4.4.1)$$

and the associated chain is the unit shift; we have a map of chains, the time map,

$$\mathrm{X}_p \to \mathbb{N}, \quad (i,j) \mapsto \max\{i,j\}\,. \quad (4.4.2)$$

We can similarly divide by the subgroup $\mathrm{B}(\mathbb{Z}_p)$, obtaining a chain intermediate between X_p and the unit shift $\mathbb{N}$. The underlying tree is

$$\mathrm{X}_p^{\square} = \mathrm{B}(\mathbb{Z}_p)\backslash\widetilde{\mathrm{X}}_p = \mathrm{B}(\mathbb{Z}_p)\backslash\mathrm{B}(\mathbb{Q}_p)/\mathrm{B}(\mathbb{Z}_p)\,, \quad (4.4.3)$$

and its boundary is

$$\partial\mathrm{X}_p^{\square} = \mathrm{B}(\mathbb{Z}_p)\backslash\mathrm{G}(\mathbb{Z}_p)/\mathrm{B}(\mathbb{Z}_p) = \mathrm{B}(\mathbb{Z}_p)\backslash\mathbb{P}^1(\mathbb{Z}_p) \cong p^{\mathbb{N}} \cup \{0\}\,. \quad (4.4.4)$$

The tree $\mathrm{X}_p^{\square}$ is isomorphic to the tree of the gamma chain

$$\mathrm{X}_p^0 = \{(i,j) \in \mathbb{N}\times\mathbb{N} \,|\, i \ge j\}\,,$$

but it will be convenient to work with a model where the state space is identified with all of $\mathbb{N}\times\mathbb{N}$.

We use the parametrization

$$\begin{aligned}
\mathrm{X}_p^{\square} &= \coprod_{N\ge 0} \mathrm{X}_{p(N)}^{\square} \cong \mathbb{N}\times\mathbb{N}\,,\\
\mathrm{X}_{p(N)}^{\square} &= \mathrm{B}(\mathbb{Z}/p^N\mathbb{Z})\backslash\mathrm{G}(\mathbb{Z}/p^N\mathbb{Z})/\mathrm{B}(\mathbb{Z}/p^N\mathbb{Z}) \\
&= \mathrm{B}(\mathbb{Z}/p^N\mathbb{Z})\backslash\mathbb{P}^1(\mathbb{Z}/p^N\mathbb{Z}) \cong \{(i,j), i+j=N\} \\
&\qquad (1:p^i) \leftrightarrow (i, N-i)\,.
\end{aligned} \quad (4.4.5)$$

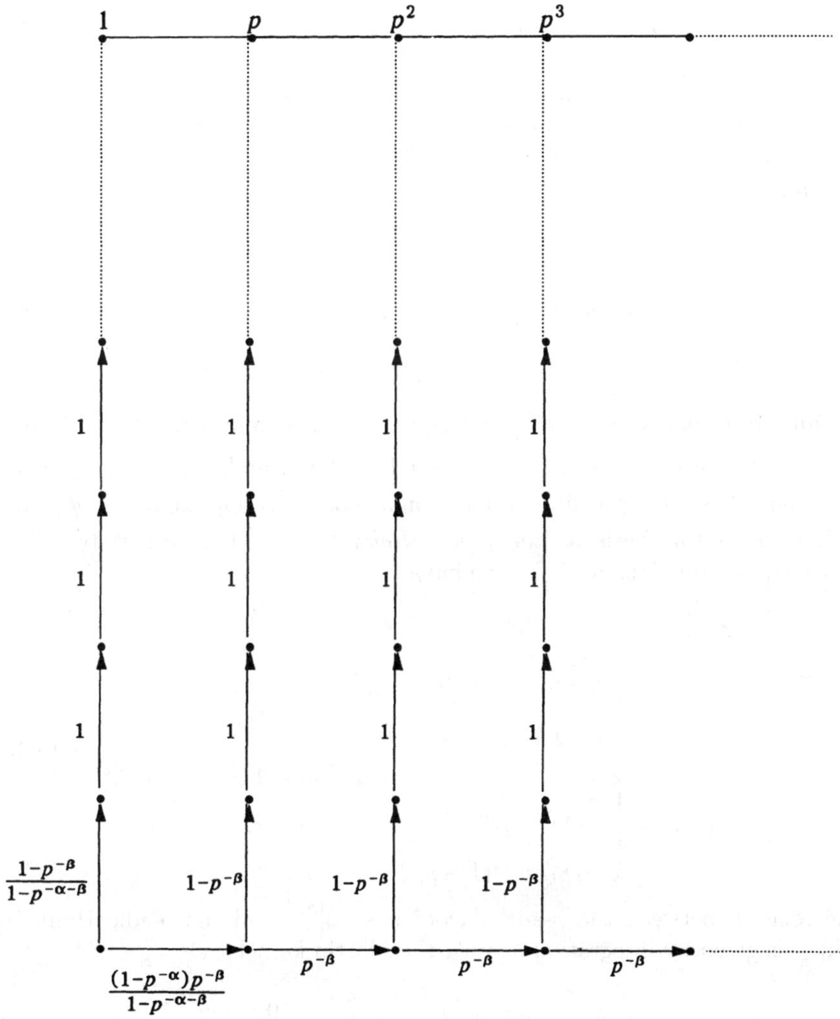

FIG. 4.3 The p-adic beta chain, $X_p^\square$.

Let $\tau_p^{(\alpha)\beta}$ denote the image of the beta measure $\tau_p^{\alpha,\beta}$ under the natural map $\partial X_p \twoheadrightarrow \partial X_p^\square$. The unique chain on $X_p^\square$ with harmonic measure $\tau_p^{(\alpha)\beta}$ will be called the (non-symmetric) *p-adic beta chain*; the transition probabilities $P_{X_p^\square}^{(\alpha)\beta}$ are given in Fig. 4.3. The time map is now

$$X_p^\square = \mathbb{N} \times \mathbb{N} \to \mathbb{N}, \quad (i,j) \mapsto i + j. \tag{4.4.6}$$

We write

$$\mathscr{H}_p^{(\alpha)\beta} = \ell_2(\partial X_p^\square, \tau_p^{(\alpha)\beta}) = \bigoplus_{m \geq 0} \mathscr{H}_p^{(\alpha)\beta}(m) \tag{4.4.7}$$

for the associated Hilbert space, and its Nth approximation

$$\mathscr{H}^{(\alpha)\beta}_{p(N)} = \ell_2\left(\mathrm{X}^{\square}_{p(N)}, \tau^{(\alpha)\beta}_{p(N)}\right) = \bigoplus_{N\geq m\geq 0} \mathscr{H}^{(\alpha)\beta}_{p}(m)\,. \tag{4.4.8}$$

Again, since the eigenspaces of the number operator $\mathscr{H}^{(\alpha)\beta)}_{p}(m)$ are one-dimensional, we have a unique (up to constant multiple) eigenfunction $\varphi^{(\alpha)\beta}_{p,m} \in \mathscr{H}^{(\alpha)\beta}_{p}(m)$. These are not too different from the Laguerre basis of (4.2); the difference occurs only for $m = 1$:

$$\begin{aligned} \varphi^{(\alpha)\beta}_{p,0} &= \mathbb{1}\,, \\ \varphi^{(\alpha)\beta}_{p,1} &= (1-p^{-\alpha})(\mathbb{1}-\phi_{p\mathbb{Z}_p}) + (1+p^{\beta})\phi_{p\mathbb{Z}_p}\,, \\ \varphi^{(\alpha)\beta}_{p,m} &= (-1)^m p^{m\beta}\phi_{p^m\mathbb{Z}_p} + (-1)^{m-1}p^{(m-1)\beta}\phi_{p^{m-1}\mathbb{Z}_p}, \quad m>1\,. \end{aligned} \tag{4.4.9}$$

It is an orthogonal basis for $\mathscr{H}^{(\alpha)\beta}_{p}$ which we call the *p-adic Jacobi basis.* Again for $0 \leq m \leq N$, we can view $\varphi^{(\alpha)\beta}_{p,m}$ as defined mod p^N, and $\{\varphi^{(\alpha)\beta}_{p,m}\}_{0\leq m\leq N}$ give an orthogonal basis for the finite-dimensional mod p^N approximation $\mathscr{H}^{(\alpha)\beta}_{p(N)}$; we shall refer to this basis as the *p-adic Hahn basis* and denote it by $\varphi^{(\alpha)\beta}_{p(N),m}$. Explicitly, as functions on $\mathrm{X}^{\square}_{p(N)}$ we have

$$\begin{aligned} \varphi^{(\alpha)\beta}_{p(N),0} &\equiv \mathbb{1}\,, \\ \varphi^{(\alpha)\beta}_{p(N),1}(i,j) &= \begin{cases} 1-p^{-\alpha} & i=0,\ j=N\,, \\ 1-p^{\beta} & i>0\,, \end{cases} \\ \varphi^{(\alpha)\beta}_{p(N),m}(i,j) &= \begin{cases} 0 & i<m-1\,, \\ (-1)^{m-1}p^{(m-1)\beta} & i=m-1, \quad m>1\,, \\ (-1)^m p^{m\beta}(1-p^{-\beta}) & i\geq m\,. \end{cases} \end{aligned} \tag{4.4.10}$$

The relation between the p-adic Jacobi base $\varphi^{(\alpha)\beta}_{p,m}$ and the p-adic Hahn base $\varphi^{(\alpha)\beta}_{p(N),m}$ is given by integration against the Martin kernel

$$\text{reduction mod}\, p^N\!: \quad \mathrm{K}^{(\alpha)\beta}_{p(N)}\varphi^{(\alpha)\beta}_{p,m} = \begin{cases} \varphi^{(\alpha)\beta}_{p(N),m} & 0\leq m\leq N\,, \\ 0 & N<m\,. \end{cases} \tag{4.4.11}$$

5

Some real hyperbolic chains

We can now return to our question of what is the real analogue of the inverse limit expressions for $\mathbb{Z}_p, \mathbb{P}^1(\mathbb{Z}_p), \mathbb{Z}_p^*$, and ask, more precisely, for the real analogue of the gamma and beta chains. In this chapter we review some of the well-known 'real hyperbolic chains', before giving what we believe is the right arithmetical approach.

5.1 THE HYPERBOLIC PLANE

The first possibility is to follow group theory, taking the classical hyperbolic plane $\widetilde{\mathrm{X}}_\eta \overset{?}{=} \mathrm{G}(\mathbb{Q}_\eta)/\mathrm{G}(\mathbb{Z}_\eta), \mathrm{G} = \mathrm{PGL}_2$, with its boundary $\partial\widetilde{\mathrm{X}}_\eta = \mathbb{P}^1(\mathbb{Q}_\eta)$; this leads to classical potential theory, from which the discrete theory of the preceding chapter was historically developed. The problem with this possibility is that the 'time' of the process is now the continuum $\mathrm{G}(\mathbb{Z}_\eta)\backslash\mathrm{G}(\mathbb{Q}_\eta)/\mathrm{G}(\mathbb{Z}_\eta) \cong [0,1)$, and the circle $\mathbb{P}^1(\mathbb{R})$ is approximated just by smaller circles; we have to insist on a discrete time chain. There are many discretizations of this model—indeed there are infinitely many (even a continuum of tree) chains with boundary $\mathbb{P}^1(\mathbb{Q}_\eta)$ and harmonic measure equal to the beta measure $\tau_\eta^{\alpha,\beta}$. The question is which is the right arithmetical one?

5.2 N-ADIC EXPANSION

One possibility is to note that the $(p+1)$-regular tree $\widetilde{\mathrm{X}}_p$ comes from expanding a p-adic number as a power series in powers of p, and hence its real analogue should be taken to be the $\mathrm{X}_\eta \overset{?}{=} (N+1)$-regular tree, using N-adic expansions for real numbers. We do indeed have a unique chain on this $(N+1)$-regular tree, whose harmonic measure is the beta measure $\tau_\eta^{\alpha,\beta}$, but there is nothing canonical about this—the choice of N is quite arbitrary.

5.3 CONTINUED FRACTION EXPANSION

Another possibility is to use the continued fraction expansion for real numbers, obtaining a chain on a binary tree. This gives a discrete and canonical chain, and a very interesting one from an arithmetical point of view. The state space is now

$$\mathrm{X}_\eta \overset{?}{=} \mathbb{Q}^+ \cong \{x = (i,j) \in \mathbb{N}^* \times \mathbb{N}^*,\ \text{g.c.d. } (i,j) = 1\}\,. \tag{5.3.1}$$

It will be convenient to remember the upper and lower bounds of each state x; we denote them by x_+ and x_-. We begin with the origin state

$$\begin{pmatrix} x_+ \\ x \\ x_- \end{pmatrix} = \begin{pmatrix} (0,1) \\ (1,1) \\ (1,0) \end{pmatrix} = `\begin{pmatrix} \infty \\ 1 \\ 0 \end{pmatrix}' \tag{5.3.2}$$

and use the branching rule for our binary tree

$$\begin{pmatrix} x_+ \\ x \\ x_- \end{pmatrix} \begin{array}{l} \overset{t_\infty}{\nearrow} \begin{pmatrix} x_+ \\ x + x_+ \\ x \end{pmatrix} \\[2ex] \underset{t_0}{\searrow} \begin{pmatrix} x \\ x + x_- \\ x_- \end{pmatrix}. \end{array} \tag{5.3.3}$$

Note that x is determined by x_-, x_+ via $x = x_- + x_+$, and that $\langle x_-, x_+ \rangle = 1$ so that $\left(\begin{smallmatrix} x_- \\ x_+ \end{smallmatrix}\right)$ is an element of the semi-group

$$\mathrm{SL}_2(\mathbb{N}) = \left\{ \begin{pmatrix} a & b \\ c & d \end{pmatrix} \in \mathrm{SL}_2(\mathbb{Z}), a,\, b,\, c,\, d \geq 0 \right\}. \tag{5.3.4}$$

Conversely, x determines x_-, x_+, and for any $x \in \mathrm{X}_\eta$ there is a *unique* $g_x = \left(\begin{smallmatrix} x_- \\ x_+ \end{smallmatrix}\right) \in \mathrm{SL}_2(\mathbb{N})$ with $(1,1)g_x = x$. The branching rule is given by

$$g_{t_\infty x} = \begin{pmatrix} 1 & 1 \\ 0 & 1 \end{pmatrix} g_x\,, \quad g_{t_0 x} = \begin{pmatrix} 1 & 0 \\ 1 & 1 \end{pmatrix} g_x\,.$$

The unique path from the origin state (1,1) to the state $x = (i,j)$ can be written as $t^{a_n} \ldots t_0^{a_3} t_\infty^{a_2} t_0^{a_1} t_\infty^{a_0}$, where $a_0 \geq 0, a_{k+1} \geq 1$ (i.e. a_0 steps to ∞ followed by a_1 steps to $0 \ldots$) and corresponds to the continued fraction expansion

$$\frac{j}{i} = a_0 + \cfrac{1}{a_1 + \cfrac{1}{a_2 + \cdots + \frac{1}{a_n + 1}}}\,. \tag{5.3.5}$$

We obtain a unique chain on this tree with harmonic measure equal to the beta measure $\tau_\eta^{\alpha,\beta}$ on the boundary $\partial\mathrm{X}_\eta = \mathbb{P}^1(\mathbb{R})/\pm 1$. It is given by

$$\mathrm{P}(x, x + x_\pm) = \frac{\tau_\eta^{\alpha,\beta}([x, x_\pm])}{\tau_\eta^{\alpha,\beta}([x_-, x_+])}\,. \tag{5.3.6}$$

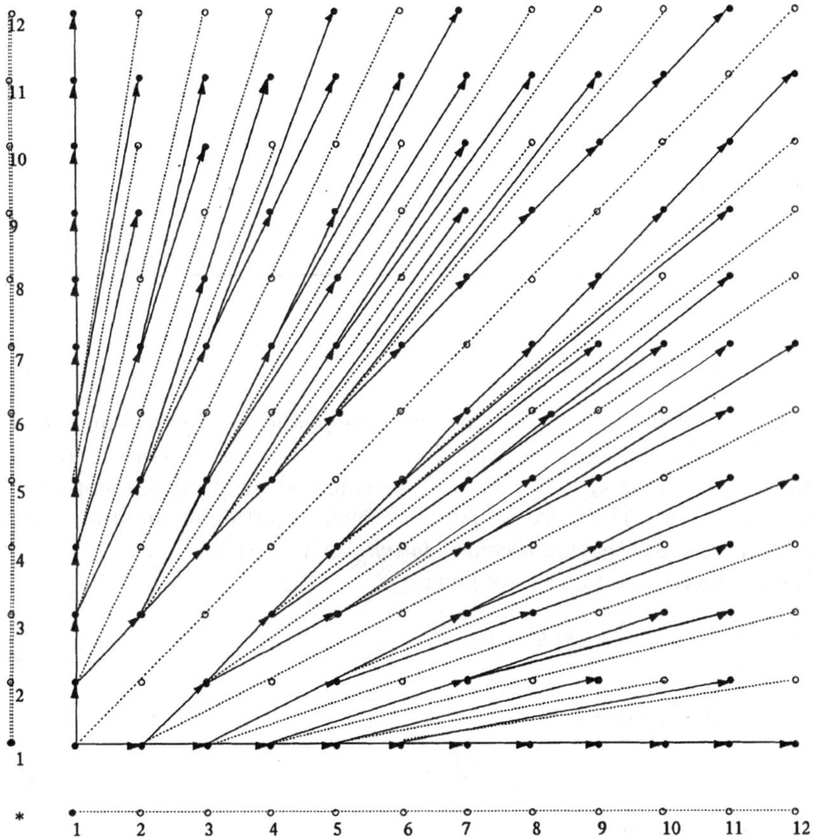

FIG. 5.1 The Farey tree.

The problem with this approach is that it is *global* (and not *local* like the p-adic beta chain): it is based on the fact that the global group $\mathrm{SL}_2(\mathbb{Z})$ is generated by the elements

$$t_\infty = \begin{pmatrix} 1 & 1 \\ 0 & 1 \end{pmatrix} \quad \text{and} \quad t_0 = \begin{pmatrix} 1 & 0 \\ 1 & 1 \end{pmatrix}$$

(with defining relations $t_\infty t_0^{-1} t_\infty = t_0^{-1} t_\infty t_0^{-1}, (t_\infty t_0^{-1} t_\infty)^4 = 1$), and, more precisely, that $\mathrm{SL}_2(\mathbb{N})$ is the *free* semi-group on t_∞, t_0. In fact, the uniform distribution (as $N \to \infty$) of the vertices of the binary tree $\{(i:j) \mid \text{g.c.d.}\,(i,j) = 1, |i|, |j| \le N\}$ within the continuum $\mathbb{P}^1(\mathbb{R})/\pm 1$ is equivalent to Riemann's hypothesis:

$$\frac{1}{\zeta(s)} = \sum_{n \ge 1} \frac{\mu(n)}{n^s} \text{has no poles in } \mathrm{Re}(s) > \frac{1}{2}$$

$$\Longleftrightarrow \sum_{\substack{1 \le j \le n \le N \\ \text{g.c.d.}(n,j)=1}} \mathrm{e}^{2\pi \mathrm{i}(j/n)} = \sum_{1 \le n \le N} \mu(n) \tag{5.3.7}$$

$$= \mathrm{O}(N^{1/2+\varepsilon}) \quad \text{for all } \varepsilon > 0.$$

6

Ramanujan's garden

6.1 THE q-ZETA FUNCTION

There is a world that interpolates between the p-adic and the real numbers; it is the world of quantum groups, basic hypergeometric series, q-series, etc. In this world, the continuum $\mathbb{Q}_\eta^*/\mathbb{Z}_\eta^* = \mathbb{R}^+$ is approximated by $q^{\mathbb{Z}}$, resembling more the p-adic $\mathbb{Q}_p^*/\mathbb{Z}_p^* = p^{\mathbb{Z}}$. The price we pay is to replace certain commuting operators X, Y by operators satisfying *q-commutativity*: $\mathrm{XY} = q\mathrm{YX}$.

The symmetric q-numbers are given by

$$\langle s\rangle_q \overset{\text{def}}{=} \frac{q^s - q^{-s}}{q - q^{-1}}, \qquad s \in \mathbb{C}, \quad q \in (0,\infty)\,; \tag{6.1.1}$$

they satisfy the symmetry

$$\langle s\rangle_q = \langle s\rangle_{q^{-1}}\,, \tag{6.1.2}$$

and have the 'real limit'

$$\lim_{q\to 1}\langle s\rangle_q = s\,. \tag{6.1.3}$$

For interpolating between the p-adic and the real numbers, it will be more useful to work with the non-symmetric q-numbers

$$[s]_q = \frac{1-q^s}{1-q}\,. \tag{6.1.4}$$

We have the 'real limit'

$$\lim_{q\to 1}[s]_q = s, \quad \text{or} \quad \lim_{N\to\infty}[s]_{q^{1/N}} = s\,, \tag{6.1.5}$$

as well as the 'p-adic limit'

$$\lim_{N\to\infty}\left[\frac{s}{N}\right]_{p^{-N}} = 1 - p^{-s}\,. \tag{6.1.6}$$

The translation between the symmetric and non-symmetric q-numbers is given by

$$\langle s\rangle_q = q^{1-s}\cdot[s]_{q^2}, \quad [s]_q = q^{(s-1)/2}\cdot\langle s\rangle_{q^{1/2}}\,. \tag{6.1.7}$$

(Thus for q a root of unity, $[n]_q$ and $\langle n\rangle_q$ are the cyclotomic units and their real counterparts respectively.)

A fundamental property of q-numbers is the following:

$$q\text{-Lie property:}\quad \langle \alpha+1\rangle_q\cdot\langle\beta\rangle_q-\langle\alpha\rangle_q\cdot\langle\beta-1\rangle_q=\langle\alpha+\beta\rangle_q\,, \tag{6.1.8}$$

or

$$[\alpha+1]_q\cdot[\beta]_q-q\cdot[\alpha]_q\cdot[\beta-1]_q=[\alpha+\beta]_q\,.$$

For us, the q-factorial $(q^s;q)_\infty$, will be of importance and its inverse will be called the *q-zeta function*:

$$\zeta_{(q)}(s)=(q^s;q)_\infty^{-1}=\prod_{n\geq 0}(1-q^{s+n})^{-1}\,. \tag{6.1.9}$$

We have the *p-adic limit*: set $q=p^{-N}$, $s:=s/N$, and let $N\to\infty$

$$\zeta_{(p^{-N})}\left(\frac{s}{N}\right)\xrightarrow[N\to\infty]{}\zeta_p(s)\,. \tag{6.1.10}$$

The 'real limit' is more involved: the Jackson q-gamma function satisfies

$$\begin{aligned}\Gamma_q(s)\overset{\text{def}}{=}\frac{\zeta_{(q)}(s)\cdot(1-q)^{-s}}{\zeta_{(q)}(1)\cdot(1-q)^{-1}}&=\prod_{n\geq 0}\frac{1-q^{1+n}}{1-q^{s+n}}\cdot\left(\frac{1-q^{2+n}}{1-q^{1+n}}\right)^{s-1}\\&\xrightarrow[q\to 1]{}\prod_{n\geq 0}\frac{1+n}{s+n}\cdot\left(\frac{2+n}{1+n}\right)^{s-1}\\&=\lim_{N\to\infty}\frac{1\cdot 2\cdots N}{s(s+1)\dots(s+N-1)}\cdot N^{s-1}\\&=\lim_{N\to\infty}\frac{\Gamma(s)\Gamma(N)}{\Gamma(s+N)}\cdot N^s\\&=\Gamma(s),\quad \text{cf. (2.1.8)}\,.\end{aligned} \tag{6.1.11}$$

Taking into account that $\Gamma(1/2)=\pi^{1/2}$, we have

$$\frac{\zeta_{(q^2)}(s/2)}{\zeta_{(q^2)}(1/2)^s}\cdot\left(\frac{\zeta_{(q^2)}(1)}{1-q^2}\right)^{s-1}=\frac{\Gamma_{q^2}(s/2)}{\Gamma_{q^2}(1/2)^s}\xrightarrow[q\to 1]{}\frac{\Gamma(s/2)}{\pi^{s/2}}=\zeta_\eta(s)\,. \tag{6.1.12}$$

We summarize the properties of $\zeta_{(q)}(s)$ as follows (cf. (2.1.3)–(2.1.5)):

$$\text{periodicity:}\quad \zeta_{(q)}\left(s+\frac{2\pi i}{\log q}\right)=\zeta_{(q)}(s),\quad \zeta_{(q)}(s+1)=(1-q^s)\cdot\zeta_{(q)}(s) \tag{6.1.13}$$

$$\text{poles:}\quad -\mathbb{N}+\frac{2\pi i}{\log q}\mathbb{Z} \tag{6.1.14}$$

$$\begin{aligned}\text{duplication:}\quad \zeta_{(q^N)}(s)&=\prod_{0\leq k<N}\zeta_{(q)}\left(s+\frac{k}{N}\cdot\frac{2\pi i}{\log q}\right),\\ \zeta_{(q^{1/N})}(Ns)&=\prod_{0\leq k<N}\zeta_{(q)}\left(s+\frac{k}{N}\right),\end{aligned} \tag{6.1.15}$$

and putting these together

$$\zeta_{(q)}(Ns) = \prod_{0 \le k_1, k_2 < N} \zeta_{(q)}\left(s + \frac{k_1}{N} + \frac{k_2}{N} \cdot \frac{2\pi \mathrm{i}}{\log q}\right).$$

To see the p-adic and real limits on an equal footing, it will be advantageous to pass to the *q-beta function*

$$\zeta_{(q)}(\alpha, \beta) \stackrel{\text{def}}{=} \frac{\zeta_{(q)}(\alpha) \cdot \zeta_{(q)}(\beta)}{\zeta_{(q)}(\alpha + \beta) \cdot \zeta_{(q)}(1)} . \tag{6.1.16}$$

We now have the simpler limits:

$$p\text{-adic limit:} \quad (1 - p^{-N}) \cdot \zeta_{(p^{-N})}\left(\frac{\alpha}{N}, \frac{\beta}{N}\right) \xrightarrow[N \to \infty]{} \zeta_p(\alpha, \beta), \tag{6.1.17}$$

$$\text{real limit:} \quad (1 - q^{2/N}) \cdot \zeta_{(q^{2/N})}\left(\frac{\alpha}{2}, \frac{\beta}{2}\right) \xrightarrow[N \to \infty]{} \zeta_\eta(\alpha, \beta), \quad \text{for any } q \in (0,1). \tag{6.1.18}$$

6.2 ELLIPTIC CURVES

Following Jacobi, we associate with the number q the elliptic curve

$$E_q = \mathbb{C}^*/q^{\mathbb{Z}} \cong \mathbb{C}/\mathcal{L}_q, \quad \mathcal{L}_q = \mathbb{Z} + \mathbb{Z} \cdot \frac{2\pi \mathrm{i}}{\log q} . \tag{6.2.1}$$

Denote the image of $s \in \mathbb{C}$ in E_q by

$$[s]_{E_q} = q^s (\mathrm{mod}\, q^{\mathbb{Z}}) = s (\mathrm{mod}\, \mathcal{L}_q) . \tag{6.2.2}$$

We have dual N-isogenies:

$$\begin{array}{ll} * \to \mu_N \to E_q \to E_{q^N} \to *, & * \to q^{\mathbb{Z}/N\mathbb{Z}} \to E_{q^N} \to E_q \to * \\ \quad [s]_{E_q} \mapsto [s]_{E_{q^N}} & \qquad [s]_{E_{q^N}} \mapsto [N \cdot s]_{E_q} . \end{array} \tag{6.2.3}$$

Thus the duplication formulae (6.1.15) express compatibility with isogenies for $\zeta_{(q)}(s)$.

The analytic theory of E_q can be based on the *theta function*

$$\sigma_{E_q}(s) = \zeta_{(q)}(s, 1-s)^{-1} = \frac{\zeta_{(q)}(1)^2}{\zeta_{(q)}(s) \cdot \zeta_{(q)}(1-s)} ; \tag{6.2.4}$$

it has simple zeros at the lattice points of $\mathcal{L}_q$, and it satisfies

$$\sigma_{E_q}\left(s + \frac{2\pi \mathrm{i}}{\log q}\right) = \sigma_{E_q}(s), \quad \sigma_{E_q}(s+1) = (-q^{-s}) \cdot \sigma_{E_q}(s) . \tag{6.2.5}$$

Given any meromorphic function, $f\colon E_q \to \mathbb{P}^1(\mathbb{C})$, write

$$f^*0 = \sum_{1\le i\le n} [\alpha_i]_{E_q}, \quad f^*\infty = \sum_{1\le i\le n} [\beta_i]_{E_q}, \tag{6.2.6}$$

where $\alpha_i,\ \beta_i \in \mathbb{C}$ can be chosen so that

$$\sum_{1\le i\le n} \alpha_i = \sum_{1\le i\le n} \beta_i\,;$$

then the theta function gives the 'prime factorization' of f:

$$f([s]_{Eq}) = \text{cons.} \prod_{1\le i\le n} \frac{\sigma_{E_q}(s-\alpha_i)}{\sigma_{E_q}(s-\beta_i)}\,. \tag{6.2.7}$$

We can now understand the p-adic and real limits more conceptually. With any $q \in (0,1)$, we associate two inverse systems of elliptic curves, $\{E_{q^N}\}_{N\ge1}$ and $\{E_{q^{1/N}}\}_{N\ge1}$, and take the inverse limits (together with their 'dense uniformizations') obtaining

$$\begin{aligned} \text{the 'adic limit': } \mathbb{C} \twoheadrightarrow \mathbb{C}^* \hookrightarrow \varprojlim_N E_{q^N} \\ s \mapsto q^s \mapsto \left[\frac{s}{N}\right]_{E_{q^N}} \end{aligned} \tag{6.2.8}$$

and

$$\begin{aligned} \text{the 'real limit': } \mathbb{C} \to \varprojlim_N E_{q^{1/N}} \\ s \mapsto [s]_{E_{q^{1/N}}} \end{aligned} \tag{6.2.9}$$

For $q = 1/p$, the adic limit yields the p-adic limits (6.1.10), (6.1.17). For any $q \in (0,1)$, the real limit yields the real limits (6.1.12), (6.1.18). More precisely, before taking the real limit, we take one step in the 'adic direction' by replacing q and s by q^2 and $s/2$ (to account for the quadratic extension $\mathbb{C}/\mathbb{Q}_\eta$; over a general number field, for a *complex* place η, this does not happen).

The p-adic and real limits thus appear to be on equal (or dual) footings replacing isogenies by their duals. But from the 'point of view of $\mathbb{C}$' these limits are rather different: the dense uniformization factorizes through $\mathbb{C}^*$ on the p-adic limit (6.2.8), while it does not in the real limit (6.2.9).

6.3 *q*-SERIES

A fundamental identity of the q-world is the

$$q\text{-beta sum:}\quad \zeta_{(q)}(\alpha,\beta) = \sum_{n\ge0} \frac{\zeta_{(q)}(\alpha+n)}{\zeta_{(q)}(1+n)} \cdot q^{\beta\cdot n} \tag{6.3.1}$$

or equivalently

$$\frac{\zeta_{(q)}(\beta)}{\zeta_{(q)}(\alpha+\beta)} = \sum_{n\geq 0} \frac{\zeta_{(q)}(\alpha+n)}{\zeta_{(q)}(\alpha)} \cdot \frac{\zeta_{(q)}(1)}{\zeta_{(q)}(1+n)} \cdot q^{\beta\cdot n}\,. \tag{6.3.2}$$

To prove it, let $\mathrm{Z}(\alpha,\beta)$ denote the sum on the right side of (6.3.2), and note that it satisfies

$$\begin{aligned} \mathrm{Z}(\alpha,\beta) - q^{\alpha}\,\mathrm{Z}(\alpha,\beta+1) &= (1-q^{\alpha})\cdot \mathrm{Z}(\alpha+1,\beta) \\ \mathrm{Z}(\alpha,\beta) - \mathrm{Z}(\alpha,\beta+1) &= (1-q^{\alpha})\cdot q^{\beta}\cdot \mathrm{Z}(\alpha+1,\beta)\,. \end{aligned} \tag{6.3.3}$$

Indeed

$$\begin{aligned} \mathrm{Z}(\alpha,\beta) - q^{\alpha}\,\mathrm{Z}(\alpha,\beta+1) &= \sum_{n\geq 0} \frac{(1-q^{\alpha})\ldots(1-q^{\alpha+n-1})}{(1-q)\ldots(1-q^{n})} \cdot q^{\beta n}\cdot(1-q^{\alpha+n}) \\ &= (1-q^{\alpha})\cdot \mathrm{Z}(\alpha+1,\beta)\,, \end{aligned}$$

and

$$\begin{aligned} \mathrm{Z}(\alpha,\beta) - \mathrm{Z}(\alpha,\beta+1) &= \sum_{n\geq 0} \frac{(1-q^{\alpha})\ldots(1-q^{\alpha+n-1})}{(1-q)\ldots(1-q^{n})} \cdot q^{\beta n}\cdot(1-q^{n}) \\ &= (1-q^{\alpha})\cdot \mathrm{Z}(\alpha+1,\beta)\cdot q^{\beta}\,. \end{aligned}$$

From (6.3.3) we obtain

$$\mathrm{Z}(\alpha,\beta) - \mathrm{Z}(\alpha,\beta+1) = [\mathrm{Z}(\alpha,\beta) - q^{\alpha}\,\mathrm{Z}(\alpha,\beta+1)]\cdot q^{\beta}\,.$$

Hence

$$\begin{aligned} \mathrm{Z}(\alpha,\beta) &= \frac{1-q^{\alpha+\beta}}{1-q^{\beta}} \cdot \mathrm{Z}(\alpha,\beta+1) = \frac{1-q^{\alpha+\beta}}{1-q^{\beta}} \cdot \frac{1-q^{\alpha+\beta+1}}{1-q^{\beta+1}} \cdot \mathrm{Z}(\alpha,\beta+2) \\ &= \cdots = \frac{\zeta_{(q)}(\beta)}{\zeta_{(q)}(\alpha+\beta)} \cdot \mathrm{Z}(\alpha,\infty) = \frac{\zeta_{(q)}(\beta)}{\zeta_{(q)}(\alpha+\beta)}\,. \end{aligned} \tag{6.3.4}$$

The q-beta sum has many consequences; in the rest of this section we shall make some of these explicit.

We can take the limit $\alpha \to \infty$ of the q-beta sum, obtaining the

$$q\text{-gamma sum:}\quad \zeta_{(q)}(\beta) = \sum_{n\geq 0} \frac{\zeta_{(q)}(1)}{\zeta_{(q)}(1+n)} \cdot q^{\beta\cdot n}\,. \tag{6.3.5}$$

Substituting the q-gamma sum for $\zeta_{(q)}(\alpha+n)$ within the q-beta sum (6.3.1), we obtain the

$$\text{symmetric } q\text{-beta sum:}\quad \zeta_{(q)}(\alpha,\beta) = \zeta_{(q)}(1)\cdot \sum_{n,m\geq 0} q^{n\cdot m} \cdot \frac{q^{\alpha m}}{\zeta_{(q)}(1+m)} \cdot \frac{q^{\beta n}}{\zeta_{(q)}(1+n)}\,. \tag{6.3.6}$$

Setting $\alpha = -N$, $\beta := \beta + N$ in the q-beta sum (6.3.2), and noting that

$$\left.\frac{\zeta_{(q)}(\alpha+k)}{\zeta_{(q)}(\alpha)}\right|_{\alpha=-N} = \frac{\zeta_{(q)}(1+N)}{\zeta_{(q)}(1+N-k)} \cdot (-1)^k \cdot q^{k(k-1)/2-Nk}, \tag{6.3.7}$$

we obtain the

$$q\text{-binom:} \quad \frac{\zeta_{(q)}(\beta+N)}{\zeta_{(q)}(\beta)} = \sum_{0\le k\le N} \left[\frac{\zeta_{(q)}(1+N)\cdot\zeta_{(q)}(1)}{\zeta_{(q)}(1+k)\cdot\zeta_{(q)}(1+N-k)}\right](-1)^k q^{k(k-1)/2}\cdot q^{\beta k}. \tag{6.3.8}$$

Conversely, as a consequence of the q-binom, we can express the powers $q^{\beta\cdot N}$ in terms of the quotients $\zeta_{(q)}(\beta+k)/\zeta_{(q)}(\beta)$:

$$q^{\beta\cdot N} = \sum_{0\le k\le N}\left[\frac{\zeta_{(q)}(1+N)\zeta_{(q)}(1)}{\zeta_{(q)}(1+k)\zeta_{(q)}(1+N-k)}\right](-1)^k q^{k(k-1)/2+k(1-N)}\cdot\frac{\zeta_{(q)}(\beta+k)}{\zeta_{(q)}(\beta)}. \tag{6.3.8$'$}$$

Letting $N\to\infty$ in the q-binom (6.3.8), we obtain the

$$q\text{-Möbius sum:} \quad \frac{1}{\zeta_{(q)}(\beta)} = \sum_{n\ge 0}\frac{\zeta_{(q)}(1)}{\zeta_{(q)}(1+n)}\cdot(-1)^n q^{n(n-1)/2}\cdot q^{\beta n}. \tag{6.3.9}$$

The q-factorial is defined by

$$[N]_q! = [N]_q\cdot[N-1]_q\cdot\ldots\cdot[1]_q = \frac{\zeta_{(q)}(1+N)}{\zeta_{(q)}(1)(1-q)^N},$$

and the q-binom coefficients by

$$\begin{bmatrix} N \\ k \end{bmatrix}_q = \frac{[N]_q!}{[k]_q![N-k]_q!} = \frac{\zeta_{(q)}(1+N)\cdot\zeta_{(q)}(1)}{\zeta_{(q)}(1+k)\cdot\zeta_{(q)}(1+N-k)}, \tag{6.3.10}$$

they satisfy the

$$\text{induction property:} \quad \begin{bmatrix} N \\ k \end{bmatrix}_q = \begin{bmatrix} N-1 \\ k \end{bmatrix}_q \cdot q^k + \begin{bmatrix} N-1 \\ k-1 \end{bmatrix}_q. \tag{6.3.11}$$

Define the subgroups of GL_N,

$$\mathrm{B}_{(M,N-M)} = {}^{M\{}\left(\begin{array}{c|c} * & * \\ \hline 0 & * \end{array}\right)_{\underbrace{}_{M}}, \qquad \mathrm{B}_{(1\ldots1)} = \begin{pmatrix} * & \cdots & * \\ & \ddots & \vdots \\ 0 & & * \end{pmatrix},$$

so that

$$\mathrm{Grass}_{M\subseteq N} = \mathrm{GL}_N/\mathrm{B}_{(M,N-M)} = \text{Grassmannian of } M\text{-dimensional subspaces in } \mathbb{A}^N, \tag{6.3.12}$$

$$\mathrm{Flag}_N = \mathrm{GL}_N/\mathrm{B}_{(1\ldots1)} = \text{flag space of } \ell_1\subseteq\ell_2\subseteq\ldots\subseteq\ell_{N-1}\subseteq\mathbb{A}^N,\ \dim\ell_j = j.$$

For q a prime power we have

$$\begin{aligned}&\begin{bmatrix} N \\ M \end{bmatrix}_q = \#\ \mathrm{Grass}_{M\subseteq N}(\mathbb{F}_q), \quad \text{e.g. } [N]_q = \#\,\mathbb{P}^{N-1}(\mathbb{F}_q)\,; \\ &[N]_q! = \#\ \mathrm{Flag}_N(\mathbb{F}_q)\,.\end{aligned} \tag{6.3.13}$$

In the real limit when $q = 1$, we obtain the usual factorial and binomial coefficients, and it is as if the category of '$\mathbb{F}_1$-vector spaces' becomes the category of sets:

$$\begin{aligned}&\begin{bmatrix} N \\ M \end{bmatrix}_1 = \binom{N}{M} = \#\{\ell \subseteq \{1,\ldots,N\}, \#\ell = M\} \\ &[N]_1! = N! = \#\{\ell_1 \subseteq \ell_2 \subseteq \ldots \subseteq \ell_{N-1} \subseteq \{1,\ldots,N\}, \#\ell_j = j\}\,.\end{aligned} \tag{6.3.14}$$

Starting from the identity

$$\left(\frac{\zeta_{(q)}(s)}{\zeta_{(q)}(s+\alpha)}\right)\cdot\left(\frac{\zeta_{(q)}(s+\alpha)}{\zeta_{(q)}(s+\alpha+\beta)}\right) = \left(\frac{\zeta_{(q)}(s)}{\zeta_{(q)}(s+\alpha+\beta)}\right),$$

and developing each of the brackets using the q-beta sum (6.3.2), then comparing coefficients of q^{ms}, we obtain

$$q\text{-Chu–Vandermonde:} \quad \frac{\zeta_{(q)}(\alpha+\beta+m)}{\zeta_{(q)}(\alpha+\beta)} = \sum_{i+j=m} \begin{bmatrix} m \\ j \end{bmatrix}_q q^{j\alpha}\,\frac{\zeta_{(q)}(\alpha+i)}{\zeta_{(q)}(\alpha)}\cdot\frac{\zeta_{(q)}(\beta+j)}{\zeta_{(q)}(\beta)}\,. \tag{6.3.15}$$

The q-zeta function satisfies the

$$\text{duality identity:} \quad \frac{\zeta_{(q)}(\beta+n)}{\zeta_{(q)}(\beta)} = (-1)^n q^{n(n-1)/2}\cdot q^{\beta n}\cdot\frac{\zeta_{(q)}(1-\beta)}{\zeta_{(q)}(1-\beta-n)}\,. \tag{6.3.16}$$

Using this, we can formulate all the above identities in a dual form. For example, we have

$$\begin{aligned}&\text{dual } q\text{-Chu–Vandermonde}: \\ &\frac{\zeta_{(q)}(\alpha+\beta+m)}{\zeta_{(q)}(\alpha+\beta)} = \sum_{0\le j\le m} \begin{bmatrix} m \\ j \end{bmatrix}_q \cdot q^{j(\beta+j)}\cdot\frac{\zeta_{(q)}(\alpha)}{\zeta_{(q)}(\alpha-j)}\cdot\frac{\zeta_{(q)}(\beta+1+m)}{\zeta_{(q)}(\beta+1+j)}\,.\end{aligned} \tag{6.3.15$'$}$$

Let $\mathbb{C}[V_{(q)}]$ denote the algebra over $\mathbb{C}$ generated by two elements X, Y satisfying

$$q\text{-commutativity:} \quad \mathrm{XY} = q\mathrm{YX}\,. \tag{6.3.17}$$

By induction on N, using (6.3.11), we have

$$\text{operator } q\text{-binom:} \quad (\mathrm{X}+\mathrm{Y})^N = \sum_{0\le k\le N} \begin{bmatrix} N \\ k \end{bmatrix}_q \mathrm{Y}^{N-k}\mathrm{X}^k\,. \tag{6.3.18}$$

The algebra $\mathbb{C}[V_{(q)}]$ is graded:

$$\mathbb{C}[V_{(q)}] = \bigoplus_{N\geq 0} \mathbb{C}[V_{(q)}]_N, \quad \dim \mathbb{C}[V_{(q)}]_{\mathrm{N}} = N+1\,. \tag{6.3.19}$$

Let $\mathbb{C}[\![V_{(q)}]\!]$ denote its adic completion

$$\mathbb{C}[\![V_{(q)}]\!] = \prod_{n\geq 0} \mathbb{C}[V_{(q)}]_N\,, \tag{6.3.20}$$

and let m (respectively j_{X}, j_{Y}) denote the embeddings of the algebra of formal power series $\mathbb{C}[\![\mathrm{Z}]\!]$ into $\mathbb{C}[\![V_{(q)}]\!]$, carrying Z into X + Y (respectively X, Y):

$$\begin{aligned} m,\ j_{\mathrm{X}},\ j_{\mathrm{Y}}\colon \mathbb{C}[\![\mathrm{Z}]\!] &\hookrightarrow \mathbb{C}[\![V_{(q)}]\!] \\ \mathrm{Z} &\mapsto \mathrm{X}+\mathrm{Y},\ \mathrm{X},\ \mathrm{Y}\,. \end{aligned} \tag{6.3.21}$$

We look at *group-like* elements $f = f(\mathrm{Z}) \in \mathbb{C}[\![\mathrm{Z}]\!]$, satisfying

$$m(f) = j_{\mathrm{Y}}(f)\cdot j_{\mathrm{X}}(f)\,, \tag{6.3.22}$$

or equivalently,

$$f(\mathrm{X}+\mathrm{Y}) = f(\mathrm{Y})\cdot f(\mathrm{X})\,. \tag{6.3.23}$$

Writing

$$f(\mathrm{Z}) = \sum_{n\geq 0} a_N \mathrm{Z}^N\,,$$

and expanding (6.3.23) using (6.3.18), we obtain $a_0 = 1$, $a_1 = \lambda, \ldots, a_N = \lambda^N/[N]_q!$ for some $\lambda \in \mathbb{C}$; thus $f(\mathrm{Z}) = \mathrm{e}_q(\lambda\mathrm{Z})$ where the q-exponential is given by

$$\mathrm{e}_q(\mathrm{Z}) = \sum_{N\geq 0} \frac{\mathrm{Z}^N}{[N]_q!}\,. \tag{6.3.24}$$

In particular, the operator q-binom (6.3.18) is equivalent to the identity

$$\mathrm{e}_q(\mathrm{X}+\mathrm{Y}) = \mathrm{e}_q(\mathrm{Y})\cdot \mathrm{e}_q(\mathrm{X})\,. \tag{6.3.25}$$

In the real limit $q = 1$, q-commutativity becomes commutativity, and $\mathrm{e}_1(z) = \mathrm{e}^z$ is the usual exponential.

The q-world often reveals new and unexpected connections. One such basic connection is the fact that for $z = q^\beta/(1-q)$ the q-exponential becomes the q-zeta (cf. (6.3.5)):

$$\mathrm{e}_q\left(\frac{q^\beta}{1-q}\right) = \zeta_{(q)}(\beta)\,. \tag{6.3.26}$$

Another important consequence (and a generalization) of the q-beta sum is Ramanujan's ${}_1\psi_1$-sum: it reads in our notation

$$\text{Ram:}\quad \sum_{n\in\mathbb{Z}} \frac{\zeta_{(q)}(\alpha+n)}{\zeta_{(q)}(\gamma+n)} q^{\beta n} = \zeta_{(q)}(\alpha,\beta)\cdot\frac{\zeta_{(q)}(1-\alpha,\gamma-\alpha-\beta)}{\zeta_{(q)}(\gamma-\alpha,1-\alpha-\beta)}, \qquad \operatorname{Re}(\gamma-\alpha)>\operatorname{Re}(\beta)>0. \tag{6.3.27}$$

Indeed at $\gamma = 1+m$ (and putting $\alpha+m$ for α)—the series on the left of (6.3.27) terminates below $n=-m$ and the identity reduces to the q-beta sum

$$\sum_n \frac{\zeta_{(q)}(\alpha+m+n)}{\zeta_{(q)}(1+m+n)} q^{\beta n} = q^{-m\beta}\cdot\zeta_{(q)}(\alpha,\beta) \quad \text{cf. (6.3.1)}$$

$$= \frac{\zeta_{(q)}(\beta)\zeta_{(q)}(\alpha+m)}{\zeta_{(q)}(1)\zeta_{(q)}(\alpha+\beta+m)}\cdot\frac{(1-q^{\alpha+\beta+m-1})\dots(1-q^{\alpha+\beta})}{(1-q^{\alpha+m-1})\dots(1-q^{\alpha})}\cdot q^{-m\beta}$$

$$= \zeta_{(q)}(\alpha+m,\beta)\cdot\frac{(1-q^{1-m-\alpha-\beta})\dots(1-q^{-\alpha-\beta})}{(1-q^{1-m-\alpha})\dots(1-q^{-\alpha})}$$

$$= \zeta_{(q)}(\alpha+m,\beta)\cdot\frac{\zeta_{(q)}(1-m-\alpha,\ 1-\alpha-\beta)}{\zeta_{(q)}(1-\alpha,\ 1-m-\alpha-\beta)}$$

The two sides of (6.3.27) are analytic functions of q^γ near 0, and agree on the sequence of q^m converging to 0, so they must be equal.

Setting $\alpha=-N$, $\beta := \beta+N$, and using (6.3.7), we obtain

$$\text{Ram}_N:\quad \sum_{N\geq n\in\mathbb{Z}} \frac{\zeta_{(q)}(1)\zeta_{(q)}(\gamma+N)}{\zeta_{(q)}(1+N-n)\zeta_{(q)}(\gamma+n)}(-1)^n q^{n(n-1)/2}q^{\beta n} = \frac{\zeta_{(q)}(\beta+N)}{\zeta_{(q)}(\beta)}\frac{\zeta_{(q)}(\gamma-\beta)}{\zeta_{(q)}(1-\beta)}. \tag{6.3.28}$$

Letting $N\to\infty$, we obtain

$$\text{Ram}_\infty:\quad \sum_{n\in\mathbb{Z}} \frac{\zeta_{(q)}(1)}{\zeta_{(q)}(\gamma+n)}(-1)^n q^{n(n-1)/2}q^{\beta n} = \frac{\zeta_{(q)}(\gamma-\beta)}{\zeta_{(q)}(\beta)\zeta_{(q)}(1-\beta)}. \tag{6.3.29}$$

When $\gamma=1$, (6.3.27), (6.3.28), (6.3.29) reduce to (6.3.1), (6.3.8), (6.3.9) respectively. When we let $\gamma=\infty$ in (6.3.29), we obtain Jacobi's

$$\text{triple product identity:}\quad \sum_{n\in\mathbb{Z}}(-1)^n q^{n(n-1)/2}q^{\beta n} = \frac{1}{\zeta_{(q)}(1)\zeta_{(q)}(\beta)\zeta_{(q)}(1-\beta)}. \tag{6.3.30}$$

When we let $\gamma = 1 + \alpha$ in (6.3.27) we obtain the *addition identity*

$$\sum_{n\in\mathbb{Z}} \frac{q^{\beta\cdot n}}{1-q^{\alpha+n}} = \frac{\zeta_{(q)}(\alpha, 1-\alpha)\cdot\zeta_{(q)}(\beta, 1-\beta)}{\zeta_{(q)}(\alpha+\beta, 1-\alpha-\beta)} = \frac{\sigma_{E_q}(\alpha+\beta)}{\sigma_{E_q}(\alpha)\sigma_{E_q}(\beta)}, \qquad 1 > \mathrm{Re}(\beta) > 0\,. \tag{6.3.31}$$

We shall denote the basic hypergeometric function by

$$ {}_r\phi_s(\alpha_1,\ldots,\alpha_r;\beta_1,\ldots,\beta_s;\gamma) = \frac{\zeta_{(q)}(1)\zeta_{(q)}(\beta_1)\ldots\zeta_{(q)}(\beta_s)}{\zeta_{(q)}(\alpha_1)\ldots\zeta_{(q)}(\alpha_r)}\,. \tag{6.3.32}$$

$$\sum_{n\geq 0}[(-1)^n q^{n(n-1)/2}]^{s+1-r}\ \frac{\zeta_{(q)}(\alpha_1+n)\ldots\zeta_{(q)}(\alpha_r+n)}{\zeta_{(q)}(1+n)\zeta_{(q)}(\beta_1+n)\ldots\zeta_{(q)}(\beta_s+n)}q^{n\cdot\gamma}\,.$$

Thus, the q-beta sum (6.3.2) reads

$$ {}_1\phi_0(\alpha;-;\beta) = \frac{\zeta_{(q)}(\beta)}{\zeta_{(q)}(\alpha+\beta)}\,;$$

the q-gamma sum (6.3.5) reads

$$ {}_1\phi_0(\infty;-;\beta) = \zeta_{(q)}(\beta)\,;$$

and the q-Möbius sum (6.3.9) reads

$$ {}_0\phi_0(-;-;\beta) = \frac{1}{\zeta_{(q)}(\beta)}\,.$$

7

The q-gamma and q-beta chains

7.1 THE q-GAMMA CHAIN

We begin by describing the *basic chain* from which all our chains will be built. Fix $q \in (0,1)$, $\beta \in (0,\infty)$. The state space is the integers $\mathbb{N}$, and the transition probabilities of the basic chain are given by

$$\mathrm{P}^{\beta}_{\mathbb{N}(q)}(i,j) = \begin{cases} q^{\beta+i} & i = j\,, \\ 1 - q^{\beta+i} & i+1 = j\,, \\ 0 & \text{otherwise}\,, \end{cases} \tag{7.1.1}$$

or schematically

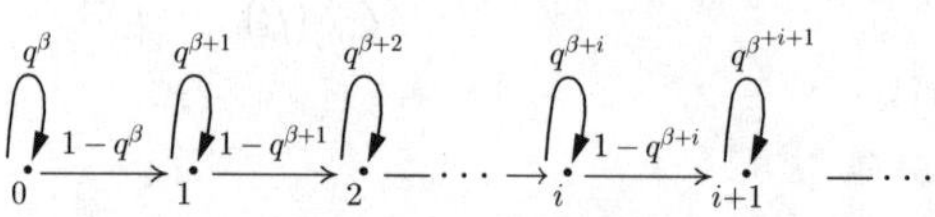

(7.1.2)

Note that this chain is not timed in the sense of (4.1) (except in the $\beta \to \infty$ limit where we obtain the unit shift).

We view $\mathrm{P}^{\beta}_{\mathbb{N}(q)}$ as an operator on $\ell_2(\mathbb{N})$, and we have

$$\left(\mathrm{P}^{\beta}_{\mathbb{N}(q)}\right)^{*} = \mathrm{X} + \mathrm{Y}, \quad \mathrm{X}f(i) = q^{\beta+i} f(i), \quad \mathrm{Y}f(i) = (1 - q^{\beta+i-1}) f(i-1)\,, \tag{7.1.3}$$

where X, Y satisfy q-commutativity $\mathrm{XY} = q\mathrm{YX}$. We have a finite probability measure $\left(\mathrm{P}^{\beta *}_{\mathbb{N}_{(q)}}\right)^{N} \delta_0 = \tau^{\beta}_{q(N)}$ supported on the set $\{0, 1, \ldots, N\}$, and using the operator q-binom (6.3.18), it is given by

$$\tau^{\beta}_{q(N)}(j) = \begin{bmatrix} N \\ j \end{bmatrix}_q \cdot \frac{\zeta_{(q)}(\beta + j)}{\zeta_{(q)}(\beta)} \cdot q^{\beta(N-j)}\,. \tag{7.1.4}$$

Another important property of the basic chain is the fact that starting the chain at j (rather than at 0) is equivalent to simply replacing β by $\beta + j$. Using this remark we obtain for the Green kernel

$$\mathrm{G}^{\beta}_{\mathbb{N}(q)}(i,j) = \begin{cases} \left(1 - q^{\beta+j}\right)^{-1} & j \geq i\,, \\ 0 & j < i\,. \end{cases} \tag{7.1.5}$$

Hence the Martin kernel is trivial

$$\mathrm{K}^{\beta}_{\mathbb{N}(q)}(i,j) = \begin{cases} 1 & j \geq i\,, \\ 0 & j < i\,, \end{cases} \tag{7.1.6}$$

and so the boundary consists of one point; $\mathbb{1}$ is the only harmonic function.

In order to get a more interesting boundary, and to have a timed chain, we pass to the product of the basic chain $\mathrm{P}^{\beta}_{\mathbb{N}(q)}$ with the unit shift $\mathrm{P}^{\infty}_{\mathbb{N}(q)}$. The state space of $\mathrm{P}^{\infty}_{\mathbb{N}(q)} \times \mathrm{P}^{\beta}_{\mathbb{N}(q)}$ is $\mathbb{N} \times \mathbb{N}$, but we are interested only with the states reachable from the origin $(0,0)$, so we set

$$\mathrm{X}^0_{(q)} = \coprod_{N \geq 0} \mathrm{X}^0_{q(N)}, \quad \mathrm{X}^0_{q(N)} = \big\{(N,j) \mid 0 \leq j \leq N\big\}. \tag{7.1.7}$$

The transition probabilities are given by

$$\mathrm{P}^{\beta}_{\mathrm{X}^0_{(q)}}((i,j),(i',j')) = \mathrm{P}^{\infty}_{\mathbb{N}(q)}(i,i') \cdot \mathrm{P}^{\beta}_{\mathbb{N}(q)}(j,j') = \begin{cases} q^{\beta+j} & i' = i+1,\ j' = j\,, \\ 1 - q^{\beta+j} & i' = i+1,\ j' = j+1\,, \\ 0 & \text{otherwise}\,, \end{cases} \tag{7.1.8}$$

or schematically

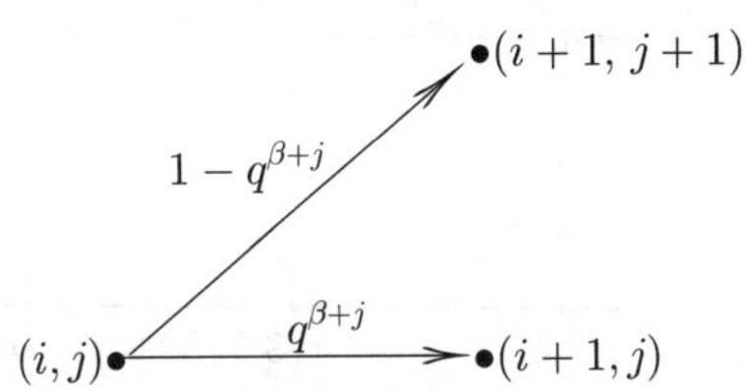

(7.1.9)

We call this chain the *q-gamma chain* (see Fig. 7.1). In the *p*-adic limit, $q = p^{-N}$, $\beta := \beta/N$, $N \to \infty$, the q-gamma chain converges to the p-adic gamma chain (4.2). In the real limit, $q := q^{2/N}$, $\beta := \beta/2$, $N \to \infty$, the q-gamma chain converges to the unit shift walking from the origin $(0,0)$ to the right (unfortunately, but in agreement with (3.3.10)!).

The probability measure $\tau^{\beta}_{q(N)} = \big(\mathrm{P}^{\beta\ *}_{\mathrm{X}^0_{(q)}}\big)^N \delta_{(0,0)}$ on $\mathrm{X}^0_{q(N)}$ is given by (7.1.4). The Green kernel is given by

$$\begin{aligned} \mathrm{G}^{\beta}_{\mathrm{X}^0_{(q)}}((i,j),(i'j')) &= \tau^{\beta+j}_{q(i'-i)}(j'-j) \\ &= \begin{bmatrix} i'-i \\ j'-j \end{bmatrix}_q \frac{\zeta_{(q)}(\beta+j')}{\zeta_{(q)}(\beta+j)} q^{(\beta+j)(i'-i-j'+j)} \quad 0 \leq j'-j \leq i'-i\,, \\ &\quad 0 \quad \text{otherwise}\,. \end{aligned} \tag{7.1.10}$$

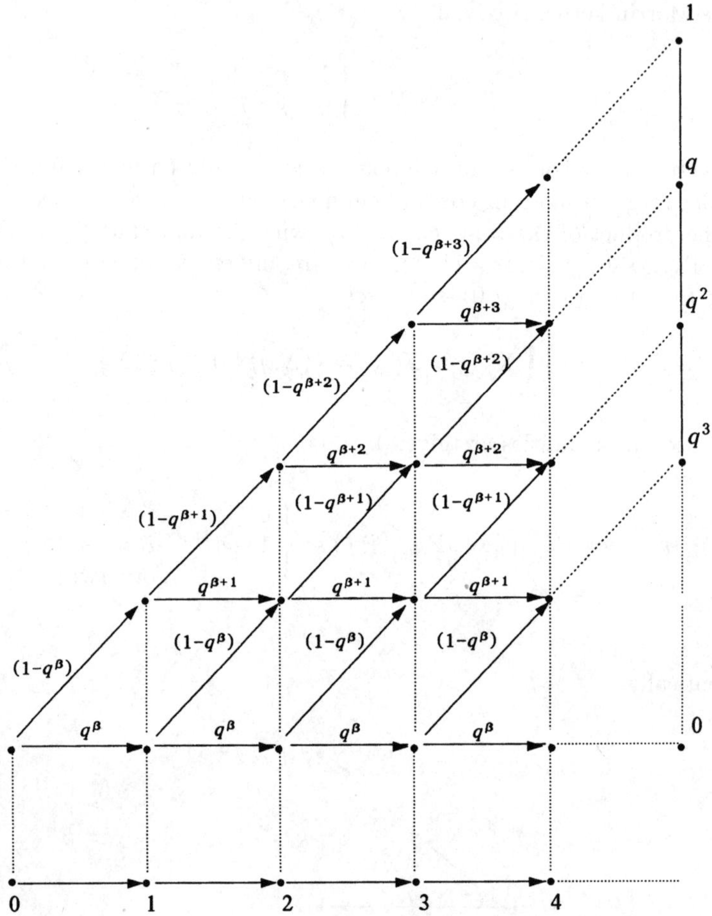

FIG. 7.1 The q-gamma chain X_q^0.

The Martin kernel is given by

$$\mathrm{K}^{\beta}_{\mathrm{X}^0_{(q)}}((i,j),(i',j')) = \frac{\begin{bmatrix} i'-i \\ j'-j \end{bmatrix}_q}{\begin{bmatrix} i' \\ j' \end{bmatrix}_q} \cdot \frac{\zeta_{(q)}(\beta)}{\zeta_{(q)}(\beta+j)} \cdot q^{j(i'-i-j'+j)-\beta(i-j)}\,. \tag{7.1.11}$$

A sequence $(i_n, j_n) \in \mathrm{X}^0_{(q)}$ converges in the Martin metric (cf. (4.1.19)) if and only if

$$\mathrm{K}^{\beta}_{\mathrm{X}^0_{(q)}}((1,0),(i_n,j_n)) = q^{-\beta} \cdot \frac{1-q^{i_n-j_n}}{1-q^{i_n}} \tag{7.1.12}$$

converges; that is, either (i_n, j_n) is eventually constant, or else $i_n \to \infty$ and $q^{i_n - j_n}$ converges in $q^{\mathbb{N}} \cup \{0 = q^{\infty}\}$ (i.e., either $i_n - j_n$ is eventually constant, or else $i_n - j_n \to \infty$). The boundary of our chain is therefore

$$\partial \mathrm{X}^0_{(q)} = g^{\mathbb{N}} \cup \{0 = g^{\infty}\} . \tag{7.1.13}$$

Here g^n is a symbol representing the equivalence class of the Cauchy sequence $\{(n+k, k) \mid k = 0, 1, 2, \ldots\}$. The extension of the Martin kernel to the boundary is given by

$$\mathrm{K}^{\beta}_{\mathrm{X}^0_{(q)}}((i,j), g^n) = \frac{\zeta_{(q)}(1+n)}{\zeta_{(q)}(1+n-i+j)} \cdot \frac{\zeta_{(q)}(\beta)}{\zeta_{(q)}(\beta+j)} \cdot q^{jn-(\beta+j)(i-j)} \quad 0 \le i-j \le n\,,$$

$$0 \quad \text{otherwise}\,, \tag{7.1.14}$$

$$\mathrm{K}^{\beta}_{\mathrm{X}^0_{(q)}}((i,j), 0) = q^{-\beta \cdot i} \quad \text{if } j = 0\,,$$

$$0 \quad \text{otherwise}\,.$$

The harmonic measure is the q-gamma measure

$$\tau^{\beta}_{\mathbb{Z}_{(q)}}(g^n) = \frac{\zeta_{(q)}(1)}{\zeta_{(q)}(\beta)} \cdot \frac{q^{\beta \cdot n}}{\zeta_{(q)}(1+n)} \,. \tag{7.1.15}$$

That this is a probability measure is just the q-gamma sum (6.3.5).

7.2 THE q-BETA CHAINS

We next consider 'position and momentum' on an equal footing, and look at the product of two basic chains $\mathrm{P}^{\alpha}_{\mathbb{N}(q)} \times \mathrm{P}^{\beta}_{\mathbb{N}(q)}$. The state space is now

$$\mathrm{X}_{(q)} = \mathbb{N} \times \mathbb{N} \tag{7.2.1}$$

and the transition probabilities are given schematically by

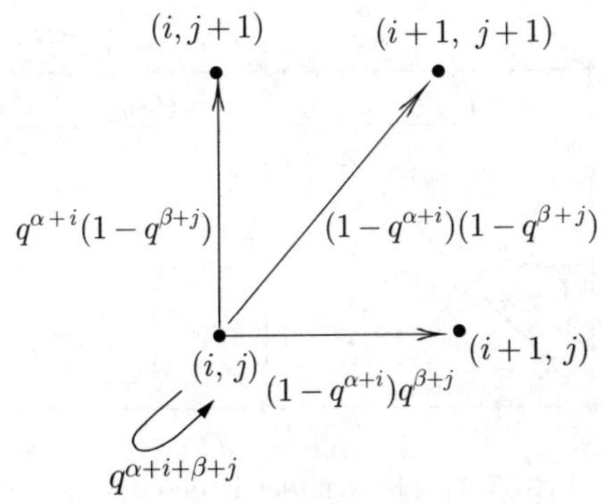

(7.2.2)

The *symmetric q-beta chain* $\mathrm{P}^{\alpha,\beta}_{\mathrm{X}_{(q)}}$ (Fig. 7.2) is the associated non-stop chain $\left(\mathrm{P}^{\alpha}_{\mathbb{N}(q)} \times \mathrm{P}^{\beta}_{\mathbb{N}(q)}\right)_!$ (cf. (4.1.9)), which is given schematically by

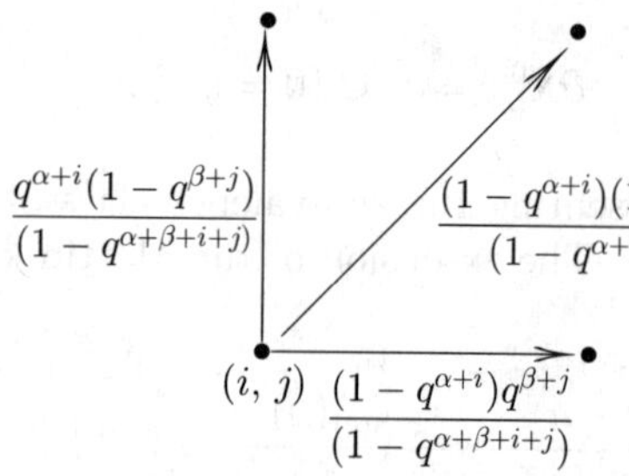

(7.2.3)

In the p-adic limit, $q = p^{-N}$, $\alpha := \alpha/N$, $\beta := \beta/N$, $N \to \infty$, the symmetric q-beta chain converges to the symmetric p-adic beta chain (4.3). In the real limit

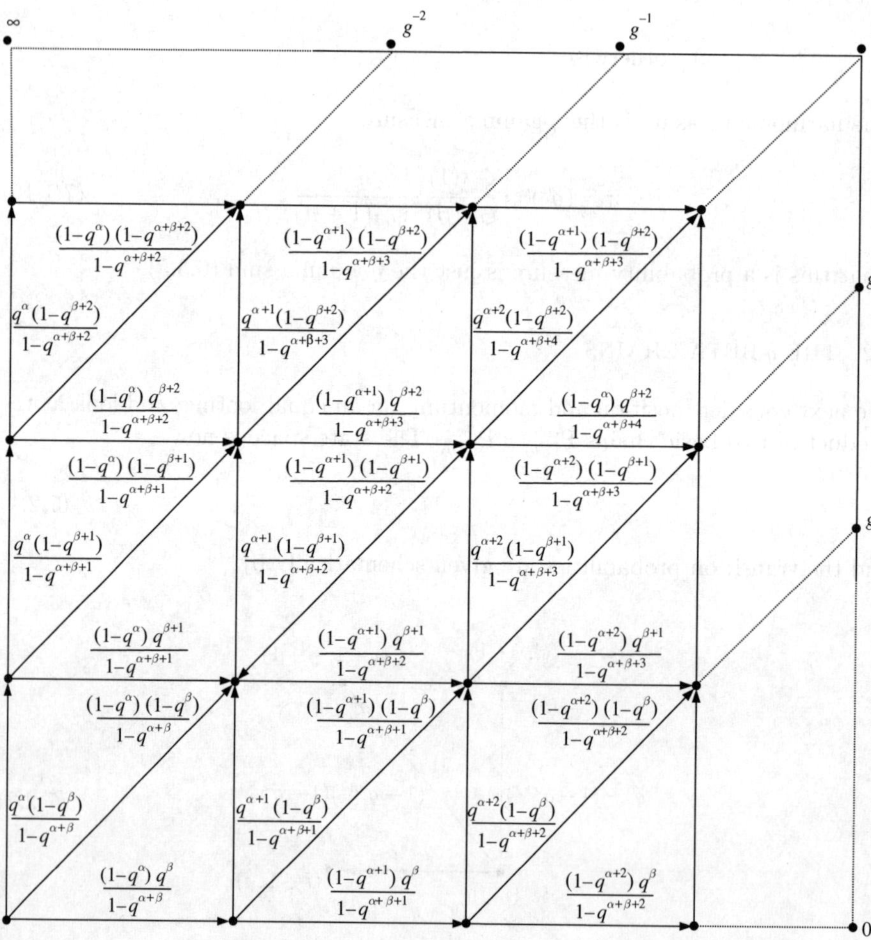

FIG. 7.2 The symmetric q-beta chain X_q.

$q := q^{2/N}$, $\alpha := \alpha/2$, $\beta := \beta/2$, $N \to \infty$, we obtain an interesting chain, the *real beta chain*, given schematically by

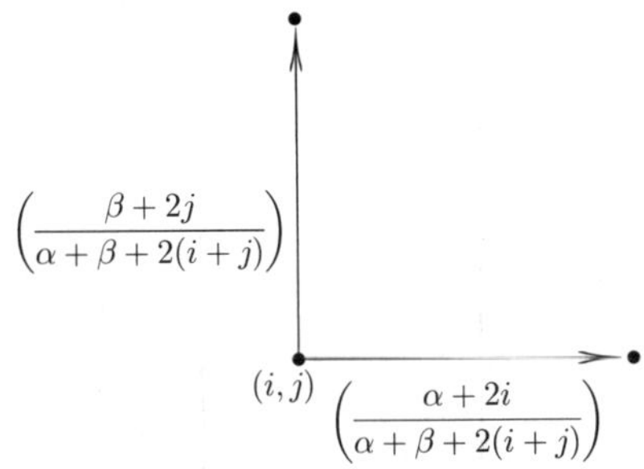

(7.2.4)

We postpone study of this chain to the next chapter.

Note that the symmetric q-beta chain $\mathrm{P}^{\alpha,\beta}_{\mathrm{X}_{(q)}}$ is not timed in the sense of (4.1). This is not surprising since $\mathrm{P}^{\alpha,\beta}_{\mathrm{X}_{(q)}}$ contains in its limits both the 'p-adic time' $(\max\{i,j\})$, and the 'real time' of the chain (7.2.4) given by $(i+j)$.

We next sacrifice the symmetry between 'position and momentum', and look at the semi-direct product $\mathrm{P}^{\beta}_{\mathbb{N}(q)} \ltimes \left(\mathrm{P}^{\alpha}_{\mathbb{N}(q)}\right)$, cf. (4.1.8). The state space is still

$$\mathrm{X}^{\square}_{q} = \mathbb{N} \times \mathbb{N}, \tag{7.2.5}$$

and the transition probabilities are given schematically by

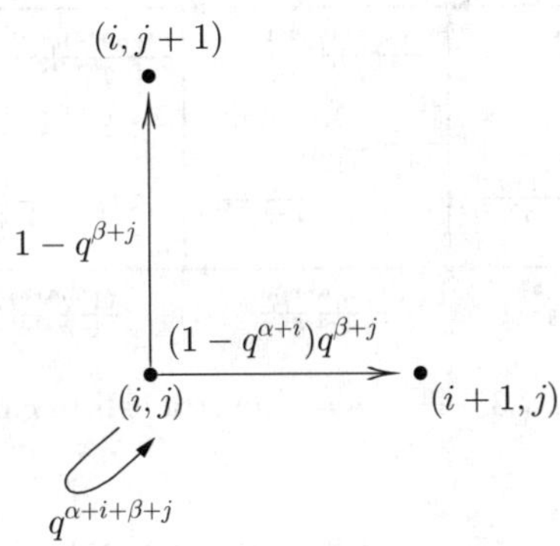

(7.2.6)

The (non-symmetric) *q-beta chain* $\mathrm{P}^{(\alpha)\beta}_{\mathrm{X}^{\square}_{q}}$ (Fig. 7.3) is the associated non-stop chain $\mathrm{P}^{(\alpha)\beta}_{\mathrm{X}^{\square}_{q}} = \left(\mathrm{P}^{\beta}_{\mathbb{N}(q)} \ltimes \left(\mathrm{P}^{\alpha}_{\mathbb{N}(q)}\right)\right)_{!}$ and its transition probabilities are given schematically by

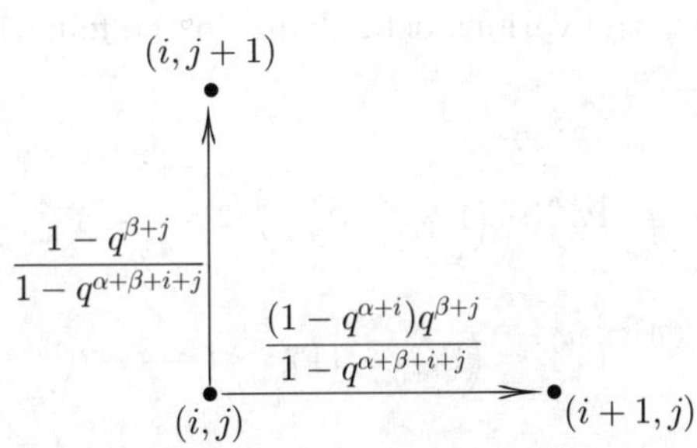

(7.2.7)

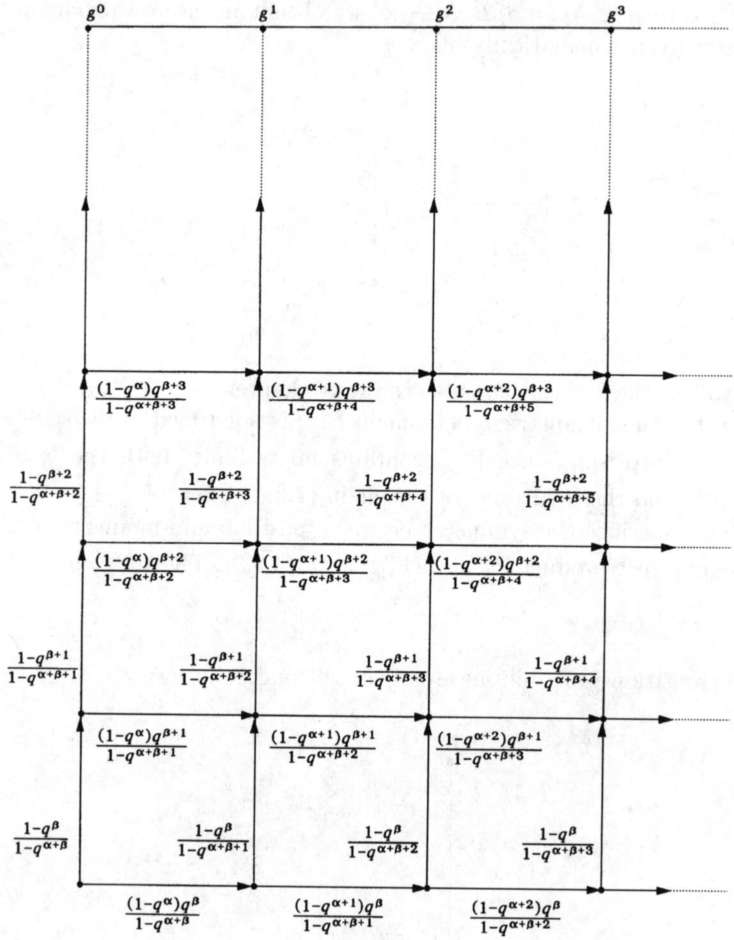

FIG. 7.3 The non-symmetric q-beta chain $\mathrm{X}_q^{\square}$.

This chain is timed with

$$\mathrm{X}_{q(N)}^{\square} = \{(i,j) \,|\, i+j=N\}. \tag{7.2.8}$$

In the p-adic limit the q-beta chain converges to the p-adic beta chain (4.4). In the real limit we obtain the same real beta chain (7.2.4); the difference between the symmetric and non-symmetric beta chains disappears in the real limit.

We can summarize the various beta chains in the following chart:

Real η | q | p-adic

symmetric: $\mathrm{P}^{\alpha,\beta}_{\mathrm{X}_{(q)}} = \left(\mathrm{P}^{\alpha}_{\mathbb{N}(q)} \times \mathrm{P}^{\beta}_{\mathbb{N}(q)}\right)_! \xrightarrow{\;ⓟ\;} \mathrm{P}^{\alpha,\beta}_{\mathrm{X}_p},\quad \mathrm{X}_p$

$\mathrm{P}^{\alpha,\beta}_{\mathrm{X}_\eta} \xleftarrow{\;ⓗ\;}$ (from both rows)

non-symmetric: $\mathrm{P}^{(\alpha)\beta}_{\mathrm{X}_q^{\square}} = \left(\mathrm{P}^{\beta}_{\mathbb{N}(q)} \ltimes \left(\mathrm{P}^{\alpha}_{\mathbb{N}(q)}\right)\right)_! \xrightarrow{\;ⓟ\;} \mathrm{P}^{(\alpha)\beta}_{\mathrm{X}_p^{\square}},\quad \mathrm{X}_p^{\square}$

$\widetilde{\mathrm{X}}_p \to \mathrm{X}_p$ (via $\mathbb{Z}_p^*$), $\mathrm{X}_p \to \mathrm{X}_p^{\square}$, $\widetilde{\mathrm{X}}_p \leftrightarrow \mathrm{X}_p^{\square}$ (via $\mathbb{Z}_p^* \ltimes \mathbb{Z}_p$)

When we view the transition probabilities of the q-Beta chain $\mathrm{P}^{(\alpha)\beta}_{\mathrm{X}^{\square}_q}$ as an operator on $\ell_2(\mathrm{X}^{\square}_q)$, we have

$$\left(\mathrm{P}^{(\alpha)\beta}_{\mathrm{X}^{\square}_q}\right)^* = \mathrm{X} + \mathrm{Y} \tag{7.2.9}$$

where

$$\mathrm{X}f(i,j) = \frac{(1-q^{\alpha+i-1})q^{\beta+j}}{1-q^{\alpha+\beta+i+j-1}} f(i-1,j), \quad \mathrm{Y}f(i,j) = \frac{1-q^{\beta+j-1}}{1-q^{\alpha+\beta+i+j-1}} f(i,j-1)$$

satisfy q-commutativity $\mathrm{XY} = q\mathrm{YX}$. Using the operator q-binom (6.3.18), we see that the probability measure $\tau^{(\alpha)\beta}_{q(N)} = \left(\mathrm{P}^{(\alpha)\beta *}_{\mathrm{X}^{\square}_q}\right)^N \delta_{(0,0)}$ is given by

$$\tau^{(\alpha)\beta}_{q(N)}(i,j) = \begin{bmatrix} N \\ i \end{bmatrix}_q \frac{\zeta_{(q)}(\alpha+i,\beta+j)}{\zeta_{(q)}(\alpha,\beta)} \cdot q^{\beta\cdot i}, \quad i+j=N. \tag{7.2.10}$$

The q-beta chain $\mathrm{P}^{(\alpha)\beta}_{\mathrm{X}^{\square}_q}$ has the property that starting the chain at the state (i,j) (rather than the origin $(0,0)$) is equivalent to replacing α and β by $\alpha+i$ and $\beta+j$. Using this remark we see that the Green kernel is given by

$$\mathrm{G}^{(\alpha)\beta}_{\mathrm{X}^{\square}_q}((i,j),(i',j')) = \begin{bmatrix} i'+j'-i-j \\ i'-i \end{bmatrix}_q \cdot \frac{\zeta_{(q)}(\alpha+i',\beta+j')}{\zeta_{(q)}(\alpha+i,\beta+j)} q^{(\beta+j)(i'-i)}, \quad i \le i',\ j \le j', \qquad 0 \quad \text{otherwise}. \tag{7.2.11}$$

The Martin kernel is therefore given by

$$\mathrm{K}^{(\alpha)\beta}_{\mathrm{X}^{\square}_q}((i,j),(i',j')) = \frac{\begin{bmatrix} i'+j'-i-j \\ i'-i \end{bmatrix}_q}{\begin{bmatrix} i'+j' \\ i' \end{bmatrix}} \cdot \frac{\zeta_{(q)}(\alpha,\beta)}{\zeta_{(q)}(\alpha+i,\beta+j)} \cdot q^{(\beta+j)(i'-i)-\beta i'}, \quad i \le i',\ j \le j', \qquad 0 \ \text{otherwise}. \tag{7.2.12}$$

A sequence $(i_n, j_n) \in \mathrm{X}^{\square}_q$ converges in the Martin metric if and only if

$$\mathrm{K}^{(\alpha)\beta}_{\mathrm{X}^{\square}_q}((1,0),(i_n,j_n)) = \frac{1-q^{i_n}}{1-q^{i_n+j_n}} \cdot \left(\frac{q^{-\beta}-q^{\alpha}}{1-q^{\alpha}}\right) \tag{7.2.13}$$

converges; that is, either (i_n, j_n) is eventually constant, or else $i_n + j_n \to \infty$ and q^{i_n} converges in $q^{\mathbb{N}} \cup \{0\}$ (i.e. either $i_n \to \infty$, or else $j_n \to \infty$ and i_n is eventually constant). We therefore identify the boundary as

$$\partial \mathrm{X}^{\square}_q = g^{\mathbb{N}} \cup \{0 = g^{\infty}\}. \tag{7.2.14}$$

The extension of the Martin kernel to the boundary is given by

$$\mathrm{K}^{(\alpha)\beta}_{\mathrm{X}^{\square}_q}((i,j),g^n) = \frac{\zeta_{(q)}(1+n)}{\zeta_{(q)}(1+n-i)} \cdot \frac{\zeta_{(q)}(\alpha,\beta)}{\zeta_{(q)}(\alpha+i,\beta+j)} q^{j(n-i)-i\beta} \quad i \le n,$$
$$0 \quad \text{otherwise}, \tag{7.2.15}$$

$$\mathrm{K}^{(\alpha)\beta}_{\mathrm{X}^{\square}_q}((i,j),0) = \frac{\zeta_{(q)}(\alpha)}{\zeta_{(q)}(\alpha+\beta)} \cdot \frac{\zeta_{(q)}(\alpha+\beta+i)}{\zeta_{(q)}(\alpha+i)} q^{-i\beta} \quad j=0,$$
$$0 \quad \text{otherwise}.$$

The harmonic measure is the q-beta measure, and is given by

$$\tau^{(\alpha)\beta}_q(g^n) = \frac{\zeta_{(q)}(\alpha+n)}{\zeta_{(q)}(1+n)} \cdot \frac{q^{\beta\cdot n}}{\zeta_{(q)}(\alpha,\beta)}\,. \tag{7.2.16}$$

That this is a probability measure is just the q-beta sum (6.3.1).

When we let $\alpha \to \infty$, the q-beta chain $\mathrm{P}^{(\alpha)\beta}_{\mathrm{X}^{\square}_q}$ converges to a chain isomorphic to the q-gamma chain $\mathrm{P}^{\beta}_{\mathrm{X}^0_q}$; the isomorphism between the state spaces $\mathrm{X}^{\square}_q = \mathbb{N}\times\mathbb{N}$ and $\mathrm{X}^0_q = \{(i,j),\ i \ge j \ge 0\}$ are given by

$$\begin{aligned} \mathrm{X}^{\square}_q &\leftrightarrow \mathrm{X}^0_q \\ (i,j) &\mapsto (i+j,j) \\ (i-j,j) &\leftarrow\!\shortmid (i,j) \end{aligned} \tag{7.2.17}$$

7.3 THE HEISENBERG RELATION AND SPECIAL BASIS

The q-beta chain on $\mathrm{X}^{\square}_q$ is timed, but its graph is not a tree—it has loops. Thus in contrast to its p-adic limit, the history of the particle walking in $\mathrm{X}^{\square}_q$ cannot be retrieved from its position at a given time (there are $\binom{i+j}{i}$ ways of getting to (i,j) from the origin). The representation of an element of the boundary $\partial\mathrm{X}^{\square}_q$ as a path in $\mathrm{X}^{\square}_q$ is far from being unique. We have of course the universal covering chain $\Omega_{\mathrm{X}^{\square}_q}$, where $\Omega_{\mathrm{X}^{\square}_q}$ is a binary tree, and every path $\omega \in \partial\Omega_{\mathrm{X}^{\square}_q} = \overset{\lim}{\longleftarrow} \mathrm{X}^{\square}_{q(N)}$ has a well-defined limit value $\lim\omega \in \partial\mathrm{X}^{\square}_q = g^{\mathbb{N}} \cup \{g^\infty = 0\}$. But the maps $\Omega_{\mathrm{X}^{\square}_q} \twoheadrightarrow \mathrm{X}^{\square}_q$, lim: $\partial\Omega_{\mathrm{X}^{\square}_q} \twoheadrightarrow \partial\mathrm{X}^{\square}_q$ are far from being one to one.

We have the Hilbert spaces

$$\begin{aligned} \mathscr{H}^{(\alpha)\beta}_{q(N)} &= \ell_2\left(\mathrm{X}^{\square}_{q(N)}, \tau^{(\alpha)\beta}_{q(N)}\right), \quad \mathrm{X}^{\square}_{q(N)} = \{(i,j), i+j=N\}, \\ &\qquad \tau^{(\alpha)\beta}_{q(N)}(i,j) = \begin{bmatrix} N \\ i \end{bmatrix}_q \frac{\zeta_{(q)}(\alpha+i,\beta+j)}{\zeta_{(q)}(\alpha,\beta)} \cdot q^{\beta i}, \\ \mathscr{H}^{(\alpha)\beta}_q &= \ell_2(\partial\mathrm{X}^{\square}_q, \tau^{(\alpha)\beta}_q), \qquad \partial\mathrm{X}^{\square}_q = g^{\mathbb{N}} \cup \{g^\infty = 0\}, \\ &\qquad \tau^{(\alpha)\beta}_q(g^i) = \frac{\zeta_{(q)}(\alpha+i)}{\zeta_{(q)}(1+i)} \cdot \frac{q^{\beta i}}{\zeta_{(q)}(\alpha,\beta)}\,. \end{aligned} \tag{7.3.1}$$

When we view the Martin kernel as an operator on these spaces, we see that K^* is not an isometry, and we cannot identify $\mathscr{H}^{(\alpha)\beta}_{q(N)}$ with a subspace of $\mathscr{H}^{(\alpha)\beta}_{q(M)}$ for any $M > N$, or of $\mathscr{H}^{(\alpha)\beta}_{q}$. We do not have a number operator and orthogonal basis, as we have in the p-adic limit due to the tree structure (cf. §4.4).

We shall look at difference operators

$$\mathrm{D}_N\colon \mathscr{H}^{(\alpha)\beta}_{q(N)} \to \mathscr{H}^{(\alpha+1)\beta+1}_{q(N-1)} \quad \text{and} \quad \mathrm{D}\colon \mathscr{H}^{(\alpha)\beta}_{q} \to \mathscr{H}^{(\alpha+1)\beta+1}_{q}$$

and use as number operators $\mathrm{D}_N^*\mathrm{D}_N$ and $\mathrm{D}^*\mathrm{D}$; their eigenfunctions will give us the desired orthogonal basis, the q-Hahn basis $\{\varphi^{(\alpha)\beta}_{q(N),m}\}$ for $\mathscr{H}^{(\alpha)\beta}_{q(N)}$, and the q-Jacobi basis $\{\varphi^{(\alpha),\beta}_{q,m}\}$ for $\mathscr{H}^{(\alpha)\beta}_{q}$. We shall see that the Martin kernel is diagonalized with respect to these bases, $\mathrm{K}_N\varphi^{(\alpha)\beta}_{q,m} = \varphi^{(\alpha)\beta}_{q(N),m}$, $m \le N$, and in the p-adic limit we obtain the basis of §4.4.

It will be useful to renormalize our measures first, and write

$$\begin{aligned} \mathrm{t}^{(\alpha)\beta}_{q(N)} &= \left(\mathrm{C}^{(\alpha)\beta}_{q(N)}\right)^{-1} \cdot \tau^{(\alpha)\beta}_{q(N)}\ \mathrm{C}^{(\alpha)\beta}_{q(N)} = \frac{1}{\zeta_{(q)}(\alpha,\beta)} \cdot \frac{\zeta_{(q)}(1+N)}{\zeta_{(q)}(\alpha+\beta+N)}, \\ \mathrm{H}^{(\alpha)\beta}_{q(N)} &= \ell_2\left(\mathrm{X}^{\square}_{q(N)}, \mathrm{t}^{(\alpha)\beta}_{q(N)}\right),\ \mathrm{t}^{(\alpha)\beta}_{q(N)}(i,j) = \frac{\zeta_{(q)}(\alpha+i)\zeta_{(q)}(\beta+j)}{\zeta_{(q)}(1+i)\zeta_{(q)}(1+j)} q^{\beta i}. \end{aligned} \tag{7.3.2}$$

For example at $\alpha = \beta = 1$, we have $\mathrm{t}^{(1)1}_{q(N)}(i,j) = q^i$.

Consider the q-difference operator

$$\nabla\colon \mathrm{H}^{(1)1}_{q(N)} \to \mathrm{H}^{(1)1}_{q(N-1)}, \quad \nabla\varphi(i,j) = \frac{\varphi(i,j+1) - \varphi(i+1,j)}{q^i - q^{i+1}}, \tag{7.3.3}$$

and its adjoint

$$\nabla^*\varphi(i,j) = \frac{\varphi(i,j-1) - \varphi(i-1,j)}{q^i - q^{i+1}}; \tag{7.3.4}$$

they satisfy q-commutativity

$$\nabla^*\nabla = q\nabla\nabla^*. \tag{7.3.5}$$

Moreover, we can write

$$\nabla^* = \mathcal{X} + \mathcal{Y}, \quad \mathcal{X}\varphi(i,j) = -\frac{\varphi(i-1,j)}{q^i(1-q)}, \quad \mathcal{Y}\varphi(i,j) = \frac{\varphi(i,j-1)}{q^i(1-q)}, \tag{7.3.6}$$

and we again have q-commutativity

$$\mathcal{X}\mathcal{Y} = q\mathcal{Y}\mathcal{X}. \tag{7.3.7}$$

We can use the operator q-binom to compute $(\nabla^*)^m = (\mathcal{X}+\mathcal{Y})^m$,

$$(\nabla^*)^m\varphi(i,j) = \frac{1}{(q^i(1-q))^m} \sum_{i_0+j_0=m} \begin{bmatrix} m \\ i_0 \end{bmatrix}_q (-1)^{i_0} q^{i_0(i_0-1)/2}\varphi(i-i_0, j-j_0). \tag{7.3.8}$$

Next let D_N denote the operator ∇, but viewed as an operator from $\mathrm{H}_{q(N)}^{(\alpha)\beta}$ to $\mathrm{H}_{q(N-1)}^{(\alpha+1)\beta+1}$:

$$\mathrm{D}_N = \nabla\colon \mathrm{H}_{q(N)}^{(\alpha)\beta} \to \mathrm{H}_{q(N-1)}^{(\alpha+1)\beta+1}, \tag{7.3.9}$$

and consider its adjoint

$$\mathrm{D}_N^* = \frac{\mathrm{t}_{q(N)}^{(1)1}}{\mathrm{t}_{q(N)}^{(\alpha)\beta}} \nabla^* \frac{\mathrm{t}_{q(N-1)}^{(\alpha+1)\beta+1}}{\mathrm{t}_{q(N-1)}^{(1)1}}. \tag{7.3.10}$$

Explicitly,

$$\mathrm{D}_N^*\varphi(i,j) = (1-q)\cdot\{[\alpha+i]_q\cdot[j]_q\cdot\varphi(i,j-1) - [i]_q\cdot[\beta+j]_q\cdot q^{-\beta}\cdot\varphi(i-1,j)\}. \tag{7.3.11}$$

The point of shifting the indices from (α,β) to $(\alpha+1,\beta+1)$, and of the renormalization of the beta measures, is that now instead of q-commutativity (7.3.5), we have the fundamental *q-Heisenberg relations*:

$$\text{‘}[\mathrm{D},\mathrm{D}^*]\text{’} = \mathrm{D}_N\mathrm{D}_N^* - \mathrm{D}_{N-1}^*\mathrm{D}_{N-1} = [\alpha+\beta]_q\cdot q^{-\beta}\cdot \mathrm{id}_{\mathrm{H}_{q(N-1)}^{(\alpha+1)\beta+1}}. \tag{7.3.12}$$

Note that the operators do not act on one space, but fit in the diagram, the ‘ladder’:

$$\mathrm{H}_{q(N)}^{(\alpha)\beta} \underset{D_N^*}{\overset{D_N}{\rightleftarrows}} \mathrm{H}_{q(N-1)}^{(\alpha+1)\beta+1} \underset{D_{N-1}^*}{\overset{D_{N-1}}{\rightleftarrows}} \mathrm{H}_{q(N-2)}^{(\alpha+2)\beta+2}. \tag{7.3.13}$$

The proof of the Heisenberg relations up the ladder (7.3.12) is straightforward. We have

(i) $\mathrm{D}_N\mathrm{D}_N^*\varphi(i,j)$

$$= \frac{1}{(1-q)q^i}\cdot\{\mathrm{D}_N^*\varphi(i,j+1) - \mathrm{D}_N^*\varphi(i+1,j)\}$$

$$= \frac{1}{(1-q)^2q^i}\left\{\begin{array}{l}(1-q^{\alpha+i})(1-q^{j+1})\varphi(i,j) - (1-q^i)(1-q^{\beta+j+1})q^{-\beta}\varphi(i-1,j+1)\\ (1-q^{i+1})(1-q^{\beta+j})q^{-\beta}\varphi(i,j) - (1-q^{\alpha+i+1})(1-q^j)\varphi(i+1,j-1)\end{array}\right\}.$$

When we calculate $\mathrm{D}_{N-1}^*\mathrm{D}_{N-1}\varphi(i,j)$ we use (7.3.11) but with α and β replaced by $\alpha+1$ and $\beta+1$. The result is

(ii) $\mathrm{D}_{N-1}^*\mathrm{D}_{N-1}\varphi(i,j)$

$$= \frac{1}{(1-q)}\Big\{(1-q^{\alpha+1+i})(1-q^j)\mathrm{D}_{N-1}\varphi(i,j-1) - (1-q^i)(1-q^{\beta+1+j})q^{-(\beta+1)}\mathrm{D}_{N-1}\varphi(i-1,j)\Big\}$$

$$= \frac{1}{(1-q)^2q^i}\left\{\begin{array}{l}(1-q^{\alpha+1+i})(1-q^j)\cdot[\varphi(i,j)-\varphi(i+1,j-1)]\\ (1-q^i)(1-q^{\beta+1+j})q^{-\beta}\cdot[\varphi(i,j)-\varphi(i-1,j+1)]\end{array}\right\}.$$

When we calculate the difference between (i) and (ii) we see that the terms with $\varphi(i+1,j-1)$ and $\varphi(i-1,j+1)$ cancel out, and we are left with a constant multiple of $\varphi(i,j)$, the constant being

$$\frac{1}{(1-q)^2 q^i}\cdot\left\{\begin{array}{l}(1-q^{\alpha+i})(1-q^{j+1})-(1-q^{\alpha+i+1})(1-q^j)\\(1-q^{i+1})(1-q^{\beta+j})q^{-\beta}-(1-q^i)(1-q^{\beta+j+1})q^{-\beta}\end{array}\right\}$$
$$=\frac{1-q^{\alpha+\beta}}{1-q}\cdot q^{-\beta},\quad \text{cf. } (6.1.8)\,.$$

We can now define for $0\le m\le N$

$$\tilde{\varphi}^{(\alpha)\beta}_{q(N),m}=\frac{q^{m(m-1)/2}}{[m]_q!}\cdot(\mathrm{D}^*)^m\mathbb{1}_{H^{(\alpha+m)\beta+m}_{q(N-m)}}=\frac{q^{m(m-1)/2}}{[m]_q!}\cdot\frac{t^{(1)1}_{q(N)}}{t^{(\alpha)\beta}_{q(N)}}(\nabla^*)^m\frac{t^{(\alpha+m)\beta+m}_{q(N-m)}}{t^{(1)1}_{q(N-m)}}\,. \tag{7.3.14}$$

Explicitly, using (7.3.8) we obtain

$$\tilde{\varphi}^{(\alpha)\beta}_{q(N),m}(i,j)=\sum_{i_0+j_0=m}\begin{bmatrix}i\\i_0\end{bmatrix}_q\cdot\begin{bmatrix}j\\j_0\end{bmatrix}_q(-1)^{i_0}q^{j_0(j_0-1)/2-i_0\beta}\frac{\zeta_{(q)}(\alpha+i+j_0)\zeta_{(q)}(\beta+j+i_0)}{\zeta_{(q)}(\alpha+i)\zeta_{(q)}(\beta+j)}\,. \tag{7.3.15}$$

We have by definition

$$\mathrm{D}^*_N\tilde{\varphi}^{(\alpha+1)\beta+1}_{q(N-1)m-1}=[m]_q\cdot q^{1-m}\cdot\tilde{\varphi}^{(\alpha)\beta}_{q(N),m}\,, \tag{7.3.16}$$

and by induction, using the q-Heisenberg relations, we obtain

$$\mathrm{D}_N\tilde{\varphi}^{(\alpha)\beta}_{q(N),m}=[m-1+\alpha+\beta]_q\cdot q^{-\beta}\cdot\tilde{\varphi}^{(\alpha+1)\beta+1}_{q(N-1),m-1}\,. \tag{7.3.17}$$

Together (7.3.16) and (7.3.17) give

$$\mathrm{D}^*_N\mathrm{D}_N\tilde{\varphi}^{(\alpha)\beta}_{q(N),m}=[m]_q\cdot[m-1+\alpha+\beta]_q\cdot q^{-(m-1+\beta)}\cdot\tilde{\varphi}^{(\alpha)\beta}_{q(N),m}\,. \tag{7.3.18}$$

Thus $\tilde{\varphi}^{(\alpha)\beta}_{q(N),m}$, $0\le m\le N$, are the eigenfunctions of the self-adjoint 'number operator' $\mathrm{D}^*_N\mathrm{D}_N$, with different eigenvalues, and so they form an orthogonal basis for $\mathrm{H}^{(\alpha)\beta}_{q(N)}$, and hence for $\mathscr{H}^{(\alpha)\beta}_{q(N)}$.

To compute the norms of $\tilde{\varphi}^{(\alpha)\beta}_{q(N),m}$, we use the q-Heisenberg relation to obtain

$$\left\|\tilde{\varphi}^{(\alpha)\beta}_{q(N),m}\right\|^2_{\mathrm{H}^{(\alpha)\beta}_{q(N)}}=\frac{1}{[m]_q}[m-1+\alpha+\beta]_q\cdot q^{m-1-\beta}\cdot\left\|\tilde{\varphi}^{(\alpha+1)\beta+1}_{q(N-1),m-1}\right\|^2_{\mathrm{H}^{(\alpha+1)\beta+1}_{q(N-1)}}\,;$$

hence

$$\left\|\tilde{\varphi}^{(\alpha)\beta}_{q(N),m}\right\|^2_{\mathrm{H}^{(\alpha)\beta}_{q(N)}}=\frac{\zeta_{(q)}(1)}{\zeta_{(q)}(1+m)}\cdot\frac{\zeta_{(q)}(\alpha+\beta+2m-1)}{\zeta_{(q)}(\alpha+\beta+m-1)}q^{-\beta m}\cdot\|\mathbb{1}\|^2_{\mathrm{H}^{(\alpha+m)\beta+m}_{q(N-m)}} \tag{7.3.19}$$

where

$$\|\mathbb{1}\|^2_{\mathrm{H}^{(\alpha+m)\beta+m}_{q(N-m)}}=\left(\mathrm{C}^{(\alpha+m)\beta+m}_{q(N-m)}\right)^{-1}.$$

All the orthogonal bases we shall encounter will similarly result from the 'Heisenberg relations up the ladder'. That is, we have a sequence of Hilbert spaces and operators

$$\mathscr{H}^0 \underset{\mathrm{D}_0^*}{\overset{\mathrm{D}_0}{\rightleftarrows}} \mathscr{H}^1 \underset{\mathrm{D}_1^*}{\overset{\mathrm{D}_1}{\rightleftarrows}} \mathscr{H}^2 \underset{\mathrm{D}_2^*}{\overset{\mathrm{D}_2}{\rightleftarrows}} \mathscr{H}^3 \rightleftarrows \cdots$$

satisfying $\mathrm{D}_N\mathrm{D}_N^*-\mathrm{D}_{N+1}^*\mathrm{D}_{N+1}=\mathrm{c}_N\cdot\mathrm{id}_{\mathscr{H}^{N+1}}$, c_N a constant, and a sequence of vectors $\mathbb{1}_N\in\mathscr{H}^N$ satisfying $\mathrm{D}_N\mathbb{1}_N=0$. It follows from above that

$$\varphi_{N,m}=(\mathrm{D}^*)^m\mathbb{1}=\mathrm{D}_N^*\mathrm{D}_{N+1}^*\cdots\mathrm{D}_{N+m-1}^*\mathbb{1}_{N+m}$$

are orthogonal in $\mathscr{H}^N$, the eigenfunctions of the number operator $\mathrm{D}_N^*\mathrm{D}_N$.

In fact the above basis $\{\tilde{\varphi}^{(\alpha)\beta}_{q(N),m}\}$, the q-Hahn basis for $\mathscr{H}^{(\alpha)\beta}_{q(N)}$, will give us all the important bases as limits: when we let $N\to\infty$ we obtain the q-Jacobi basis $\{\varphi^{(\alpha)\beta}_{q,m}\}$ for the boundary space $\mathscr{H}^{(\alpha)\beta}_{q}$, and when we let $\alpha\to\infty$ we obtain the q-Laguerre basis $\{\varphi^{\beta}_{\mathbb{Z}_q,m}\}$ for the Hilbert space associated with the q-gamma measure $\mathscr{H}^{\beta}_{\mathbb{Z}_q}$. In the p-adic limit we will retrieve our basis of Chapter 4 that resulted from the tree structure (the Heisenberg relations up the ladder collapse in the p-adic limit; the space $\mathscr{H}^1$ becomes infinitely far from $\mathscr{H}^0$). In the real limit the Heisenberg relation up the ladder survives, and will serve as a substitute for the lack of a tree structure. To get the p-adic and real limits on equal footings, we will have to be somewhat pedantic about our normalizations.

Using the q-Chu–Vandermonde (6.3.15) identity, we can transform our formula (7.3.15) for $\tilde{\varphi}^{(\alpha)\beta}_{q(N),m}$ into a more useful formula,

$$\tilde{\varphi}^{(\alpha)\beta}_{q(N),m}(i,j)=\sum_{i_0+j_0=m}\begin{bmatrix} i\\ i_0\end{bmatrix}_q\begin{bmatrix} j\\ j_0\end{bmatrix}_q(-1)^{i_0}q^{i\cdot j_0+j_0(j_0-1)/2-i_0(\beta+j_0)}\cdot \frac{\zeta_{(q)}(\alpha+m)}{\zeta_{(q)}(\alpha+i_0)}\cdot\frac{\zeta_{(q)}(\beta+m)}{\zeta_{(q)}(\beta+j_0)}. \tag{7.3.20}$$

The difference between (7.3.15) and (7.3.20) is small, and is important only in order to compare our basis with those appearing in the literature. However, it is not trivial, and so we give the details (this should be omitted on a first reading).

We start with the identity

$$\sum_{i_0+j_0=m} \begin{bmatrix} m \\ i_0 \end{bmatrix}_q (-1)^{i_0} q^{i_0(i_0-1)/2} \cdot q^{i_0 x + j_0 y}$$

$$= q^{my} \cdot \frac{\zeta_{(q)}(x-y+m)}{\zeta_{(q)}(x-y)} \text{ by (6.3.8)}$$

$$= q^{my} \sum_{i_0+j_0=m} \begin{bmatrix} m \\ i_0 \end{bmatrix}_q q^{i_0 x} \cdot \frac{\zeta_{(q)}(x+j_0)}{\zeta_{(q)}(x)} \cdot \frac{\zeta_{(q)}(-y+i_0)}{\zeta_{(q)}(-y)} \text{ by (6.3.15)}$$

$$= \sum_{i_0+j_0=m} \begin{bmatrix} m \\ i_0 \end{bmatrix}_q (-1)^{i_0} q^{i_0(i_0-1)/2} \cdot q^{i_0 x} \frac{\zeta_{(q)}(x+j_0)}{\zeta_{(q)}(x)} \cdot q^{j_0 y} \frac{\zeta_{(q)}(1+y)}{\zeta_{(q)}(1+y-i_0)} \text{ by (6.3.16)}.$$

We multiply it by

$$\frac{\zeta_{(q)}(x)}{\zeta_{(q)}(x+\alpha+m)} \frac{\zeta_{(q)}(1+y-\beta-m)}{\zeta_{(q)}(1+y)},$$

and expand using (6.3.2), obtaining

$$\sum_{i_0+j_0=m} \begin{bmatrix} m \\ i_0 \end{bmatrix}_q (-1)^{i_0} q^{i_0(i_0-1)/2} \cdot q^{i_0 x + j_0 y} \cdot \sum_{i \geq 0} \frac{\zeta_{(q)}(1)}{\zeta_{(q)}(1+i)} \frac{\zeta_{(q)}(\alpha+m+i)}{\zeta_{(q)}(\alpha+m)} \cdot q^{ix} \cdot$$

$$\sum_{j \geq 0} \frac{\zeta_{(q)}(1)}{\zeta_{(q)}(1+j)} \frac{\zeta_{(q)}(\beta+m+j)}{\zeta_{(q)}(\beta+m)} q^{j(1+y-\beta-m)} = \sum_{i_0+j_0=m} \begin{bmatrix} m \\ i_0 \end{bmatrix}_q (-1)^{i_0} q^{i_0(i_0-1)/2} \cdot q^{i_0 x + j_0 y} \cdot$$

$$\sum_{i \geq 0} \frac{\zeta_{(q)}(1)}{\zeta_{(q)}(1+i)} \frac{\zeta_{(q)}(\alpha+i_0+i)}{\zeta_{(q)}(\alpha+i_0)} \cdot q^{i(x+j_0)} \cdot \sum_{j \geq 0} \frac{\zeta_{(q)}(1)}{\zeta_{(q)}(1+j)} \frac{\zeta_{(q)}(\beta+j_0+j)}{\zeta_{(q)}(\beta+j_0)} q^{j(1+y-\beta-m)}.$$

Comparing coefficients of q^{ix+jy} we obtain (after multiplying by $(1-q)^{i+j-m}$)

$$\sum_{i_0+j_0=m} \begin{bmatrix} m \\ i_0 \end{bmatrix}_q (-1)^{i_0} q^{i_0(i_0-1)/2} \cdot \frac{\zeta_{(q)}(\alpha+j_0+i)}{[i-i_0]!\,\zeta_{(q)}(\alpha+m)} \cdot \frac{\zeta_{(q)}(\beta+i_0+j)}{[j-j_0]!\,\zeta_{(q)}(\beta+m)} \cdot q^{(j-j_0)(1-\beta-m)}$$

$$= \sum_{i_0+j_0=m} \begin{bmatrix} m \\ i_0 \end{bmatrix}_q (-1)^{i_0} q^{i_0(i_0-1)/2} \cdot \frac{\zeta_{(q)}(\alpha+i) q^{(i-i_0)j_0}}{[i-i_0]!\,\zeta_{(q)}(\alpha+i_0)} \cdot \frac{\zeta_{(q)}(\beta+j)}{[j-j_0]!\,\zeta_{(q)}(\beta+j_0)} \cdot q^{(j-j_0)(1-\beta-m)}.$$

Multiplying this by

$$\frac{[i]_q!\,[j]_q!}{[m]_q!} \frac{\zeta_{(q)}(\alpha+m)}{\zeta_{(q)}(\alpha+i)} \frac{\zeta_{(q)}(\beta+m)}{\zeta_{(q)}(\beta+j)} q^{-m(m-1)/2-m\beta-j(1-\beta-m)}$$

we finally see that (7.3.15) equals (7.3.20).

Returning to the 'probability one' spaces $\mathscr{H}^{(\alpha)\beta}_{q(N)}$, it will be useful to renormalize the basis $\tilde{\varphi}^{(\alpha)\beta}_{q(N),m}$, and define the *q-Hahn basis* $\{\varphi^{(\alpha)\beta}_{q(N),m}\}$, $0 \le m \le N$, for the space $\mathscr{H}^{(\alpha)\beta}_{q(N)}$, by

$$\varphi^{(\alpha)\beta}_{q(N),m} = \frac{\zeta_{(q)}(\alpha+\beta+N)}{\zeta_{(q)}(\alpha+\beta+N+m)} \cdot \tilde{\varphi}^{(\alpha)\beta}_{q(N),m} \,. \tag{7.3.21}$$

We can now rewrite (7.3.15) to (7.3.20) in terms of $\varphi^{(\alpha)\beta}_{q(N),m}$, in the following subsections.

q-Hahn basis

$$\begin{aligned}\varphi^{(\alpha)\beta}_{q(N),m}(i,j) &= \sum_{i_0+j_0=m} \begin{bmatrix} i \\ i_0 \end{bmatrix}_q \begin{bmatrix} j \\ j_0 \end{bmatrix}_q (-1)^{i_0} q^{j_0(j_0-1)/2 - i_0\beta} \cdot \frac{\zeta_{(q)}(\alpha+i+j_0, \beta+j+i_0)}{\zeta_{(q)}(\alpha+i, \beta+j)} \\ &= \frac{\zeta_{(q)}(\alpha+\beta+N)}{\zeta_{(q)}(\alpha+\beta+N+m)} \sum_{i_0+j_0=m} \begin{bmatrix} i \\ i_0 \end{bmatrix}_q \begin{bmatrix} j \\ j_0 \end{bmatrix}_q (-1)^{i_0} q^{j_0(j_0-1)/2 + (i-i_0)j_0 - i_0\beta} \cdot \\ &\qquad \frac{\zeta_{(q)}(\alpha+m)}{\zeta_{(q)}(\alpha+i_0)} \cdot \frac{\zeta_{(q)}(\beta+m)}{\zeta_{(q)}(\beta+j_0)} \,. \end{aligned} \tag{7.3.22}$$

$$\begin{aligned} (1-q^{\alpha+\beta+N})^{-1} \cdot \mathrm{D}^*_N \varphi^{(\alpha+1)\beta+1}_{q(N-1),m-1} &= [m]_q q^{1-m} \cdot \varphi^{(\alpha)\beta}_{q(N),m} \,, \\ (1-q^{\alpha+\beta+N}) \cdot \mathrm{D}_N \varphi^{(\alpha)\beta}_{q(N),m} &= [m-1+\alpha+\beta]_q \cdot q^{-\beta} \cdot \varphi^{(\alpha+1)\beta+1}_{q(N-1),m-1} \,, \end{aligned} \tag{7.3.23}$$

$$\mathrm{D}^*_N \mathrm{D}_N \varphi^{(\alpha)\beta}_{q(N),m} = [m]_q \cdot [m-1+\alpha+\beta]_q \cdot q^{-(m-1+\beta)} \cdot \varphi^{(\alpha)\beta}_{q(N),m} \,.$$

$$\begin{aligned} \left\| \varphi^{(\alpha)\beta}_{q(N),m} \right\|^2_{\mathscr{H}^{(\alpha)\beta}_{q(N)}} &= \begin{bmatrix} N \\ m \end{bmatrix}_q \cdot \frac{\zeta_{(q)}(\alpha+m, \beta+m)}{\zeta_{(q)}(\alpha,\beta)} \cdot \\ &\qquad \frac{\zeta_{(q)}(\alpha+\beta+2m-1, \alpha+\beta+N)}{\zeta_{(q)}(\alpha+\beta+m-1, \alpha+\beta+N+m)} \cdot q^{-\beta m} \,. \end{aligned} \tag{7.3.24}$$

q-Jacobi basis

We can now let $N \to \infty$ to obtain the limit operators D and D*, and the limit basis $\{\varphi^{(\alpha)\beta}_{q,m}\}$, $m \ge 0$, for the boundary space $\mathscr{H}^{(\alpha)\beta}_{q}$. We write

$$\tau^{(\alpha)\beta}_q = \frac{1}{\zeta_{(q)}(\alpha,\beta)} \cdot \mathrm{t}^{(\alpha)\beta}_q, \quad \mathrm{t}^{(\alpha)\beta}_q(g^i) = \frac{\zeta_{(q)}(\alpha+i)}{\zeta_{(q)}(1+i)} \cdot q^{\beta i}, \quad \mathrm{H}^{(\alpha)\beta}_q = \ell_2(g^{\mathbb{N}}, \mathrm{t}^{(\alpha)\beta}_q) \,. \tag{7.3.25}$$

Consider the q-difference operator

$$\mathrm{D}\colon \mathrm{H}^{(\alpha)\beta}_q \to \mathrm{H}^{(\alpha+1)\beta+1}_q, \quad \mathrm{D}\varphi(g^i) = \frac{\varphi(g^i) - \varphi(g^{i+1})}{q^i - q^{i+1}} \,, \tag{7.3.26}$$

and its adjoint D^*, given explicitly by

$$\mathrm{D}^*\varphi(g^i) = [\alpha + i]_q \cdot \varphi(g^i) - [i]_q \cdot q^{-\beta} \cdot \varphi(g^{i-1}) . \tag{7.3.27}$$

They satisfy the q-Heisenberg relations

$$\mathrm{DD}^* - \mathrm{D}^*\mathrm{D} = [\alpha + \beta]_q \cdot q^{-\beta} \cdot \mathrm{id}_{\mathrm{H}_q^{(\alpha+1)\beta+1}} . \tag{7.3.28}$$

The ('little') q-*Jacobi polynomials* are given by

$$\varphi_{q,m}^{(\alpha)\beta} = \frac{q^{m(m-1)/2}}{[m]_q!} (\mathrm{D}^*)^m \mathbb{1}_{\mathrm{H}_q^{(\alpha+m)\beta+m}} . \tag{7.3.29}$$

Explicitly

$$\begin{aligned}
\varphi_{q,m}^{(\alpha)\beta}(g^i) &= \lim_{j\to\infty} \varphi_{q(i+j),m}^{(\alpha)\beta}(i,j) \qquad\qquad (7.3.30)\\
&= \sum_{i_0+j_0=m} \begin{bmatrix} i \\ i_0 \end{bmatrix}_q (-1)^{i_0} q^{j_0(j_0-1)/2 - i_0\beta} \frac{\zeta_{(q)}(1)}{\zeta_{(q)}(1+j_0)} \frac{\zeta_{(q)}(\alpha+i+j_0)}{\zeta_{(q)}(\alpha+i)} \\
&\qquad \text{from (7.3.15)} \\
&= \sum_{i_0+j_0=m} \begin{bmatrix} i \\ i_0 \end{bmatrix}_q (-1)^{i_0} q^{ij_0 + j_0(j_0-1)/2 - i_0(\beta+j_0)} \frac{\zeta_{(q)}(1)}{\zeta_{(q)}(1+j_0)} \frac{\zeta_{(q)}(\alpha+m)}{\zeta_{(q)}(\alpha+i_0)} \frac{\zeta_{(q)}(\beta+m)}{\zeta_{(q)}(\beta+j_0)} \\
&\qquad \text{from (7.3.20).}
\end{aligned}$$

Note that

$$\varphi_{q,m}^{(\alpha)\beta}(0) = \lim_{i\to\infty} \varphi_{q,m}^{(\alpha)\beta}(g^i) = (-1)^m q^{-m\beta} \frac{\zeta_{(q)}(1)}{\zeta_{(q)}(1+m)} \frac{\zeta_{(q)}(\beta+m)}{\zeta_{(q)}(\beta)} ;$$

hence

$$\frac{\varphi_{q,m}^{(\alpha)\beta}(g^i)}{\varphi_{q,m}^{(\alpha)\beta}(0)} = \sum_{i_0+j_0=m} \begin{bmatrix} m \\ j_0 \end{bmatrix}_q (-1)^{j_0} q^{j_0(j_0-1)/2 + j_0(\beta+i-i_0)} \frac{\zeta_{(q)}(\beta)}{\zeta_{(q)}(\beta+j_0)} \frac{\zeta_{(q)}(\alpha+m)}{\zeta_{(q)}(\alpha+i_0)} \frac{\zeta_{(q)}(1+i)}{\zeta_{(q)}(1+i-i_0)} .$$

Using the q-binom (6.3.8) to expand $\zeta_{(q)}(1+i)/\zeta_{(q)}(1+i-i_0)$, this is equal to

$$\sum_{0\le k\le m} \begin{bmatrix} m \\ k \end{bmatrix}_q (-1)^k q^{k(k-1)/2 + k(1+i-m)} \frac{\zeta_{(q)}(\beta)}{\zeta_{(q)}(\beta+k)} \sum_{0\le j_0\le k} \begin{bmatrix} k \\ j_0 \end{bmatrix}_q \frac{\zeta_{(q)}(\beta+k)}{\zeta_{(q)}(\beta+j_0)} \frac{\zeta_{(q)}(\alpha+m)}{\zeta_{(q)}(\alpha+m-j_0)} q^{j_0(\beta+j_0-1)} ,$$

and using the dual q-Chu–Vandermonde (6.3.15)$'$ the inner sum is equal to $\zeta_{(q)}(\alpha+\beta+m+k-1)/\zeta_{(q)}(\alpha+\beta+m-1)$. Thus we have another expression for $\varphi_{q,m}^{(\alpha)\beta}(g^i)$, showing more clearly that they are polynomials in q^i, namely

$$\begin{aligned}\frac{\varphi_{q,m}^{(\alpha)\beta}(g^i)}{\varphi_{q,m}^{(\alpha)\beta}(0)} &= \sum_{0\le k\le m} \begin{bmatrix} m \\ k \end{bmatrix}_q (-1)^k q^{k(k-1)/2-km} \frac{\zeta_{(q)}(\beta)}{\zeta_{(q)}(\beta+k)} \frac{\zeta_{(q)}(\alpha+\beta+m+k-1)}{\zeta_{(q)}(\alpha+\beta+m-1)} q^{(1+i)k} \\ &= {}_2\phi_1(-m, \alpha+\beta+m-1;\ \beta;\ 1+i) \text{ using (6.3.7).}\end{aligned} \tag{7.3.31}$$

We now have directly from the q-Heisenberg relation (7.3.28) and the definition (7.3.29)

$$\begin{aligned}\mathrm{D}^*\varphi_{q,m-1}^{(\alpha+1)\beta+1} &= [m]_q \cdot q^{1-m} \cdot \varphi_{q,m}^{(\alpha)\beta}\,, \\ \mathrm{D}\varphi_{q,m}^{(\alpha)\beta} &= [m-1+\alpha+\beta]_q \cdot q^{-\beta} \cdot \varphi_{q,m-1}^{(\alpha+1)\beta+1}\,, \\ \mathrm{D}^*\mathrm{D}\varphi_{q,m}^{(\alpha)\beta} &= [m]_q \cdot [m-1+\alpha+\beta]_q \cdot q^{-(m-1+\beta)} \cdot \varphi_{q,m}^{(\alpha)\beta}\,.\end{aligned}$$

The $\{\varphi_{q,m}^{(\alpha)\beta}\}, m \ge 0$, form an orthogonal base for $\mathrm{H}_q^{(\alpha)\beta}$, and hence for $\mathscr{H}_q^{(\alpha)\beta}$, where their norm is given by

$$\left\|\varphi_{q,m}^{(\alpha)\beta}\right\|^2_{\mathscr{H}_q^{(\alpha)\beta}} = \frac{\zeta_{(q)}(1)}{\zeta_{(q)}(1+m)} \cdot \frac{\zeta_{(q)}(\alpha+m,\beta+m)}{\zeta_{(q)}(\alpha,\beta)} \cdot \frac{\zeta_{(q)}(\alpha+\beta+2m-1)}{\zeta_{(q)}(\alpha+\beta+m-1)} \cdot q^{-\beta m}\,. \tag{7.3.32}$$

There is compatibility between the operators D_N and D_N^* and the limit operators D and D* on the boundary, given by the Martin kernel. Letting $\mathrm{K}_{q(N)}^{(\alpha)\beta}$ denote the operator from $\mathscr{H}_q^{(\alpha)\beta}$ to $\mathscr{H}_{q(N)}^{(\alpha)\beta}$ given by the Martin kernel, we have explicitly by (7.2.15)

$$\begin{aligned}\mathrm{K}_{q(N)}^{(\alpha)\beta}\varphi(i,j) &= \int \mathrm{K}_q^{(\alpha)\beta}((i,j),g^m)\varphi(g^m)\tau_q^{(\alpha)\beta}(g^m) \\ &= \frac{1}{\zeta_{(q)}(\alpha+i,\beta+j)} \sum_{m\ge 0} \frac{\zeta_{(q)}(\alpha+i+m)}{\zeta_{(q)}(1+m)} q^{m(\beta+j)} \varphi(g^{i+m})\,.\end{aligned} \tag{7.3.33}$$

Considering the operators in the diagram

$$\begin{array}{ccc} \mathscr{H}_q^{(\alpha)\beta} & \overset{\mathrm{D}}{\underset{\mathrm{D}^*}{\rightleftarrows}} & \mathscr{H}_q^{(\alpha+1)\beta+1} \\ \Big\downarrow \mathrm{K}_{q(N)}^{(\alpha)\beta} & & \Big\downarrow \mathrm{K}_{q(N-1)}^{(\alpha+1)\beta+1} \\ \mathscr{H}_{q(N)}^{(\alpha)\beta} & \overset{\mathrm{D}_N}{\underset{\mathrm{D}_N^*}{\rightleftarrows}} & \mathscr{H}_{q(N-1)}^{(\alpha+1)\beta+1} \end{array} \tag{7.3.34}$$

we have

$$(1-q^{\alpha+\beta+N})^{-1}\cdot \mathrm{D}_N^*\circ \mathrm{K}_{q(N-1)}^{(\alpha+1)\beta+1} = \mathrm{K}_{q(N)}^{(\alpha)\beta}\circ \mathrm{D}^*\,,$$
$$(1-q^{\alpha+\beta+N})\cdot \mathrm{D}_N\circ \mathrm{K}_{q(N)}^{(\alpha)\beta} = \mathrm{K}_{q(N-1)}^{(\alpha+1)\beta+1}\circ \mathrm{D}\,. \tag{7.3.35}$$

From this we obtain

$$\mathrm{K}_{q(N)}^{(\alpha)\beta}\,\varphi_{q,m}^{(\alpha)\beta} = \begin{cases}\varphi_{q(N),m}^{(\alpha)\beta} & 0\le m\le N\,,\\ 0 & N<m\,.\end{cases} \tag{7.3.36}$$

We have thus diagonalized $\mathrm{K}_{q(N)}^{(\alpha)\beta}$, and we see that it is not an orthogonal projection because of the difference in the norms between (7.3.24) and (7.3.32).

In the p-adic limit the $\varphi_{q,m}^{(\alpha)\beta}$ and $\varphi_{q(N),m}^{(\alpha)\beta}$ converge to the p-adic Jacobi and Hahn basis of §4.4,

$$\lim_{d\to\infty}\varphi_{(p^{-d}),m}^{(\alpha/d)\beta/d}(g^i) = \varphi_{p,m}^{(\alpha)\beta}(p^i\mathbb{Z}_p^*),\qquad \lim_{d\to\infty}\varphi_{(p^{-d})(N),m}^{(\alpha/d)\beta/d}(i,j) = \varphi_{p(N),m}^{(\alpha)\beta}(i,j)\,. \tag{7.3.37}$$

We see that $\mathrm{K}_{q(N)}^{(\alpha)\beta}$ is the q-analogue of the reduction mod p^N map, comparing (7.3.36) with (4.4.11).

q-Laguerre basis

We can now let $\alpha\to\infty$ to obtain the theory for the q-gamma chain. First we write

$$\mathscr{H}_{q(N)}^{\beta} = \ell_2(\mathrm{X}_{q(N)}^{\square},\tau_{q(N)}^{\beta})$$
$$\tau_{q(N)}^{\beta}(i,j) = \begin{bmatrix}N\\ j\end{bmatrix}_q\cdot\frac{\zeta_{(q)}(\beta+j)}{\zeta_{(q)}(\beta)}\cdot q^{\beta i} = \mathrm{C}_{q(N)}^{\beta}\cdot \mathrm{t}_{q(N)}^{\beta} \tag{7.3.38}$$

where

$$\mathrm{C}_{q(N)}^{\beta} = \frac{\zeta_{(q)}(1)\zeta_{(q)}(1+N)}{\zeta_{(q)}(\beta)}\quad\text{and}\quad \mathrm{t}_{q(N)}^{\beta}(i,j) = \frac{\zeta_{(q)}(\beta+j)}{\zeta_{(q)}(1+j)}\cdot\frac{q^{\beta i}}{\zeta_{(q)}(1+i)}\,.$$

Put

$$\mathrm{H}_{q(N)}^{\beta} = \ell_2\left(\mathrm{X}_{q(N)}^{\square},\mathrm{t}_{q(N)}^{\beta}\right).$$

Letting

$$\mathrm{D}_N = \nabla\colon \mathrm{H}_{q(N)}^{\beta}\to \mathrm{H}_{q(N-1)}^{\beta+1},\quad \mathrm{D}_N\varphi(i,j) = \frac{\varphi(i,j+1)-\varphi(i+1,j)}{q^i-q^{i+1}}\,, \tag{7.3.39}$$

we have for its adjoint

$$\mathrm{D}_N^* = \frac{\mathrm{t}_{q(N)}^{(1)1}}{\mathrm{t}_{q(N)}^{\beta}}\nabla^*\frac{\mathrm{t}_{q(N-1)}^{\beta+1}}{\mathrm{t}_{q(N-1)}^{(1)1}}$$
$$\mathrm{D}_N^*\varphi(i,j) = [j]_q\cdot\varphi(i,j-1) - [i]_q\cdot(1-q^{\beta+j})\cdot q^{-\beta}\cdot\varphi(i-1,j)\,. \tag{7.3.40}$$

The q-Heisenberg relations take the form

$$\mathrm{D}_N\mathrm{D}_N^* - \mathrm{D}_{N-1}^*\mathrm{D}_{N-1} = \frac{q^{-\beta}}{1-q}\cdot \mathrm{id}_{\mathrm{H}^{\beta+1}_{q(N-1)}}\,. \tag{7.3.41}$$

Setting for $0 \le m \le N$

$$\begin{aligned}\varphi^{\beta}_{q(N),m} &= \frac{q^{m(m-1)/2}}{[m]_q!}(\mathrm{D}^*)^m \mathbb{1}_{\mathrm{H}^{\beta+m}_{q(N-m)}} \\ &= \frac{q^{m(m-1)/2}}{[m]_q!}\frac{\mathrm{t}^{(1)1}_{q(N)}}{\mathrm{t}^{\beta}_{q(N)}}(\nabla^*)^m \frac{\mathrm{t}^{\beta+m}_{q(N-m)}}{\mathrm{t}^{(1)1}_{q(N-m)}}\,,\end{aligned} \tag{7.3.42}$$

we have

$$\begin{aligned}\varphi^{\beta}_{q(N),m}(i,j) &= \sum_{i_0+j_0=m}\begin{bmatrix} i \\ i_0\end{bmatrix}_q\cdot\begin{bmatrix} j \\ j_0\end{bmatrix}_q\cdot(-1)^{i_0}\cdot q^{j_0(j_0-1)/2-i_0\beta}\frac{\zeta_{(q)}(\beta+j+i_0)}{\zeta_{(q)}(\beta+j)} \\ &\qquad\qquad\text{from (7.3.15)} \\ &= \sum_{i_0+j_0=m}\begin{bmatrix} i \\ i_0\end{bmatrix}_q\cdot\begin{bmatrix} j \\ j_0\end{bmatrix}_q\cdot(-1)^{i_0}\cdot q^{ij_0+j_0(j_0-1)/2-i_0(\beta+j_0)}\frac{\zeta_{(q)}(\beta+m)}{\zeta_{(q)}(\beta+j_0)} \\ &\qquad\qquad\text{from (7.3.20)}\,.\end{aligned} \tag{7.3.43}$$

We have

$$\begin{aligned}\mathrm{D}_N^*\varphi^{\beta+1}_{q(N-1),m-1} &= [m]_q\cdot q^{1-m}\cdot\varphi^{\beta}_{q(N),m}\,, \\ \mathrm{D}_N\varphi^{\beta}_{q(N),m} &= \frac{q^{-\beta}}{(1-q)}\cdot\varphi^{\beta+1}_{q(N-1),m-1}\,, \\ \mathrm{D}_N^*\mathrm{D}_N\varphi^{\beta}_{q(N),m} &= [m]_q\cdot\frac{q^{-(m-1+\beta)}}{(1-q)}\cdot\varphi^{\beta}_{q(N),m}\,.\end{aligned} \tag{7.3.44}$$

The $\{\varphi^{\beta}_{q(N),m}\}$, $0\le m\le N$, form an orthogonal basis for $\mathscr{H}^{\beta}_{q(N)}$; their norm is given by

$$\left\|\varphi^{\beta}_{q(N),m}\right\|^2_{\mathscr{H}^{\beta}_{q(N)}} = \begin{bmatrix} N \\ m\end{bmatrix}_q\cdot\frac{\zeta_{(q)}(\beta+m)}{\zeta_{(q)}(\beta)}q^{-\beta m}\,. \tag{7.3.45}$$

Going over to the boundary limit $N\to\infty$, we set

$$\begin{aligned}&\mathscr{H}^{\beta}_{\mathbb{Z}_q} = \ell_2\left(\partial\mathrm{X}^{\square}_q,\tau^{\beta}_{\mathbb{Z}_q}\right),\quad \mathrm{H}^{\beta}_{\mathbb{Z}_q} = \ell_2\left(\partial\mathrm{X}^{\square}_q,\mathrm{t}^{\beta}_{\mathbb{Z}_q}\right), \\ &\tau^{\beta}_{\mathbb{Z}_q}(g^i) = \frac{\zeta_{(q)}(1)}{\zeta_{(q)}(\beta)}\cdot\frac{q^{\beta i}}{\zeta_{(q)}(1+i)},\quad \mathrm{t}^{\beta}_{\mathbb{Z}_q}(g^i) = \frac{q^{\beta i}}{\zeta_{(q)}(1+i)}\,.\end{aligned} \tag{7.3.46}$$

Consider the q-difference operator

$$\mathrm{D}\colon \mathrm{H}^{\beta}_{\mathbb{Z}_q} \to \mathrm{H}^{\beta+1}_{\mathbb{Z}_q}, \quad \mathrm{D}\varphi(g^i) = \frac{\varphi(g^i) - \varphi(g^{i+1})}{q^i - q^{i+1}}. \tag{7.3.47}$$

Its adjoint is given by

$$\mathrm{D}^*\varphi(g^i) = \frac{1}{(1-q)}\left[\varphi(g^i) - (1-q^i)q^{-\beta}\varphi(g^{i-1})\right], \tag{7.3.48}$$

retaining the q-Heisenberg relations

$$\mathrm{DD}^* - \mathrm{D}^*\mathrm{D} = \frac{q^{-\beta}}{(1-q)}\mathrm{id}_{\mathrm{H}^{\beta+1}_{\mathbb{Z}_q}}. \tag{7.3.49}$$

We define the q-*Laguerre basis* for $\mathscr{H}^{\beta}_{\mathbb{Z}_q}$ by

$$\begin{aligned}\varphi^{\beta}_{\mathbb{Z}_q,m}(g^i) &= \frac{q^{m(m-1)/2}}{[m]_q!}(\mathrm{D}^*)^m \mathbb{1}_{\mathrm{H}^{\beta+m}_{\mathbb{Z}_q}} \\ &= \lim_{\alpha\to\infty} \varphi^{(\alpha)\beta}_{q,m}(g^i) \\ &= \lim_{j\to\infty} \varphi^{\beta}_{q(i+j),m}(i,j).\end{aligned} \tag{7.3.50}$$

Explicitly

$$\begin{aligned}\varphi^{\beta}_{\mathbb{Z}_q,m}(g^i) &= \sum_{i_0+j_0=m} \begin{bmatrix} i \\ i_0 \end{bmatrix}_q (-1)^{i_0} q^{j_0(j_0-1)/2 - i_0\beta} \frac{\zeta_{(q)}(1)}{\zeta_{(q)}(1+j_0)} \\ &= \sum_{i_0+j_0=m} \begin{bmatrix} i \\ i_0 \end{bmatrix}_q (-1)^{i_0} q^{j_0(j_0-1)/2 - i_0\beta + j_0(i-i_0)} \frac{\zeta_{(q)}(1)}{\zeta_{(q)}(1+j_0)} \frac{\zeta_{(q)}(\beta+m)}{\zeta_{(q)}(\beta+j_0)}.\end{aligned} \tag{7.3.51}$$

We have

$$\begin{aligned}\mathrm{D}^*\varphi^{\beta+1}_{\mathbb{Z}_q,m-1} &= [m]_q \cdot q^{1-m} \cdot \varphi^{\beta}_{\mathbb{Z}_q,m}, \\ \mathrm{D}\varphi^{\beta}_{\mathbb{Z}_q,m} &= \frac{q^{-\beta}}{(1-q)} \cdot \varphi^{\beta+1}_{\mathbb{Z}_q,m-1}, \\ \mathrm{D}^*\mathrm{D}\varphi^{\beta}_{\mathbb{Z}_q,m} &= [m]_q \cdot \frac{q^{-(m-1+\beta)}}{(1-q)} \cdot \varphi^{\beta}_{\mathbb{Z}_q,m}.\end{aligned} \tag{7.3.52}$$

The norms are given by

$$\left\|\varphi^{\beta}_{\mathbb{Z}_q,m}\right\|^2_{\mathscr{H}^{\beta}_{\mathbb{Z}_q}} = \frac{\zeta_{(q)}(1)}{\zeta_{(q)}(1+m)} \cdot \frac{\zeta_{(q)}(\beta+m)}{\zeta_{(q)}(\beta)} \cdot q^{-\beta m} \stackrel{\text{def}}{=} \mathrm{C}^{\beta}_{(q)}(m). \tag{7.3.53}$$

Note also that

$$\varphi^{\beta}_{\mathbb{Z}_q,m}(0) = \lim_{i\to\infty} \varphi^{\beta}_{\mathbb{Z}_q,m}(g^i) = (-1)^m \cdot \mathrm{C}^{\beta}_{(q)}(m), \tag{7.3.54}$$

and

$$\frac{\varphi^{\beta}_{\mathbb{Z}_q,m}(g^i)}{\varphi^{\beta}_{\mathbb{Z}_q,m}(0)} = \sum_{0\le j_0\le m} \begin{bmatrix} m \\ j_0 \end{bmatrix}_q (-1)^{j_0} q^{j_0(j_0-1)/2-j_0 m} \frac{\zeta_{(q)}(\beta)}{\zeta_{(q)}(\beta+j_0)} q^{(1+i)j_0} \tag{7.3.55}$$
$$= {}_2\phi_1(-m,\infty;\beta;1+i),$$

a polynomial of degree m in q^{1+i}.

Consider the operator $\mathrm{K}^{\beta}_{q(N)}$ from $\mathscr{H}^{\beta}_{\mathbb{Z}_q}$ to $\mathscr{H}^{\beta}_{q(N)}$ given by the Martin kernel, given explicitly by

$$\mathrm{K}^{\beta}_{q(N)}\varphi(i,j) = \frac{\zeta_{(q)}(1)}{\zeta_{(q)}(\beta+j)} \sum_{k\ge 0} \frac{q^{k(\beta+j)}}{\zeta_{(q)}(1+k)} \varphi(g^{i+k}), \quad \text{cf.(7.1.14)}. \tag{7.3.56}$$

We have compatibility

$$\mathrm{D}^*_N \mathrm{K}^{\beta+1}_{q(N-1)} = \mathrm{K}^{\beta}_{q(N)} \mathrm{D}^*,$$
$$\mathrm{D}_N \mathrm{K}^{\beta}_{q(N)} = \mathrm{K}^{\beta+1}_{q(N-1)} \mathrm{D}, \tag{7.3.57}$$

from which we deduce that

$$\mathrm{K}^{\beta}_{q(N)} \varphi^{\beta}_{\mathbb{Z}_q,m} = \begin{cases} \varphi^{\beta}_{q(N),m} & 0 \le m \le N, \\ 0 & N < m. \end{cases} \tag{7.3.58}$$

In the p-adic limit we obtain the p-adic Laguerre basis of §4.2,

$$\lim_{d\to\infty} \varphi^{\beta/d}_{\mathbb{Z}_{(p^{-d})},m}(g^i) = \varphi^{\beta}_{\mathbb{Z}_p,m}\left(p^i \mathbb{Z}^*_p\right), \quad \lim_{d\to\infty} \varphi^{\beta/d}_{p^{-d}(N),m}(i,j) = \varphi^{\beta}_{p(N),m}(i,j). \tag{7.3.59}$$

We see that $\mathrm{K}^{\beta}_{q(N)}$ is the q-analogue of the 'reduction mod p^N' map, comparing (7.3.58) with (4.2.14).

The properties (4.2.15)–(4.2.20) are all limits of analogous properties of the q-Laguerre basis, for example

Mellin transform: $$\tau^s_{\mathbb{Z}_q}(\varphi^{\beta}_{\mathbb{Z}_q,m}) = (-1)^m \mathrm{C}^{\beta-s}_{(q)}(m). \tag{7.3.60}$$

$\beta'-\beta$ relation: $$\varphi^{\beta'}_{\mathbb{Z}_q,m} = \sum_{0\le j\le m} (-1)^j \mathrm{C}^{\beta'-\beta}_{(q)}(j) \cdot \varphi^{\beta}_{\mathbb{Z}_q,m-j}. \tag{7.3.61}$$

The best way to verify such formulae is by induction on m using (7.3.52) and (7.3.49).

8

The real beta chains

8.1 THE BETA CHAIN

The real beta chain is obtained as the real limit ($\alpha := \alpha/2$, $\beta := \beta/2$, $q := q^{2/N}$, $N \to \infty$) of both the symmetric and the non-symmetric q-beta chains. The state space is given by

$$\mathrm{X}_\eta = \mathbb{N} \times \mathbb{N} = \coprod_{N \geq 0} \mathrm{X}_{\eta(N)}, \quad \mathrm{X}_{\eta(N)} = \{(i,j) \,|\, i+j=N\}, \tag{8.1.1}$$

and the transition probabilities are given schematically by

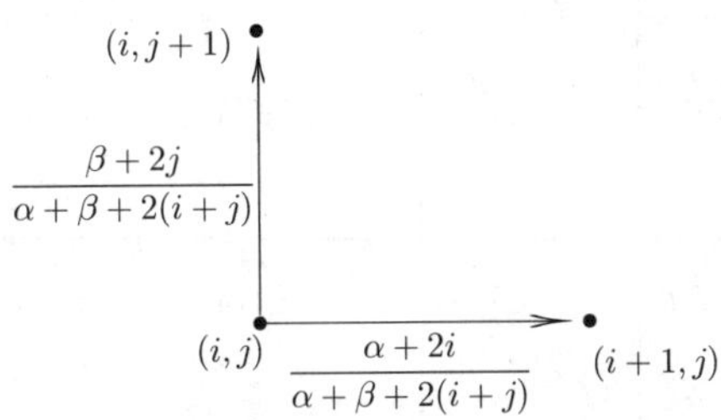

(8.1.2)

(cf. also Fig. 8.1).

This is the *discrete, canonical,* chain we were looking for. It is *local*—the transition probabilities are just the factors of the beta function $\zeta_\eta(\alpha,\beta)^{-1}$—and as we shall soon see, it has boundary

$$\partial \mathrm{X}_\eta = \mathbb{Z}_\eta^* \backslash \mathbb{P}^1(\mathbb{Q}_\eta) = \pm 1 \backslash \mathbb{P}^1(\mathbb{R}), \tag{8.1.3}$$

with harmonic measure equal to the real beta measure $\tau_\eta^{\alpha,\beta}$.

On the finite set $\mathrm{X}_{\eta(N)}$ we have the probability measure $\tau_{\eta(N)}^{\alpha,\beta} = (\mathrm{P}_{\alpha,\beta}^*)^N \delta_{(0,0)}$; it is the negative hypergeometric distribution, explicitly given by

$$\tau_{\eta(N)}^{\alpha,\beta}(i,j) = \frac{N!}{i!\,j!} \frac{\zeta_\eta(\alpha+2i, \beta+2j)}{\zeta_\eta(\alpha,\beta)}, \quad i+j=N. \tag{8.1.4}$$

For example, for $\alpha = \beta = 1$,

$$\tau_{\eta(N)}^{1,1}(i,j) = \frac{(2i!)!(2j)!}{(2^N \cdot i! \cdot j!)^2}. \tag{8.1.5}$$

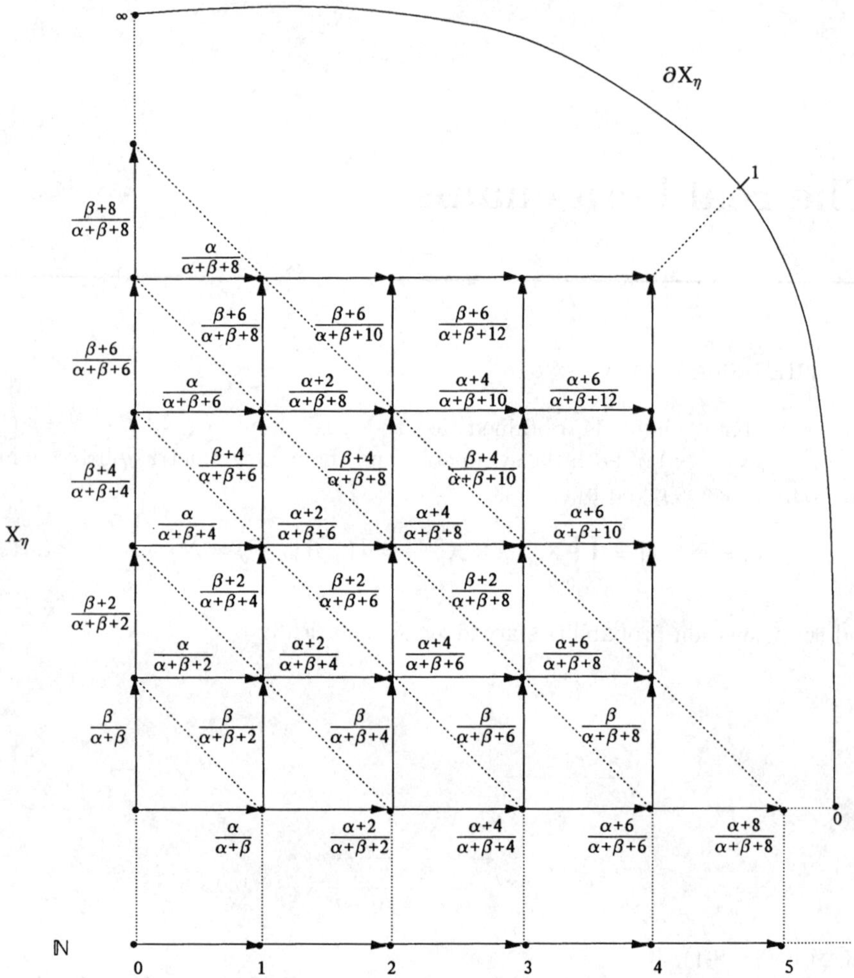

FIG. 8.1 The real beta chain X_η.

For example, for $\alpha = \beta = 2$ (or equivalently, for $\alpha = \beta = 1$ and a *complex* prime η of a number field),

$$\tau^{2,2}_{\eta(N)}(i,j) = \frac{1}{N+1}, \quad \text{independent of } i,j, \quad i+j=N. \tag{8.1.6}$$

Note that starting the chain at (i,j), rather than at the origin $(0,0)$, is equivalent to replacing α, β by $\alpha + 2i, \beta + 2j$, respectively. Using this remark we obtain from (8.1.4) that the Green kernel is given by

$$\mathrm{G}^{\alpha,\beta}_\eta((i,j),(i',j')) = \begin{cases} \dfrac{((i'+j')-(i+j))!}{(i'-i)!(j'-j)!}\dfrac{\zeta_\eta(\alpha+2i',\beta+2j')}{\zeta_\eta(\alpha+2i,\beta+2j)} & i' \geq i,\ j' \geq j\,, \\ 0 & \text{otherwise}\,. \end{cases} \tag{8.1.7}$$

The Martin kernel is therefore given by

$$\mathrm{K}_\eta^{\alpha,\beta}((i,j),(i',j')) = \frac{((i'+j')-(i+j))!}{(i'-i)!(j'-j)!}\,\frac{i'!\,j'!}{(i'+j')!}\,\frac{\zeta_\eta(\alpha,\beta)}{\zeta_\eta(\alpha+2i,\beta+2j)} \qquad i'\geq i,\ j'\geq j\,. \tag{8.1.8}$$

It follows that a sequence $(i_n, j_n) \in \mathrm{X}_\eta$ is a Cauchy sequence with respect to the Martin metric if and only if $\mathrm{K}_\eta^{\alpha,\beta}((1,0),(i_n,j_n))$ and $\mathrm{K}_\eta^{\alpha,\beta}((0,1),(i_n,j_n))$ are Cauchy sequences, or, equivalently, either (i_n, j_n) is a fixed point of X_η for $n \gg 0$, or else $i_n + j_n \to \infty$ and $(i_n : j_n)$ converges in $\partial\mathrm{X}_\eta = \pm1\backslash\mathbb{P}^1(\mathbb{R})$.

The extension of the Martin kernel to the boundary is given by

$$\begin{aligned}\mathrm{K}_\eta^{\alpha,\beta}((i,j),x) &= \lim_{\substack{i_n+j_n\to\infty\\ j_n/i_n\to x^2}} \mathrm{K}_\eta^{\alpha,\beta}((i,j),(i_n,j_n))\\ &= \rho_{\infty\eta}(x)^{2i}\cdot\rho_{0\eta}(x)^{2j}\cdot\frac{\zeta_\eta(\alpha,\beta)}{\zeta_\eta(\alpha+2i,\beta+2j)} = \frac{\tau_\eta^{\alpha+2i,\beta+2j}(x)}{\tau_\eta^{\alpha,\beta}(x)}\,.\end{aligned} \tag{8.1.9}$$

It follows from this that the harmonic measure is indeed the real beta measure $\tau_\eta^{\alpha,\beta}$.

Recalling the p-adic analogue, the intuitive meaning of (8.1.9) is the 'characteristic function' of the set of real numbers $x \in \mathbb{R}$ such that $\pm x$ is 'congruent to $(N-j,j)$ modulo η^N' is given by

$$\rho_{\infty\eta}(x)^{2(N-j)}\rho_{0\eta}(x)^{2j} = \frac{x^{2j}}{(1+x^2)^N}\,.$$

Again every point of the boundary is extreme, and we have a bijection between $\mathrm{P}_\eta^{\alpha,\beta}$-harmonic functions and probability measures on $\partial\mathrm{X}_\eta$. For example, the measure $\tau_\eta^{\alpha',\beta'}$ corresponds to the $\mathrm{P}_\eta^{\alpha,\beta}$-harmonic function

$$\varphi_{\alpha,\beta}^{\alpha',\beta'}(i,j) = \frac{\zeta_\eta(\alpha'+2i,\beta'+2j)}{\zeta_\eta(\alpha+2i,\beta+2j)}\cdot\frac{\zeta_\eta(\alpha,\beta)}{\zeta_\eta(\alpha',\beta')}\,, \tag{8.1.10}$$

and the chains $\mathrm{P}_\eta^{\alpha,\beta}$, for different values of α, β, are all conjugate, in the sense that as operators on $\ell_2(\mathrm{X}_\eta)$

$$\mathrm{P}_\eta^{\alpha',\beta'} = \left(\varphi_{\alpha,\beta}^{\alpha',\beta'}\right)^{-1}\mathrm{P}_\eta^{\alpha,\beta}\varphi_{\alpha,\beta}^{\alpha',\beta'}\,. \tag{8.1.11}$$

8.2 THE HEISENBERG RELATIONS AND THE SPECIAL BASIS

The real beta chain, unlike its p-adic analogue, is not a tree chain, and the representation of a real number as a path in X_η is far from being unique. This is now compensated for by the fact that the product of transition probabilities along any path in X_η does not depend on the path itself but only on its beginning and end points: from (i,j) to (i',j'), we obtain

$$\frac{\zeta_\eta(\alpha+2i',\beta+2j')}{\zeta_\eta(\alpha+2i,\beta+2j)}\,.$$

Integration against the Martin kernel does not give embeddings or projections. We have

$$\begin{gathered}\mathscr{H}^{\alpha,\beta}_{\eta(N)} = \ell_2\Big(\mathrm{X}_{\eta(N)}, \tau^{\alpha,\beta}_{\eta(N)}\Big) \underset{\left(\mathrm{K}^{\alpha,\beta}_{\eta(N)}\right)^*}{\overset{\mathrm{K}^{\alpha,\beta}_{\eta(N)}}{\leftrightarrows}} \mathscr{H}^{\alpha,\beta}_{\eta} = \mathrm{L}_2\Big(\partial\mathrm{X}_\eta, \tau^{\alpha,\beta}_\eta\Big)\\ \mathrm{K}^{\alpha,\beta}_{\eta(N)}\varphi(i,j) = \tau^{\alpha+2i,\beta+2j}_\eta(\varphi)\,,\\ \left(\mathrm{K}^{\alpha,\beta}_{\eta(N)}\right)^*\varphi(x) = \sum_{i+j=N}\frac{N!}{i!\,j!}\varphi(i,j)\cdot\rho_\infty(x)^{2i}\cdot\rho_0(x)^{2j}\,,\end{gathered} \tag{8.2.1}$$

and $\mathrm{K}^{\alpha,\beta}_{\eta(N)} \circ \left(\mathrm{K}^{\alpha,\beta}_{\eta(N)}\right)^*$ is given by the kernel

$$\frac{\zeta_\eta(\alpha+2(i+i'),\beta+2(j+j'))\cdot\zeta_\eta(\alpha,\beta)}{\zeta_\eta(\alpha+2i,\beta+2j)\cdot\zeta_\eta(\alpha+2i',\beta+2j')}\,.$$

To solve this problem, we have to take the real limit of the theory developed in §7.3. First, we have the ordinary difference operators

$$\begin{aligned}\nabla\colon \ell_2(\mathrm{X}_{\eta(N)}) \to \ell_2(\mathrm{X}_{\eta(N-1)}),\quad & \nabla\varphi(i,j) = \varphi(i,j+1)-\varphi(i+1,j)\,,\\ \nabla^*\colon \ell_2(\mathrm{X}_{\eta(N-1)}) \to \ell_2(\mathrm{X}_{\eta(N)}),\quad & \nabla^*\varphi(i,j) = \varphi(i,j-1)-\varphi(i-1,j)\,,\end{aligned} \tag{8.2.2}$$

where now $\ell_2(\mathrm{X}_{\eta(N)})$ denotes ℓ_2 with respect to the counting measure, and ∇ and ∇^* satisfy ordinary commutativity $\nabla\nabla^* = \nabla^*\nabla$. Next, we write

$$\tau^{\alpha,\beta}_{\eta(N)} = \mathrm{C}^{\alpha,\beta}_{\eta(N)}\cdot \mathrm{t}^{\alpha,\beta}_{\eta(N)},\quad \mathrm{C}^{\alpha,\beta}_{\eta(N)} = \frac{N!}{\zeta_\eta(\alpha,\beta)}\cdot\frac{1}{\Gamma((\alpha+\beta)/2+N)}\,, \tag{8.2.3}$$

$$\mathrm{t}^{\alpha,\beta}_{\eta(N)}(i,j) = \frac{\Gamma(\alpha/2+i)\Gamma(\beta/2+j)}{i!\cdot j!},\quad \mathrm{H}^{\alpha,\beta}_{\eta(N)} = \ell_2\Big(\mathrm{X}_{\eta(N)}, \mathrm{t}^{\alpha,\beta}_{\eta(N)}\Big)\,, \tag{8.2.4}$$

and we let $\widetilde{\mathrm{D}}_N$ denote the operator ∇ viewed as an operator from $\mathrm{H}^{\alpha,\beta}_{\eta(N)}$ to $\mathrm{H}^{\alpha+2,\beta+2}_{\eta(N-1)}$. Its adjoint is given by

$$\begin{gathered}\widetilde{\mathrm{D}}^*_N\colon \mathrm{H}^{\alpha+2,\beta+2}_{\eta(N-1)} \to \mathrm{H}^{\alpha,\beta}_{\eta(N)}\\ \widetilde{\mathrm{D}}^*_N\varphi(i,j) = j\cdot\left(\frac{\alpha}{2}+i\right)\varphi(i,j-1) - i\cdot\left(\frac{\beta}{2}+j\right)\varphi(i-1,j)\,.\end{gathered} \tag{8.2.5}$$

Concerning the operators

$$\mathrm{H}^{\alpha,\beta}_{\eta(N)} \underset{\widetilde{\mathrm{D}}^*_N}{\overset{\widetilde{\mathrm{D}}_N}{\rightleftarrows}} \mathrm{H}^{\alpha+2,\beta+2}_{\eta(N-1)} \underset{\widetilde{\mathrm{D}}^*_{N-1}}{\overset{\widetilde{\mathrm{D}}_{N-1}}{\rightleftarrows}} \mathrm{H}^{\alpha+4,\beta+4}_{\eta(N-2)} \tag{8.2.6}$$

we have the Heisenberg relations up the ladder

$$\widetilde{\mathrm{D}}_N\widetilde{\mathrm{D}}^*_N - \widetilde{\mathrm{D}}^*_{N-1}\widetilde{\mathrm{D}}_{N-1} = \left(\frac{\alpha+\beta}{2}\right)\cdot \mathrm{id}_{\mathrm{H}^{\alpha+2,\beta+2}_{\eta(N-1)}}\,. \tag{8.2.7}$$

Returning to the 'probability one' spaces $\mathscr{H}^{\alpha,\beta}_{\eta(N)}$, we set

$$\begin{aligned} \mathrm{D}_N &= \left(\frac{\alpha+\beta}{2}+N\right)\cdot\nabla\colon \mathscr{H}^{\alpha,\beta}_{\eta(N)} \longrightarrow \mathscr{H}^{\alpha+2,\beta+2}_{\eta(N-1)}\,, \\ \mathrm{D}_N^{+} &= \left(\frac{\alpha+\beta}{2}+N\right)^{-1}\cdot\widetilde{\mathrm{D}}_N^{*}\colon \mathscr{H}^{\alpha+2,\beta+2}_{\eta(N-1)} \longrightarrow \mathscr{H}^{\alpha,\beta}_{\eta(N)}\,, \end{aligned} \tag{8.2.8}$$

preserving the Heisenberg relations

$$\mathrm{D}_N\mathrm{D}_N^{+} - \mathrm{D}_{N-1}^{+}\mathrm{D}_{N-1} = \left(\frac{\alpha+\beta}{2}\right)\cdot \mathrm{id}_{\mathscr{H}^{\alpha+2,\beta+2}_{\eta(N-1)}}\,. \tag{8.2.9}$$

It follows that the number operator

$$\begin{aligned} \mathrm{N}^{\alpha,\beta}_{\eta(N)} &= \mathrm{D}_N^{+}\mathrm{D}_N \quad \text{on } \mathscr{H}^{\alpha,\beta}_{\eta(N)}\,, \\ \mathrm{N}^{\alpha,\beta}_{\eta(N)}\varphi(i,j) &= \left[i\left(\frac{\beta}{2}+j\right)+\left(\frac{\alpha}{2}+i\right)j\right]\varphi(i,j) \\ &\quad - i\left(\frac{\beta}{2}+j\right)\varphi(i-1,j+1) - \left(\frac{\alpha}{2}+i\right)j\varphi(i+1,j-1)\,, \end{aligned} \tag{8.2.10}$$

has as eigenvectors the orthogonal basis of $\mathscr{H}^{\alpha,\beta}_{\eta(N)}$ given by the *real Hahn polynomials*:

$$\begin{aligned} \varphi^{\alpha,\beta}_{\eta(N),m}(i,j) &= \frac{1}{m!}(\mathrm{D}^{+})^{m}\mathbb{1} \\ &= \lim_{q\to 1}\varphi^{(\alpha/2)\beta/2}_{q^2(N),m}(i,j) \\ &= \sum_{i_0+j_0=m}\binom{i}{i_0}\binom{j}{j_0}(-1)^{i_0}\frac{\zeta_\eta(\alpha+2(i+j_0),\beta+2(j+i_0))}{\zeta_\eta(\alpha+2i,\beta+2j)} \quad \text{from (7.3.22)} \\ &= \frac{\zeta_\eta(\alpha+\beta+4m,\alpha+\beta+2N)\zeta_\eta(\alpha+2m,\beta+2m)}{\zeta_\eta(\alpha+\beta+2m,\alpha+\beta+2N+2m)} \\ &\quad \sum_{i_0+j_0=m}\binom{i}{i_0}\binom{j}{j_0}(-1)^{i_0}\frac{1}{\zeta_\eta(\alpha+2i_0,\beta+2j_0)}\,. \end{aligned} \tag{8.2.11}$$

We have the basic relations

$$\begin{aligned} \mathrm{D}_N^{+}\varphi^{\alpha+2,\beta+2}_{\eta(N-1),m-1} &= m\cdot\varphi^{\alpha,\beta}_{\eta(N),m}\,, \\ \mathrm{D}_N\varphi^{\alpha,\beta}_{\eta(N),m} &= \left(m-1+\frac{\alpha+\beta}{2}\right)\cdot\varphi^{\alpha+2,\beta+2}_{\eta(N-1),m-1}\,, \\ \mathrm{N}^{\alpha,\beta}_{\eta(N)}\varphi^{\alpha,\beta}_{\eta(N),m} &= m\cdot\left(m-1+\frac{\alpha+\beta}{2}\right)\cdot\varphi^{\alpha,\beta}_{\eta(N),m}\,. \end{aligned} \tag{8.2.12}$$

The norms are given by

$$\begin{aligned} &\left\|\varphi^{\alpha,\beta}_{\eta(N),m}\right\|^2_{\mathscr{H}^{\alpha,\beta}_{\eta(N)}} \\ &= \binom{N}{m}\cdot\frac{\zeta_\eta(\alpha+2m,\beta+2m)}{\zeta_\eta(\alpha,\beta)}\cdot\frac{\zeta_\eta(\alpha+\beta+4m-2,\alpha+\beta+2N)}{\zeta_\eta(\alpha+\beta+2m-2,\alpha+\beta+2N+2m)}\,. \end{aligned} \tag{8.2.13}$$

The real Jacobi basis

We can now pass to the limit $N \to \infty$, but in order to describe the limit operators of D_N and D_N^+ and the limit polynomials of $\varphi^{\alpha,\beta}_{\eta(N),m}$, we recall the Cayley transform (cf. (3.3.3))

$$z \stackrel{\text{def}}{=} \rho_\infty(x)^2 - \rho_0(x)^2 = \frac{1-x^2}{1+x^2} \in [-1,1]\,. \tag{8.2.14}$$

We get a bijection and a 'dictionary':

$$\begin{aligned}
\pm1\backslash\mathbb{P}^1(\mathbb{R}) &\leftrightarrow [-1,1]\\
x &\leftrightarrow z\\
\infty,1,0 &\leftrightarrow -1,0,1\\
\rho_\infty(x)^2,\rho_0(x)^2 &\leftrightarrow \frac{1+z}{2},\frac{1-z}{2}\\
\mathrm{d}^*x = \frac{\mathrm{d}x}{|x|_\eta} &\leftrightarrow \frac{2\mathrm{d}z}{1-z^2}\\
x\frac{\partial}{\partial x} &\leftrightarrow -(1-z^2)\frac{\partial}{\partial z}\\
r\varphi(x) = \varphi(x^{-1}) &\leftrightarrow r\varphi(z) = \varphi(-z)
\end{aligned} \tag{8.2.15}$$

and an isomorphism:

$$\mathscr{H}^{\alpha,\beta}_\eta \cong \mathrm{L}_2\left([-1,1],(1+z)^{\alpha/2-1}\cdot(1-z)^{\beta/2-1}\mathrm{d}z\frac{2^{1-(\alpha+\beta)/2}}{\zeta_\eta(\alpha,\beta)}\right).$$

Note that when $j/i \to x^2$, we have

$$\rho_\infty(i,j) \stackrel{\text{def}}{=} \left(\frac{i}{i+j}\right)^{1/2} \to \rho_\infty(x), \quad \rho_0(i,j) \stackrel{\text{def}}{=} \left(\frac{j}{i+j}\right)^{1/2} \to \rho_0(x) \tag{8.2.16}$$

and

$$z(i,j) \stackrel{\text{def}}{=} \frac{i-j}{i+j} \to z\,.$$

The limits of the operators D_N and D_N^+ are given by

$$\begin{aligned}
&\mathrm{D}\colon \mathscr{H}^{\alpha,\beta}_\eta \to \mathscr{H}^{\alpha+2,\beta+2}_\eta, \quad \mathrm{D} = -2\frac{\partial}{\partial z} = \rho_\infty(x)^{-2}\rho_0(x)^{-2}\frac{x}{2}\frac{\partial}{\partial x}\\
&\mathrm{D}^+\colon \mathscr{H}^{\alpha+2,\beta+2}_\eta \to \mathscr{H}^{\alpha,\beta}_\eta, \quad \mathrm{D}^+ = -\rho_\infty(x)^{-\alpha}\rho_0(x)^{-\beta}\frac{x}{2}\frac{\partial}{\partial x}\rho_\infty(x)^{\alpha}\rho_0(x)^{\beta}\,.
\end{aligned} \tag{8.2.17}$$

We retain the Heisenberg relation up the ladder

$$\mathrm{D}\mathrm{D}^+ - \mathrm{D}^+\mathrm{D} = \left(\frac{\alpha+\beta}{2}\right)\cdot \mathrm{id}_{\mathscr{H}^{\alpha+2,\beta+2}_\eta}\,. \tag{8.2.18}$$

As before, it follows that the number operator

$$\mathrm{N}_\eta^{\alpha,\beta} = \mathrm{D}^+\mathrm{D} \text{ on } \mathscr{H}_\eta^{\alpha,\beta},$$

$$\mathrm{N}_\eta^{\alpha,\beta} = -\rho_\infty(x)^{-2}\rho_0(x)^{-2}\left(\frac{x}{2}\frac{\partial}{\partial x}\right)^2 + \left[\left(\frac{\alpha}{2}-1\right)\rho_\infty(x)^{-2} - \left(\frac{\beta}{2}-1\right)\rho_0(x)^{-2}\right]\frac{x}{2}\frac{\partial}{\partial x}, \tag{8.2.19}$$

has as eigenvectors the orthogonal basis of $\mathscr{H}_\eta^{\alpha,\beta}$, given by the real *Jacobi polynomials*:

$$\begin{aligned}
\varphi_{\eta,m}^{\alpha,\beta}(x) &= \frac{1}{m!}(\mathrm{D}^+)^m\mathbb{1} \\
&= \rho_\infty(x)^{2-\alpha}\rho_0(x)^{2-\beta}\frac{1}{m!}\left[-\rho_\infty(x)^{-2}\rho_0(x)^{-2}\frac{x}{2}\frac{\partial}{\partial x}\right]^m \\
&\quad \rho_\infty(x)^{\alpha+2(m-1)}\rho_0(x)^{\beta+2(m-1)} \\
&= \lim_{\substack{i,j\to\infty \\ j/i\to x^2}} \varphi_{\eta(N),m}^{\alpha,\beta}(i,j) \\
&= \sum_{i_0+j_0=m}\frac{\pi^m}{i_0!\,j_0!}\frac{\zeta_\eta(\alpha+2m)\zeta_\eta(\beta+2m)}{\zeta_\eta(\alpha+2i_0)\zeta_\eta(\beta+2j_0)}(-1)^{i_0}\rho_\infty(x)^{2i_0}\rho_0(x)^{2j_0} \\
&= \mathrm{P}_m^{(\beta/2-1,\alpha/2-1)}(z)
\end{aligned} \tag{8.2.20}$$

where $\mathrm{P}_m^{(\alpha,\beta)}(z)$ are the classical Jacobi polynomials (in z).

For example,

$$\varphi_{\eta,0}^{\alpha,\beta} = \mathbb{1}, \quad -\varphi_{\eta,1}^{\alpha,\beta} = \left(\frac{\alpha+\beta}{4}\right)\frac{1-x^2}{1+x^2} + \left(\frac{\beta-\alpha}{4}\right).$$

We have the basic relations

$$\begin{aligned}
\mathrm{D}^+\varphi_{\eta,m-1}^{\alpha+2,\beta+2} &= m\cdot\varphi_{n,m}^{\alpha,\beta}, \\
\mathrm{D}\varphi_{\eta,m}^{\alpha,\beta} &= \left(m-1+\frac{\alpha+\beta}{2}\right)\cdot\varphi_{n,m-1}^{\alpha+2,\beta+2}, \\
\mathrm{N}_\eta^{\alpha,\beta}\varphi_{\eta,m}^{\alpha,\beta} &= m\cdot\left(m-1+\frac{\alpha+\beta}{2}\right)\cdot\varphi_{\eta,m}^{\alpha,\beta}.
\end{aligned} \tag{8.2.21}$$

The norms are given by

$$\left\|\varphi_{\eta,m}^{\alpha,\beta}\right\|_{\mathscr{H}_\eta^{\alpha,\beta}}^2 = \frac{\zeta_\eta(\alpha+2m,\beta+2m)}{\zeta_\eta(\alpha,\beta)}\cdot\frac{\zeta_\eta(\alpha+\beta+4m-2)}{\zeta_\eta(\alpha+\beta+2m-2)}\cdot\frac{\pi^m}{m!}. \tag{8.2.22}$$

The compatibility between the operators D_N and D_N^+ and the limit operators D and D^+ on the boundary is now given by the commutativity of the diagrams

$$\begin{array}{ccc} \mathscr{H}_\eta^{\alpha,\beta} & \xrightarrow{\ \mathrm{D}\ } & \mathscr{H}_\eta^{\alpha+2,\beta+2} \\ {\scriptstyle \mathrm{K}_{\eta(N)}^{\alpha,\beta}}\downarrow & & \downarrow{\scriptstyle \mathrm{K}_{\eta(N-1)}^{\alpha+2,\beta+2}} \\ \mathscr{H}_{\eta(N)}^{\alpha,\beta} & \xrightarrow{\ \mathrm{D}_N\ } & \mathscr{H}_{\eta(N-1)}^{\alpha+2,\beta+2} \end{array} \qquad \begin{array}{ccc} \mathscr{H}_\eta^{\alpha,\beta} & \xleftarrow{\ \mathrm{D}^+\ } & \mathscr{H}_\eta^{\alpha+2,\beta+2} \\ {\scriptstyle \mathrm{K}_{\eta(N)}^{\alpha,\beta}}\downarrow & & \downarrow{\scriptstyle \mathrm{K}_{\eta(N-1)}^{\alpha+2,\beta+2}} \\ \mathscr{H}_{\eta(N)}^{\alpha,\beta} & \xleftarrow{\ \mathrm{D}_N^+\ } & \mathscr{H}_{\eta(N-1)}^{\alpha+2,\beta+2} \end{array} \tag{8.2.23}$$

From this we obtain the 'reduction mod η^N' formula, relating the real Jacobi and Hahn basis (cf. (4.4.11), (7.3.36)):

$$\mathrm{K}_{\eta(N)}^{\alpha,\beta}\varphi_{\eta,m}^{\alpha,\beta} = \begin{cases} \varphi_{\eta(N),m}^{\alpha,\beta} & m \leq N\,, \\ 0 & m > N\,. \end{cases} \tag{8.2.24}$$

We have thus diagonalized $\mathrm{K}_{\eta(N)}^{\alpha,\beta}$, and it is not a projection because of the difference in the norms (8.2.13) and (8.2.22).

Given a measure μ on the boundary $\partial\mathrm{X}_\eta$, let $f_\eta^{\alpha,\beta} = \mathrm{K}_\eta^{\alpha,\beta}\mu$ denote the associated (α,β)-harmonic function on X_η (cf. (4.1.20)), and let $f_{\eta(N)}^{\alpha,\beta} \in \mathscr{H}_{\eta(N)}^{\alpha,\beta}$ denote its restriction to $\mathrm{X}_{\eta(N)}$. We can expand $f_{\eta(N)}^{\alpha,\beta}$ using our orthogonal base $\varphi_{\eta(N),m}^{\alpha,\beta}$:

$$\begin{aligned} f_{\eta(N)}^{\alpha,\beta}(i,j) &= \sum_{0\leq m\leq N} \mathrm{C}_m^{(N)}\cdot\varphi_{\eta(N),m}^{\alpha,\beta}(i,j), \quad i+j=N \\ \mathrm{C}_m^{(N)} &= \frac{1}{\left\|\varphi_{\eta(N),m}^{\alpha,\beta}\right\|^2}\cdot\left(f_{\eta(N)}^{\alpha,\beta},\varphi_{\eta(N),m}^{\alpha,\beta}\right)_{\mathscr{H}_{\eta(N)}^{\alpha,\beta}}. \end{aligned} \tag{8.2.25}$$

Our normalizations (and the terminology 'reduction mod η^N') are further justified by the fact that the Fourier coefficients $\mathrm{C}_m^{(N)}$ are *independent of N*. In fact we have for all $N \geq m$

$$\mathrm{C}_m^{(N)} = \frac{1}{\left\|\varphi_{\eta,m}^{\alpha,\beta}\right\|^2}\cdot\int\mu(\mathrm{d}x)\varphi_{\eta,m}^{\alpha,\beta}(x)\,, \tag{8.2.26}$$

that is the $\mathrm{C}_m = \mathrm{C}_m^{(N)}$ are the Fourier coefficients of the measure μ with respect to the orthogonal base $\left\{\varphi_{\eta,m}^{\alpha,\beta}\right\}$ of $\mathscr{H}_\eta^{\alpha,\beta}$. This phenomenon holds also for the q-analogue, and of course for its p-adic limit.

The real Laguerre basis

When we let $\alpha \to \infty$, the real beta chain converges to the unit shift walking from the origin straight to $0 \in \partial\mathrm{X}_\eta$, in agreement with (3.3.10). We do not have a simple gamma chain for the real prime! The problem is that the point $0 \in \partial\mathrm{X}_\eta$ is one of our reference points in the real beta chain (via $\rho_0(x)$); when we concentrate on the 'infinitesimal neighbourhoods' of the point $1 \in \partial\mathrm{X}_\eta$ we will have a simple limit chain, cf. §8.3.

We can, however, take the appropriately normalized limit $\alpha \to \infty$ of the operators D and D^+, and of the Jacobi polynomials, to obtain the Heisenberg relations leading to the Laguerre polynomials. We shall describe this next. We have

$$\mathrm{D}\colon \mathscr{H}^{\beta}_{\mathbb{Z}_\eta} \longrightarrow \mathscr{H}^{\beta+2}_{\mathbb{Z}_\eta}, \quad \mathrm{D} = \lim_{\alpha\to\infty} \left(\frac{\alpha}{2}\right)^{-1} \cdot \pi\left(\left(\frac{\alpha}{2\pi}\right)^{1/2}\right) \circ \mathrm{D} \circ \pi\left(\left(\frac{\alpha}{2\pi}\right)^{-1/2}\right), \tag{8.2.27}$$

where $\pi(a)\varphi(x) = \varphi(a^{-1}x)$, and where the D on the right is from (8.2.17).

Explicitly, this turns out to be the simple operator

$$\mathrm{D} = \frac{1}{\pi x^2}\frac{x}{2}\frac{\partial}{\partial x}.$$

Similarly, we have

$$\begin{aligned}
\mathrm{D}^+ = \mathrm{D}^+_\beta &= \left(\frac{\beta}{2\pi}\right)^{-1} \cdot \mathrm{D}^*\colon \mathscr{H}^{\beta+2}_{\mathbb{Z}_\eta} \to \mathscr{H}^{\beta}_{\mathbb{Z}_\eta}, \\
\mathrm{D}^+ &= \lim_{\alpha\to\infty} \pi\left(\left(\frac{\alpha}{2\pi}\right)^{1/2}\right) \circ \mathrm{D}^+ \circ \pi\left(\left(\frac{\alpha}{2\pi}\right)^{-1/2}\right) \quad \text{the } \mathrm{D}^+ \text{ from (8.2.17)} \\
&= \phi_{\mathbb{Z}_\eta}(x)^{-1}|x|_\eta^{-\beta}\left(\frac{-x}{2}\frac{\partial}{\partial x}\right)\phi_{\mathbb{Z}_\eta}(x)|x|_\eta^{\beta} = -\frac{x}{2}\frac{\partial}{\partial x} + \pi x^2 - \frac{\beta}{2}.
\end{aligned} \tag{8.2.28}$$

The Heisenberg relation up the ladder reads

$$\mathrm{D}\mathrm{D}^+_\beta - \mathrm{D}^+_{\beta+2}\mathrm{D} = \mathrm{id}_{\mathscr{H}^{\beta+2}_{\mathbb{Z}_\eta}}. \tag{8.2.29}$$

It follows that the number operator

$$\mathrm{N}^{\beta}_{\mathbb{Z}_\eta} = \mathrm{D}^+\mathrm{D} \text{ on } \mathscr{H}^{\beta}_{\mathbb{Z}_\eta}, \quad \mathrm{N}^{\beta}_{\mathbb{Z}_\eta} = -\frac{1}{4\pi}\frac{\partial^2}{\partial x^2} + \left(1 - \frac{\beta-1}{2\pi x^2}\right)\frac{x}{2}\frac{\partial}{\partial x}, \tag{8.2.30}$$

has as eigenvectors the orthogonal base of $\mathscr{H}^{\beta}_{\mathbb{Z}_\eta}$ given by the *Laguerre polynomials*:

$$\begin{aligned}
\varphi^{\beta}_{\mathbb{Z}_\eta,m}(x) &= \frac{1}{m!}(\mathrm{D}^+)^m \mathbb{1} \\
&= \lim_{\alpha\to\infty} \varphi^{\alpha,\beta}_{\eta,m}\left(\left(\frac{\alpha}{2\pi}\right)^{-1/2} x\right) \\
&= \phi_{\mathbb{Z}_\eta}(x)^{-1}|x|_\eta^{2-\beta}\frac{1}{m!}\left(-\frac{1}{2x}\frac{\partial}{\partial x}\right)^m \phi_{\mathbb{Z}_\eta}(x)|x|_\eta^{\beta+2(m-1)} \\
&= \frac{\pi^m}{m!}\zeta_\eta(\beta+2m)\sum_{i_0+j_0=m}\binom{m}{j_0}\frac{(-1)^{i_0}x^{2j_0}}{\zeta_\eta(\beta+2j_0)} \\
&= \mathrm{L}^{(\beta/2-1)}_m(\pi x^2)
\end{aligned} \tag{8.2.31}$$

where $\mathrm{L}^{(\beta)}_m(x)$ are the classical Laguerre polynomials. For example,

$$\varphi^{\beta}_{\mathbb{Z}_\eta,0} = \mathbb{1}, \quad \varphi^{\beta}_{\mathbb{Z}_\eta,1}(x) = \pi x^2 - \beta/2.$$

We have the basic relations

$$\begin{aligned} \mathrm{D}^+\varphi^{\beta+2}_{\mathbb{Z}_\eta,m-1} &= m\cdot\varphi^{\beta}_{\mathbb{Z}_\eta,m}\,,\\ \mathrm{D}\varphi^{\beta}_{\mathbb{Z}_\eta,m} &= \varphi^{\beta+2}_{\mathbb{Z}_\eta,m-1}\,,\\ \mathbb{N}^{\beta}_{\mathbb{Z}_\eta}\,\varphi^{\beta}_{\mathbb{Z}_\eta,m} &= m\cdot\varphi^{\beta}_{\mathbb{Z}_\eta,m}\,. \end{aligned} \tag{8.2.32}$$

The norms are given by

$$\left\|\varphi^{\beta}_{\mathbb{Z}_\eta,m}\right\|^2_{\mathscr{H}^{\beta}_{\mathbb{Z}\eta}} = \frac{\pi^m}{m!}\,\frac{\zeta_\eta(\beta+2m)}{\zeta_\eta(\beta)}\,. \tag{8.2.33}$$

In the q-beta chain, as well as in its p-adic limit, we had a simple 'vertical' convergence to the boundary ($j\to\infty$, $i=$ fixed), which is not the case for the real limit ($i,\ j\to\infty$, $j/i\to x^2$). We can now complete our limit relations, and summarize them in the following diagram. We let $\textcircled{p}_\ell$ denote the p-adic substitution $q=p^{-\ell}$, $\alpha:=\alpha/\ell$, $\beta:=\beta/\ell$, and we let $\textcircled{\eta}_\ell$ denote the real substitution $q:=q^{2/\ell}$, $\alpha:=\alpha/2$, $\beta:=\beta/2$.

$$\begin{array}{lccccc}
 & \underline{p\neq\eta} & & \underline{q} & & \underline{\eta}\\[2ex]
\text{Hahn:} & \varphi^{(\alpha)\beta}_{p(N),m}(i,j) & \xleftarrow[\infty\leftarrow\ell]{\textcircled{p}_\ell} & \varphi^{(\alpha)\beta}_{q(N),m}(i,j) & \xrightarrow[\ell\to\infty]{\textcircled{\eta}_\ell} & \varphi^{\alpha,\beta}_{\eta(N),m}(i,j)\\
 & \Big\downarrow{\scriptstyle j\to\infty} & & \Big\downarrow{\scriptstyle j\to\infty} & & \Big\downarrow{\scriptstyle i,j\to\infty,\ j/i\to x^2}\\
\text{Jacobi:} & \varphi^{(\alpha)\beta}_{p,m}(p^i) & \xleftarrow[\infty\leftarrow\ell]{\textcircled{p}_\ell} & \varphi^{(\alpha)\beta}_{q,m}(g^i) & \xrightarrow[i,\ell\to\infty,\ q^{i/\ell}\to\rho_0(x)]{\textcircled{\eta}_\ell} & \varphi^{\alpha,\beta}_{\eta,m}(x)\\
 & \Big\downarrow{\scriptstyle \alpha\to\infty} & & \Big\downarrow{\scriptstyle \alpha\to\infty} & & \Big\downarrow{\scriptstyle x:=\left(\frac{2\pi}{\alpha}\right)^{1/2}x,\ \alpha\to\infty}\\
\text{Laguerre:} & \varphi^{\beta}_{\mathbb{Z}_p,m}(p^i) & \xleftarrow[\infty\leftarrow\ell]{\textcircled{p}_\ell} & \varphi^{\beta}_{\mathbb{Z}_q,m}(g^i) & \xrightarrow[i,\ell\to\infty,\ \frac{q^{2i/\ell}}{1-q^{2/\ell}}\to\pi x^2]{\textcircled{\eta}_\ell} & \varphi^{\beta}_{\mathbb{Z}_\eta,m}(x)
\end{array} \tag{8.2.34}$$

Note that in the limit relation for the real Laguerre polynomials

$$\lim_{i,\ell\to\infty}\varphi^{\beta/2}_{\mathbb{Z}_{q^{2/\ell}},m}(g^i) = \varphi^{\beta}_{\mathbb{Z}_\eta,m}(x)\,, \tag{8.2.35}$$

the condition imposed on i, ℓ is

$$\frac{q^{2i/\ell}}{1-q^{2/\ell}} \to \pi x^2 . \tag{8.2.36}$$

It can also be written, using $\Gamma_{q^{2/\ell}}(1/2) \xrightarrow[\ell\to\infty]{} \Gamma(1/2) = \sqrt{\pi}$, as

$$(1-q)^{-1} \cdot \tau^{1}_{(q)}(g^i)\Big|_{\eta_\ell} = (1-q^{2/\ell})^{-1} \cdot \tau^{1/2}_{(q^{2/\ell})}(g^i) = \frac{\zeta_{(q^{2/\ell})}(1)}{\zeta_{(q^{2/\ell})}(1/2)} \frac{q^{i/\ell}}{1-q^{2/\ell}} \xrightarrow[\ell\to\infty]{} |x| , \tag{8.2.37}$$

with $\tau^{\beta}_{(q)}(g^i) = \dfrac{\zeta_{(q)}(1)}{\zeta_{(q)}(\beta)} \cdot q^{\beta i}$ the q-Tate distribution (cf. §10.1).

Any properties of the q-Jacobi polynomials, or the q-Laguerre polynomials, now give the corresponding properties in both the p-adic and real limits. We list the following sample of properties of the real Laguerre basis, analogous to the p-adic properties (4.2.15) to (4.2.20):

$$\text{norm:} \quad \left\|\varphi^{\beta}_{\mathbb{Z}_\eta,m}\right\|^2_{\mathscr{H}^{\beta}_{\mathbb{Z}_\eta}} = \frac{1}{m!}\frac{\beta}{2}\left(\frac{\beta}{2}+1\right)\cdots\left(\frac{\beta}{2}+m-1\right) \stackrel{\text{def}}{=} \mathrm{C}^{\beta}_{\eta}(m) \tag{8.2.38}$$

$$\text{special value:} \quad \varphi^{\beta}_{\mathbb{Z}_\eta,m}(0) = (-1)^m \mathrm{C}^{\beta}_{\eta}(m) \tag{8.2.39}$$

$$\text{Mellin transform:} \quad \tau^{s}_{\mathbb{Z}_\eta}\left(\varphi^{\beta}_{\mathbb{Z}_\eta,m}\right) = (-1)^m \mathrm{C}^{\beta-s}_{\eta}(m) \tag{8.2.40}$$

$$\beta'\!-\!\beta \text{ relation:} \quad \varphi^{\beta'}_{\mathbb{Z}_\eta,m} = \sum_{0\le j\le m} (-1)^j \mathrm{C}^{\beta'-\beta}_{\eta}(j)\varphi^{\beta}_{\mathbb{Z}_\eta,m-j}. \tag{8.2.41}$$

The generating function for the Laguerre basis is

$$\sum_{m\ge0} \frac{\varphi^{\beta}_{\mathbb{Z}_\eta,m}(x)}{\mathrm{C}^{\beta}_{\eta}(m)} \cdot \frac{\pi^m y^{2m}}{m!} \mathrm{e}^{-\pi y^2} = \sum_{m\ge0} \frac{\zeta_\eta(\beta)(-1)^m}{\zeta_\eta(\beta+2m)} \frac{\pi^m (xy)^{2m}}{m!} = \mathcal{F}_\beta(xy) . \tag{8.2.42}$$

Here $\mathcal{F}_\beta$ is the Bessel function of the first kind; see §10.2 below. Note that (8.2.42) depends only on $x \cdot y$.

Another generating function for the Laguerre basis is

$$\sum_{m\ge0} \varphi^{\beta}_{\mathbb{Z}_\eta,m}(x) \cdot y^{2m} = (1+y^2)^{-\beta/2} \mathrm{e}^{-\pi x^2 y^2/1+y^2} = \rho_\infty(y)^\beta \phi_{\mathbb{Z}_\eta}(\rho_0(y)x) . \tag{8.2.43}$$

The heat kernel is the well-known Hardy–Hille kernel

$$\begin{aligned} t^{\mathrm{N}^{\beta}_{\mathbb{Z}_\eta}} &= \sum_{m\ge0} \left\|\varphi^{\beta}_{\mathbb{Z}_\eta,m}\right\|^{-2} \cdot \varphi^{\beta}_{\mathbb{Z}_\eta,m}(x) \cdot \varphi^{\beta}_{\mathbb{Z}_\eta,m}(y) \cdot t^m \\ &= (1-t)^{-\beta/2} \cdot \mathrm{e}^{-\pi(x^2+y^2)t/(1-t)} \cdot \mathcal{I}_\beta\left(\frac{t^{1/2}xy}{1-t}\right), \end{aligned} \tag{8.2.44}$$

with the modified Bessel function

$$\mathcal{I}_\beta(x) = \mathcal{F}_\beta(\mathrm{i}x) = \sum_{m\geq 0} \frac{\zeta_\eta(\beta)}{\zeta_\eta(\beta+2m)} \cdot \frac{\pi^m x^{2m}}{m!}. \tag{8.2.45}$$

At $\beta=1$, we have $\mathcal{I}_1(x) = (\mathrm{e}^{2\pi x}+\mathrm{e}^{-2\pi x})/2 = \cosh(2\pi x)$, and we obtain the Mehler–Fock kernel ('made even'):

$$t^{\mathrm{N}^1_{\mathbb{Z}_\eta}} = (1-t)^{-1/2}\mathrm{e}^{-\pi(x^2+y^2)t/(1-t)} \cdot \cosh\left(2\pi xy \frac{t^{1/2}}{1-t}\right). \tag{8.2.46}$$

We can make the algebraic sum $s_q^{(\alpha)\beta} = \bigoplus_{m\geq 0} \mathbb{C}\varphi_{q,m}^{(\alpha)\beta}\left(\text{respectively } s_{q(N)}^{(\alpha)\beta} = \mathscr{H}_{q(N)}^{(\alpha)\beta}\right)$ into an algebra by decreeing that $e_{q,m}^{(\alpha)\beta} \overset{\text{def}}{=} \frac{\varphi_{q,m}^{(\alpha)\beta}(0)}{\|\varphi_{q,m}^{(\alpha)\beta}\|^2} \cdot \varphi_{q,m}^{(\alpha)\beta}$ $\Big($respectively $e_{q(N),m}^{(\alpha)\beta}$ formed similarly with $\varphi_{q(N),m}^{(\alpha)\beta}$, $m=0,\dots,N\Big)$ satisfy $e_{m_1}\cdot e_{m_2} = \delta_{m_1,m_2}\cdot e_{m_i}$; in the p-adic and real limits (respectively for $p\neq\eta$) at $\alpha=\beta=1$ this is the same product as convolution of functions on $B(\mathbb{Z}_p)\setminus \mathrm{PGL}_2(\mathbb{Z}_p)/B(\mathbb{Z}_p)$ and $\{\pm1\}\setminus \mathrm{PO}_2/\{\pm1\}$ $\Big($respectively $B(\mathbb{Z}/pN)\setminus\mathrm{PGL}_2(\mathbb{Z}/pN)/B(\mathbb{Z}/pN)\Big)$. We can pass to the $\alpha\to\infty$ limit and similarly make $s_{\mathbb{Z}_q}^{\beta} = \bigoplus_{m\geq 0} \mathbb{C}\varphi_{\mathbb{Z}_q,m}^{\beta}$ $\left(\text{respectively } s_{q(N)}^{\beta} = \mathscr{H}_{q(N)}^{\beta}\right)$ into an algebra using $e_{\mathbb{Z}_q,m}^{\beta} = (-1)^m\cdot\varphi_{\mathbb{Z}_q,m}^{\beta}$ $\left(\text{respectively } (-1)^m\cdot\varphi_{q(N),m}^{\beta}\right)$ as units; in the p-adic limit at $\beta=1$ this is the same product as convolution of functions in $s(\mathbb{Z}_p)^{\mathbb{Z}_p^*}$ $\left(\text{respectively } s(\mathbb{Z}/pN)^{(\mathbb{Z}/pN)^*}\right)$.

8.3 THE REAL UNITS

To obtain the p-adic beta chain on a tree which is independent of the prime p, we had to divide by the action of the units $\mathbb{Z}_p^*$, passing from the hyperbolic plane $\widetilde{\mathrm{X}}_p$ to X_p, but the chain on $\widetilde{\mathrm{X}}_p$ is just a random walk over the chain on X_p. We can similarly add the real units $\mathbb{Z}_\eta^* = \{\pm1\}$ into the picture. We let

$$\begin{aligned}\widetilde{\mathrm{X}}_\eta &= \coprod_{N\geq 0} \widetilde{\mathrm{X}}_{\eta(N)},\\ \widetilde{\mathrm{X}}_{\eta(N)} &= \{(0,N),(\pm1,N-1),\dots,(\pm i,N-i),\dots,(\pm(N-1),1),(N,0)\},\end{aligned} \tag{8.3.1}$$

with its natural graph structure, so that we have a map of graphs

$$\widetilde{\mathrm{X}}_\eta \to \mathrm{X}_\eta,\quad (\pm i,j)\mapsto(i,j). \tag{8.3.2}$$

The random walk on $\widetilde{\mathrm{X}}_\eta$ over the real beta chain $\mathrm{P}_\eta^{\alpha,\beta}$ on X_η (cf. (4.1.4)) gives us the beta chain $\widetilde{\mathrm{P}}_\eta^{\alpha,\beta}$ on $\widetilde{\mathrm{X}}_\eta$ as in Fig. 8.2. The harmonic measure is now $\widetilde{\tau}_\eta^{\alpha,\beta}$, the real beta measure on the circle

$$\partial\widetilde{\mathrm{X}}_\eta = \mathbb{P}^1(\mathbb{Q}_\eta) = \mathbb{P}^1(\mathbb{R}). \tag{8.3.3}$$

We see that $\partial\widetilde{\mathrm{X}}_\eta$ is approximated by

$$\widetilde{\mathrm{X}}_{\eta(N)} = \text{`}\mathbb{P}^1(\mathbb{Z}/\eta^N)\text{'} = \text{the } 2N\text{-gon}. \tag{8.3.4}$$

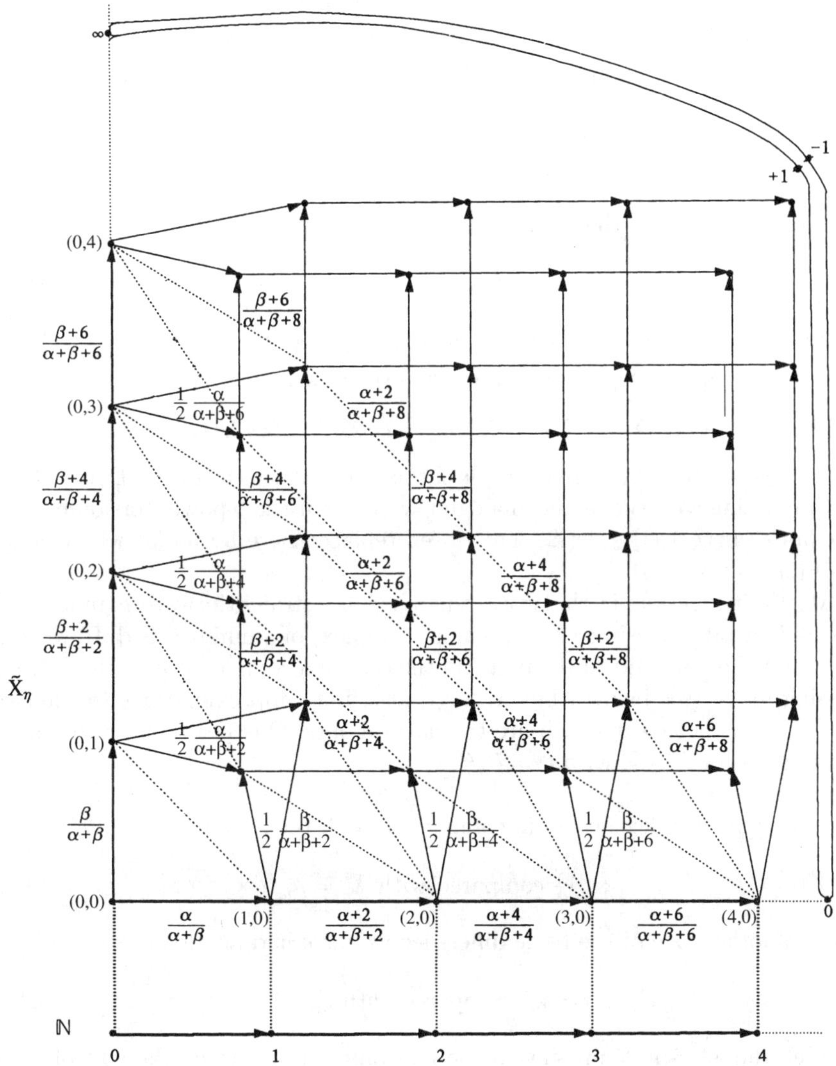

FIG. 8.2 The real beta chain (with ±1) $\widetilde{X}_\eta$.

The graph $\widetilde{X}_p$ for $p \neq \eta$ has many symmetries; indeed $\mathrm{PGL}_2(\mathbb{Z}_p)$ acts on it by graph automorphisms, and if we forget its orientation (going from the origin to the boundary), the unoriented graph is even two-point homogeneous under the action of $\mathrm{PGL}_2(\mathbb{Q}_p)$. In contrast, the graph $\widetilde{X}_\eta$ (oriented or not!) has only $\mathbb{Z}_\eta^* = \{\pm 1\}$ as symmetries. Some more symmetries reveal themselves if we ascend one step further. We have a graph over $\widetilde{X}_\eta$ (in analogy with (4.3.16)),

$$\widetilde{\widetilde{X}}_\eta = V(\mathbb{Z}) = \coprod_{N\geq 0} \widetilde{\widetilde{X}}_{\eta(N)}, \quad \widetilde{\widetilde{X}}_{\eta(N)} = \{(i,j) \in \mathbb{Z}\times\mathbb{Z}, |i|+|j| = N\} = `V^*(\mathbb{Z}/\eta^N)\text{'}. \tag{8.3.5}$$

The units $\mathbb{Z}_\eta^* = \{\pm 1\}$ act on $\widetilde{\widetilde{\mathrm{X}}}_\eta$ by graph automorphisms, $\pm 1 \cdot (i,j) = (\pm i, \pm j)$, and the quotient is $\widetilde{\mathrm{X}}_\eta$. We obtain the beta chain on $\widetilde{\widetilde{\mathrm{X}}}_\eta$ as the random walk on $\widetilde{\widetilde{\mathrm{X}}}$ over the beta chain on $\widetilde{\mathrm{X}}_\eta$. The Martin boundary is now the circle in the plane

$$\partial \widetilde{\widetilde{\mathrm{X}}}_\eta = `V^*(\mathbb{Z}_\eta)' = \left\{ (x,y) \in V(\mathbb{Q}_\eta) \,\middle|\, |x,y|_\eta = 1 \right\}. \tag{8.3.6}$$

In summary, we have the tower of chains

$$\widetilde{\widetilde{\mathrm{X}}}_\eta \to \widetilde{\mathrm{X}}_\eta \to \mathrm{X}_\eta \tag{8.3.7}$$

where each arrow is obtained by dividing by the action of $\mathbb{Z}_\eta^* = \{\pm 1\}$. The underlying graph of $\widetilde{\widetilde{\mathrm{X}}}_\eta$ has as symmetries the dihedral group

$$\mathrm{D}_4 = \mathrm{GL}_2(\mathbb{Z}) \cap \mathrm{O}_2 = (\mathbb{Z}/2\mathbb{Z}) \ltimes (\mathbb{Z}/4\mathbb{Z}), \tag{8.3.8}$$

and the associated unoriented graph has as symmetries the group $\mathrm{D}_4 \ltimes V(\mathbb{Z})$.

Comparing with the p-adic analogue, we see that the p-power torsion points of an elliptic curve $V(\mathbb{Q}_p)/V(\mathbb{Z}_p)$ have been replaced here by its lattice of periods $V(\mathbb{Z})$ (cf. end of §4.3).

Most of the results in this work generalize in a straightforward manner to the case of an arbitrary prime (finite, real, or complex) of a number field. However, it is not clear how to 'add to the units' in the case of a complex prime: there are the 'missing units' (see below), but we need also finite approximations to the circle $\mathbb{C}^{(1)}$, and it is not clear which are the natural ones. One possibility is to use the chain (8.3.5) on $\mathbb{Z} \times \mathbb{Z}$ with $\alpha = \beta = 1$.

The real units are degenerate in comparison with their p-adic analogue,

$$\mathbb{Z}_\eta^* = \{\pm 1\} \text{ compared with } \mathbb{Z}_p^* = \mu_{p-1} \times \mathrm{e}^{p\mathbb{Z}_p}. \tag{8.3.9}$$

This is another face of the basic difference we have had all along,

$$\mathbb{Q}_\eta^*/\mathbb{Z}_\eta^* = \mathbb{R}^+ \text{ compared with } \mathbb{Q}_p^*/\mathbb{Z}_p^* = p^{\mathbb{Z}}. \tag{8.3.10}$$

We seek the '$\mathrm{e}^{p\mathbb{Z}_p}$-part' missing in the real units; it should be thought of as the 'union of infinitesimal neighbourhoods of 1 in $\mathbb{Q}_\eta^*$'. Bearing in mind the p-adic analogue (cf. (4.3.11), (4.3.12)), we take the limit $\alpha = \beta \to \infty$ in the real beta chain. We obtain a random walk on the graph X_η, with probability 1/2 for walking from (i,j) to either $(i+1,j)$ or $(i,j+1)$. We get

$$\tau_{\eta(N)}^{\infty,\infty}(i,j) = \lim_{\alpha\to\infty} \tau_{\eta(N)}^{\alpha,\alpha}(i,j) = \frac{N!}{i!\,j!}\frac{1}{2^N}. \tag{8.3.11}$$

The Green kernel is

$$\mathrm{G}_\eta^{\infty,\infty}((i,j),(i',j')) = \begin{cases} \dfrac{(i'+j'-i-j)!}{(i'-i)!(j'-j)!}\dfrac{2^{i+j}}{2^{i'+j'}} & i' \geq i, j' \geq j, \\ 0 & \text{otherwise}, \end{cases} \tag{8.3.12}$$

and the Martin kernel is therefore

$$\mathrm{K}_\eta^{\infty,\infty}((i,j),(i',j')) = \begin{cases} \dfrac{(i'+j'-i-j)!}{(i'-i)!(j'-j)!}\dfrac{i'!j'!}{(i'+j')!}2^{i+j} & i' \geq i, j' \geq j\,, \\ 0 & \text{otherwise}\,. \end{cases} \tag{8.3.13}$$

It follows that a sequence (i_n, j_n) converges in the Martin metric if and only if $\mathrm{K}_\eta^{\infty,\infty}((0,1),(i_n,j_n)) = 2j_n/(i_n+j_n)$ and $\mathrm{K}_\eta^{\infty,\infty}((1,0),(i_n,j_n)) = 2i_n/(i_n+j_n)$ converge, which is equivalent again to j_n/i_n converging in $[0,\infty] = \{\pm\}\backslash\mathbb{P}^1(\mathbb{R})$, or, using the Cayley transform, this is equivalent to $(i_n-j_n)/(i_n+j_n)$ converging in $[-1,1]$.

The extension of the Martin kernel to the boundary is

$$\mathrm{K}_\eta^{\infty,\infty}((i,j),x) = \lim_{j_n/i_n \to x^2} \mathrm{K}_\eta^{\infty,\infty}((i,j),(i_n,j_n)) = 2^{i+j}\rho_\infty(x)^{2i}\rho_0(x)^{2j}\,, \tag{8.3.14}$$

or in terms of the $z = \rho_\infty^2 - \rho_0^2$ variable,

$$\mathrm{K}_\eta^{\infty,\infty}((i,j),z) = \lim_{\frac{i_n-j_n}{i_n+j_n}\to z} \mathrm{K}_\eta^{\infty,\infty}((i,j),(i_n,j_n)) = (1+z)^i \cdot (1-z)^j\,. \tag{8.3.15}$$

It follows that the harmonic measure is the δ-measure supported at $x=1$ (or $z=0$). We can similarly take the limit $\alpha=\beta\to\infty$ of the beta chain on $\widetilde{\mathrm{X}}_\eta$. The boundary is then the circle $\mathbb{P}^1(\mathbb{R})$, and the harmonic measure is $\frac{1}{2}(\delta_1+\delta_{-1})$ which is the Haar measure on $\mathbb{Z}_\eta^*$, in analogy with the p-adic case (4.3.12), and in agreement with (3.3.11).

In looking for the missing $\mathrm{e}^{p\mathbb{Z}_p}$ part of $\mathbb{Z}_\eta^*$, we shall concentrate on X_η. Let

$$\mathscr{H}_{\eta(N)}^{\infty,\infty} = \ell_2\left(\mathrm{X}_{\eta(N)}, \tau_{\eta(N)}^{\infty,\infty}\right). \tag{8.3.16}$$

Let $\widetilde{\mathrm{D}}_N = \frac{1}{2}\nabla$ be the difference operator

$$\widetilde{\mathrm{D}}_N \colon \mathscr{H}_{\eta(N)}^{\infty,\infty} \to \mathscr{H}_{\eta(N-1)}^{\infty,\infty}, \quad \widetilde{\mathrm{D}}_N\varphi(i,j) = \frac{1}{2}(\varphi(i,j+1)-\varphi(i+1,j))\,, \tag{8.3.17}$$

and let $\widetilde{\mathrm{D}}_N^+ = N\cdot\widetilde{\mathrm{D}}_N^*$,

$$\widetilde{\mathrm{D}}_N^+ \colon \mathscr{H}_{\eta(N-1)}^{\infty,\infty} \to \mathscr{H}_{\eta(N)}^{\infty,\infty}, \quad \widetilde{\mathrm{D}}_N^+\varphi(i,j) = j\varphi(i,j-1) - i\varphi(i-1,j)\,. \tag{8.3.18}$$

We have the Heisenberg relations

$$\widetilde{\mathrm{D}}_N\widetilde{\mathrm{D}}_N^+ - \widetilde{\mathrm{D}}_{N-1}^+\widetilde{\mathrm{D}}_{N-1} = \mathrm{id}_{\mathscr{H}_{\eta(N-1)}^{\infty,\infty}}\,. \tag{8.3.19}$$

We define

$$\tilde{\varphi}_{\eta(N),m}^{\infty,\infty}(i,j) = \frac{1}{m!}(\widetilde{\mathrm{D}}^+)^m\mathbb{1} = 2^m \cdot \lim_{\alpha\to\infty}\varphi_{\eta(N),m}^{\alpha,\alpha}(i,j) = \sum_{i_0+j_0=m}\binom{i}{i_0}\binom{j}{j_0}(-1)^{i_0}\,, \tag{8.3.20}$$

and we have from the Heisenberg relation

$$\begin{aligned} \widetilde{\mathrm{D}}_N^+ \tilde{\varphi}^{\infty,\infty}_{\eta(N-1),m-1} &= m \cdot \tilde{\varphi}^{\infty,\infty}_{\eta(N),m} \,, \\ \widetilde{\mathrm{D}}_N \tilde{\varphi}^{\infty,\infty}_{\eta(N),m} &= \tilde{\varphi}^{\infty,\infty}_{\eta(N-1),m-1} \,, \\ \widetilde{\mathrm{D}}_N^+ \widetilde{\mathrm{D}}_N \tilde{\varphi}^{\infty,\infty}_{\eta(N),m} &= m \cdot \tilde{\varphi}^{\infty,\infty}_{\eta(N),m} \,. \end{aligned} \tag{8.3.21}$$

Thus $\{\tilde{\varphi}^{\infty,\infty}_{\eta(N),m}\}$, $0 \le m \le N$, form an orthogonal basis for $\mathscr{H}^{\infty,\infty}_{\eta(N)}$, and are eigenvectors of the number operator $\widetilde{\mathrm{N}}^{\infty,\infty}_{\eta(N)} = \widetilde{\mathrm{D}}_N^+ \widetilde{\mathrm{D}}_N$,

$$\widetilde{\mathrm{N}}^{\infty,\infty}_{\eta(N)} \varphi(i,j) = \frac{1}{2}[N\varphi(i,j) - j\varphi(i+1,j-1) - i\varphi(i-1,j+1)] \,. \tag{8.3.22}$$

We have

$$\left\| \tilde{\varphi}^{\infty,\infty}_{\eta(N),m} \right\|^2 = \frac{N}{m} \cdot \left\| \tilde{\varphi}^{\infty,\infty}_{\eta(N-1),m-1} \right\|^2 ,$$

and so the norms are given by

$$\left\| \tilde{\varphi}^{\infty,\infty}_{\eta(N),m} \right\|^2_{\mathscr{H}^{\infty,\infty}_{\eta(N)}} = \binom{N}{m} . \tag{8.3.23}$$

Consider the universal covering chain $\Omega_{\mathrm{X}_\eta} \to \mathrm{X}_\eta$ as in (4.1.11), which is the random walk on the binary tree. We can denote elements of $\Omega^N_{\mathrm{X}_\eta}$ by sequences $\varepsilon = (\varepsilon_1, \dots, \varepsilon_N)$, $\varepsilon_i = \pm 1$, and we have $\tau_{\Omega^N_{\mathrm{X}_\eta}}(\varepsilon) = 1/2^N$ for all ε. The associated Hilbert space can be written as a tensor product

$$\ell_2\Big(\Omega^N_{\mathrm{X}_\eta}, \tau_{\Omega^N_{\mathrm{X}_\eta}}\Big) = \mathscr{H}^{\infty,\infty}_{\eta(1)} \otimes \dots \otimes \mathscr{H}^{\infty,\infty}_{\eta(1)} \qquad (N \text{ times}) \,. \tag{8.3.24}$$

An orthogonal basis for this space is given by all ‘monomials’

$$x_1^{e_1} \dots x_N^{e_N} , \quad e_i = 0, 1, \quad x_i(\varepsilon) = \varepsilon_i \,. \tag{8.3.25}$$

The symmetric group Σ_N acts naturally on this space, and the subspace of Σ_N-invariant vectors can be identified with our space (8.3.16)

$$\ell_2\Big(\Omega^N_{\mathrm{X}_\eta}, \tau_{\Omega^N_{\mathrm{X}_\eta}}\Big)^{\Sigma_N} \cong \mathscr{H}^{\infty,\infty}_{\eta(N)} \,. \tag{8.3.26}$$

Thus an orthogonal basis for $\mathscr{H}^{\infty,\infty}_{\eta(N)}$ is obtained by the elementary symmetric functions

$$\sigma_{N,m}(x_1, \dots, x_N) = (-1)^m \sum_{j_1 < \dots < j_m} x_{j_1} \dots x_{j_m} , \quad 0 \le m \le N \,. \tag{8.3.27}$$

Indeed, these are just our basis elements $\tilde{\varphi}^{\infty,\infty}_{\eta(N),m}$ in disguise:

$$\tilde{\varphi}^{\infty,\infty}_{\eta(N),m}(i,j) = \sigma_{N,m}(\underbrace{1, \dots, 1}_{i}, \underbrace{-1, \dots, -1}_{j}) \,. \tag{8.3.28}$$

Using the generating function for the $\sigma_{N,m}$

$$\prod_{1\le k\le N}(1-tx_k)=\sum_{0\le m\le N}t^m\sigma_{N,m}(x_1,\dots,x_N)\,, \tag{8.3.29}$$

we have

$$(1-t)^i(1+t)^j=\sum_{0\le m\le N}t^m\tilde{\varphi}^{\infty,\infty}_{\eta(N),m}(i,j)\,. \tag{8.3.30}$$

But our operators $\widetilde{\mathrm{D}}_N$ and $\widetilde{\mathrm{D}}^+_N$, and the basis $\tilde{\varphi}^{\infty,\infty}_{\eta(N),m}$, do not converge in the limit $N\to\infty$ without further normalizations. We put

$$\mathrm{D}_N=\sqrt{2\pi N}\,\widetilde{\mathrm{D}}_N,\quad \text{so its adjoint is } \mathrm{D}^*_N=\sqrt{\frac{2\pi}{N}}\widetilde{\mathrm{D}}^+_N\,, \tag{8.3.31}$$

and the Heisenberg relations read

$$\mathrm{D}_N\mathrm{D}^*_N-\mathrm{D}^*_{N-1}\mathrm{D}_{N-1}=2\pi\cdot\mathrm{id}_{\mathscr{H}^{\infty,\infty}_{\eta(N-1)}}\,. \tag{8.3.32}$$

We obtain the orthogonal basis for $\mathscr{H}^{\infty,\infty}_{\eta(N)}$

$$\varphi^{\infty,\infty}_{\eta(N),m}=\frac{1}{m!}(\mathrm{D}^*)^m\mathbb{1}=\frac{(2\pi)^{m/2}}{(N(N-1)\dots(N-m+1))^{1/2}}\tilde{\varphi}^{\infty,\infty}_{\eta(N),m} \tag{8.3.33}$$

and (8.3.21), (8.3.22), and (8.3.23) now read

$$\begin{aligned}
\mathrm{D}^*_N\varphi^{\infty,\infty}_{\eta(N-1),m-1}&=m\cdot\varphi^{\infty,\infty}_{\eta(N),m}\,,\\
\mathrm{D}_N\varphi^{\infty,\infty}_{\eta(N),m}&=2\pi\cdot\varphi^{\infty,\infty}_{\eta(N-1),m-1}\,,\\
\mathrm{N}^{\infty,\infty}_{\eta(N)}&=\mathrm{D}^*_N\mathrm{D}_N=2\pi\cdot\widetilde{\mathrm{N}}^{\infty,\infty}_{\eta(N)}\,,\\
\mathrm{N}^{\infty,\infty}_{\eta(N)}\;\varphi^{\infty,\infty}_{\eta(N),m}&=2\pi m\cdot\varphi^{\infty,\infty}_{\eta(N),m}\,,\\
\left\|\varphi^{\infty,\infty}_{\eta(N),m}\right\|^2_{\mathscr{H}^{\infty,\infty}_{\eta(N)}}&=\frac{(2\pi)^m}{m!}\,.
\end{aligned} \tag{8.3.34}$$

The essential thing in (8.3.31) is the factor $\sqrt{N}$; the factor $\sqrt{2\pi}$ is inessential. If we associate the point (i,j) to the point

$$w(i,j)\overset{\mathrm{def}}{=}\frac{i-j}{\sqrt{2\pi(i+j)}}=\sqrt{\frac{N}{2\pi}}\cdot z(i,j)\in\left[-\left(\frac{N}{2\pi}\right)^{1/2},\left(\frac{N}{2\pi}\right)^{1/2}\right],\quad i+j=N\,, \tag{8.3.35}$$

then the measure $\tau^{\infty,\infty}_{\eta(N)}$ corresponds to the measure on $\mathbb{R}$ given by

$$\left(\frac{1}{2}\left(\delta_{1/\sqrt{2\pi N}}+\delta_{-1/\sqrt{2\pi N}}\right)\right)^{*N}=\frac{1}{2^N}\sum_{i+j=N}\frac{N!}{i!\,j!}\delta_{(i-j)/\sqrt{2\pi N}}\,. \tag{8.3.36}$$

By the central limit theorem this converges as $N \to \infty$ to the probability measure $e^{-\pi w^2} dw$. The operators D_N and D_N^* now converge to the operators

$$D = -\frac{\partial}{\partial w}, \quad D^* = e^{\pi w^2} \frac{\partial}{\partial w} e^{-\pi w^2} = \frac{\partial}{\partial w} - 2\pi w. \tag{8.3.37}$$

They satisfy the ordinary Heisenberg relations

$$DD^* - D^*D = 2\pi \cdot \mathrm{id}_{\mathscr{H}_\eta^{\infty,\infty}} \text{ on } \mathscr{H}_\eta^{\infty,\infty} = L_2(\mathbb{R}, e^{-\pi w^2} dw). \tag{8.3.38}$$

The basis $\varphi_{\eta(N),m}^{\infty,\infty}$ now converge to the *Hermite polynomials*:

$$\begin{aligned}
\varphi_{\eta,m}^{\infty,\infty}(w) &= \frac{1}{m!}(D^*)^m \mathbb{1} \\
&= \lim_{\substack{i,j\to\infty \\ w(i,j)\to w}} \varphi_{\eta(i+j),m}^{\infty,\infty}(i,j) \\
&= \lim_{\alpha\to\infty} \varphi_{\eta,m}^{\alpha,\alpha}(z)|_{z=(2\pi/\alpha)w} \\
&= e^{\pi w^2} \frac{1}{m!} \left(\frac{\partial}{\partial w}\right)^m e^{-\pi w^2} \\
&= \frac{(-1)^m (2\pi)^m}{m!} \int (w + \mathrm{i}y)^m e^{-\pi y^2} dy \\
&= \sum_{i_0+2j_0=m} \frac{(-1)^{i_0}}{i_0!\, j_0!} (2\pi w)^{i_0} (-\pi)^{j_0}.
\end{aligned} \tag{8.3.39}$$

We again have the basic relations, the limit $N \to \infty$ of (8.3.34),

$$\begin{aligned}
D^* \varphi_{\eta,m-1}^{\infty,\infty} &= m \varphi_{\eta,m}^{\infty,\infty}, \\
D \varphi_{\eta,m}^{\infty,\infty} &= 2\pi\, \varphi_{\eta,m-1}^{\infty,\infty}, \\
\mathbb{N}_\eta^{\infty,\infty} \varphi_{\eta,m}^{\infty,\infty} &= 2\pi m \cdot \varphi_{\eta,m}^{\infty,\infty}, \\
\mathbb{N}_\eta^{\infty,\infty} = D^* D &= 2\pi w \frac{\partial}{\partial w} - \frac{\partial^2}{\partial w^2}, \\
\left\| \varphi_{\eta,m}^{\infty,\infty} \right\|_{\mathscr{H}_\eta^{\infty,\infty}}^2 &= \frac{(2\pi)^m}{m!}.
\end{aligned} \tag{8.3.40}$$

The generating function of the Hermite polynomials is

$$e^{-\pi y^2 - 2\pi y w} = \sum_{m\geq 0} y^m \cdot \varphi_{\eta,m}^{\infty,\infty}(w). \tag{8.3.41}$$

The heat kernel is the Mehler kernel

$$t^{(1/2\pi)\mathbb{N}_\eta^{\infty,\infty}} = \sum_{m\geq 0} \frac{\varphi_{\eta,m}^{\infty,\infty}(x) \cdot \varphi_{\eta,m}^{\infty,\infty}(y)}{\|\varphi_{\eta,m}^{\infty,\infty}\|^2} \cdot t^m = (1-t^2)^{-1/2} e^{-\pi(x^2+y^2)t^2/(1-t^2) + 2\pi x y t/(1-t^2)}. \tag{8.3.42}$$

The effective, and best possible, estimates for all our limit relations, relating the various bases (cf. (8.3.39) and (8.2.34)), are obtained by comparing their recursion relations. Using these estimates to approximate the Hermite basis $\varphi_{\eta,m}^{\infty,\infty}(w)$ in terms of the elementary symmetric functions $\varphi_{\eta(N),m}^{\infty,\infty}(i,j)$, Beckner (1975) proves a deep inequality for the Fourier transform $\mathcal{F}_1$ on the real line $\mathbb{R} = \mathbb{Q}_\eta$, namely the $p=\eta$ case of the norm inequality (the $p \neq \eta$ case is trivial):

$$\begin{aligned} \|\mathcal{F}_1\|_{L_r(\mathbb{Q}_p)\to L_{r'}(\mathbb{Q}_p)} &\overset{\text{def}}{=} \sup_{\varphi\neq 0} \frac{\|\mathcal{F}_1\varphi\|_{L_{r'}}}{\|\varphi\|_{L_r}} \quad \left(\frac{1}{r}+\frac{1}{r'}=1\right) \\ &\overset{(!)}{=} \frac{\left\|\phi_{\mathbb{Z}_p}\right\|_{L_{r'}}}{\left\|\phi_{\mathbb{Z}_p}\right\|_{L_r}} = \begin{cases} 1 & p\neq\eta\,, \\ r^{r/2}/r'^{\,r'/2} & p=\eta\,. \end{cases} \end{aligned} \tag{8.3.43}$$

For $p=\eta$ this is related to the hypercontractivity of the harmonic oscillator $\mathrm{N}_\eta^{\infty,\infty}$. We end this section by stating the corresponding properties of $\mathrm{N}_\eta^{\alpha,\beta}$; the proof is deeper and depends on the delicate L_2–L_∞ behaviour of the eigenfunctions $\varphi_{\eta,m}^{\alpha,\beta}$. The case $\alpha=\beta$ is treated in Mueller and Weissler (1982). The case $\alpha\neq\beta$ can be treated similarly, but special care is needed at the boundary points $0,\ \infty \in \partial \mathrm{X}_\eta$. The point ∞ is an exit (respectively regular, entrance) boundary point, according to $\alpha=0$ (respectively $\alpha\in(0,2),\ \alpha\geq 2$). In the regular case one should impose the 'reflecting barrier' condition on the domain of $\mathrm{N}_\eta^{\alpha,\beta}$, see Trotter (1958):

$$\rho_\infty(x)^{\alpha-2}\rho_0(x)^{\beta-2}x\,\frac{\partial}{\partial x}\varphi(x) \underset{x\to\infty}{\longrightarrow} 0\,;$$

similarly for $x\to 0$, with α replaced by β. The inequalities then read:

log Sobolev inequality: for $\varphi \in \mathrm{Dom\ D}, \varphi$ real valued,

$$\int |\varphi(x)|^2 \log \frac{|\varphi(x)|}{\|\varphi\|_{\mathscr{H}_\eta^{\alpha,\beta}}}\,\tau_\eta^{\alpha,\beta}(\mathrm{d}x) \leq \left(\frac{\alpha+\beta}{2}\right)^{-1} \left\| \left(\mathrm{N}_\eta^{\alpha,\beta}\right)^{1/2} \varphi \right\|_{\mathscr{H}_\eta^{\alpha,\beta}}^2 . \tag{8.3.44}$$

hypercontractivity: for $\alpha>\beta>0,\ t>0,\ 1<r_1<r_2<\infty$,

$$\left\| \mathrm{e}^{-t\mathrm{N}_\eta^{\alpha,\beta}} \right\|_{L_{r_1}\left(\tau_\eta^{\alpha,\beta}\right)\to L_{r_2}\left(\tau_\eta^{\alpha,\beta}\right)} \leq 1 \Longleftrightarrow \mathrm{e}^{-t(\alpha+\beta)} \leq \frac{r_1-1}{r_2-1}\,. \tag{8.3.45}$$

9

Global 'chains' and higher dimensions

9.1 RESTRICTED DIRECT PRODUCTS OF CHAINS

Let $\mathcal{P}$ be a set, and for each $p \in \mathcal{P}$ let X_p be a timed chain with state space

$$\mathrm{X}_p = \coprod_{N \geq 0} \mathrm{X}_{p(N)}, \quad \mathrm{X}_{p(0)} = \{1_p\}, \tag{9.1.1}$$

and transition probabilities P_p describing how our particle walks from states in $\mathrm{X}_{p(N)}$ to states in $\mathrm{X}_{p(N+1)}$. Let G_p denote the Green kernel, let K_p denote the Martin kernel of X_p, $p \in \mathcal{P}$, and let $\partial \mathrm{X}_p$ denote the Martin boundary. We form the *restricted* direct product of the X_p: the state space is given by

$$\mathrm{X} = \left\{ x = (x_p) \in \prod_{p \in \mathcal{P}} \mathrm{X}_p \,\middle|\, x_p = 1_p \text{ for all but finitely many } p \right\}. \tag{9.1.2}$$

Let $\mathcal{N}(\mathcal{P})$ denote the free abelian monoid generated by $\mathcal{P}$. Its elements are

$$\mathcal{N}(\mathcal{P}) = \left\{ p_1^{N_1} \dots p_\ell^{N_\ell},\ p_i \in \mathcal{P},\ N_i \in \mathbb{N} \right\} \tag{9.1.3}$$

so we can multiply them, and we have a partial ordering on $\mathcal{N}(\mathcal{P})$ given by

$$N \mid M \Leftrightarrow N = p_1^{N_1} \dots p_\ell^{N_\ell}, \quad M = p_1^{M_1} \dots p_\ell^{M_\ell}, \quad \text{and} \quad N_i \leq M_i \quad \text{for } i = 1, \dots, \ell. \tag{9.1.4}$$

We can now write X as a disjoint union

$$\mathrm{X} = \coprod_{N \in \mathcal{N}(\mathcal{P})} \mathrm{X}_N, \tag{9.1.5}$$

where for $N = p_1^{N_1} \dots p_\ell^{N_\ell}$ we set

$$\mathrm{X}_N = \prod_{1 \leq i \leq \ell} \mathrm{X}_{p_i(N_i)} \times \prod_{p \in \mathcal{P} \setminus \{p_1, \dots, p_\ell\}} \{1_p\}.$$

Note that we do *not* have a natural chain structure on X; for this we will need a way of choosing a *finite* subset of $\mathcal{P}$ at each step of our particle walk (or a probability measure on the collection of all finite subsets of $\mathcal{P}$). Thus we do *not* have a single operator $\mathrm{P_X}$ acting on functions on X. Instead, we have the collection of commuting operators $\{\mathrm{P}_p\}$, $p \in \mathcal{P}$, acting through the x_p-variable, and generating an action of $\mathcal{N}(\mathcal{P})$ on the functions on X. A function

$f\colon \mathrm{X} \to [0,\infty)$, $f(1_{\mathrm{X}}) = 1$, will be called *harmonic* if $\mathrm{P}_p f = f$ for all $p \in \mathcal{P}$ (or, equivalently, if $\mathrm{P}_N f = f$ for all $N \in \mathcal{N}(\mathcal{P})$). The Green kernel is now defined by

$$\mathrm{G}_{\mathrm{X}}(x,y) = \prod_{p\in\mathcal{P}} \mathrm{G}_p(x_p,y_p) = `\prod_{p\in\mathcal{P}}(1-\mathrm{P}_p)^{-1}' = \sum_{N\in\mathcal{N}(\mathcal{P})} \mathrm{P}_N(x,y)\,, \tag{9.1.6}$$

and the Martin kernel is defined by

$$\mathrm{K}_{\mathrm{X}}(x,y) = \frac{\mathrm{G}_{\mathrm{X}}(x,y)}{\mathrm{G}_{\mathrm{X}}(1_{\mathrm{X}},y)} = \prod_{p\in\mathcal{P}} \mathrm{K}_p(x_p,y_p)\,. \tag{9.1.7}$$

We can now define the Martin metric on X, so that a sequence $x_n \in \mathrm{X}$ is a Cauchy sequence if and only if $\mathrm{K}_{\mathrm{X}}(x,x_n)$ is a Cauchy sequence (of real numbers) for all $x \in \mathrm{X}$. Completing with respect to this metric we obtain the compact space $\mathrm{X} \cup \partial\mathrm{X}$, where the boundary $\partial\mathrm{X}$ is the product of the boundaries $\partial\mathrm{X}_p$, $p \in \mathcal{P}$:

$$\partial\mathrm{X} = \prod_{p\in\mathcal{P}} \partial\mathrm{X}_p\,. \tag{9.1.8}$$

The Martin kernel $\mathrm{K}_{\mathrm{X}}(x,\,y)$ extends by continuity to the boundary

$$\mathrm{K}_{\mathrm{X}}(x,y) = \lim_{x_n\to y} \mathrm{K}_{\mathrm{X}}(x,x_n) = \prod_{p\in\mathcal{P}} \mathrm{K}_p(x_p,y_p), \quad x\in\mathrm{X}, \quad y\in\partial\mathrm{X}\,. \tag{9.1.9}$$

Integration against K_{X} induces a one-to-one correspondence between harmonic functions f on X and probability measures μ on the boundary $\partial\mathrm{X}$:

$$\begin{gathered} f\colon \mathrm{X} \to [0,\infty),\ f(1_{\mathrm{X}}) = 1,\ \mathrm{P}_N f = f \quad \text{for all } N \in \mathcal{N}(\mathcal{P}) \\ \Leftrightarrow f(x) = \int_{\partial\mathrm{X}} \mathrm{K}_{\mathrm{X}}(x,y)\mu(dy) \end{gathered} \tag{9.1.10}$$

(we are assuming that all points of $\partial\mathrm{X}_p$ are extreme for all $p \in \mathcal{P}$, as will be the case in all our applications).

The *harmonic measure* τ_{X} on $\partial\mathrm{X}$ corresponds to the constant harmonic function $\mathbb{1}$; it is just the product of the harmonic measures τ_p of the chains X_p, $p \in \mathcal{P}$.

Taking $\mathcal{P}$ to be the collection of finite primes $\mathcal{N}(\mathcal{P}) = \mathbb{N}^*$ is the monoid of natural numbers under multiplication. Taking for each $p \in \mathcal{P}$ the p-adic beta chain $\widetilde{\mathrm{X}}_p$, the restricted direct product gives us the beta 'chain' over the ring of finite adeles $\mathbb{A}_{\mathrm{fin}} = \prod_{p\in\mathcal{P}} \mathbb{Q}_p \colon \mathbb{Z}_p$. Its state space is given by

$$\widetilde{\mathrm{X}}_{\mathbb{A}_{\mathrm{fin}}} = \mathrm{PGL}_2(\mathbb{A}_{\mathrm{fin}})/\mathrm{PGL}_2(\widehat{\mathbb{Z}}) = \coprod_{N\geq 1} \mathbb{P}^1(\mathbb{Z}/N\mathbb{Z}) \tag{9.1.11}$$

and its boundary is given by

$$\partial\widetilde{\mathrm{X}}_{\mathbb{A}_{\mathrm{fin}}} = \varprojlim_N \mathbb{P}^1(\mathbb{Z}/N\mathbb{Z}) = \mathbb{P}^1(\widehat{\mathbb{Z}}) = \prod_{p\neq\eta} \mathbb{P}^1(\mathbb{Q}_p)\,. \tag{9.1.12}$$

We can similarly take the quotients X_p or $\mathrm{X}_p^{\square}$ of these beta chains, or the covering $\widetilde{\widetilde{\mathrm{X}}}_p$; forming the restricted direct product gives the analogous 'chain' over $\mathbb{A}_{\mathrm{fin}}$.

The situation is summarized in the following diagram:

$$
\begin{array}{ccccc}
\underline{\text{State space}} & & & & \underline{\text{Boundary}} \\
\widetilde{\widetilde{\mathrm{X}}}_{\mathbb{A}_{\mathrm{fin}}} = V(\mathbb{A}_{\mathrm{fin}})/V(\widehat{\mathbb{Z}}) & = & \coprod\limits_{N\geq 1} V^*(\mathbb{Z}/N\mathbb{Z}) \Longrightarrow \partial\widetilde{\widetilde{\mathrm{X}}}_{\mathbb{A}_{\mathrm{fin}}} & = & V^*(\widehat{\mathbb{Z}}) = \prod\limits_{p\neq\eta} V^*(\mathbb{Z}_p) \\
\downarrow & & \downarrow \qquad\qquad \downarrow & & \downarrow \\
\widetilde{\mathrm{X}}_{\mathbb{A}_{\mathrm{fin}}} = \mathrm{PGL}_2(\mathbb{A}_{\mathrm{fin}})/\mathrm{PGL}_2(\widehat{\mathbb{Z}}) & = & \coprod\limits_{N\geq 1} \mathbb{P}^1(\mathbb{Z}/N\mathbb{Z}) \Longrightarrow \partial\widetilde{\mathrm{X}}_{\mathbb{A}_{\mathrm{fin}}} & = & \mathbb{P}^1(\widehat{\mathbb{Z}}) = \prod\limits_{p\neq\eta} \mathbb{P}^1(\mathbb{Z}_p) \\
\downarrow & & \downarrow \qquad\qquad \downarrow & & \downarrow \\
\mathrm{X}_{\mathbb{A}_{\mathrm{fin}}} = \widehat{\mathbb{Z}}^*\backslash\widetilde{\mathrm{X}}_{\mathbb{A}_{\mathrm{fin}}} & = & \coprod\limits_{N\geq 1} (\mathbb{Z}/N\mathbb{Z})^*\backslash\mathbb{P}^1(\mathbb{Z}/N\mathbb{Z}) \Longrightarrow \partial\mathrm{X}_{\mathbb{A}_{\mathrm{fin}}} & = & \widehat{\mathbb{Z}}^*\backslash\mathbb{P}^1(\widehat{\mathbb{Z}}) = \prod\limits_{p\neq\eta} \mathbb{Z}_p^*\backslash\mathbb{P}^1(\mathbb{Z}_p) \\
\downarrow & & \downarrow \qquad\qquad \downarrow & & \downarrow \\
\mathrm{X}^{\square}_{\mathbb{A}_{\mathrm{fin}}} = \mathrm{B}(\widehat{\mathbb{Z}})\backslash\widetilde{\mathrm{X}}_{\mathbb{A}_{\mathrm{fin}}} & = & \coprod\limits_{N\geq 1} \mathrm{B}(\mathbb{Z}/N\mathbb{Z})\backslash\mathbb{P}^1(\mathbb{Z}/N\mathbb{Z}) \Longrightarrow \partial\mathrm{X}^{\square}_{\mathbb{A}_{\mathrm{fin}}} & = & \mathrm{B}(\widehat{\mathbb{Z}})\backslash\mathbb{P}^1(\widehat{\mathbb{Z}}) = \prod\limits_{p\neq\eta} \mathrm{B}(\mathbb{Z}_p)\backslash\mathbb{P}^1(\mathbb{Z}_p) \\
\downarrow & & \downarrow \qquad\qquad \downarrow & & \downarrow \\
\mathbb{N}^* = \mathrm{PGL}_2(\widehat{\mathbb{Z}})\backslash\widetilde{\mathrm{X}}_{\mathbb{A}_{\mathrm{fin}}} & = & \coprod\limits_{N\geq 1} \{*\} \Longrightarrow \partial\mathbb{N}^* & = & \mathrm{PGL}_2(\widehat{\mathbb{Z}})\backslash\mathbb{P}^1(\widehat{\mathbb{Z}}) = \{*\}
\end{array}
\tag{9.1.13}
$$

We can now put together the picture (9.1.13) of the beta chains over $\mathbb{A}_{\mathrm{fin}}$, and the analogous picture for the real prime, cf. §8.3. The beta 'chains' over the adeles $\mathbb{A}$ are the product of the 'chains' over $\mathbb{A}_{\mathrm{fin}}$, and the chains over the real prime η: $\mathrm{X}_{\mathbb{A}} = \mathrm{X}_{\mathbb{A}_{\mathrm{fin}}} \times \mathrm{X}_{\eta}$. Alternatively, $\mathrm{X}_{\mathbb{A}}$ is the restricted direct product of the X_p over all $p \geqslant \eta$. We summarize the situation in the following diagram:

$$
\begin{array}{ccccc}
\widetilde{\widetilde{\mathrm{X}}}_{\mathbb{A}} = V(\mathbb{Z}) \times (V(\mathbb{Q})/V(\mathbb{Z})) & = & \coprod\limits_{n\in\mathcal{N}} \text{`}V^*(\mathbb{Z}/n)\text{'} & \Longrightarrow & \partial\widetilde{\widetilde{\mathrm{X}}}_{\mathbb{A}} = \prod\limits_{p\geq\eta} V^*(\mathbb{Z}_p) \\
\downarrow & & \downarrow & & \downarrow \\
\widetilde{\mathrm{X}}_{\mathbb{A}} = \{\pm 1\}\widehat{\mathbb{Z}}^*\backslash\widetilde{\widetilde{\mathrm{X}}}_{\mathbb{A}} & = & \coprod\limits_{n\in\mathcal{N}} \text{`}\mathbb{P}^1(\mathbb{Z}/n)\text{'} & \Longrightarrow & \partial\widetilde{\widetilde{\mathrm{X}}}_{\mathbb{A}} = \prod\limits_{p\geq\eta} \mathbb{P}^1(\mathbb{Q}_p) \\
\downarrow & & \downarrow & & \downarrow \\
\mathrm{X}_{\mathbb{A}} = \{\pm 1\}\widehat{\mathbb{Z}}^*\backslash\widetilde{\mathrm{X}}_{\mathbb{A}} & = & \coprod\limits_{n\in\mathcal{N}} \mathrm{X}_{\mathbb{A}(n)} & \Longrightarrow & \partial\mathrm{X}_{\mathbb{A}} = \prod\limits_{p\geq\eta} \mathbb{Z}_p^*\backslash\mathbb{P}^1(\mathbb{Q}_p) \\
\downarrow & & \downarrow & & \downarrow \\
\mathcal{N} = \bigoplus\limits_{p\geq\eta} p^{\mathbb{N}} = \mathbb{N}^* \times \eta^{\mathbb{N}} & = & \coprod\limits_{n\in\mathcal{N}} \{*\} & \Longrightarrow & \partial\mathcal{N} = \{*\}
\end{array}
\tag{9.1.14}
$$

9.2 HIGHER DIMENSIONAL BETA CHAINS

The beta chains have natural generalizations to higher dimensions. We can consider the symmetric and non-symmetric beta chains by forming the product and semi-direct product of r basic chains, $\mathrm{P}^{\alpha_1}_{\mathbb{N}(q)} \times \cdots \times \mathrm{P}^{\alpha_r}_{\mathbb{N}(q)}$ and $\mathrm{P}^{\alpha_1}_{\mathbb{N}(q)} \ltimes \cdots \ltimes \mathrm{P}^{\alpha_r}_{\mathbb{N}(q)}$. Their associated non-stop chains converge in both the real and p-adic limits. Let us concentrate on the non-symmetric q-beta chain $\mathrm{P}^{\bar{\alpha}}_q = \left(\mathrm{P}^{\alpha_1}_{\mathbb{N}(q)} \ltimes \cdots \ltimes \mathrm{P}^{\alpha_2}_{\mathbb{N}(q)}\right)_!$; its state space is $\mathrm{X}^r_q = \mathbb{N}^r = \{\overline{m} = (m_1, \ldots, m_r)\}$. It is timed with time map

$$|\cdot|\colon \mathrm{X}^r_q \to \mathbb{N},\ |\overline{m}| = m_1 + \cdots + m_r\,, \tag{9.2.1}$$

so that

$$\mathrm{X}^r_{q(N)} = \left\{\overline{m} \in \mathbb{N}^r \middle| |\overline{m}| = N\right\}, \quad \text{and} \quad \#\mathrm{X}^r_{q(N)} = \binom{N+r-1}{N}.$$

We use the notation $\overline{m}(i+) = (m_1, \ldots, m_{i-1}, m_i + 1, m_{i+1}, \ldots, m_r)$ and similarly $\overline{m}(i-)$, $\overline{m}(i+,j-)$, $\overline{m}(i++)$, etc. The q-beta chain $\mathrm{P}^{\bar{\alpha}}_q$ is given explicitly by

$$\begin{aligned} \mathrm{P}^{\bar{\alpha}}_q(\overline{m}, \bar{n}) = \frac{1 - q^{\alpha_i + m_i}}{1 - q^{|\alpha| + |m|}} \cdot q^{\sum_{j<i}(\alpha_j + m_j)} \quad & \text{if } \bar{n} = \overline{m}(i+), i = 1, \ldots, r\,, \\ 0 \quad & \text{otherwise}\,. \end{aligned} \tag{9.2.2}$$

On $\mathrm{X}^r_{q(N)}$ we have the probability measure $\tau^{\bar{\alpha}}_{q(N)} = \left(\mathrm{P}^{\bar{\alpha}*}_q\right)^N \delta_{\bar{o}}$

We can write

$$\begin{aligned} \mathrm{P}^{\bar{\alpha}*}_q &= \mathrm{A}_1 + \cdots + \mathrm{A}_r\,, \\ \mathrm{A}_i\varphi(\overline{m}) &= \frac{1 - q^{\alpha_i + m_i - 1}}{1 - q^{|\alpha| + |m| - 1}} \cdot q^{\sum_{j<i}(\alpha_j + m_j)} \varphi(\overline{m}(i-))\,, \end{aligned} \tag{9.2.3}$$

and we have q-commutativity

$$\mathrm{A}_i \mathrm{A}_j = q \mathrm{A}_j \mathrm{A}_i \quad \text{for } j < i\,.$$

Using the (r-dimensional) operator q-binom we obtain

$$\tau^{\bar{\alpha}}_{q(N)}(\overline{m}) = \begin{bmatrix} N \\ \overline{m} \end{bmatrix}_q \cdot \frac{\zeta_{(q)}(\bar{\alpha} + \overline{m})}{\zeta_{(q)}(\bar{\alpha})} \cdot q^{\sum_{j<i\le r} \alpha_j m_i} \tag{9.2.4}$$

where the higher q-binomial coefficients are given by

$$\begin{bmatrix} N \\ \overline{m} \end{bmatrix}_q = \frac{[N]_q!}{[m_1]_q! \ldots [m_r]_q!}, \quad m_1 + \cdots + m_r = N\,, \tag{9.2.5}$$

and the higher q-beta function is given by

$$\zeta_{(q)}(\bar{\alpha}) = \frac{\zeta_{(q)}(\alpha_1) \ldots \zeta_{(q)}(\alpha_r)}{\zeta_{(q)}(\alpha_1 + \cdots + \alpha_r)} \zeta_{(q)}(1)^{1-r}\,. \tag{9.2.6}$$

Note that starting the chain at $\overline{m}$, rather than at $\bar{o}$, is equivalent to replacing $\bar{\alpha}$ by $\bar{\alpha} + \overline{m}$. Hence the Green kernel is given by

$$\begin{aligned} \mathrm{G}^{\bar{\alpha}}(\overline{m}, \bar{n}) = \tau^{\bar{\alpha} + \bar{m}}(\bar{n} - \overline{m}) \quad & \text{if } n_i \ge m_i, i = 1, \ldots, r\,, \\ 0 \quad & \text{otherwise}\,. \end{aligned} \tag{9.2.7}$$

The Martin kernel is therefore

$$\mathrm{K}_q^{\bar\alpha}(\overline{m},\bar n)=\frac{\tau^{\bar\alpha+\bar m}(\bar n-\overline{m})}{\tau^{\bar\alpha}(\bar n)}=\frac{\begin{bmatrix}|\bar n|-|\overline{m}|\\ \bar n-\overline{m}\end{bmatrix}_q}{\begin{bmatrix}|\bar n|\\ \bar n\end{bmatrix}_q}\cdot\frac{\zeta_{(q)}(\bar\alpha)}{\zeta_{(q)}(\bar\alpha+\overline{m})}\cdot q^{\sum_{j<i}(m_jn_i-(m_j+\alpha_j)m_i)}. \tag{9.2.8}$$

It follows that the boundary is given by

$$\partial\mathrm{X}_q^r=\mathbb{N}^{r-1}\amalg\mathbb{N}^{r-2}\amalg\cdots\amalg\mathbb{N}^1\amalg\mathbb{N}^0=\{\infty\}, \tag{9.2.9}$$

where the points of $\mathbb{N}^{r-j}$ are denoted $(\infty,m_{j+1},\ldots,m_r)$ and are limits of points of X_q^r for 'vertical' sequences:

$$(\infty,m_{j+1},\ldots,m_r)=\lim_{m_j\to\infty}(x_1,\ldots,x_{j-1},m_j,m_{j+1},\ldots,m_r) \tag{9.2.10}$$

with $x_1,\ldots,x_{j-1}\in\mathbb{N}$ arbitrary.

The 'main part' of the boundary is $\mathbb{N}^{r-1}$, where the extension of the Martin kernel is given by

$$\begin{aligned}&\mathrm{K}_q^{\bar\alpha}(\bar m,(\infty,n_2,\ldots,n_r))\\&=\frac{\zeta_{(q)}(1+n_2)\ldots\zeta_{(q)}(1+n_r)}{\zeta_{(q)}(1+n_2-m_2)\ldots\zeta_{(q)}(1+n_r-m_r)}\cdot\frac{\zeta_{(q)}(\bar\alpha)}{\zeta_{(q)}(\bar\alpha+\overline{m})}\cdot q^{\sum_{j<i}(m_jn_i-(m_j+\alpha_j)m_i)},\\&\qquad\qquad n_i\ge m_i,\quad i=2,\ldots,r,\\&\qquad\qquad 0\quad\text{otherwise}.\end{aligned} \tag{9.2.11}$$

The harmonic measure $\tau_q^{\bar\alpha}$ is supported on $\mathbb{N}^{r-1}$, and is given by

$$\tau_q^{\bar\alpha}(\infty,n_2,\ldots,n_r)=\frac{1}{\zeta_{(q)}(\bar\alpha)}\cdot\frac{\zeta_{(q)}(\alpha_2+n_2)\ldots\zeta_{(q)}(\alpha_r+n_r)}{\zeta_{(q)}(1+n_2)\ldots\zeta_{(q)}(1+n_r)}\cdot q^{\sum_{j<i}\alpha_jn_i}. \tag{9.2.12}$$

We have the Hilbert space

$$\mathscr{H}_q^{\bar\alpha}=\ell_2\left(\mathbb{N}^{r-1},\tau_q^{\bar\alpha}\right) \tag{9.2.13}$$

and its finite-dimensional 'approximations'

$$\mathscr{H}_{q(N)}^{\bar\alpha}=\ell_2\left(\mathrm{X}_{q(N)}^r,\tau_{q(N)}^{\bar\alpha}\right), \tag{9.2.14}$$

but again the Martin kernel does not give injections or projections.

Consider the difference operators

$$\mathrm{D}_N\colon\mathscr{H}_{q(N)}^{\bar\alpha}\to\mathscr{H}_{q(N-1)}^{\bar\alpha(1+,2+)},\quad \mathrm{D}_N\varphi(\overline{m})=q^{-(m_2+\cdots+m_r)}\frac{\varphi(\overline{m}(2+))-\varphi(\overline{m}(1+))}{1-q} \tag{9.2.15}$$

and let

$$\mathrm{D}_N^+ = \frac{[|\bar{\alpha}|+N]_q [N]_q [\alpha_1]_q [\alpha_2]_q}{[|\bar{\alpha}|]_q \cdot [|\bar{\alpha}|+1]_q} (1-q)^2 \cdot \mathrm{D}_N^* .$$

Explicitly

$$\begin{aligned} \mathrm{D}_N^+ \varphi(\overline{m}) = \ & (1-q) q^{N-m_1-m_2} \Big\{ q^{-\alpha_1} [m_1+\alpha_1]_q [m_2]_q \varphi(\overline{m}(2-)) \\ & - [m_1]_q [m_2+\alpha_2]_q \varphi(\overline{m}(1-)) \Big\} . \end{aligned} \tag{9.2.16}$$

We have the Heinsenberg relations up the ladder:

$$\begin{gathered} \mathscr{H}^{\bar{\alpha}}_{q(N)} \underset{\mathrm{D}_N^+}{\overset{\mathrm{D}_N}{\rightleftarrows}} \mathscr{H}^{\bar{\alpha}(1+,2+)}_{q(N-1)} \underset{\mathrm{D}_{N-1}^+}{\overset{\mathrm{D}_{N-1}}{\rightleftarrows}} \mathscr{H}^{\bar{\alpha}(1++,2++)}_{q(N-2)} \\ \mathrm{D}_N \mathrm{D}_N^+ - \mathrm{D}_{N-1}^+ \mathrm{D}_{N-1} = q^{-\alpha_1} \cdot [\alpha_1+\alpha_2]_q \cdot \mathrm{id}_{\mathscr{H}^{\bar{\alpha}(1+,2+)}_{q(N-1)}} . \end{gathered} \tag{9.2.17}$$

We have an injection from the $(r-1)$-dimensional spaces to the r-dimensional ones,

$$\begin{gathered} \rho_{r-1} \colon \mathscr{H}^{(\alpha_1+\alpha_2,\alpha_3,\ldots,\alpha_r)}_{q(N)} \hookrightarrow \mathscr{H}^{\bar{\alpha}}_{q(N)} \\ \rho_{r-1}\varphi(\overline{m}) = \varphi(m_1+m_2, m_3, \ldots, m_r) . \end{gathered} \tag{9.2.18}$$

The dual projection is given by averaging

$$\rho_{r-1}^* \varphi(\overline{m}) = \sum_{i+j=m_1} \tau^{(\alpha_1,\alpha_2)}_{q(m_1)}(i,j) \varphi(i,j,m_2,\ldots,m_{r-1}) . \tag{9.2.19}$$

We have

$$\text{image } \rho_{r-1} = \text{kernel } \mathrm{D}_N . \tag{9.2.20}$$

It follows from (9.2.17), (9.2.20) that we obtain an orthogonal basis for $\mathscr{H}^{\bar{\alpha}}_{q(N)}$ by setting

$$\begin{gathered} \varphi^{\bar{\alpha}}_{q(N),\bar{v}} \stackrel{\text{def}}{=} \frac{q^{\binom{|\bar{v}|}{2}}}{[v_1]_q! \cdots [v_{r-1}]_q!} (\mathrm{D}^+)^{v_1} \rho_{r-1} (\mathrm{D}^+)^{v_2} \rho_{r-2} \cdots \rho_2 (\mathrm{D}^+)^{v_{r-1}} \mathbb{1} , \\ \bar{v} = (v_1, \ldots, v_{r-1}), \quad |\bar{v}| = v_1 + \cdots + v_{r-1} \le N . \end{gathered} \tag{9.2.21}$$

For $r=2$ these are the q-Hahn polynomials,

$$\varphi^{\alpha_1,\alpha_2}_{q(N),v_1}(m_1,m_2) = (q^{\alpha_1}(1-q))^{-v_1} \cdot \tilde{\varphi}^{(\alpha_2)\alpha_1}_{q(N),v_1}(m_1,m_2) , \tag{9.2.22}$$

cf. (7.3.14, 7.3.15). For $r>2$ it is a combination of the two-dimensional basis

$$\varphi^{\bar{\alpha}}_{q(N),\bar{v}}(\overline{m}) = \prod_{1\le j<r} q^{v_j V_{j-1}} \varphi^{(\mathrm{A}_j - 2V_{j-1},\alpha_{j+1})}_{q(\mathrm{M}_{j+1}-V_{j-1}),v_j} (\mathrm{M}_j - V_{j-1}, m_{j+1}) , \tag{9.2.23}$$

with $V_j = v_1 + \cdots + v_j$, $\mathrm{M}_j = m_1 + \cdots + m_j$, $\mathrm{A}_j = \alpha_1 + \cdots + \alpha_j$.

The basis $\varphi^{\bar\alpha}_{q(N),\bar v}$ for different N diagonalize the Martin kernel

$$\mathrm{K}^{\bar\alpha}_{N,N'}\varphi^{\bar\alpha}_{q(N'),\bar v} = \frac{\begin{bmatrix}|\bar\alpha| + N' + |\bar v| - 1 \\ |\bar v|\end{bmatrix}_q}{\begin{bmatrix}|\bar\alpha| + N + |\bar v| - 1 \\ |\bar v|\end{bmatrix}_q} \cdot \varphi^{\bar\alpha}_{q(N),\bar v} \quad \text{for } |\bar v| \le N,\ 0 \text{ for } |\bar v| > N\,. \tag{9.2.24}$$

We can now take the limit $N \to \infty$ to the boundary spaces. The difference operator is

$$\mathrm{D}\colon \mathscr{H}^{\bar\alpha}_q \to \mathscr{H}^{\bar\alpha(1+,2+)}_q, \quad \mathrm{D}\varphi(\bar n) = q^{-|\bar n|} \cdot \frac{\varphi(\bar n(2+)) - \varphi(\bar n)}{1-q}, \tag{9.2.25}$$
$$\bar n = (\infty, n_2, \dots, n_r), \quad |\bar n| = n_2 + \cdots + n_r\,,$$

and its 'adjoint'

$$\mathrm{D}^+ = \frac{[\alpha_1]_q [\alpha_2]_q}{[|\bar\alpha|]_q \cdot [|\bar\alpha| + 1]_q} \mathrm{D}^*$$

is given explicitly by

$$\mathrm{D}^+\varphi(\bar n) = q^{|\bar n| - n_2} \cdot \left\{ q^{-\alpha_1} [n_2]_q \varphi(\bar n(2-)) - [\alpha_2 + n_2]_q \varphi(\bar n) \right\}. \tag{9.2.26}$$

They satisfy the Heisenberg relations up the ladder

$$\mathrm{DD}^+ - \mathrm{D}^+\mathrm{D} = q^{-\alpha_1} [\alpha_1 + \alpha_2]_q \cdot \mathrm{id}_{\mathscr{H}^{\bar\alpha(1+,2+)}_q}\,. \tag{9.2.27}$$

We have embeddings

$$\rho_{r-1}\colon \mathscr{H}^{(\alpha_1+\alpha_2,\alpha_3,\dots,\alpha_r)}_q \hookrightarrow \mathscr{H}^{\bar\alpha}_q$$
$$\rho_{r-1}\varphi(\infty, n_2, \dots, n_r) = \varphi(\infty, n_3, \dots, n_r) \tag{9.2.28}$$

with
$$\mathrm{image}\rho_r - 1 = \mathrm{kernel}\ \mathrm{D}\,.$$

We get the orthogonal basis for $\mathscr{H}^{\bar\alpha}_q$ given by

$$\varphi^{\bar\alpha}_{q,\bar v} = \frac{q^{\binom{|\bar v|}{2}}}{[v_1]_q! \dots [v_{r-1}]_q!} (\mathrm{D}^+)^{v_1} \rho_{r-1} \dots \rho_2 (\mathrm{D}^+)^{v_{r-1}} \mathbb{1}, \quad \bar v = (v_1, \dots, v_{r-1}) \in \mathbb{N}^{r-1}\,. \tag{9.2.29}$$

For $r = 2$ these are the little q-Jacobi polynomials (7.3.29), and for $r > 2$ these are combinations of the two-dimensional basis. Explicitly

$$\varphi^{\bar\alpha}_{q,\bar v}(\infty, n_2, \dots, n_r)$$
$$= \varphi^{\bar\alpha}_{q,\bar v}(\infty) \cdot \prod_{1 \le j < r} {}_2\phi_1(-v_j, \mathrm{A}_{j+1} + 2V_j - v_j - 1; \mathrm{A}_j + 2V_{j-1}; 1 + n_{j+1})\,. \tag{9.2.30}$$

The compatibility between the operators D, D^+ and ρ, and their finite-dimensional approximations with respect to the Martin kernel, yields the compatibility of our basis,

$$\mathrm{K}^{\bar\alpha}_N \varphi^{\bar\alpha}_{q,\bar v} = \frac{\zeta_{(q)}(|\bar\alpha| + N)}{\zeta_{(q)}(|\bar\alpha| + N - |\bar v|)} \cdot \varphi^{\bar\alpha}_{q(N),\bar v} \quad \text{for } |\bar v| \le N\,, \tag{9.2.31}$$
$$0 \quad \text{for } |\bar v| > N\,.$$

In the p-adic limit ⓟ: $q = p^{-\ell}$, $\alpha_j := \alpha_j/\ell$, $\ell \to \infty$ we obtain the chain $\mathrm{P}_p^{\bar{\alpha}}$ with state space $\mathrm{X}_p^r = \mathbb{N}^r$, given by

$$\mathrm{P}_p^{\bar{\alpha}}(\overline{m}, \bar{n}) = \begin{cases} \dfrac{1-p^{-\alpha_i}}{1-p^{-|\bar{\alpha}|}} \cdot p^{-\sum_{j<i} \alpha_j} & \overline{m} = \bar{o},\ \bar{n} = \bar{o}(i+)\,, \\ (1-p^{-\alpha_i}) \cdot p^{-\sum_{j<i} \alpha_j} & \overline{m} = (0, \ldots, 0, m_{i+1}, \ldots m_r) \neq \bar{o},\ \bar{n} = \overline{m}(i+)\,, \\ p^{-\sum_{j<i} \alpha_j} & \overline{m} = (0, \ldots, m_i, \ldots m_r),\ m_i \neq 0,\ \bar{n} = \overline{m}(i+)\,, \\ 0 & \text{otherwise}\,. \end{cases} \tag{9.2.32}$$

This is a tree chain, where the unique path from $\bar{o}$ to $\overline{m}$ is given by

$$\begin{aligned} &\bar{o} \to (\bar{o}, 1) \to (\bar{o}, 2) \to \cdots \to (\bar{o}, m_r) \to (\bar{o}, 1, m_r) \to \cdots \\ &\to (\bar{o}, m_{r-1}, m_r) \to \cdots \to (0, m_2, \ldots, m_r) \to \cdots \to \overline{m}\,. \end{aligned} \tag{9.2.33}$$

It has the following interpretation. Consider the algebraic space

$$\mathrm{X} = \mathrm{B}_{(1,\ldots,1)} \backslash \mathrm{GL}_r / \mathrm{B}_{(1,r-1)} = \mathrm{B}_{(1,\ldots,1)} \backslash \mathbb{P}^{r-1} \tag{9.2.34}$$

where $\mathrm{B}_{(1,\ldots,1)}$ is the minimal Borel subgroup of GL_r corresponding to the partition $r = 1 + \cdots + 1$, and $\mathrm{B}_{(1,r-1)}$ is the maximal Borel subgroup corresponding to $r = 1 + (r-1)$. We have a natural identification

$$\begin{aligned} \mathrm{X}_{p(N)} = \mathrm{X}(\mathbb{Z}/p^N\mathbb{Z}) = \mathrm{B}_{(1\ldots1)}(\mathbb{Z}/p^N\mathbb{Z}) \backslash \mathbb{P}^{r-1}(\mathbb{Z}/p^N\mathbb{Z}) &\xrightarrow{\sim} \{\overline{m} \in \mathbb{N}^r \mid |\overline{m}| = N\} \\ (1 : p^{m_r} : p^{m_r+m_{r-1}} : \cdots : p^{m_r+\cdots+m_2}) &\leftrightarrow \overline{m} \end{aligned} \tag{9.2.35}$$

and we see that the underlining tree of the p-adic chain $\mathrm{P}_p^{\bar{\alpha}}$ is identical with the tree structure on $\coprod_{N\geq 0} \mathrm{X}(\mathbb{Z}/p^N\mathbb{Z})$ resulting from reductions mod p^N. Similarly, we have an identification of the boundary

$$\begin{aligned} \partial \mathrm{X}_p = \mathrm{X}(\mathbb{Z}_p) = \varprojlim_N \mathrm{X}_{p(N)} = \mathrm{B}_{(1\ldots1)}(\mathbb{Z}_p) \backslash \mathbb{P}^{r-1}(\mathbb{Z}_p) &\xrightarrow{\sim} \mathbb{N}^{r-1} \amalg \cdots \amalg \mathbb{N}^1 \amalg \{\infty\} \\ (1 : p^{m_r} : p^{m_r+m_{r-1}} : \cdots : p^{m_r+\cdots+m_2}) &\longleftarrow (\infty, m_2, \ldots, m_r)\,. \end{aligned} \tag{9.2.36}$$

The Martin kernel now give embeddings and projections identifying $\mathscr{H}_{p(N)}^{\bar{\alpha}}$ with the subspace of $\mathscr{H}_p^{\bar{\alpha}}$ of functions defined mod p^N. The p-adic limit of our basis $\{\varphi_{p(N),\bar{v}}^{\bar{\alpha}} \mid |\bar{v}| = N\}$ forms a basis for the orthogonal complement of $\mathscr{H}_{p(N-1)}^{\bar{\alpha}}$ inside $\mathscr{H}_{p(N)}^{\bar{\alpha}}$. Since we have a tree chain, it is described completely by the associated harmonic measure $\tau_p^{\bar{\alpha}}$, which under the identification (9.2.36) is given by the image on $\partial \mathrm{X}_p$ of the p-adic beta measure on $\mathbb{P}^{r-1}(\mathbb{Z}_p)$; that is, the image of the product of r gamma measures: $\tau_p^{\bar{\alpha}} = \mathrm{pr}_*(\tau_{\mathbb{Z}_p}^{\alpha_1} \otimes \cdots \otimes \tau_{\mathbb{Z}_p}^{\alpha_r})$.

In the real limit ⓝ: $q := q^{2/\ell}$, $\alpha_j := \alpha_j/2$, $\ell \to \infty$, we obtain the real chain $\mathrm{P}_\eta^{\bar{\alpha}}$ with

$$\mathrm{P}_\eta^{\bar{\alpha}}(\bar{m}, \bar{n}) = \frac{\alpha_i + 2m_i}{|\bar{\alpha}| + 2|\bar{m}|} \quad \text{for } \bar{n} = \bar{m}(i+),\ i = 1, \ldots, r\,, \qquad 0 \quad \text{otherwise}\,. \tag{9.2.37}$$

It is also the real limit of the symmetric q-beta chain, and is symmetric in $\{\alpha_1, \ldots, \alpha_r\}$. It is timed with $\mathrm{X}_{\eta(N)} = \{\overline{m} \in \mathbb{N}^r \mid |\overline{m}| = N\}$. The finite approximations of the harmonic measure are given by

$$\tau^{\bar{\alpha}}_{\eta(N)}(\overline{m}) = (\mathrm{P}^{\bar{\alpha}*}_{\eta})^N \delta_{\bar{o}}(\overline{m}) = \binom{N}{\overline{m}} \cdot \frac{\zeta_\eta(\bar{\alpha} + 2\overline{m})}{\zeta_\eta(\bar{\alpha})}. \tag{9.2.38}$$

Starting the chain at $\overline{m}$ instead of $\bar{o}$ is equivalent to replacing $\bar{\alpha}$ by $\bar{\alpha} + 2\overline{m}$. The Green kernel is therefore

$$\mathrm{G}^{\bar{\alpha}}_{\eta}(\overline{m}, \bar{n}) = \tau^{\bar{\alpha}+2\overline{m}}_{\eta}(\bar{n} - \overline{m}) = \binom{|\bar{n}| - |\overline{m}|}{\bar{n} - \overline{m}} \cdot \frac{\zeta_\eta(\bar{\alpha} + 2\bar{n})}{\zeta_\eta(\bar{\alpha} + 2\overline{m})}. \tag{9.2.39}$$

The Martin kernel is

$$\mathrm{K}^{\bar{\alpha}}_{\eta}(\overline{m}, \bar{n}) = \frac{\binom{|\bar{n}| - |\overline{m}|}{\bar{n} - \overline{m}}}{\binom{|\bar{n}|}{\bar{n}}} \cdot \frac{\zeta_\eta(\bar{\alpha})}{\zeta_\eta(\bar{\alpha} + 2\overline{m})}, \quad n_i \geq m_i, \; i = 1, \ldots, r, \tag{9.2.40}$$

$$0 \quad \text{otherwise}.$$

It follows that a sequence $\overline{m}^{(k)} = \left(m_1^{(k)}, \ldots, m_r^{(k)}\right)$ converges in the Martin metric if it is eventually constant or else $|\overline{m}^{(k)}| = m_1^{(k)} + \cdots + m_r^{(k)} \underset{k \to \infty}{\longrightarrow} \infty$, and

$$x_j^2 \overset{\text{def}}{=} \lim_{k \to \infty} \frac{m_j^{(k)}}{|\overline{m}^{(k)}|} = \lim_{k \to \infty} \frac{\alpha_j}{|\bar{\alpha}|} \mathrm{K}^{\bar{\alpha}}_{\eta}(\bar{o}(j+), \overline{m}^{(k)}) \in [0,1] \quad \text{exists for } j = 1, \ldots, r. \tag{9.2.41}$$

We can thus identify the real boundary as

$$\partial \mathrm{X}_\eta = \{\bar{x} \in [0,1]^r \mid x_1^2 + \cdots + x_r^2 = 1\} = \{\pm 1\}^r \backslash \mathrm{S}^{r-1} = \{\pm 1\}^{r-1} \backslash \mathbb{P}^{r-1}(\mathbb{R}). \tag{9.2.42}$$

The extension of the Martin kernel to the boundary is given by

$$\mathrm{K}^{\bar{\alpha}}_{\eta}(\overline{m}, \bar{x}) = \frac{\zeta_\eta(\bar{\alpha})}{\zeta_\eta(\bar{\alpha} + 2\overline{m})} x_1^{2m_1} \cdots x_r^{2m_r}. \tag{9.2.43}$$

The harmonic measure is

$$\tau^{\bar{\alpha}}_{\eta}(\bar{x}) = |x_1|^{\alpha_1 - 1} \cdots |x_r|^{\alpha_r - 1} \frac{\mathrm{d}^0 \bar{x}}{\zeta_\eta(\bar{\alpha})}, \tag{9.2.44}$$

where $\mathrm{d}^0\bar{x}$ is the measure on the $(r-1)$-sphere induced from the Haar measure on $\mathrm{O}_r(\mathbb{R}) = \mathrm{GL}_r(\mathbb{Z}_\eta)$, with $\mathrm{d}^0\bar{x}(\mathrm{S}^{r-1}) = \zeta_\eta(1, \ldots, 1) = 1/\zeta_\eta(r)$.

The measure $\tau_\eta^{\bar\alpha}$ can also be described as the image of the product of r gamma measures, $\tau_\eta^{\bar\alpha} = \mathrm{pr}_*\left(\tau_{\mathbb{Z}_\eta}^{\alpha_1} \otimes \cdots \otimes \tau_{\mathbb{Z}_\eta}^{\alpha_r}\right)$.

The q-operators $\mathrm{D}_N, \mathrm{D}_N^+$ and ρ_{r-1} converge in the real limit. We have

$$\mathrm{D}_N\colon \mathscr{H}_{\eta(N)}^{\bar\alpha} \longrightarrow \mathscr{H}_{\eta(N-1)}^{\bar\alpha(1++,2++)}, \quad \mathrm{D}_N\varphi(\overline{m}) = \varphi(\overline{m}(2+)) - \varphi(\overline{m}(1+)) \tag{9.2.45}$$

and its adjoint

$$\mathrm{D}_N^+ = \frac{\alpha_1 \cdot \alpha_2 \cdot N \cdot (N + |\bar\alpha|/2)}{|\bar\alpha| \cdot (|\bar\alpha| + 2)} \mathrm{D}_N^*$$

is given by

$$\mathrm{D}_N^+\varphi(\overline{m}) = \left(m_1 + \frac{\alpha_1}{2}\right) m_2 \varphi(\overline{m}(2-)) - m_1\left(m_2 + \frac{\alpha_2}{2}\right)\varphi(\overline{m}(1-)).$$

They satisfy the Heisenberg relations up the ladder

$$\mathrm{D}_N\mathrm{D}_N^+ - \mathrm{D}_{N-1}^+\mathrm{D}_{N-1} = \left(\frac{\alpha_1 + \alpha_2}{2}\right) \cdot \mathrm{id}_{\mathscr{H}_{\eta(N-1)}^{\bar\alpha(1++,2++)}}. \tag{9.2.46}$$

We have the injection $\rho_{r-1}\colon \mathscr{H}_{\eta(N)}^{(\alpha_1+\alpha_2,\alpha_3,\ldots,\alpha_r)} \hookrightarrow \mathscr{H}_{\eta(N)}^{\bar\alpha}$,

$$\rho_{r-1}\varphi(\overline{m}) = \varphi(m_1 + m_2, m_3, \ldots, m_r) \tag{9.2.47}$$

satisfying $\qquad \mathrm{image}\rho_{r-1} = \mathrm{kernel}\ \mathrm{D}_N$.

The resulting basis for $\mathscr{H}_{\eta(N)}^{\bar\alpha}$ is

$$\varphi_{\eta(N),\bar v}^{\bar\alpha} = \frac{1}{v_1!\ldots v_{r-1}!}(\mathrm{D}^+)^{v_1}\rho_{r-1}\ldots\rho_2(\mathrm{D}^+)^{v_{r-1}}\mathbb{1} \tag{9.2.48}$$
$$\bar v = (v_1, \ldots, v_{r-1}), \quad |\bar v| \le N.$$

For $r = 2$ these are the real Hahn polynomials (8.2.11), and for $r > 2$ they are combinations of such:

$$\varphi_{\eta(N),\bar v}^{\bar\alpha}(\overline{m}) = \prod_{1\le j<r} \varphi_{\eta,v_j}^{(\mathrm{A}_j+2V_{j-1},\alpha_{j+1})}(\mathrm{M}_j - V_{j-1}, m_{j+1}) \tag{9.2.49}$$
$$\mathrm{M}_j = m_1 + \cdots + m_j, \quad V_j = v_1 + \cdots + v_j, \quad \mathrm{A}_j = \alpha_1 + \cdots + \alpha_j.$$

The operators $\mathrm{D}_N, \mathrm{D}_N^+$ and ρ_{r-1} are compatible with the Martin kernel and our basis therefore diagonalizes it; for $N \le N'$:

$$\mathrm{K}_{N,N'}^{\bar\alpha}\varphi_{\eta(N'),\bar v}^{\bar\alpha} = \begin{cases} \dfrac{\dbinom{\frac12|\bar\alpha| + N' + |\bar v| - 1}{|\bar v|}}{\dbinom{\frac12|\bar\alpha| + N + |\bar v| - 1}{|\bar v|}}\varphi_{\eta(N),\bar v}^{\bar\alpha} & \text{for } |\bar v| \le N, \\ 0 & \text{for } |\bar v| > N. \end{cases} \tag{9.2.50}$$

We can now describe the $N \to \infty$ limit theory for the boundary space $\mathscr{H}_\eta^{\bar{\alpha}} = L_2(\{\pm 1\}^r \backslash S^{r-1}, \tau_\eta^{\bar{\alpha}})$, and it will be useful to introduce spherical coordinates on S^{r-1}:

$$\begin{aligned} x_1 &= \sin\theta_1 \dots \sin\theta_{r-2} \cdot \sin\theta \\ x_2 &= \sin\theta_1 \dots \sin\theta_{r-2} \cdot \cos\theta \\ x_3 &= \sin\theta_1 \dots \cos\theta_{r-2} \\ &\vdots \\ x_r &= \cos\theta_1, \quad 0 \le \theta \le 2\pi, \quad 0 \le \theta_j \le \pi. \end{aligned} \tag{9.2.51}$$

The difference operators D_N converge to the operator

$$\begin{gathered} D\colon \mathscr{H}_\eta^{\bar{\alpha}} \longrightarrow \mathscr{H}_\eta^{\bar{\alpha}(1++,2++)}, \\ D\varphi(\bar{x}) = \frac{-2}{x_1^2 + x_2^2} \frac{\partial}{\partial \cos 2\theta} \varphi(\bar{x}) = \frac{-1}{2(x_1^2+x_2^2)} \frac{1}{\cos\theta} \frac{\partial}{\partial\cos\theta} \varphi(\bar{x}). \end{gathered} \tag{9.2.52}$$

Its adjoint

$$D^+ = \frac{\alpha_1 \cdot \alpha_2}{|\bar{\alpha}| \cdot (|\bar{\alpha}| + 2)} D^* = \lim_{N\to\infty} D_N^+$$

is given by

$$\begin{aligned} D^+ = & \frac{x_1^2 + x_2^2}{2} \cdot (1 - \cos 2\theta)^{1-\alpha_1/2} (1 + \cos 2\theta)^{1-\alpha_2/2} \\ & \frac{\partial}{\partial \cos 2\theta} (1 - \cos 2\theta)^{\alpha_1/2} (1 + \cos 2\theta)^{\alpha_2/2}. \end{aligned} \tag{9.2.53}$$

They satisfy the Heisenberg relation up the ladder

$$DD^+ - D^+D = \left(\frac{\alpha_1 + \alpha_2}{2}\right) \cdot \mathrm{id}_{\mathscr{H}_\eta^{\bar{\alpha}(1++,2++)}}. \tag{9.2.54}$$

We have the injections

$$\begin{gathered} \rho_{r-1}\colon \mathscr{H}_\eta^{(\alpha_1+\alpha_2,\alpha_3,\dots,\alpha_r)} \hookrightarrow \mathscr{H}_\eta^{\bar{\alpha}} \\ \rho_{r-1}\varphi(\bar{x}) = \varphi\left(\sqrt{x_1^2 + x_2^2}, x_3, \dots, x_r\right). \end{gathered} \tag{9.2.55}$$

The adjoint projections are given by averaging

$$\rho_{r-1}^* \varphi(x_1, \dots, x_{r-1}) = \tau_\eta^{(\alpha_1,\alpha_2)}(\varphi(-,-,x_2,\dots,x_{r-1})), \tag{9.2.56}$$

and image $\rho_{r-1} = $ kernel D. We obtain the orthogonal basis for $\mathscr{H}_\eta^{\bar{\alpha}}$ given by

$$\varphi_{\eta,\bar{v}}^{\bar{\alpha}} = \frac{1}{v_1! \dots v_{r-1}!} (D^+)^{v_1} \rho_{r-1} \dots \rho_2 (D^+)^{v_{r-1}} 1. \tag{9.2.57}$$

$$\bar{v} = (v_1, \dots, v_{r-1}) \in \mathbb{N}^{r-1}.$$

For $r = 2$ these are the Jacobi polynomials, and for $r > 2$ combinations of these given explicitly by

$$\varphi^{\bar{\alpha}}_{\eta,\bar{v}}(\bar{x}) = \prod_{1 \le j < r} \mathrm{S}^{2v_{j-1}}_{r-j} \cdot (-1)^{v_j} \cdot \mathrm{P}^{(\frac{1}{2}\mathrm{A}_j + 2V_{j-1} - 1, \frac{1}{2}\alpha_{j+1} - 1)}_{v_j} (\mathrm{S}^2_{r-j-1} \cdot \cos 2\theta_{r-j}) \qquad (9.2.58)$$

where $\mathrm{S}_{r-j} = \sin\theta_1 \ldots \sin\theta_{r-j}$, $V_j = v_1 + \cdots + v_j$, $\mathrm{A}_j = \alpha_1 + \cdots + \alpha_j$, and $\mathrm{P}^{(\alpha,\beta)}_v(z)$ are the classical Jacobi polynomials.

We have the diagram

$$\begin{array}{ccc} \mathscr{H}^{\bar{\alpha}}_{\eta} & \underset{\mathrm{D}^+}{\overset{\mathrm{D}}{\rightleftarrows}} & \mathscr{H}^{\bar{\alpha}(1++,2++)}_{\eta} \\ \mathrm{K}^{\bar{\alpha}*}_N \uparrow \downarrow \mathrm{K}^{\bar{\alpha}}_N & & \mathrm{K}^{\bar{\alpha}(+)*}_{(N-1)} \uparrow \downarrow \mathrm{K}^{\bar{\alpha}(+)}_{(N-1)} \\ \mathscr{H}^{\bar{\alpha}}_{\eta(N)} & \underset{\mathrm{D}^+_N}{\overset{\mathrm{D}_N}{\rightleftarrows}} & \mathscr{H}^{\bar{\alpha}(1++,2++)}_{\eta(N-1)} \end{array} \qquad (9.2.59)$$

and the compatibilities (we write $\bar{\alpha}(+)$ for $\bar{\alpha}(1++,2++)$):

$$\begin{aligned} \mathrm{K}^{\bar{\alpha}(+)}_{N-1}\mathrm{D} &= \left(\frac{|\bar{\alpha}|}{2} + N\right)\mathrm{D}_N\mathrm{K}^{\bar{\alpha}}_N , \\ N \cdot \mathrm{D}^+\mathrm{K}^{\bar{\alpha}(+)*}_{N-1} &= \mathrm{K}^{\bar{\alpha}*}_N\mathrm{D}^+_N , \end{aligned} \qquad (9.2.60)$$

$$\begin{aligned} \mathrm{D}\mathrm{K}^{\bar{\alpha}*}_N &= N \cdot \mathrm{K}^{\bar{\alpha}(+)*}_{N-1}\mathrm{D}_N , \\ \left(\frac{|\bar{\alpha}|}{2} + N\right)\mathrm{K}^{\bar{\alpha}}_N\mathrm{D}^+ &= \mathrm{D}^+_N\mathrm{K}^{\bar{\alpha}(+)}_{N-1} , \end{aligned} \qquad (9.2.61)$$

and also

$$\mathrm{K}^{\bar{\alpha}*}_N \rho_{r-1} = \rho_{r-1}\mathrm{K}^{(\alpha_1+\alpha_2,\alpha_3,\ldots,\alpha_r)*}_N . \qquad (9.2.62)$$

We obtain the compatibility of our basis

$$\mathrm{K}^{\bar{\alpha}}_N\varphi^{\bar{\alpha}}_{\eta,\bar{v}} = \frac{\Gamma(|\bar{\alpha}|/2 + N + |\bar{v}|)}{\Gamma(|\bar{\alpha}|/2 + N)}\varphi^{\bar{\alpha}}_{\eta(N),\bar{v}} \quad \text{for } |\bar{v}| \le N, \quad 0 \text{ for } |\bar{v}| > N , \qquad (9.2.63)$$

as well as

$$\mathrm{K}^{\bar{\alpha}*}_N \varphi^{\bar{\alpha}}_{\eta(N),\bar{v}} = \frac{N!}{(N - |\bar{v}|)!}\varphi^{\bar{\alpha}}_{\eta,\bar{v}} . \qquad (9.2.64)$$

We have thus diagonalized the Martin kernel, and it is not an embedding or a projection, but it replaces the tree structure, and gives the real analogue of reduction mod p^N.

We have the limit relations, generalizing (8.2.34),

$$\begin{array}{ccccccc} & \underline{p\neq\eta} & & \underline{q} & & \underline{p=\eta} & \\ & \varphi^{\bar\alpha}_{p(N),\bar v} & \xleftarrow{\;(p)\;} & \varphi^{\bar\alpha}_{q(N),\bar v} & \xrightarrow{\;(\eta)\;} & \varphi^{\bar\alpha}_{\eta(N),\bar v} & \\ N\to\infty & \downarrow & & \downarrow & & \downarrow & \qquad (9.2.65) \\ & \varphi^{\bar\alpha}_{p,\bar v} & \xleftarrow[\;(p)\;]{} & \varphi^{\bar\alpha}_{q,\bar v} & \xrightarrow[\;(\eta)\;]{} & \varphi^{\bar\alpha}_{\eta,\bar v} & \end{array}$$

For $p\neq\eta$ we have coverings of chains

$$\begin{array}{lcl} \widetilde{\widetilde{\widetilde{\mathrm{X}}}}_p = \coprod\limits_{N\geq 0} V_r^*(\mathbb{Z}/p^N) \simeq (\mathbb{Q}_p/\mathbb{Z}_p)^r & \Longrightarrow & \partial\widetilde{\widetilde{\widetilde{\mathrm{X}}}}_p = V_r^*(\mathbb{Z}_p) = \{x\in\mathbb{Q}_p^{\oplus r} \mid |x|_p = 1\} \\ \downarrow & & \downarrow \\ \widetilde{\widetilde{\mathrm{X}}}_p = \coprod\limits_{N\geq 0} \mathbb{P}^{r-1}(\mathbb{Z}/p^N) & \Longrightarrow & \partial\widetilde{\widetilde{\mathrm{X}}}_p = \mathbb{P}^{r-1}(\mathbb{Z}_p) = \mathbb{Z}_p^*\backslash V_r^*(\mathbb{Z}_p) \\ \downarrow & & \downarrow \\ \widetilde{\mathrm{X}}_p = \coprod\limits_{N\geq 0} ((\mathbb{Z}/p^N)^*)^{r-1}\backslash \mathbb{P}^{r-1}(\mathbb{Z}/p^N) & \Longrightarrow & \partial\widetilde{\mathrm{X}}_p = (\mathbb{Z}_p^*)^{r-1}\backslash\mathbb{P}^{r-1}(\mathbb{Z}_p) \\ \downarrow & & \downarrow \\ \mathrm{X}_p = \coprod\limits_{N\geq 0} \mathrm{B}_{(1,\ldots,1)}(\mathbb{Z}/p^N)\backslash\mathbb{P}^{r-1}(\mathbb{Z}/p^N) & \Longrightarrow & \partial\mathrm{X}_p = \mathrm{B}_{(1,\ldots,1)}(\mathbb{Z}_p)\backslash\mathbb{P}^{r-1}(\mathbb{Z}_p) \\ \downarrow & & \downarrow \\ \mathbb{N} = \coprod\limits_{N\geq 0}\{*\} & \Longrightarrow & \{*\}\,. \end{array} \qquad (9.2.66)$$

X_p is the non-symmetric p-adic beta chain described above; $\widetilde{\mathrm{X}}_p$ is the symmetric version obtained as the p-adic limit of the symmetric q-beta chain; and $\widetilde{\widetilde{\mathrm{X}}}_p$ and $\widetilde{\widetilde{\widetilde{\mathrm{X}}}}_p$ are the random walks over $\widetilde{\mathrm{X}}_p$ (the underlining trees of $\widetilde{\widetilde{\mathrm{X}}}_p$ and $\widetilde{\widetilde{\widetilde{\mathrm{X}}}}_p$ depend on p, but the transition probabilities 'come' from $\widetilde{\mathrm{X}}_p$).

For the real prime $p=\eta$, there is no difference between X_η and $\widetilde{\mathrm{X}}_\eta$: the real chain is the limit of both the symmetric and non-symmetric q-beta chains (we can write $\mathrm{B}_{(1,\ldots,1)} = \mathbb{G}_m^r \ltimes N$, with N the unipotent radical, $\mathrm{B}_{(1,\ldots,1)}(\mathbb{Z}_\eta) = \mathbb{G}_m^r(\mathbb{Z}_\eta) = \{\pm1\}^r$, and $N(\mathbb{Z}_\eta)$ disappear). We can similarly add the real units $\{\pm1\}^{r-1}$ or $\{\pm1\}^r$ into the game by considering the real chains $\widetilde{\widetilde{\mathrm{X}}}_\eta$ or $\widetilde{\widetilde{\widetilde{\mathrm{X}}}}_\eta$ as the random walks over X_η, where

$$\begin{array}{lcl} \widetilde{\widetilde{\widetilde{\mathrm{X}}}}_\eta = \mathbb{Z}^r & \Longrightarrow & \partial\widetilde{\widetilde{\widetilde{\mathrm{X}}}}_\eta = \mathrm{S}^{r-1} = \text{`}V_r^*(\mathbb{Z}_\eta)\text{'} = \{x\in\mathbb{R}^{\oplus r} \mid |x|_\eta = 1\} \\ \downarrow & & \downarrow \\ \widetilde{\widetilde{\mathrm{X}}}_\eta = \{\pm1\}\backslash\mathbb{Z}^r & \Longrightarrow & \partial\widetilde{\widetilde{\mathrm{X}}}_\eta = \mathbb{P}^{r-1}(\mathbb{R}) \\ \downarrow & & \downarrow \\ \mathrm{X}_\eta = \{\pm1\}^r\backslash\mathbb{Z}^r = \mathbb{N}^r & \Longrightarrow & \partial\mathrm{X}_\eta = \{\pm1\}^{r-1}\backslash\mathbb{P}^{r-1}(\mathbb{R})\,. \end{array} \qquad (9.2.67)$$

One moral is that the '$\mathbb{Z}/\eta^N$' points of $\mathbb{P}^{r-1}$ are given by

$$\text{'}\mathbb{P}^{r-1}(\mathbb{Z}/\eta^N)\text{'} = \{\pm 1\}\backslash\{\overline{m} \in \mathbb{Z}^r, |\overline{m}| = |m_1| + \cdots + |m_r| = N\}. \tag{9.2.68}$$

For all $p \geq \eta$, when we set all the parameters α_i equal to 1 the harmonic measure $\tau_p^{(\bar{1})}$ is the unique $\mathrm{GL}_r(\mathbb{Z}_p)$-invariant probability measure on $\mathbb{P}^{r-1}(\mathbb{Q}_p)$, or up on $V_r^*(\mathbb{Z}_p)$ (with $\mathrm{GL}_r(\mathbb{Z}_\eta) = \mathrm{O}_r$ and $V_r^*(\mathbb{Z}_\eta) = \mathrm{S}^{r-1}$ the $(r-1)$-dimensional sphere). The Hilbert space $\mathrm{L}_2(V_r^*(\mathbb{Z}_p), \tau_p^{(\bar{1})})$ is naturally a unitary representation of $\mathrm{GL}_r(\mathbb{Z}_p)$; it is the induction of the trivial representation of $\mathrm{H}_p = \mathrm{GL}_{r-1}(\mathbb{Z}_p) \ltimes \mathbb{Z}_p^{r-1}$ (respectively $\mathrm{H}_\eta = \mathrm{O}_{r-1}$ for $p = \eta$). It decomposes as the direct sum of $\mathrm{GL}_r(\mathbb{Z}_p)$-representations containing an H_p-invariant vector. For $p = \eta$ this decomposition is given by

$$\mathrm{L}_2\left(\mathrm{S}^{r-1}, \tau_\eta^{(\bar{1})}\right) = \mathrm{Ind}_{\mathrm{O}_{r-1}}^{\mathrm{O}_r}(\mathbb{1}) = \bigoplus_{m\geq 0} V_m^r \tag{9.2.69}$$

with

$$V_m^r = \ker\left\{\Delta = \sum_{1\leq j\leq r} \partial^2/\partial x_j^2 : \mathbb{C}[x_1, \ldots, x_r]_{\deg=m} \longrightarrow \mathbb{C}[x_1, \ldots, x_r]_{\deg=m-2}\right\}$$

the harmonic polynomials of degree m. Lifting our basis from $\partial \mathrm{X}_\eta = \{\pm 1\}^r \backslash \mathrm{S}^{r-1}$ to S^{r-1} we obtain an orthogonal basis for $\mathrm{L}_2(\mathrm{S}^{r-1})$ consisting of eigenfunctions under the diagonal matrices $\{\pm 1\}^r$; these are the spherical harmonics and they give the refinement of (9.2.69) into the simultaneous decomposition of $\mathrm{L}_2(\mathrm{S}^{r-1})$ under the action of $\mathrm{O}_r \supseteq \mathrm{O}_{r-1} \supseteq \cdots \supseteq \mathrm{O}_1 = \{\pm 1\}$.

10
The Fourier transform

10.1 THE TATE DISTRIBUTION AND THE BETA FUNCTION AT IMAGINARY ARGUMENT

Consider the *q-Tate distribution* on $g^{\mathbb{Z}}$:

$$\tau^{\beta}_{(q)}(g^i) = \frac{\zeta_{(q)}(1)}{\zeta_{(q)}(\beta)} q^{i \cdot \beta} . \tag{10.1.1}$$

It is normalized by $\tau^{\beta}_{(q)}(\phi_{\mathbb{Z}_q}) = 1$, where the *q-vacuum* $\phi_{\mathbb{Z}_q}$ is the function on $g^{\mathbb{Z}}$ given by

$$\phi_{\mathbb{Z}_q}(g^i) = \frac{1}{\zeta_{(q)}(1+i)} = \begin{cases} \prod_{k>i}(1-q^k) & i \geq 0 , \\ 0 & i < 0 . \end{cases} \tag{10.1.2}$$

The normalization $\tau^{\beta}_{(q)}(\phi_{\mathbb{Z}_q}) = 1$ is just the q-gamma sum (6.3.5).

Note that when $q \to 0$, the $\phi_{\mathbb{Z}_q}$ converge to the characteristic function of $g^{\mathbb{N}}$, and so in the p-adic limit

$$\lim_{Ⓟ} \phi_{\mathbb{Z}_q} = \phi_{\mathbb{Z}_p} = \text{characteristic function of } \mathbb{Z}_p . \tag{10.1.3}$$

In the real limit Ⓡ we let $q \to 1$ and $i \to \infty$, in such a way that $q^i/(1-q) \to \pi x^2$, cf. (8.2.36), (8.2.37). To see the limit Ⓡ of $\phi_{\mathbb{Z}_q}$ we expand it using the q-Möbius sum (6.3.9):

$$\begin{aligned} \phi_{\mathbb{Z}_q}(g^i) &= \sum_{n \geq 0} \frac{\zeta_{(q)}(1)}{\zeta_{(q)}(1+n)} (-1)^n q^{n(n-1)/2} q^{n(1+i)} \\ &= \sum_{n \geq 0} \frac{(1-q)^n}{(1-q)\dots(1-q^n)} (-1)^n q^{n(n+1)/2} \left(\frac{q^i}{1-q}\right)^n \end{aligned} \tag{10.1.4}$$

and so

$$\lim_{Ⓡ} \phi_{\mathbb{Z}_q} = \sum_{n \geq 0} \frac{1}{n!} \cdot (-1)^n \cdot (\pi x^2)^n = \mathrm{e}^{-\pi x^2} = \phi_{\mathbb{Z}_\eta} . \tag{10.1.5}$$

For $\beta > 0$, $\tau^{\beta}_{(q)}$ is a positive measure, and we have the associated Hilbert space

$$\mathscr{H}^{\beta}_{q} = \ell_2\left(g^{\mathbb{Z}}, \tau^{\beta}_{(q)}\right) . \tag{10.1.6}$$

The unitary representation of $g^{\mathbb{Z}}$ on $\mathscr{H}_q^{\beta}$, is given by

$$\pi_q^{\beta}(g^j)\varphi(g^i) = q^{-j\beta/2}\varphi(g^{i-j})\,. \tag{10.1.7}$$

We view $\phi_{\mathbb{Z}_q}$ as an element of $\mathscr{H}_q^{\beta}$; it is a cyclic vector, and the associated positive-definite function on $g^{\mathbb{Z}}$ is given by (cf. (3.2.7)′)

$$\tilde{\rho}_{(q)}^{\beta}(g^i) = \left(\phi_{\mathbb{Z}_q}, \pi_q^{\beta}(g^i)\phi_{\mathbb{Z}_q}\right)_{\mathscr{H}_q^{\beta}} = \frac{\zeta_{(q)}(1)}{\zeta_{(q)}(\beta)} \sum_{k\geq 0, i} \frac{q^{(k-i/2)\beta}}{\zeta_{(q)}(1+k)\zeta_{(q)}(1+k-i)}\,. \tag{10.1.8}$$

We identify the dual group of $g^{\mathbb{Z}}$ with $\mathrm{i}\mathbb{R}/(2\pi\mathrm{i}/\log q)\mathbb{Z}$, and for a function f on $g^{\mathbb{Z}}$, we write its q-Mellin transform as

$$\hat{f}(s) = \int_{g^{\mathbb{Z}}} f(a)|a|_{(q)}^{s}\mathrm{d}^*a = \sum_{j\in\mathbb{Z}} f(g^j)q^{js}, \quad \text{where } |g^j|_{(q)}^{s} = q^{js}\,. \tag{10.1.9}$$

The inverse Mellin transform is given by

$$f(g^j) = \int_{\mathrm{i}\mathbb{R}/(2\pi\mathrm{i}/\log q)\mathbb{Z}} \hat{f}(s)q^{-js}\mathrm{d}_q s\,, \tag{10.1.10}$$

with the normalized Haar measure

$$\mathrm{d}_q s = \mathrm{d}t \cdot \left|\frac{\log q}{2\pi}\right|, \quad \text{where } s = \mathrm{i}t\,.$$

The Mellin transform of the function $\tilde{\rho}_{(q)}^{\beta}$ is given by (cf. (3.2.8)′)

$$\begin{aligned}\sum_{j\in\mathbb{Z}} \tilde{\rho}_{(q)}^{\beta}(g^j)\cdot q^{js} &= \frac{\zeta_{(q)}(1)}{\zeta_{(q)}(\beta)} \sum_{k,j\geq 0} \frac{q^{k(\beta/2+s)+j(\beta/2-s)}}{\zeta_{(q)}(1+k)\zeta_{(q)}(1+j)} \\ &= \frac{\zeta_{(q)}(\beta/2+s)\zeta_{(q)}(\beta/2-s)}{\zeta_{(q)}(\beta)\zeta_{(q)}(1)} = \zeta_{(q)}(\beta/2+s, \beta/2-s), \text{ by (6.3.5)}\,.\end{aligned} \tag{10.1.11}$$

We obtain isomorphic descriptions of $\mathscr{H}_q^{\beta}$:

$$\begin{array}{ccc} \text{'multiplicative'} & & \text{'additive'} \\ \mathscr{H}_q^{\beta,*} & \xrightarrow{\ \sim\ } & \mathscr{H}_q^{\beta} \\ & \searrow{\scriptstyle\wr} \quad {\scriptstyle\wr}\swarrow & \\ & \text{'analytic': } \widehat{\mathscr{H}}_q^{\beta} = \mathrm{L}_2\left(\mathrm{i}\mathbb{R}\Big/\dfrac{2\pi\mathrm{i}}{\log q}\mathbb{Z}, \zeta_{(q)}(\beta/2+s, \beta/2-s)\mathrm{d}_q s\right) & \end{array} \tag{10.1.12}$$

The correspondence between elements of these spaces is given by

$$\begin{array}{ccc} \mathscr{H}_q^{\beta,*} \ni f & \longmapsto & \pi_q^\beta(f)\phi_{\mathbb{Z}_q} = \varphi \in \mathscr{H}_q^\beta \\ & \searrow \quad \swarrow & \\ & \hat{f}(s) = \tau_{(q)}^{\beta/2+s}(\varphi) \in \widehat{\mathscr{H}}_q^\beta & \end{array} \tag{10.1.13}$$

Here $\mathscr{H}_q^{\beta,*}$ is the Hilbert space of functions f on $g^{\mathbb{Z}}$ satisfying

$$(f,f)_{\mathscr{H}_q^{\beta,*}} = \left(\pi_q^\beta(f)\phi_{\mathbb{Z}_q}, \pi_q^\beta(f)\phi_{\mathbb{Z}_q}\right)_{\mathscr{H}_q^\beta} = \int |\hat{f}(s)|^2 \zeta_{(q)}(\beta/2+s, \beta/2-s)\mathrm{d}_q s < \infty .$$

For a polynomial $\hat{f}(s) = \mathrm{F}(q^s, q^{-s}) \in \mathbb{C}[q^s, q^{-s}]$, we can invert the right arrow in (10.1.13) via

$$\varphi = \mathrm{F}\left(\pi_q^\beta(g^1), \pi_q^\beta(g^{-1})\right)\phi_{\mathbb{Z}_q} .$$

In both the p-adic and real limits, the positive-definite function $\tilde{\rho}_{(q)}^\beta$ converges to its counterpart $\tilde{\rho}_p^\beta$ and the dual measure converges to its counterpart $\zeta_p(\beta/2+s,\ \beta/2-s)\mathrm{d}_p s$, where we identify the dual group of $\mathbb{Q}_p^*/\mathbb{Z}_p^*$ by

$$(\mathbb{Q}_p^*/\mathbb{Z}_p^*)^{\vee} = \begin{cases} \mathrm{i}\mathbb{R}\Big/\dfrac{2\pi\mathrm{i}}{\log p}\mathbb{Z} & p \neq \eta \\ \mathrm{i}\mathbb{R} & p = \eta, \end{cases} \qquad \mathrm{d}_p s = \begin{cases} \dfrac{\log p}{2\pi\mathrm{i}}\mathrm{d}s & p \neq \eta , \\ \dfrac{1}{4\pi\mathrm{i}}\mathrm{d}s & p = \eta . \end{cases} \tag{10.1.14}$$

The isomorphic description of $\mathscr{H}_p^\beta = \mathrm{L}_2(\mathbb{Q}_p^*/\mathbb{Z}_p^*, \tau_p^\beta), \tau_p^\beta(\mathrm{d}x) = |x|_p^\beta \mathrm{d}^*x/\zeta_p(\beta)$, persists in both the p-adic and real limits, the correspondence given by the $p \geq \eta$ counterpart of (10.1.13). The only difference is that in the real case, the multiplicative space $\mathscr{H}_\eta^{\beta,*}$ does not consist merely of functions on $\mathbb{Q}_\eta^*/\mathbb{Z}_\eta^* = \mathbb{R}^+$, but also contains certain distributions.

We can invert the right arrow in (10.1.13) in the real case, for a polynomial $\hat{f}(s) \in \mathbb{C}[s]$, by putting

$$\varphi = \hat{f}\left(x\frac{\partial}{\partial x} + \frac{\beta}{2}\right)\phi_{\mathbb{Z}_\eta} . \tag{10.1.15}$$

Here $x\partial/\partial x + \beta/2$ is the (skew-adjoint) infinitesimal generator of the unitary representation π_η^β of $\mathbb{R}^+$:

$$\frac{\partial}{\partial w}\bigg|_{w=0} \pi_\eta^\beta(\mathrm{e}^{-w})\varphi(x) = \frac{\partial}{\partial w}\bigg|_{w=0} \mathrm{e}^{w\beta/2}\varphi(\mathrm{e}^w x) = \left(\frac{\beta}{2} + x\frac{\partial}{\partial x}\right)\varphi(x) .$$

For all $p \geq \eta$, at $\beta = 1$ $\tau_p^1(\mathrm{d}x) = \mathrm{d}x$ is the additive Haar measure and $\mathscr{H}_p^1 = \mathrm{L}_2(\mathbb{Q}_p, \mathrm{d}x)^{\mathbb{Z}_p^*}$.

We can use polar coordinates on r-dimensional space and identify

$$\mathbb{Q}_p^r\backslash\{0\} \simeq \mathrm{p}^{\mathbb{Z}} \times V_r^*(\mathbb{Z}_p), \quad \mathbb{Q}_\eta^r\backslash\{0\} = \mathbb{R}^+ \times \mathrm{S}^{r-1} \tag{10.1.16}$$

The r-dimensional additive Haar measure $\mathrm{d}\mathbf{x}$ becomes identified with $\tau_p^r \otimes \tilde{\tau}_p^{(1,\dots,1)}$, where $\tau_p^r(\mathrm{d}a) = |a|_p^r \, \mathrm{d}^* a/\zeta_p(r)$ is the Tate measure at $\beta = r$, and $\tilde{\tau}_p^{(1,\dots,1)}$ is the $\mathrm{GL}_r(\mathbb{Z}_p)$-invariant measure on $V_r^*(\mathbb{Z}_p)$. Using the spaces $\mathscr{H}_q^{\bar{\alpha}} = \ell_2(\partial \mathrm{X}_q, \tau_q^{\bar{\alpha}})$ of (9.2.13) with $\bar{\alpha} = (\alpha_1, \dots, \alpha_r)$, and the spaces $\mathscr{H}_q^{|\bar{\alpha}|} = \ell_2(g^{\mathbb{Z}}, \tau_q^{|\bar{\alpha}|})$ of (10.1.6) with $|\bar{\alpha}| = \alpha_1 + \cdots + \alpha_r$, we can form their tensor product $\mathscr{H}_q^{|\bar{\alpha}|} \otimes \mathscr{H}_q^{\bar{\alpha}}$; in the real limit at $\alpha_j = 1$ we obtain the space $\mathrm{L}_2(\mathbb{R}^r, \mathrm{d}\mathbf{x})^{\{\pm 1\}^r}$, and in the p-adic limit at $\alpha_j = 1$ we obtain $\mathrm{L}_2(\mathbb{Q}_p^r, \mathrm{d}\mathbf{x})^{\mathrm{B}_{(1\dots1)}(\mathbb{Z}_p)}$.

10.2 THE FOURIER–BESSEL TRANSFORM

In the realizations $\mathscr{H}_q^{\beta,*}$ (and $\widehat{\mathscr{H}_q^{\beta}}$) we have the operator $\mathcal{F}_q^{\beta,*}$ (and the corresponding operator $\widehat{\mathcal{F}_q^{\beta}}$) given by

$$\mathcal{F}_q^{\beta,*} f(a) = f(a^{-1}) \quad (\text{and } \widehat{\mathcal{F}}_q^{\beta} \hat{f}(s) = \hat{f}(-s)). \tag{10.2.1}$$

These correspond to the *Fourier–Bessel transform* $\mathcal{F}_q^{\beta}$ on $\mathscr{H}_q^{\beta}$. From the way we have defined it, it clearly satisfies:

$$\begin{aligned}
&\text{(i)} \quad \mathcal{F}_q^{\beta} \text{ is unitary self-adjoint:} \\
&\qquad \mathcal{F}_q^{\beta} = (\mathcal{F}_q^{\beta})^* = (\mathcal{F}_q^{\beta})^{-1} \quad (\text{`Hankel's inversion'}) \\
&\text{(ii)} \quad \mathcal{F}_q^{\beta} \pi_q^{\beta}(a) = \pi_q^{\beta}(a^{-1}) \mathcal{F}_q^{\beta} \\
&\text{(iii)} \quad \mathcal{F}_q^{\beta} \phi_{\mathbb{Z}_q} = \phi_{\mathbb{Z}_q} \quad (\text{`Weber's First formula'}) \\
&\text{(iv)} \quad \tau_q^s(\mathcal{F}_q^{\beta} \varphi) = \tau_q^{\beta - s}(\varphi).
\end{aligned} \tag{10.2.2}$$

The opeator $\mathcal{F}_q^{\beta}$ is given by a kernel $\mathcal{F}_q^{\beta}(x, y)$ on $g^{\mathbb{Z}}$, but by (10.2.2)(ii) $\mathcal{F}_q^{\beta}(x, y)$ depends only on $x \cdot y$ and we can write

$$\mathcal{F}_q^{\beta} \varphi(y) = \int_{g^{\mathbb{Z}}} \varphi(x) \mathcal{F}_q^{\beta}(x \cdot y) \tau_{(q)}^{\beta}(\mathrm{d}x). \tag{10.2.3}$$

By (10.2.2)(iv) the kernel $\mathcal{F}_q^{\beta}(x)$ satisfies (and is characterized by)

$$\tau_{(q)}^s(\mathcal{F}_q^{\beta}) = \frac{\zeta_{(q)}(\beta)}{\zeta_{(q)}(\beta - s)}.$$

It is given explicitly by

$$\mathcal{F}_q^{\beta}(g^i) = {}_1\phi_1(\infty; \beta; 1 + i) \tag{10.2.4}$$

$$= \sum_{k \geq 0} (-1)^k q^{k(k-1)/2} \frac{\zeta_{(q)}(\beta)}{\zeta_{(q)}(1 + i + k)} \frac{\zeta_{(q)}(1)}{\zeta(q)(1 + k)} q^{\beta k} \tag{10.2.4$_p$}$$

$$= \sum_{k \geq 0} (-1)^k q^{k(k-1)/2} \frac{\zeta(q)(\beta)}{\zeta_{(q)}(\beta + k)} \frac{\zeta_{(q)}(1)}{\zeta_{(q)}(1 + k)} q^{(1+i)k}. \tag{10.2.4$_\eta$}$$

Indeed we only have to check that $(*)$ is satisfied:

$$\begin{aligned}\tau^s_{(q)}({}_1\phi_1(\infty;\beta;1+i)) &= \frac{\zeta_{(q)}(1)}{\zeta_{(q)}(s)}\sum_i q^{is}\sum_k(-1)^k q^{k(k-1)/2}\frac{\zeta_{(q)}(\beta)}{\zeta_{(q)}(1+i+k)}\frac{\zeta_{(q)}(1)}{\zeta_{(q)}(1+k)}q^{\beta k}\\ &= \frac{\zeta_{(q)}(\beta)}{\zeta_{(q)}(s)}\sum_i q^{is}\frac{\zeta_{(q)}(1)}{\zeta_{(q)}(1+i)}\sum_k(-1)^k q^{k(k-1)/2}\frac{\zeta_{(q)}(1)}{\zeta_{(q)}(1+k)}q^{k(\beta-s)}\\ &= \frac{\zeta_{(q)}(\beta)}{\zeta_{(q)}(\beta-s)},\ \text{by (6.3.5), (6.3.9)}.\end{aligned}$$

The equivalence of the above expansions $(10.2.4)_p$ and $(10.2.4)_\eta$- follows from the symmetry between α and β in the function (using the q-Möbius sum (6.3.9)):

$$\begin{aligned}\frac{\zeta_{(q)}(1)}{\zeta_{(q)}(\beta)}\mathcal{F}^\beta_q(g^{\alpha-1}) &= \frac{\zeta_{(q)}(1)}{\zeta_{(q)}(\beta)}{}_1\phi_1(\infty;\beta;\alpha)\\ &= \zeta_{(q)}(1)\sum_{k,j\geq 0} q^{kj}\cdot(-1)^k q^{k(k-1)/2}\frac{\zeta_{(q)}(1)}{\zeta_{(q)}(1+k)}q^{\beta k}\\ &\qquad (-1)^j q^{j(j-1)/2}\frac{\zeta_{(q)}(1)}{\zeta_{(q)}(1+j)}q^{\alpha j}.\end{aligned}\tag{10.2.5}$$

Note that

$$\lim_{\beta\to\infty}\mathcal{F}^\beta_q(g^i) = \frac{1}{\zeta_{(q)}(1+i)} = \phi_{\mathbb{Z}_q}(g^i).\tag{10.2.6}$$

The operator $\mathcal{F}^\beta_q$ converges in both the p-adic and real limits, and we obtain the operators $\mathcal{F}^\beta_p$ on $\mathscr{H}^\beta_p$, satisfying the analogous properties of (10.2.2), $p\geq\eta$. At $\beta=1$, $\mathcal{F}^\beta_p$ is the ordinary Fourier transform applied to $\mathbb{Z}^*_p$-invariant functions in $\mathrm{L}_2(\mathbb{Q}_p, dx)$, $p\geq\eta$. The kernel $\mathcal{F}^\beta_q(x)$ also converges in both the p-adic and real limits; to identify the limit kernel, we use $(10.2.4)_p$ and $(10.2.4)_\eta$ respectively to obtain:

$$\mathcal{F}^\beta_p(x) = \zeta_p(\beta)\cdot(\phi_{\mathbb{Z}_p}(x) - p^{-\beta}\phi_{\mathbb{Z}_p}(px)),\quad p\neq\eta,\tag*{$(10.2.7)_p$}$$

$$\mathcal{F}^\beta_\eta(x) = \sum_{k\geq 0}(-1)^k\frac{\zeta_\eta(\beta)}{\zeta_\eta(\beta+2k)}\cdot\frac{1}{k!}\cdot(\pi x^2)^k = \zeta_\eta(\beta)|x|^{1-\beta/2}\pi\mathrm{J}_{\beta/2-1}(2\pi x)\tag*{$(10.2.7)_\eta$}$$

where $\mathrm{J}_\beta(x)$ is the Bessel function.

We have the limits

$$\lim_{\beta\to\infty}\mathcal{F}^\beta_p(x) = \phi_{\mathbb{Z}_p}(x),\quad p\neq\eta,\tag*{$(10.2.8)_p$}$$

$$\lim_{\beta\to\infty}\mathcal{F}^\beta_\eta\left(\left(\frac{\beta}{2\pi}\right)^{1/2}x\right) = \phi_{\mathbb{Z}_\eta}(x).\tag*{$(10.2.8)_\eta$}$$

For the reals, using polar coordinates (10.1.16) and the decomposition (9.2.69) of $\mathrm{L}_2(\mathrm{S}^{r-1})$, we obtain a decomposition

$$\mathrm{L}_2(\mathbb{R}^r, \mathbf{dx}) \simeq \bigoplus_{m\geq 0}\mathscr{H}^r_\eta\otimes V^r_m.\tag{10.2.9}$$

The r-dimensional Fourier transform $\mathcal{F}_{\mathbb{R}^r}$ similarly decomposes as a direct sum of Fourier–Bessel transforms:

$$\mathcal{F}_{\mathbb{R}^r} \simeq \bigoplus_{m\geq 0} i^m | \,|_\eta^m \mathcal{F}_\eta^{r+2m} | \,|_\eta^{-m} \otimes \mathrm{id}_{V_m^r}\,; \tag{10.2.10}$$

this follows from the Funk–Hecke theorem (Erdélyi *et al.* 1953, p. 247).

For $p \geq \eta$, the function $\mathcal{F}_p^{1+\beta}(x)$ is the (ordinary) Fourier transform of a compactly supported probability measure $\mathcal{F}_p^{1+\beta} = \mathcal{F}_p^1\, \tilde{g}_p^\beta$. Here $\tilde{g}_p^\beta(\mathrm{d}x) = \frac{\zeta_p(1+\beta)}{\zeta_p(\beta)} \cdot g_p^\beta(x)\mathrm{d}x$, with

$$g_p^\beta(x) = \begin{cases} 0 & |x|_p > 1 \\ (1-p^{-\beta})^{-1} & |x|_p = 1, p \neq \eta; \\ 1 & |x|_p < 1 \end{cases} \quad g_\eta^\beta(x) = \begin{cases} 0 & |x|_\eta \geq 1\,, \\ (1-x^2)^{\frac{\beta}{2}-1} & |x|_\eta \leq 1\,. \end{cases} \tag{10.2.11}$$

The functions g_p^β satisfy

$$g_p^\beta(x) \underset{\beta\to\infty}{\longrightarrow} \phi_{\mathbb{Z}_p}(x), \quad p \neq \eta; \quad g_\eta^\beta\left(\left(\frac{\beta}{2\pi}\right)^{-\frac{1}{2}} x\right) \underset{\beta\to\infty}{\longrightarrow} \phi_{\mathbb{Z}_\eta}(x)\,. \tag{10.2.12}$$

$$\int g_p^\beta(x)|x|_p^\alpha \mathrm{d}^* x = \zeta_p(\alpha, \beta), \quad p \geq \eta\,. \tag{10.2.13}$$

10.3 SYMMETRIC CONVOLUTION

Define

$$\delta_q^\beta(x_1, \ldots, x_n) = \int_{g^{\mathbb{Z}}} \mathcal{F}_q^\beta(x_1 y) \ldots \mathcal{F}_q^\beta(x_n y) \tau_{(q)}^\beta(\mathrm{d}y)\,. \tag{10.3.1}$$

This is symmetric in the x_i and homogeneous:

$$\delta_q^\beta(a\cdot x_1, \ldots, a \cdot x_n) = \delta_q^\beta(x_1, \ldots, x_n) \cdot |a|_{(q)}^{-\beta}, \quad \text{where } |g^i|_{(q)}^\beta = q^{i\beta}\,. \tag{10.3.2}$$

For $n = 2$, using the fact that $\mathcal{F}_q^\beta \mathcal{F}_q^\beta = \mathrm{id}_{\mathscr{H}_q^\beta}$, we see that

$$\begin{aligned}(\varphi(x), \delta_q^\beta(x,y))_{\mathscr{H}_q^\beta} &= \int_{g^{\mathbb{Z}}} \tau_{(q)}^\beta(\mathrm{d}x)\varphi(x) \int_{g^{\mathbb{Z}}} \tau_{(q)}^\beta(\mathrm{d}z) \mathcal{F}_q^\beta(xz)\mathcal{F}_q^\beta(yz) \\ &= \int_{g^{\mathbb{Z}}} \tau_{(q)}^\beta(\mathrm{d}z) \mathcal{F}_q^\beta\varphi(z)\mathcal{F}_q^\beta(yz) = \mathcal{F}_q^\beta\mathcal{F}_q^\beta\varphi(y) = \varphi(y)\,. \end{aligned} \tag{10.3.3}$$

Thus $\delta_q^\beta(x,y)$ is the reproducing kernel of $\mathscr{H}_q^\beta$, and it is given explicitly by

$$\delta_q^\beta(g^i, g^j) = \delta_{i,j} \frac{\zeta_{(q)}(\beta)}{\zeta_{(q)}(1)} \cdot q^{-i\beta}\,. \tag{10.3.4}$$

We set

$$\Delta_q^\beta \varphi(x,y) = \varphi(x \oplus_\beta y) = \int_{g^{\mathbb{Z}}} \varphi(z)\delta_q^\beta(x,y,z)\tau_{(q)}^\beta(\mathrm{d}z)\,. \tag{10.3.5}$$

This operation is co-commutative $\varphi(x \oplus_\beta y) = \varphi(y \oplus_\beta x)$, and also co-associative $\varphi((x \oplus_\beta y) \oplus_\beta z) = \varphi(x \oplus_\beta (y \oplus_\beta z))$; in fact

$$\varphi(x_1 \oplus_\beta \ldots \oplus_\beta x_n) = \int_{g^{\mathbb{Z}}} \varphi(z)\delta_q^\beta(x_1,\ldots,x_n,z)\tau_{(q)}^\beta(\mathrm{d}z)\,. \tag{10.3.6}$$

We obtain a dual commutative and associative operation, the *β-convolution*, given by

$$\varphi_1 *_\beta \varphi_2(y) = \iint_{g^{\mathbb{Z}}} \varphi_1(x_1)\varphi_2(x_2)\delta_q^\beta(x_1,x_2,y)\tau_{(q)}^\beta(\mathrm{d}x_1)\tau_{(q)}^\beta(\mathrm{d}x_2)\,. \tag{10.3.7}$$

The duality between the β-convolution and the co-convolution Δ_q^β is given by

$$(\varphi_1 *_\beta \varphi_2, \varphi)_{\mathscr{H}_q^\beta} = (\varphi_1 \otimes \varphi_2, \Delta_q^\beta \varphi)_{\mathscr{H}_q^\beta \otimes \mathscr{H}_q^\beta} \tag{10.3.8}$$

The operation $\mathcal{F}_q^\beta$ interchanges β-convolution and multiplication,

$$\mathcal{F}_q^\beta(\varphi_1 *_\beta \varphi_2) = \left(\mathcal{F}_q^\beta \varphi_1\right) \cdot \left(\mathcal{F}_q^\beta \varphi_2\right) \tag{10.3.9}$$

or in terms of the kernel $\mathcal{F}_q^\beta(x)$:

$$\mathcal{F}_q^\beta(x \oplus_\beta y) = \mathcal{F}_q^\beta(x) \cdot \mathcal{F}_q^\beta(y)\,. \tag{10.3.10}$$

The operation of β-(co)-convolution converges in the p-adic and real limits, and the above formulae remain valid. The only difference is that in the real limit $\delta_\eta^\beta(x,y)$ is the discontinuous integral of Weber and Schafheitlin, cf. Watson (1922). That is, (10.3.4) does not make sense in the real limit, and $\delta_\eta^\beta(x,y)$ is best thought of as the reproducing *distribution* for $\mathscr{H}_\eta^\beta$; (10.3.3) is still valid for 'nice' $\varphi \in \mathscr{H}_\eta^\beta$. The kernel $\delta_p^\beta(x,y,z)$ appearing in the formulae for β-(co)-convolution can be evaluated explicitly:

$$\delta_p^\beta(x,y,z) = \begin{cases} (1-2p^{-\beta}) \cdot \zeta_p(\beta)^2 \cdot |z|_p^{-\beta} & |x|_p = |y|_p = |z|_p\,, \\ \zeta_p(\beta) \cdot |z|_p^{-\beta}, & |x|_p < |y|_p = |z|_p \quad p \neq \eta\,, \\ 0 & |x|_p, |y|_p < |z|_p\,. \end{cases} \tag{10.3.11$_p$}$$

$$\delta_\eta^\beta(x,y,z) = \frac{1}{4}\frac{\zeta_\eta(\beta)^2}{\zeta_\eta(\beta-1)} \cdot \left|\frac{xyz}{2}\right|_\eta^{2-\beta} \cdot \Delta(x,y,z)^{\beta-3}\,. \tag{10.3.11$_\eta$}$$

(cf. Watson 1922, p. 411, (3)), where $\Delta(x,y,z)$ is the area of the triangle of sides $|x|, |y|, |z|$ (and 0 if there is no such triangle, e.g. $|x| + |y| < |z|$). Similarly we have an explicit formula for β-co-convolution:

$$\varphi(x \oplus_\beta y) = \varphi(|x,y|_p) + \phi_{\mathbb{Z}_p^*}(x/y) \int_{|z|_p<1} (\varphi(zy) - \varphi(y))|z|_p^\beta \mathrm{d}^*z, \quad p \neq \eta\,, \tag{10.3.12$_p$}$$

$$\varphi(x \oplus_\beta y) = \int_0^\pi \varphi((x^2+y^2-2xy\cos\theta)^{1/2}) \cdot |\sin\theta|^{\beta-2} \frac{\mathrm{d}\theta}{\zeta_\eta(\beta-1,1)}, \quad p = \eta\,. \tag{10.3.12$_\eta$}$$

Indeed, for $(10.3.12)_\eta$ setting $z = (x^2 + y^2 - 2xy\cos\theta)^{1/2}$ we have by $(10.3.11)_\eta$

$$\begin{aligned}
\delta_\eta^\beta(x,y,z)\cdot\tau_\eta^\beta(\mathrm{d}z) &= \frac{1}{2}\frac{\zeta_\eta(\beta)}{\zeta_\eta(\beta-1)}\left|\frac{xyz}{2}\right|^{2-\beta}\Delta(x,y,z)^{\beta-3}|z|^{\beta-1}\,\mathrm{d}z \\
&= \frac{1}{2}\frac{\zeta_\eta(\beta)}{\zeta_\eta(\beta-1)}\left|\frac{xy}{2}\right|^{2-\beta}\cdot\left[\frac{((x+y)^2-z^2)(z^2-(x-y)^2)}{2^4}\right]^{(\beta-3)/2} z\,\mathrm{d}z \\
&= \frac{1}{2}\frac{\zeta_\eta(\beta)}{\zeta_\eta(\beta-1)}\left|\frac{xy}{2}\right|^{2-\beta}\left[\frac{xy(1+\cos\theta)\cdot xy(1-\cos\theta)}{2^2}\right]^{(\beta-3)/2} xy\,|\sin\theta|\,\mathrm{d}\theta \\
&= |\sin\theta|^{\beta-2}\frac{\mathrm{d}\theta}{\zeta_\eta(\beta-1,1)}\,.
\end{aligned}$$

Note that we have the limit relation

$$\lim_{\beta\to\infty}\varphi(x\oplus_\beta y) = \varphi(|x,y|_p), \quad p \geq \eta\,. \tag{10.3.13}$$

10.4 THE BASIC BASIS AND THE LAGUERRE BASIS

The 'half' of the space $\mathscr{H}_q^\beta$ consisting of functions supported on $g^{\mathbb{N}}$ is closely related to the space $\mathscr{H}_{\mathbb{Z}_q}^\beta$ we encountered in (7.3.46). We have dense inclusions

$$\begin{array}{ccccc}
\phi_{\mathbb{Z}_q}\cdot\mathscr{H}_{\mathbb{Z}_q}^\beta & & & & \mathscr{H}_{\mathbb{Z}_q}^\beta \\
\| & & & & \| \\
\ell_2\left(g^{\mathbb{N}},\zeta_{(q)}(1+i)q^{\beta\cdot i}\dfrac{\zeta_{(q)}(1)}{\zeta_{(q)}(\beta)}\right) & \hookrightarrow & \ell_2\left(g^{\mathbb{N}},q^{\beta\cdot i}\dfrac{\zeta_{(q)}(1)}{\zeta_{(q)}(\beta)}\right) & \hookrightarrow & \ell_2\left(g^{\mathbb{N}},\dfrac{q^{\beta\cdot i}}{\zeta_{(q)}(1+i)}\dfrac{\zeta_{(q)}(1)}{\zeta_{(q)}(\beta)}\right).
\end{array} \tag{10.4.1}$$

In the p-adic limit these inclusions are equalities. In the real limit we obtain the dense inclusions

$$\begin{array}{ccccc}
\phi_{\mathbb{Z}_\eta}\cdot\mathscr{H}_{\mathbb{Z}_\eta}^\beta & & \mathscr{H}_\eta^\beta & & \mathscr{H}_{\mathbb{Z}_\eta}^\beta \\
\| & & \| & & \| \\
\mathrm{L}_2\left(\mathbb{R},\mathrm{e}^{\pi x^2}\dfrac{|x|^\beta\mathrm{d}^*x}{\zeta_\eta(\beta)}\right)^{\pm1} & \hookrightarrow & \mathrm{L}_2\left(\mathbb{R},\dfrac{|x|^\beta\mathrm{d}^*x}{\zeta_\eta(\beta)}\right)^{\pm1} & \hookrightarrow & \mathrm{L}_2\left(\mathbb{R},\mathrm{e}^{-\pi x^2}\dfrac{|x|^\beta\mathrm{d}^*x}{\zeta_\eta(\beta)}\right)^{\pm1}.
\end{array} \tag{10.4.1$_\eta$}$$

Define the *q-basic basis* $\{\phi_{(q),m}\}, m \geq 0$, to consist of the functions

$$\phi_{(q),m}(g^i) = (-1)^m\frac{\zeta_{(q)}(1)}{\zeta_{(q)}(1+m)}\cdot\frac{q^{(m+i)m}}{\zeta_{(q)}(1+m+i)}\,. \tag{10.4.2}$$

Note that $\phi_{(q),0}(g^i) = \zeta_{(q)}(1+i)^{-1} = \phi_{\mathbb{Z}_q}(g^i)$ and that $\phi_{(q),m}$ is supported on g^i, $i \geq -m$. Further the $\phi_{(q),m}$ are independent of the parameter β.

We let $\mathscr{H}_{\mathbb{Q}_q/\mathbb{Z}_q}^\beta$ denote the Hilbert space with orthogonal base given by the $\phi_{(q),m}$ with

$$\|\,\phi_{(q),m}\,\|_{\mathscr{H}_{\mathbb{Q}_q/\mathbb{Z}_q}^\beta}^2 = \mathrm{C}_{(q)}^\beta(m) = \frac{\zeta_{(q)}(1)}{\zeta_{(q)}(1+m)}\cdot\frac{\zeta_{(q)}(\beta+m)}{\zeta_{(q)}(\beta)}\cdot q^{-m\beta}, \quad \text{cf. (7.3.53)}\,. \tag{10.4.3}$$

Thus for two functions in $\mathscr{H}^{\beta}_{\mathbb{Q}_q/\mathbb{Z}_q}$,

$$\varphi = \sum_{m\geq 0} a_m \cdot \phi_{(q),m}, \quad \varphi' = \sum_{m\geq 0} a'_m \cdot \phi_{(q),m},$$

we have

$$(\varphi, \varphi')_{\mathscr{H}^{\beta}_{\mathbb{Q}_q/\mathbb{Z}_q}} = \sum_{m\geq 0} a_m \cdot \bar{a}'_m \cdot \mathrm{C}^{\beta}_{(q)}(m). \tag{10.4.4}$$

Comparing with the norms of the q-Laguerre basis (7.3.53), we see that $\mathscr{H}^{\beta}_{\mathbb{Q}_q/\mathbb{Z}_q}$ is isometric to the Hilbert space $\mathscr{H}^{\beta}_{\mathbb{Z}_q}$, and hence also to the space $\phi_{\mathbb{Z}_q} \cdot \mathscr{H}^{\beta}_{\mathbb{Z}_q}$,

$$\begin{array}{ccc} \mathscr{H}^{\beta}_{\mathbb{Q}_q/\mathbb{Z}_q} & \stackrel{\sim}{\longleftrightarrow} & \phi_{\mathbb{Z}_q} \cdot \mathscr{H}^{\beta}_{\mathbb{Z}_q} \\ \Cup & & \Cup \\ \phi_{(q),m} & \longleftrightarrow & \phi_{\mathbb{Z}_q} \cdot \varphi^{\beta}_{\mathbb{Z}_q,m}\,. \end{array} \tag{10.4.5}$$

From the q-gamma sum (6.3.5) and the definition (10.4.2) we obtain

$$\tau^{s}_{(q)}(\phi_{(q),m}) = (-1)^m \mathrm{C}^{s}_{(q)}(m). \tag{10.4.6}$$

Comparing this with the analogous result for $\phi_{\mathbb{Z}_q} \cdot \varphi^{\beta}_{\mathbb{Z}_{q,m}}$, (7.3.60), we see that

$$\tau^{\beta/2+s}_{(q)}(\phi_{(q),m}) = (-1)^m \mathrm{C}^{\beta/2+s}_{(q)}(m) = \tau^{\beta/2-s}_{(q)}\left(\phi_{\mathbb{Z}_q} \cdot \varphi^{\beta}_{\mathbb{Z}_q,m}\right). \tag{10.4.7}$$

Thus the isomorphism (10.4.5) is implemented by the Fourier–Bessel transform $\mathcal{F}^{\beta}_{q}$:

$$\mathcal{F}^{\beta}_{q}(\phi_{(q),m}) = \phi_{\mathbb{Z}_q} \cdot \varphi^{\beta}_{\mathbb{Z}_{q,m}}, \quad \phi_{(q),m} = \mathcal{F}^{\beta}_{q}(\phi_{\mathbb{Z}_q} \cdot \varphi^{\beta}_{\mathbb{Z}_q,m}). \tag{10.4.8}$$

We get the new expression for the Bessel kernel

$$\mathcal{F}^{\beta}_{q}(x \cdot y) = \sum_{m\geq 0} \frac{\zeta_{(q)}(1+m)}{\zeta_{(q)}(1)} \frac{\zeta_{(q)}(\beta)}{\zeta_{(q)}(\beta+m)} q^{m\beta} \cdot \phi_{\mathbb{Z}_q}(x) \varphi^{\beta}_{\mathbb{Z}_q,m}(x) \cdot \phi_{(q),m}(y). \tag{10.4.9}$$

Note that the image in the analytic space $\widehat{\mathscr{H}}^{\beta}_{q}$ under $\tau^{\beta/2+s}_{(q)}$ of the linear span of the $\phi_{\mathbb{Z}_q} \cdot \varphi^{\beta}_{\mathbb{Z}_q,m}$ (respectively $\phi_{(q),m}$) consists of all polynomials in q^s (respectively q^{-s}); hence the closures of the images of $\phi_{\mathbb{Z}_q}\mathscr{H}^{\beta}_{\mathbb{Z}_q}$ and $\mathscr{H}^{\beta}_{\mathbb{Q}_q/\mathbb{Z}_q}$ are the Hardy subspaces of $\widehat{\mathscr{H}}^{\beta}_{q}$, and in particular,

$$\mathscr{H}^{\beta}_{\mathbb{Q}_q/\mathbb{Z}_q} \cap \phi_{\mathbb{Z}_q}\mathscr{H}^{\beta}_{\mathbb{Z}_q} = \mathbb{C} \cdot \phi_{\mathbb{Z}_q}, \quad \mathscr{H}^{\beta}_{\mathbb{Q}_q/\mathbb{Z}_q} + \phi_{\mathbb{Z}_q}\mathscr{H}^{\beta}_{\mathbb{Z}_q} = \mathscr{H}^{\beta}_{q}.$$

We can summarize the above discussions in the diagram

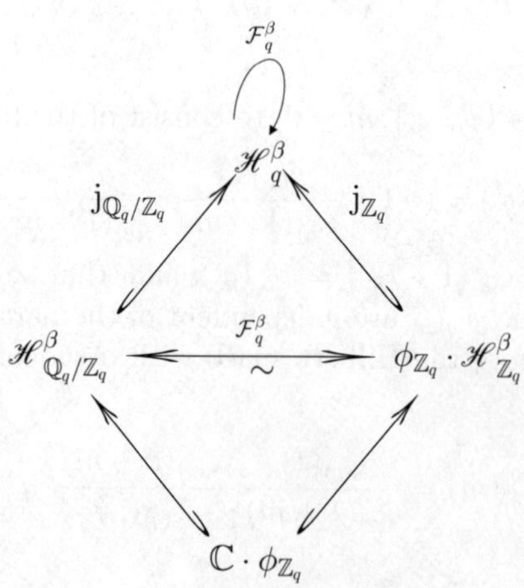

(10.4.10)

The adjoint of $\mathrm{j}_{\mathbb{Z}_q}$ is given by multiplication by $\phi_{\mathbb{Z}_q}$; the adjoint of $\mathrm{j}_{\mathbb{Q}_q/\mathbb{Z}_q}$ is given by β-convolution with $\phi_{\mathbb{Z}_q}$. The space $\mathscr{H}^{\beta}_{\mathbb{Q}_q/\mathbb{Z}_q}$ has a reproducing kernel given by $\phi_{\mathbb{Z}_q}(x \oplus_\beta y)$, and we can write

$$\phi_{\mathbb{Z}_q}(x \oplus_\beta y) = \sum_{m \geq 0} \frac{\zeta_{(q)}(1+m)}{\zeta_{(q)}(1)} \frac{\zeta_{(q)}(\beta)}{\zeta_{(q)}(\beta+m)} q^{m\beta} \cdot \phi_{(q),m}(x) \cdot \phi_{(q),m}(y) \qquad (10.4.11)$$

Consider next the p-adic limit. We have already noticed that $\phi_{(q),0} = \phi_{\mathbb{Z}_q}$ converges to $\phi_{\mathbb{Z}_p}$. For $m > 0$, when $q \to 0$, $\phi_{(q),m}$ converges to $(-1)^m \delta_{g^{-m}}$; hence in the p-adic limit $\phi_{(q),m}$ converges to $(-1)^m \phi_{p^{-m}\mathbb{Z}_p^*}$, (where $\phi_{p^{-m}\mathbb{Z}_p^*}$ is the characteristic function of $p^{-m}\mathbb{Z}_p^*$). We see that $\mathscr{H}^{\beta}_{\mathbb{Q}_p/\mathbb{Z}_p}$, the p-adic limit of $\mathscr{H}^{\beta}_{\mathbb{Q}_q/\mathbb{Z}_q}$, consists of the subspace of $\mathscr{H}^{\beta}_{p}$ of functions invariant under $\mathbb{Z}_p$-translation. The image in the analytic space $\widehat{\mathscr{H}}^{\beta}_{p}$ under $\tau_p^{\beta/2+s}$ of the subspace $\mathscr{H}^{\beta}_{\mathbb{Z}_p}$ (respectively $\mathscr{H}^{\beta}_{\mathbb{Q}_p/\mathbb{Z}_p}$) is precisely the Hardy space of boundary values of analytic functions in $\mathrm{Re}(s) > 0$ (respectively $\mathrm{Re}(s) < 0$). The p-adic limit of the expansion (10.4.9) for the Bessel kernel reads

$$\mathcal{F}^{\beta}_{p}(x \cdot y) = \phi_{\mathbb{Z}_p}(x)\phi_{\mathbb{Z}_p}(y) + \sum_{m \geq 1} \frac{p^{-m\beta}}{1-p^{-\beta}} \cdot \varphi^{\beta}_{\mathbb{Z}_p,m}(x) \cdot \phi_{p^{-m}\mathbb{Z}_p^*}(y) \,. \qquad (10.4.9)_p$$

The expansion of the reproducing kernel $\phi_{\mathbb{Z}_p}(x \oplus_\beta y)$ of $\mathscr{H}^{\beta}_{\mathbb{Q}_p/\mathbb{Z}_p}$ is simply

$$\phi_{\mathbb{Z}_p}(x \oplus_\beta y) = \phi_{\mathbb{Z}_p}(x)\phi_{\mathbb{Z}_p}(y) + \sum_{m \geq 1} \frac{p^{-m\beta}}{1-p^{-\beta}} \cdot \phi_{p^{-m}\mathbb{Z}_p^*}(x) \cdot \phi_{p^{-m}\mathbb{Z}_p^*}(y) \,. \qquad (10.4.11)_p$$

Consider next the real limit ⓡ: $q \to 1$, $i \to \infty$, $q^i/(1-q) \to \pi x^2$. We have already noted that $\phi_{(q),0} = \phi_{\mathbb{Z}_q}$ converges to $\phi_{\mathbb{Z}_\eta} = \mathrm{e}^{-\pi x^2}$. Similarly,

$$\begin{aligned} \lim_{ⓡ} \phi_{(q),m} &= \lim_{ⓡ} (-1)^m \cdot \frac{\zeta_{(q)}(1)(1-q)^m}{\zeta_{(q)}(1+m)} \cdot q^{m^2} \cdot \left(\frac{q^i}{1-q}\right)^m \cdot \frac{1}{\zeta_{(q)}(1+m+i)} \\ &= (-1)^m \cdot \frac{1}{m!} \cdot (\pi x^2)^m \cdot \mathrm{e}^{-\pi x^2} \stackrel{\text{def}}{=} \phi_{\eta,m}(x) \,. \end{aligned} \qquad (10.4.2)_\eta$$

We see that $\mathscr{H}^{\beta}_{\mathbb{Q}_\eta/\mathbb{Z}_\eta}$, the real limit of $\mathscr{H}^{\beta}_{\mathbb{Q}_q/\mathbb{Z}_q}$, consists of (the restriction to $\mathbb{R}$ of entire even) functions of the form

$$\varphi(x) = \sum_{m \geq 0} a_m \frac{(\pi x^2)^m}{m!} \cdot \mathrm{e}^{-\pi x^2}$$

with

$$\| \varphi \|^2_{\mathscr{H}^{\beta}_{\mathbb{Q}_\eta/\mathbb{Z}_\eta}} = \sum_{m \geq 0} |a_m|^2 \cdot \frac{\pi^m}{m!} \frac{\zeta_\eta(\beta+2m)}{\zeta_\eta(\beta)} < \infty \,. \qquad (10.4.4)_\eta$$

$\mathscr{H}^{\beta}_{\mathbb{Q}_\eta/\mathbb{Z}_\eta}$ is not a closed subspace of $\mathscr{H}^{\beta}_{\eta}$, but we have a dense inclusion $\mathrm{j}_{\mathbb{Q}_\eta/\mathbb{Z}_\eta}: \mathscr{H}^{\beta}_{\mathbb{Q}_\eta/\mathbb{Z}_\eta} \hookrightarrow \mathscr{H}^{\beta}_{\eta}$, its adjoint given by β-convolution with $\phi_{\mathbb{Z}_\eta}$. The images of $\mathscr{H}^{\beta}_{\mathbb{Q}_\eta/\mathbb{Z}_\eta}$, and of $\phi_{\mathbb{Z}_\eta} \cdot \mathscr{H}^{\beta}_{\mathbb{Z}_\eta}$, in $\mathscr{H}^{\beta}_{\eta}$ correspond to each other under the Fourier–Bessel transform: $\mathcal{F}^{\beta}_{\eta}$ interchanges the basic basis $\{\phi_{\eta,m}(x)\}$ and the Laguerre basis $\{\phi_{\mathbb{Z}_\eta}(x)\varphi^{\beta}_{\mathbb{Z}_\eta,m}(x)\}$. We have the expansion for the Bessel kernel

$$\mathcal{F}_\eta^\beta(x\cdot y)=\sum_{m\geq 0}\frac{\zeta_\eta(\beta)}{\zeta_\eta(\beta+2m)}(-1)^m\cdot\varphi^\beta_{\mathbb{Z}_\eta,m}(x)\cdot y^{2m}\cdot \mathrm{e}^{-\pi(x^2+y^2)}\,. \tag{10.4.9$_\eta$}$$

The space $\mathscr{H}^\beta_{\mathbb{Q}_\eta/\mathbb{Z}_\eta}$ has a reproducing kernel $\phi_{\mathbb{Z}_\eta}(x\oplus_\beta y)$, and we can write

$$\begin{aligned}\phi_{\mathbb{Z}_\eta}(x\oplus_\beta y)&=\sum_{m\geq 0}\frac{\zeta_\eta(\beta)}{\zeta_\eta(\beta+2m)}\cdot\frac{(\pi(xy)^2)^m}{m!}\cdot \mathrm{e}^{-\pi(x^2+y^2)}\\&=\mathcal{F}_\eta^\beta\left(\sqrt{-1}xy\right)\cdot \mathrm{e}^{-\pi(x^2+y^2)}\quad\text{('Weber's second formula').}\end{aligned} \tag{10.4.11$_\eta$}$$

Define the following operators on $\mathscr{H}^\beta_{\mathbb{Q}_q/\mathbb{Z}_q}$:

$$\begin{aligned}\mathrm{A}_+\varphi(g^i)&=-\frac{q^{i+1}}{1-q}\varphi(g^{i+1})\,,\\ \mathrm{A}_-\varphi(g^i)&=-[(1-q^{1-i})\varphi(g^{i-1})+q^{1-i}\varphi(g^{i-2})]\\ \mathrm{N}\varphi(g^i)&=q^i\varphi(g^i)+\varphi(g^{i-1})\,.\end{aligned} \tag{10.4.12}$$

Applying these operators to the basis $\phi_{(q),m}$, we obtain

$$\begin{aligned}\mathrm{A}_+\phi_{(q),m}&=[m+1]_q q^{-m}\phi_{(q),m+1}\,,\\ \mathrm{A}_-\phi_{(q),m}&=\phi_{(q),m-1}\,,\\ \mathrm{N}\phi_{(q),m}&=q^{-m}\cdot\phi_{(q),m}\,.\end{aligned} \tag{10.4.13}$$

The operators A_+, A_-, N generate the *q-Heisenberg algebra* with relations

$$\mathrm{A}_\pm\mathrm{N}=q^{\pm 1}\cdot\mathrm{N}\mathrm{A}_\pm \tag{10.4.14}$$

$$\mathrm{A}_-\mathrm{A}_+=-q\cdot\frac{1-q^{-1}\mathrm{N}}{1-q}\,,\quad \mathrm{A}_+\mathrm{A}_-=-q\cdot\frac{1-\mathrm{N}}{1-q}\,, \tag{10.4.15}$$

and consequently,

$$\mathrm{A}_-\mathrm{A}_+-q^{-1}\mathrm{A}_+\mathrm{A}_-=1\,, \tag{10.4.16}$$

$$\mathrm{A}_-\mathrm{A}_+-\mathrm{A}_+\mathrm{A}_-=\mathrm{N}\,. \tag{10.4.17}$$

Note that N is self-adjoint on any space where the $\phi_{(q),m}$ are an orthogonal basis. For example, letting $\mathscr{H}^{(!)}_{\mathbb{Q}_q/\mathbb{Z}_q}$ be the Hilbert space with $\phi_{(q),m}$ an orthogonal basis, but with the norms

$$\|\phi_{(q),m}\|^2_{\mathscr{H}^{(!)}_{\mathbb{Q}_q/\mathbb{Z}_q}}=\frac{q^{m(m-1)/2}}{[m]_q!}\,, \tag{10.4.18}$$

then as operators on $\mathscr{H}^{(!)}_{\mathbb{Q}_q/\mathbb{Z}_q}$ we have

$$\mathrm{N}^*=\mathrm{N},\quad \mathrm{A}_\pm^*=\mathrm{A}_\mp\,. \tag{10.4.19}$$

The Heisenberg operators up the ladder, $\mathrm{D}\colon\mathscr{H}^\beta_{\mathbb{Z}_q}\to\mathscr{H}^{\beta+1}_{\mathbb{Z}_q}$ of (7.3.47), and $\mathrm{D}^+=(1-q^\beta)\mathrm{D}^*$ of (7.3.48), when transformed from the spaces $\mathscr{H}^\beta_{\mathbb{Z}_q}$ to the spaces $\mathscr{H}^\beta_{\mathbb{Q}_q/\mathbb{Z}_q}$, are given in terms of the operators $\mathrm{A}_\pm$ by

$$\mathrm{D}^+:=\mathcal{F}_q^\beta\phi_{\mathbb{Z}_q}\mathrm{D}^+\phi_{\mathbb{Z}_q}^{-1}\mathcal{F}_q^{\beta+1}=\mathrm{A}_+\text{ viewed as an operator }\mathscr{H}^{\beta+1}_{\mathbb{Q}_q/\mathbb{Z}_q}\to\mathscr{H}^\beta_{\mathbb{Q}_q/\mathbb{Z}_q}\,, \tag{10.4.20}$$

$$\mathrm{D}:=\mathcal{F}_q^{\beta+1}\phi_{\mathbb{Z}_q}\mathrm{D}\phi_{\mathbb{Z}_q}^{-1}\mathcal{F}_q^{\beta}=\frac{q^{-\beta}}{1-q}\cdot\mathrm{A}_-\text{ viewed as an operator }\mathscr{H}^{\beta}_{\mathbb{Q}_q/\mathbb{Z}_q}\to\mathscr{H}^{\beta+1}_{\mathbb{Q}_q/\mathbb{Z}_q}\,. \tag{10.4.21}$$

Note that now D^+ is independent of β, and D is $q^{-\beta}$ times an operator independent of β. The Heisenberg relation up the ladder (cf. (7.3.49))

$$\mathrm{DD}^+ - \mathrm{D}^+\mathrm{D} = \frac{q^{-\beta}}{1-q} \cdot \mathrm{id}_{\mathscr{H}^{\beta+1}_{\mathbb{Q}_q/\mathbb{Z}_q}} \tag{10.4.22}$$

are equivalent via (10.4.20) and (10.4.21) to the q-Heisenberg relation (10.4.16).

Note that we can write

$$\mathrm{N} = \mathcal{X} + \mathcal{Y}, \quad \mathcal{X}\varphi(g^i) = q^i \cdot \varphi(g^i), \quad \mathcal{Y}\varphi(g^i) = \varphi(g^{i-1}) \tag{10.4.23}$$

where $\mathcal{X}, \mathcal{Y}$ q-commute, $\mathcal{XY} = q \cdot \mathcal{YX}$. It follows that we have

$$\mathrm{N}^t = (\mathcal{X} + \mathcal{Y})^t = \sum_{k\geq 0} \begin{bmatrix} t \\ k \end{bmatrix}_q \mathcal{Y}^k \mathcal{X}^{t-k}, \tag{10.4.24}$$

$$\mathrm{N}^t\varphi(g^i) = \sum_{k\geq 0} \begin{bmatrix} t \\ k \end{bmatrix}_q q^{(t-k)(i-k)} \varphi(g^{i-k}),$$

$$\mathrm{N}^t \phi_{(q),m} = q^{-tm} \phi_{(q),m}\,. \tag{10.4.25}$$

Note that we can express all the operators of the q-Heisenberg algebra in terms of the q-commuting operators $\mathcal{X}, \mathcal{Y}$ of (10.4.23):

$$\mathrm{A}_+ = -\frac{1}{1-q} \mathcal{Y}^{-1} \cdot \mathcal{X}, \tag{10.4.26}$$

$$\mathrm{A}_- = -\mathcal{Y}[1 - \mathcal{X}^{-1}(1 - \mathcal{Y})]\,. \tag{10.4.27}$$

We can transform the action of the q-Heisenberg alegebra to the space $\phi_{\mathbb{Z}_q} \cdot \mathscr{H}^{\beta}_{\mathbb{Z}_q}$ by conjugating by the Fourier–Bessel transform $\mathcal{F}^{\beta}_q$. To this end, note that we have

$$\mathcal{F}^{\beta}_q \mathcal{Y} \mathcal{F}^{\beta}_q = q^{\beta} \cdot \mathcal{Y}^{-1}, \quad \mathcal{Y}^{-1}\varphi(g^i) = \varphi(g^{i+1})\,, \tag{10.4.28}$$

and a computation using the expression $(10.2.4)_\eta$ for $\mathcal{F}^{\beta}_q$ gives

$$\mathcal{F}^{\beta}_q \mathcal{X} \mathcal{F}^{\beta}_q = (1 - q^{\beta}\mathcal{Y}^{-1})\mathcal{X}^{-1}(1 - \mathcal{Y}), \quad \mathcal{X}^{-1}\varphi(g^i) = q^{-i}\varphi(g^i)\,. \tag{10.4.29}$$

Thus the action of the q-Heisenberg algebra when transformed to $\phi_{\mathbb{Z}_q} \cdot \mathscr{H}^{\beta}_{\mathbb{Z}_q}$ is given by

$$\mathrm{N} = \mathcal{F}^{\beta}_q(\mathcal{X} + \mathcal{Y})\mathcal{F}^{\beta}_q = (1 - q^{\beta}\mathcal{Y}^{-1})\mathcal{X}^{-1}(1 - \mathcal{Y}) + q^{\beta}\mathcal{Y}^{-1}, \tag{10.4.30}$$

$$\mathrm{A}_+ = -\frac{1}{1-q}\mathcal{F}^{\beta}_q \mathcal{Y}^{-1}\mathcal{X}\mathcal{F}^{\beta}_q = \frac{1}{1-q} \cdot (1 - q^{-\beta}\mathcal{Y})\mathcal{X}^{-1}(1 - \mathcal{Y})\,, \tag{10.4.31}$$

$$\mathrm{A}_- = -\mathcal{F}^{\beta}_q \mathcal{Y}[1 - \mathcal{X}^{-1}(1 - \mathcal{Y})]\mathcal{F}^{\beta}_q = -q^{\beta}\mathcal{Y}^{-1}[1 - (1 - \mathcal{Y})^{-1}\mathcal{X}]\,, \tag{10.4.32}$$

where

$$(1 - \mathcal{Y})^{-1}\varphi(g^i) = \sum_{k\geq 0} \mathcal{Y}^k \varphi(g^i) = \sum_{k\geq 0} \varphi(g^{i-k})$$

is well defined on functions supported on $g^{\mathbb{N}}$.

Finally we can transform the action of the q-Heisenberg algebra to the space $\mathscr{H}^{\beta}_{\mathbb{Z}_q}$ by conjugating by (the operator of multiplication) $\phi_{\mathbb{Z}_q}$. Note that we have

$$\phi_{\mathbb{Z}_q}^{-1} \mathcal{X} \phi_{\mathbb{Z}_q} = \mathcal{X}\,, \tag{10.4.33}$$

$$\phi_{\mathbb{Z}_q}^{-1} \mathcal{Y} \phi_{\mathbb{Z}_q} = (1 - \mathcal{X})\mathcal{Y}\,. \tag{10.4.34}$$

From this, and from (10.4.30)–(10.4.32), we obtain the action of the q-Heisenberg algebra on $\mathscr{H}^{\beta}_{\mathbb{Z}_q}$:

$$\mathrm{N} = \phi_{\mathbb{Z}_q}^{-1}[(1-q^{\beta}\mathcal{Y}^{-1})\mathcal{X}^{-1}(1-\mathcal{Y}) + q^{\beta}\mathcal{Y}^{-1}]\phi_{\mathbb{Z}_q}$$
$$= (1-q^{\beta}\mathcal{Y}^{-1})\mathcal{X}^{-1}(1-\mathcal{Y}) + \mathcal{Y}, \qquad (10.4.35)$$

$$\mathrm{A}_+ = \frac{1}{1-q}\phi_{\mathbb{Z}_q}^{-1}(1-q^{-\beta}\mathcal{Y})\mathcal{X}^{-1}(1-\mathcal{Y})\phi_{\mathbb{Z}_q}$$
$$= \frac{1}{1-q}(1-q^{-\beta}(1-\mathcal{X})\mathcal{Y})\mathcal{X}^{-1}(1-(1-\mathcal{X})\mathcal{Y}), \qquad (10.4.36)$$

$$\mathrm{A}_- = -q^{\beta}\phi_{\mathbb{Z}_q}^{-1}\mathcal{Y}^{-1}\left[1 - \sum_{k\geq 0}\mathcal{Y}^k\mathcal{X}\right]\phi_{\mathbb{Z}_q}$$
$$= -q^{\beta}\left[\mathcal{Y}^{-1} - \sum_{k\geq 0}((1-\mathcal{X})\mathcal{Y})^k\mathcal{X}\right], \qquad (10.4.37)$$

or explicitly

$$\mathrm{N}\varphi(g^i) = q^{-i}[(1+q^{\beta-1})\varphi(g^i) - (1-q^i)\varphi(g^{i-1}) - q^{\beta-1}\varphi(g^{i+1})], \qquad (10.4.38)$$

$$\mathrm{A}_+\varphi(g^i) = \frac{q^{-i}}{1-q}[\varphi(g^i) - (1-q^i)(1+q^{1-\beta})\varphi(g^{i-1})$$
$$+ (1-q^i)(1-q^{i-1})q^{1-\beta}\varphi(g^{i-2})], \qquad (10.4.39)$$

$$\mathrm{A}_-\varphi(g^i) = -q^{\beta}\left[\varphi(g^{i+1}) - \sum_{0\leq k\leq i}\frac{\zeta_{(q)}(1+i)}{\zeta_{(q)}(1+i-k)}q^{i-k}\varphi(g^{i-k})\right]. \qquad (10.4.40)$$

10.5 THE BETA MEASURE

The last four sections have analogues in the context of the beta measure. We indicate here the p-adic and real versions briefly. We have the unitary representations (3.2.10) for $p \geq \eta$:

$$\pi_p^{\alpha,\beta}\colon \mathbb{Q}_p^*/\mathbb{Z}_p^* \to \mathrm{U}(\mathscr{H}_p^{\alpha,\beta}) = \text{group of unitary operators on } \mathscr{H}_p^{\alpha,\beta} = \mathrm{L}_2(\mathbb{P}^1(\mathbb{Q}_p), \tau_p^{\alpha,\beta})^{\mathbb{Z}_p^*}. \qquad (10.5.1)$$

The vector $\mathbb{1} \in \mathscr{H}_p^{\alpha,\beta}$ is a cyclic vector. The associated positive-definite function $(\pi_p^{\alpha,\beta}(a)\mathbb{1},\mathbb{1})_{\mathscr{H}_p^{\alpha,\beta}} = \tau_p^{\alpha,\beta}(\pi_p^{\alpha,\beta}(a)\mathbb{1})$ on $\mathbb{Q}_p^*/\mathbb{Z}_p^*$ gives rise to the positive measure on the dual group $(\mathbb{Q}_p^*/\mathbb{Z}_p^*)^{\vee}$. It is given for $p \geq \eta$ by

$$\hat{\tau}_p^{\alpha,\beta}(s) = \int \mathrm{d}^*a|a|_p^s(\pi_p^{\alpha,\beta}(a)\mathbb{1},\mathbb{1})_{\mathscr{H}_p^{\alpha,\beta}}$$
$$= \frac{\zeta_p(\alpha/2-s,\alpha/2+s)\cdot\zeta_p(\beta/2-s,\beta/2+s)}{\zeta_p((\alpha+\beta)/2,(\alpha+\beta)/2)}$$
$$= \frac{\zeta_p(\alpha/2-s,\beta/2+s)\zeta_p(\alpha/2+s,\beta/2-s)}{\zeta_p(\alpha,\beta)}. \qquad (10.5.2)$$

We obtain isomorphic descriptions of the space $\mathscr{H}_p^{\alpha,\beta}$:

$$\begin{array}{ccc} \text{`multiplicative'}\ \mathscr{H}_p^{\alpha,\beta,*} & \xrightarrow{\sim} & \mathscr{H}_p^{\alpha,\beta}\ \text{`additive'} \\ & \searrow\wr \quad \wr\swarrow & \\ & \text{`analytic'}: \widehat{\mathscr{H}}_p^{\alpha,\beta} = \mathrm{L}_2\Big(\big(\mathbb{Q}_p^*/\mathbb{Z}_p^*\big)^{\mathrm{v}}, \hat{\tau}_p^{\alpha,\beta}(s)\mathrm{d}_p s\Big) & \end{array} \tag{10.5.3}$$

It is given by

$$\begin{array}{ccc} f & \longmapsto & \pi_p^{\alpha,\beta}(f)\mathbb{1} = \varphi \\ & \searrow \quad \swarrow & \\ & \hat{f}(s) = \tau_p^{\alpha/2-s,\beta/2+s}(\varphi) & \end{array} \tag{10.5.4}$$

In the realizations $\mathscr{H}_p^{\alpha,\beta,*}$ and $\widehat{\mathscr{H}}_p^{\alpha,\beta}$ we have the unitary self-adjoint operators given by

$$\mathcal{F}_p^{\alpha,\beta,*} f(a) = f(a^{-1}) \quad \text{and} \quad \widehat{\mathcal{F}}_p^{\alpha,\beta} \hat{f}(s) = \hat{f}(-s)\,. \tag{10.5.5}$$

These correspond under the isomorphism (10.5.3) to the *Fourier–Jacobi operator* $\mathcal{F}_p^{\alpha,\beta}$ on $\mathscr{H}_p^{\alpha,\beta}$. It satisfies:

$$\begin{aligned} &\text{(i) } \mathcal{F}_p^{\alpha,\beta} = \big(\mathcal{F}_p^{\alpha,\beta}\big)^* = \big(\mathcal{F}_p^{\alpha,\beta}\big)^{-1} \\ &\text{(ii) } \mathcal{F}_p^{\alpha,\beta}\pi_p^{\alpha,\beta}(a) = \pi_p^{\alpha,\beta}(a^{-1})\mathcal{F}_p^{\alpha,\beta} \\ &\text{(iii) } \mathcal{F}_p^{\alpha,\beta}\mathbb{1} = \mathbb{1} \\ &\text{(iv) } \tau_p^{\alpha/2-s,\beta/2+s}\big(\mathcal{F}_p^{\alpha,\beta}\varphi\big) = \tau_p^{\alpha/2+s,\beta/2-s}(\varphi) \\ &\text{(v) for } \alpha = \beta, \mathcal{F}_p^{\alpha,\beta}\varphi(x:y) = \varphi(y:x)\,. \end{aligned} \tag{10.5.6}$$

The Fourier–Jacobi operator can be written in the form

$$\mathcal{F}_p^{\alpha,\beta}\varphi(y) = \int \varphi(x)\mathcal{F}_p^{\alpha,\beta}(x \cdot y)\tau_p^{\alpha,\beta}(\mathrm{d}x)\,. \tag{10.5.7}$$

The kernel $\mathcal{F}_p^{\alpha,\beta}(x \cdot y)$ has the following expansions:

$$\mathcal{F}_p^{\alpha,\beta}(x) = \begin{cases} \dfrac{\zeta_p(\alpha,\beta)\zeta_p((\alpha+\beta)/2)}{\zeta_p(\alpha)\zeta_p((\beta-\alpha)/2)} & |x|_p > 1\,, \\ \zeta_p(\alpha,\beta)\zeta_p((\alpha+\beta)/2)\cdot(1 - p^{-\alpha} - p^{-\beta} + p^{-(\alpha+\beta)/2}) & |x|_p = 1,\ p \neq \eta\,, \\ \dfrac{\zeta_p(\alpha,\beta)\zeta_p((\alpha+\beta)/2)}{\zeta_p(\beta)\zeta_p((\alpha-\beta)/2)} & |x|_p < 1\,, \end{cases} \tag*{(10.5.8)$_p$}$$

$$\mathcal{F}_\eta^{\alpha,\beta}(x) = \begin{cases} |1, x^{-1}|_\eta^{(\alpha+\beta)/2} \sum_{n \geq 0} \frac{\zeta_\eta(\alpha,\beta)\zeta_\eta((\alpha+\beta)/2+2n)}{\zeta_\eta(\alpha+2n)\zeta_\eta((\beta-\alpha)/2+2n)} \frac{(-\pi x^{-2})^n}{n!} & |x|_\eta > 1\,, \\ |1, x|_\eta^{(\alpha+\beta)/2} \sum_{n \geq 0} \frac{\zeta_\eta(\alpha,\beta)\zeta_\eta((\alpha+\beta)/2+2n)}{\zeta_\eta(\beta+2n)\zeta_\eta((\alpha-\beta)/2+2n)} \frac{(-\pi x^{2})^n}{n!} & |x|_\eta < 1\,. \end{cases} \tag{10.5.8$_\eta$}$$

As in 10.3 we obtain an operation of (co-)convolution for $p \geq \eta$:

$$\begin{aligned} \Delta_p^{\alpha,\beta}\varphi(x,y) &= \varphi(x \bigoplus_{\alpha,\beta} y) = \int \varphi(z)\delta_p^{\alpha,\beta}(x,y,z)\tau_p^{\alpha,\beta}(\mathrm{d}z)\,, \\ \varphi_1 *_{\alpha,\beta} \varphi_2(z) &= \int\!\!\int \varphi_1(x)\varphi_2(y)\delta_p^{\alpha,\beta}(x,y,z)\tau_p^{\alpha,\beta}(\mathrm{d}x)\tau_p^{\alpha,\beta}(\mathrm{d}y)\,, \\ \delta_p^{\alpha,\beta}(x,y,z) &= \int \mathcal{F}_p^{\alpha,\beta}(x\cdot t)\mathcal{F}_p^{\alpha,\beta}(y\cdot t)\mathcal{F}_p^{\alpha,\beta}(z\cdot t)\tau_p^{\alpha,\beta}(\mathrm{d}t)\,, \\ (\Delta_p^{\alpha,\beta}\varphi, \varphi_1\otimes\varphi_2)_{\mathscr{H}_p^{\alpha,\beta}\otimes\mathscr{H}_p^{\alpha,\beta}} &= (\varphi, \varphi_1 *_{\alpha,\beta} \varphi_2)_{\mathscr{H}_p^{\alpha,\beta}}\,. \end{aligned} \tag{10.5.9}$$

10.6 THE PURE GAMMA BASIS AND THE CUT-OFF BASIS

Let us return to the analytic space $\widehat{\mathscr{H}}_q^\beta$. We can decompose it into the ± 1 eigenspaces of the operator $\widehat{\mathcal{F}}_q^\beta$, $\widehat{\mathcal{F}}_q^\beta \hat{f}(s) = \hat{f}(-s)$,

$$\widehat{\mathscr{H}}_q^\beta = \widehat{\mathscr{H}}_{q,+}^\beta \oplus \widehat{\mathscr{H}}_{q,-}^\beta\,. \tag{10.6.1}$$

Let us write

$$\mathrm{X} = (q^s + q^{-s})/2 = \cos((\log q)\cdot t), \quad \mathrm{Y} = (q^s - q^{-s})/2\mathrm{i} = \sin((\log q)\cdot t), \quad s = \mathrm{i}t\,, \tag{10.6.2}$$

and view X, Y as commuting self-adjoint operators on $\widehat{\mathscr{H}}_q^\beta$ via multiplication. We have

$$\mathrm{X}^2 + \mathrm{Y}^2 = 1, \quad \widehat{\mathcal{F}}_q^\beta \mathrm{X} = \mathrm{X}\widehat{\mathcal{F}}_q^\beta, \quad \widehat{\mathcal{F}}_q^\beta \mathrm{Y} = -\mathrm{Y}\widehat{\mathcal{F}}_q^\beta\,. \tag{10.6.3}$$

Since $\{\mathrm{X}^n\mathbb{1}\}_{n\geq 0}$ are dense in $\widehat{\mathscr{H}}_{q,+}^\beta$, they can be orthogonalized to a basis $\{\hat{f}_{q,+,n}^\beta(s) = \hat{f}_{q,+,n}^\beta(\mathrm{X})\mathbb{1}\}_{n\geq 0}$ of $\widehat{\mathscr{H}}_{q,+}^\beta$. Similarly, since $\{\mathrm{X}^n\mathrm{Y}\mathbb{1}\}_{n\geq 0}$ are dense in $\widehat{\mathscr{H}}_{q,-}^\beta$, they can be orthogonalized to a basis $\{\hat{f}_{q,-,n}^\beta(s) = \hat{f}_{q,-,n-1}^\beta(\mathrm{X})\mathrm{Y}\mathbb{1}\}_{n\geq 1}$ of $\widehat{\mathscr{H}}_{q,-}^\beta$. The $\hat{f}_{q,\pm,n}^\beta(\mathrm{X})$ are just the orthogonal polynomials with respect to the weight function on $[-1,1]$ given by

$$\prod_{n\geq 0} \frac{(1-q^{1+n})(1-q^{\beta+n})}{(1-2\mathrm{X}q^{\beta/2+n}+q^{\beta+2n})} \cdot (1-\mathrm{X}^2)^{\mp 1/2}\mathrm{d}\mathrm{X}\,. \tag{10.6.4}$$

The polynomials $\hat{f}_{q,\pm,n}^\beta(\mathrm{X})$ have n zeros in the interval $[-1,1]$; hence in terms of the variable s, $\hat{f}_{q,\pm,n}^\beta(s)$ have $2n$ zeros in $\mathrm{i}\mathbb{R}/(2\pi\mathrm{i}/\log q)\mathbb{Z}$. We shall refer to them as the *pure basis*.

The polynomials $\hat{f}^{\beta}_{q,\pm,n}(s)$ are given in terms of the moments (10.1.8),

$$\begin{aligned}\tilde{\rho}^{\beta}_{(q)}(g^i) &= \int_{\mathrm{i}\mathbb{R}/\frac{2\pi\mathrm{i}}{\log q}\mathbb{Z}} q^{is}\zeta_{(q)}(\beta/2+s,\beta/2-s)\mathrm{d}_q s \\ &= q^{i\cdot\beta}{}_1\phi_1(\beta;\infty;1+i) = \frac{q^{i\cdot\beta}}{\zeta_{(q)}(1+i)\zeta_{(q)}(\beta)}{}_2\phi_1(\infty,\infty;1+i;\beta) \qquad (10.6.5) \\ &= \frac{q^{i\beta}}{\zeta_{(q)}(1+i)}{}_0\phi_1(-;1+i;1+i+\beta)\,,\end{aligned}$$

as follows (see Szegö 1959):

$$\begin{aligned}\hat{f}^{\beta}_{q,+,n}(s) &= q^{-ns}\mathrm{P}^{\beta}_{2n}(q^s) + q^{ns}\mathrm{P}^{\beta}_{2n}(q^{-s})\,, \\ \hat{f}^{\beta}_{q,-,n+1}(s) &= q^{-ns}\mathrm{P}^{\beta}_{2n+1}(q^s) - q^{ns}\mathrm{P}^{\beta}_{2n+1}(q^{-s})\,,\end{aligned} \qquad (10.6.6)$$

with $\mathrm{P}^{\beta}_{n}(q^s)$, the associated basis for the Hardy space, given by

$$\mathrm{P}^{\beta}_{n}(q^s) = \det\begin{bmatrix} \tilde{\rho}^{\beta}_{(q)}(g^0) & \tilde{\rho}^{\beta}_{(q)}(g^1) & \cdots & \tilde{\rho}^{\beta}_{(q)}(g^n) \\ \tilde{\rho}^{\beta}_{(q)}(g^1) & \tilde{\rho}^{\beta}_{(q)}(g^0) & \cdots & \tilde{\rho}^{\beta}_{(q)}(g^{n-1}) \\ \vdots & & & \vdots \\ \tilde{\rho}^{\beta}_{(q)}(g^{n-1}) & \tilde{\rho}^{\beta}_{(q)}(g^{n-2}) & \cdots & \tilde{\rho}^{\beta}_{(q)}(g^1) \\ 1 & q^s & \cdots & q^{ns}\end{bmatrix}. \qquad (10.6.7)$$

Unfortunately, the moments $\tilde{\rho}^{\beta}_{(q)}(g^i)$ cannot be summed, and so one cannot get a hold on the polynomials $\hat{f}_{q,\pm,n}(s)$. However, in the p-adic and real limits we do get bases which are explicitly given.

The pure gamma basis

For $p \neq \eta$, we obtain the basis for $\widehat{\mathscr{H}}^{\beta}_{p}$ given by

$$\hat{f}^{\beta}_{p,+,0}(s) = 1, \quad \hat{f}^{\beta}_{p,\pm,n}(s) = \mathrm{C}^{\beta/2+s}_{p}(n) \pm \mathrm{C}^{\beta/2-s}_{p}(n), \quad n \geq 1\,, \qquad (10.6.8)$$

and the corresponding base of $\mathscr{H}^{\beta}_{p}$ given by

$$\varphi^{\beta}_{p,+,0} = \phi_{\mathbb{Z}_p}, \quad \varphi^{\beta}_{p,\pm,n} = (-1)^n \cdot \left(\phi_{p^{-n}\mathbb{Z}_p^*} \pm \varphi^{\beta}_{\mathbb{Z}_p,n}\right), \quad n \geq 1\,. \qquad (10.6.9)$$

For the real prime we obtain the basis for $\widehat{\mathscr{H}}^{\beta}_{\eta}$ given by the Meixner–Pollaczeck polynomials

$$\begin{aligned}\hat{f}^{\beta}_{\eta,n}(s) &= (-1)^n\frac{(\beta/2)_n}{n!}{}_2\mathrm{F}_1\left(-n,\frac{\beta}{4}+\frac{s}{2};\frac{\beta}{2};2\right) \\ &= \frac{(-\pi)^n}{n!}\sum_{0\leq k\leq n}\binom{n}{k}(-1)^k\frac{\zeta_\eta(\beta+2n)}{\zeta_\eta(\beta+2k)}\cdot\frac{\zeta_\eta(\beta/2+s+2k)}{\zeta_\eta(\beta/2+s)}\cdot 2^k\,. \qquad (10.6.10)\end{aligned}$$

The $\hat{f}^{\beta}_{\eta,n}(s)$ are orthogonal on $\mathrm{i}\mathbb{R}$ with respect to the measure $\zeta_\eta(\beta/2+s,\beta/2-s)\mathrm{d}_\eta s$ and the parity is now given by

$$\hat{f}^{\beta}_{\eta,n}(-s) = (-1)^n\cdot\hat{f}^{\beta}_{\eta,n}(s)\,. \qquad (10.6.11)$$

The corresponding base $\{\varphi^{\beta}_{\eta,n} = \hat{f}^{\beta}_{\eta,n}(x\partial/\partial x + \beta/2)\phi_{\mathbb{Z}_\eta}\}$ for the space $\mathscr{H}^{\beta}_{\eta}$ is closely related to the Laguerre base for the space $\phi_{\mathbb{Z}_\eta} \cdot \mathscr{H}^{\beta}_{\mathbb{Z}_\eta}$ (cf. $(10.4.1)_\eta$), and is explicitly given by

$$\varphi^{\beta}_{\eta,n}(x) = \phi_{\mathbb{Z}_\eta}(x) \cdot \varphi^{\beta}_{\mathbb{Z}_\eta,n}(\sqrt{2}x)\,. \tag{10.6.12}$$

The innocent looking $\sqrt{2}$ on the right of (10.6.12) has remarkable consequences:

$$\int \varphi^{\beta}_{\mathbb{Z}_\eta,n}(x)\phi_{\mathbb{Z}_\eta}(x)|x|^s \frac{\mathrm{d}^*x}{\zeta_\eta(s)} = (-1)^n \mathrm{C}^{\beta-s}_{\eta}(n) \tag{10.6.13}$$

(cf. (8.2.40)) a polynomial in $\beta - s$ with zeros at $\beta - s \in \{0, -2, -4, \ldots, -2(n-1)\}$;

$$\int \varphi^{\beta}_{\mathbb{Z}_\eta,n}(\sqrt{2}x)\phi_{\mathbb{Z}_\eta}(x)|x|^s \frac{\mathrm{d}^*x}{\zeta_\eta(s)} = \hat{f}^{\beta}_{\eta,n}\left(s - \frac{\beta}{2}\right) \tag{10.6.14}$$

a polynomial in s with n zeros on the line $\mathrm{Re}(s) = \beta/2$.

Thus the $\sqrt{2}$ has the effect of shifting the poles of $\zeta_\eta(s)$ into the critical line.

Was Riemann aware of this phenomenon in some form when he conjectured his famous hypothesis? (Recall that on the eight pages of his draft on the zeta function you can find $\sqrt{2}$ expanded to 37 decimals!).

The Mexiner–Pollaczeck polynomials $\hat{f}^{\beta}_{\eta,n}(s)$ satisfy the recursion relation

$$s\hat{f}^{\beta}_{\eta,n}(s) = (n+1)\cdot \hat{f}^{\beta}_{\eta,n+1}(s) - (n-1+\beta/2)\cdot \hat{f}^{\beta}_{\eta,n-1}(s)\,, \tag{10.6.15}$$

which is just the Mellin transform of the relation

$$-(x\partial/\partial x + \beta/2)\varphi^{\beta}_{\eta,n}(x) = (n+1)\cdot \varphi^{\beta}_{\eta,n+1}(x) - (n-1+\beta/2)\cdot \varphi^{\beta}_{\eta,n-1}(x)\,. \tag{10.6.16}$$

It follows that we can interpret $\hat{f}^{\beta}_{\eta,m}(s)$ as the characteristic polynomial of the Jacobi operator, which in the realization $\mathscr{H}^{\beta}_{\eta}$ is just the infinitesimal generator of the unitary action of the multiplicative group:

$$\begin{aligned} n!\hat{f}^{\beta}_{\eta,n}(s) &= \det\begin{bmatrix} s & \beta/2 & & & & \\ 1 & s & 1+\beta/2 & & & \\ & 2 & s & & & \\ & & & \ddots & & \\ & & & & s & n-2+\beta/2 \\ & & & & n-1 & s \end{bmatrix} \\ &= \det\bigl(\Phi^{\beta}_{n}(s - \beta/2 - x\partial/\partial x)\Phi^{\beta}_{n}\bigr)\,, \end{aligned} \tag{10.6.17}$$

where Φ^{β}_{n} is the projection onto the subspace of $\mathscr{H}^{\beta}_{\eta}$ spanned by $\{(x\partial/\partial x + \beta/2)^j \phi_{\mathbb{Z}_\eta}\}$, $0 \le j \le n-1$.

Define the operators

$$\mathrm{D}^{+}_{\beta}\colon \mathscr{H}^{\beta+2}_{\eta} \longrightarrow \mathscr{H}^{\beta}_{\eta}, \quad \mathrm{D}^{+}_{\beta} = -\frac{x}{2}\frac{\partial}{\partial x} - \frac{\beta}{2} + \pi x^2 \tag{10.6.18}$$

(this is the same operator as in (8.2.28), but now considered on spaces without the subscript $\mathbb{Z}_\eta$),

$$\mathrm{D}^- = \frac{1}{\beta}\left(\mathrm{D}_\beta^+\right)^*: \mathscr{H}_\eta^\beta \longrightarrow \mathscr{H}_\eta^{\beta+2}, \quad \mathrm{D}^- = \frac{1}{2}\left(1 + \frac{1}{2\pi x}\frac{\partial}{\partial x}\right). \tag{10.6.19}$$

We have the Heisenberg relations up the ladder

$$\mathrm{D}^-\mathrm{D}_\beta^+ - \mathrm{D}_{\beta+2}^+\mathrm{D}^- = \mathrm{id}_{\mathscr{H}_\eta^{\beta+2}}, \tag{10.6.20}$$

and we have

$$\varphi_{\eta,n}^\beta = \frac{1}{n!}(\mathrm{D}^+)^n \phi_{\mathbb{Z}_\eta} = \frac{1}{n!}\mathrm{D}_\beta^+\mathrm{D}_{\beta+2}^+ \dots \mathrm{D}_{\beta+2(n-1)}^+ \phi_{\mathbb{Z}_\eta} \tag{10.6.21}$$

(i.e. the same operator as in (8.2.31), but now applied to $\phi_{\mathbb{Z}_\eta}$ rather than to $\mathbb{1}$). From this we obtain

$$\begin{aligned}
\mathrm{D}_\beta^+\varphi_{\eta,n-1}^{\beta+2} &= n \cdot \varphi_{\eta,n}^\beta, \\
\mathrm{D}^-\varphi_{\eta,n}^\beta &= \varphi_{\eta,n-1}^{\beta+2}, \\
\mathrm{N}_\eta^\beta\varphi_{\eta,n}^\beta &= n \cdot \varphi_{\eta,n}^\beta, \\
\mathrm{N}_\eta^\beta = \mathrm{D}_\beta^+\mathrm{D}^- &= -\frac{1}{2}\left(\frac{1}{4\pi}\frac{\partial^2}{\partial x^2} + \left(\frac{\beta-1}{4\pi x}\right)\frac{\partial}{\partial x} + \frac{\beta}{2} - \pi x^2\right).
\end{aligned} \tag{10.6.22}$$

We can translate these operators to the analytic spaces:

$$\begin{aligned}
\widehat{\mathrm{D}}_\beta^+: \widehat{\mathscr{H}}_\eta^{\beta+2} \to \widehat{\mathscr{H}}_\eta^{\beta+2}, \quad & \widehat{\mathrm{D}}_\beta^+\hat{f}(s) = \tfrac{1}{2}[(\beta/2+s)\hat{f}(s+1) - (\beta/2-s)\hat{f}(s-1)], \\
\widehat{\mathrm{D}}^-: \widehat{\mathscr{H}}_\eta^\beta \to \widehat{\mathscr{H}}_{\hat{\eta}}^{\beta+2}, \quad & \widehat{\mathrm{D}}^-\hat{f}(s) = \tfrac{1}{2}[\hat{f}(s+1) - \hat{f}(s-1)],
\end{aligned} \tag{10.6.23}$$

preserving the Heisenberg relations up the ladder, and

$$\hat{f}_{\eta,n}^\beta(s) = \frac{1}{n!}(\widehat{\mathrm{D}}^+)^n\mathbb{1} = \frac{1}{n!}\widehat{\mathrm{D}}_\beta^+\widehat{\mathrm{D}}_{\beta+2}^+ \dots \widehat{\mathrm{D}}_{\beta+2(n-1)}^+\mathbb{1}. \tag{10.6.24}$$

From the fact that $\hat{f}_{\eta,n}^\beta(s) = (1/n)\widehat{\mathrm{D}}_\beta^+\hat{f}_{\eta,n-1}^{\beta+2}(s)$ we can see directly (by induction on n) that $\hat{f}_{\eta,n}^\beta(s)$ is a polynomial of degree n in s with all its zeros on the line $\mathrm{Re}(s) = 0$ (without recourse to any general facts of orthogonal polynomials).

By changing the measure $\zeta_{(q)}(\beta/2+s, \beta/2-s)\mathrm{d}_q s$ slightly (or, equivalently, by changing the cyclic vector $\phi_{\mathbb{Z}_q}$), we shall give below (§10.8) the more manageable q-analogue of the basis $\hat{f}_{\eta,n}^\beta$. Note that the $\hat{f}_{\eta,n}^\beta$ alternates in sign (10.6.11), and as for its q-analogue, it will be necessary to orthogonalize $\{\mathrm{Y}^n\mathbb{1}\}_{n\geq 0}$; the operator $(1-q)^{-1}\cdot\mathrm{Y}$ converges in the real limit to multiplication by $i \cdot s$ which corresponds to the infinitesimal generator of the action of the multiplicative group.

The cut-off basis (or the generalized prolate spheroidal wave functions)

For a finite prime $p \neq \eta$, we have the following finite-dimensional approximations to the Hilbert space $\mathscr{H}_p^\beta$:

$$\mathscr{H}^\beta_{p(N)} = \bigoplus_{\substack{0 \le n \le N \\ +,-}} \mathbb{C}\varphi^\beta_{p,n,\pm} \tag{10.6.25}$$

$$(\text{analytic}) = \left\{\varphi \in \mathscr{H}^\beta_p \mid \tau_p^{\beta/2+s}(\varphi) \in \bigoplus_{|n|\le N} \mathbb{C}\cdot p^{ns}\right\}$$

$$(\text{multiplicative}) = \left\{\pi^\beta_p(f)\phi_{\mathbb{Z}_p} \mid \operatorname{supp} f \subseteq \{p^{-N} \le |a| \le p^N\}\right\} = \bigoplus_{|n|\le N} \mathbb{C}\pi^\beta_p(p^n)\phi_{\mathbb{Z}_p}$$

$$(\text{additive}) = \left\{\varphi \in \mathscr{H}^\beta_p \mid \operatorname{supp} \varphi,\ \operatorname{supp} \mathcal{F}^\beta_p\varphi \subseteq \{|x| \le p^N\}\right\}.$$

We can ask for the real analogue $\mathscr{H}^\beta_{\eta(N)}$, and it turns out that each of the characterizations in (10.6.25) gives rise to a different analogue. The analytic analogue is

$$\begin{aligned}\mathscr{H}^\beta_{\eta(N)} &= \bigoplus_{0\le n\le N} \mathbb{C}\varphi^\beta_{\eta,n} \\ &= \left\{\varphi \in \mathscr{H}^\beta_\eta \,\middle|\, \tau_\eta^{\beta/2+s}(\varphi) \in \bigoplus_{0\le n\le N} \mathbb{C}\cdot s^n\right\} \\ &= \operatorname{span}\left\{\pi^\beta_\eta((a\partial/\partial a)^j)\phi_{\mathbb{Z}_\eta} = (x\partial/\partial x + \beta/2)^j \mathrm{e}^{-\pi x^2},\ 0 \le j \le N\right\}.\end{aligned} \tag{10.6.26}$$

The last line in (10.6.26) gives $\mathscr{H}^\beta_{\eta(N)}$ a multiplicative interpretation, but note that it is very different from the (infinite-dimensional) space $\{\pi^\beta_\eta(f)\phi_{\mathbb{Z}_\eta} \mid \operatorname{supp} f \subseteq [c^{-1}, c]\}$.

In looking for the real analogue of the additive space in (10.6.25), we are confronted by a problem: a compactly supported function φ has an entire holomorphic Fourier transform $\mathcal{F}\varphi$, so $\mathcal{F}\varphi$ is never compactly supported. This problem was analysed in Slepian and Pollak (1961), and Landau and Pollak (1962), with the following surprising solution.

Let

$$\mathrm{B}_c(x) = \begin{cases} 1 & |x|_\eta \le c\,, \\ 0 & |x|_\eta > c\,, \end{cases} \tag{10.6.27}$$

and view B_c as a projection on $\mathscr{H}^\beta_\eta$ via multiplication, with image

$$\mathscr{H}^\beta_{\eta,c} = \mathrm{B}_c\mathscr{H}^\beta_\eta\,. \tag{10.6.28}$$

Let $\widehat{\mathrm{B}}_c = \mathcal{F}^\beta_\eta \mathrm{B}_c \mathcal{F}^\beta_\eta$; it is another projection on $\mathscr{H}^\beta_\eta$, with image

$$\mathscr{H}^\beta_{\eta,\hat{c}} = \mathcal{F}^\beta_\eta\mathscr{H}^\beta_{\eta,c} = \widehat{\mathrm{B}}_c\mathscr{H}^\beta_\eta\,. \tag{10.6.29}$$

The projections $\mathrm{B}_c, \widehat{\mathrm{B}}_c$ do not commute, and as indicated above we have trivial intersection $\mathscr{H}^\beta_{\eta,c} \cap \mathscr{H}^\beta_{\eta,\hat{c}} = \{0\}$.

We can consider the integral operator

$$\mathrm{B}_c\widehat{\mathrm{B}}_c \colon \mathscr{H}^\beta_{\eta,c} \to \mathscr{H}^\beta_{\eta,c}\,. \tag{10.6.30}$$

The associated decomposition into eigenfunctions is

$$\mathscr{H}^{\beta}_{\eta,c} = \bigoplus_{n\geq 0} \mathbb{C}\varphi^{\beta}_{\eta,c,n} \tag{10.6.31}$$

with

$$\mathrm{B}_c\widehat{\mathrm{B}}_c\varphi^{\beta}_{\eta,c,n} = \lambda_n \cdot \varphi^{\beta}_{\eta,c,n}, \quad \lambda_n = \lambda^{\beta}_{c,n}\,.$$
$$1 > \lambda_0 > \lambda_1 > \ldots \downarrow 0\,.$$

It follows that

$$\mathscr{H}^{\beta}_{\eta,\hat{c}} = \bigoplus_{n=0} \mathbb{C}\widehat{\mathrm{B}}_c\varphi^{\beta}_{\eta,c,n} \tag{10.6.31$'$}$$

and $\|\widehat{\mathrm{B}}_c\varphi^{\beta}_{\eta,c,n}\|^2 = \lambda_n \cdot \|\varphi^{\beta}_{\eta,c,n}\|^2$ or, equivalently, $\|\mathrm{B}_c\mathcal{F}^{\beta}_{\eta}\varphi^{\beta}_{\eta,c,n}\|^2 = \lambda_n \cdot \|\mathcal{F}^{\beta}_{\eta}\varphi^{\beta}_{\eta,c,n}\|^2$. Thus for λ_n near $1, \mathcal{F}^{\beta}_{\eta}\varphi^{\beta}_{\eta,c,n}$ has most of its 'energy' supported in $[-c,c]$, and for λ_n near $0, \mathcal{F}^{\beta}_{\eta}\varphi^{\beta}_{\eta,c,n}$ has most of its 'energy' supported outside $[-c,c]$.

The main result of Landau, Pollak, and Slepian is that the λ_n for $n \leq 2c^2$ are near 1, then they decrease sharply from 1 to 0 in an interval of length $\mathrm{O}(\log c)$, and λ_n are near 0 for $n \geq 2c^2 + \mathrm{O}(\log c)$. This can be seen most easily by combining the case $\beta = 1$ of even functions on $\mathrm{L}_2(\mathbb{R})$ with the case $\beta = 3$ of odd functions. In this case $\mathrm{B}_c\widehat{\mathrm{B}}_c$ is an integral operator on $[-c,c]$ with kernel

$$\delta_c(x,y) = \frac{\sin 2\pi c(x,y)}{\pi(x,y)}$$

and its eigenfunctions are alternatingly even and odd. We have

$$\sum_{n\geq 0} \lambda_n = \mathrm{tr}(\mathrm{B}_c\widehat{\mathrm{B}}_c) = \int_{-c}^{c} \mathrm{d}x\, \delta_c(x,x) = 4c^2\,,$$
$$\sum_{n\geq 0} \lambda_n^2 = \mathrm{tr}\Big((\mathrm{B}_c\widehat{\mathrm{B}}_c)^2\Big) = \int_{-c}^{c}\!\!\int \mathrm{d}x\,\mathrm{d}y\, \delta_c(x,y)^2 \geq 4c^2 - \frac{\log \pi c^2}{\pi^2} - 1\,.$$

It follows from the inequality

$$\lambda_N \leq \frac{\Sigma\lambda_n - \Sigma\lambda_n^2}{N - \Sigma\lambda_n}$$

(valid for $N > \Sigma\lambda_n$, and any $1 \geq \lambda_0 \geq \lambda_1 \geq \cdots \geq 0$) that

$$\lambda_N \leq \frac{(\log \pi c^2)/\pi^2 + 1}{N - 4c^2} \quad \text{for } N > 4c^2\,.$$

Thus for $N \geq 4c^2 + \frac{1}{\varepsilon}\left(\frac{\log \pi c^2}{\pi^2} + 1\right)$, we have $\lambda_N \leq \varepsilon$.

The 'intersection' of $\mathscr{H}^{\beta}_{\eta,c}$ and $\mathscr{H}^{\beta}_{\eta,\hat{c}}$ can thus be taken to be (for all $\beta > 0$)

$$\mathscr{H}^{\beta}_{\eta}\langle c\rangle \stackrel{\mathrm{def}}{=} \bigoplus_{0\leq n\leq 2c^2} \mathbb{C}\varphi^{\beta}_{\eta,c,n}\,. \tag{10.6.32}$$

The $\varphi^{\beta}_{\eta,c,n}$ are also eigenfunctions of a differential operator $\mathrm{H}^{\beta}_{\eta,c}$ that commutes with $\mathrm{B}_c, \widehat{\mathrm{B}}_c$, and $\mathcal{F}^{\beta}_{\eta}$. For $\beta = 1$ this is the well-known operator

$$\begin{aligned}\mathrm{H}^{1}_{\eta,c} &= -\frac{\partial}{\partial x}(c^2 - x^2)\frac{\partial}{\partial x} + (2\pi c x)^2 \\ &= c^2\left((2\pi x)^2 - \frac{\partial^2}{\partial x^2}\right) + \frac{\partial}{\partial x}x^2\frac{\partial}{\partial x}\end{aligned} \qquad (10.6.33)$$

and thus $\varphi^{1}_{\eta,c,n}$ are the (even) prolate spheroidal wave functions.

We shall give next a brief description of the q-analogue, which will give us in the real limit the operators $\mathrm{H}^{\beta}_{\eta,c}$ for all β, and clarify some of the algebra behind these operators. We consider

$$\mathrm{B}_c(g^k) = \begin{cases} 1 & k \geq c \\ 0 & k < c \end{cases} \qquad c \in \mathbb{Z}, \qquad (10.6.34)$$

as a projection on $\mathscr{H}^{\beta}_{q}$ with image $\mathrm{B}_c\mathscr{H}^{\beta}_{q} = \mathscr{H}^{\beta}_{q,c}$, and consider the dual projection

$$\widehat{\mathrm{B}}_c = \mathcal{F}^{\beta}_{q}\mathrm{B}_c\mathcal{F}^{\beta}_{q}\,. \qquad (10.6.35)$$

Together they give rise to the integral operator

$$\mathrm{B}_c\widehat{\mathrm{B}}_c\colon\ \mathscr{H}^{\beta}_{q,c} \longrightarrow \mathscr{H}^{\beta}_{q,c}\,, \qquad (10.6.36)$$

with kernel

$$\delta^{\beta}_{q,c}(x,y) = \int\limits_{|z|_{(q)}\leq q^c} \tau^{\beta}_{q}(\mathrm{d}z)\mathcal{F}^{\beta}_{q}(zx)\mathcal{F}^{\beta}_{q}(zy)\,. \qquad (10.6.37)$$

As in (10.6.31) we have the associated decomposition into eigenfunctions

$$\begin{aligned}\mathscr{H}^{\beta}_{q,c} &= \bigoplus_{n\geq 0}\mathbb{C}\varphi^{\beta}_{q,c,n} \\ \mathrm{B}_c\widehat{\mathrm{B}}_c\varphi^{\beta}_{q,c,n} &= \lambda_n \cdot \varphi^{\beta}_{q,c,n}, \qquad \lambda_n = \lambda^{\beta}_{q,c,n}\,.\end{aligned} \qquad (10.6.38)$$

In the analytic model $\widehat{\mathscr{H}}^{\beta}_{q}$, the operator B_c is the projection onto

$$\widehat{\mathscr{H}}^{\beta}_{q,c} = \bigoplus_{n\geq c}\mathbb{C}\cdot\frac{q^{ns}}{\zeta_{(q)}(\beta/2+s)}\,, \qquad (10.6.39)$$

and $\widehat{\mathrm{B}}_c$ is the projection onto

$$\widehat{\mathrm{B}}_c\widehat{\mathscr{H}}^{\beta}_{q} = \widehat{\mathcal{F}}^{\beta}_{q}\widehat{\mathscr{H}}^{\beta}_{q,c} = \bigoplus_{n\geq c}\mathbb{C}\cdot\frac{q^{-ns}}{\zeta_{(q)}(\beta/2-s)}\,. \qquad (10.6.40)$$

In particular, for $c \leq 0$, $\mathrm{B}_c\mathscr{H}^{\beta}_{q} \cap \widehat{\mathrm{B}}_c\mathscr{H}^{\beta}_{q}$ is a $(2|c|+1)$-dimensional space, whose p-adic limit gives the space $\mathscr{H}^{\beta}_{p(|c|)}$ of (10.6.25). On the other hand, when $c > 0$, $\mathrm{B}_c\mathscr{H}^{\beta}_{q}$ and $\widehat{\mathrm{B}}_c\mathscr{H}^{\beta}_{q}$ have trivial intersection. To describe the differential operator $\mathrm{H} = \mathrm{H}^{\beta}_{q,c}$, recall the q-commuting operators $\mathcal{X}, \mathcal{Y}$ of §10.4:

$$\mathcal{X}\varphi(g^n) = q^n\cdot\varphi(g^n), \quad \mathcal{Y}\varphi(g^n) = \varphi(g^{n-1})\,; \qquad (10.6.41)$$

as operators on $\mathscr{H}^{\beta}_{q}$:

$$\mathcal{X}^* = \mathcal{X}, \quad \mathcal{Y}^* = q^{\beta}\cdot\mathcal{Y}^{-1} = \mathcal{F}^{\beta}_{q}\mathcal{Y}\mathcal{F}^{\beta}_{q}\,. \qquad (10.6.42)$$

$$\mathcal{F}_q^\beta \mathcal{X} \mathcal{F}_q^\beta = (1-\mathcal{Y}^*)\mathcal{X}^{-1}(1-\mathcal{Y}), \quad \text{cf. (10.6.29)}. \tag{10.6.43}$$

We let H be the self-adjoint operator on $\mathscr{H}_{q,c}^\beta$ given by

$$\mathrm{H} = \mathrm{H}_{q,c}^\beta = q^c \cdot \left[\mathcal{X} + \mathcal{F}_q^\beta \mathcal{X} \mathcal{F}_q^\beta\right] - (1-\mathcal{Y}^*)(1-\mathcal{Y}). \tag{10.6.44}$$

It is clear that H commutes with $\mathcal{F}_q^\beta$. Using (10.6.43) we can write

$$\mathrm{H} = q^c \cdot \mathcal{X} + (1-\mathcal{Y}^*)(q^c\mathcal{X}^{-1} - 1)(1-\mathcal{Y}). \tag{10.6.45}$$

Integration by parts, using $(q^c\mathcal{X}^{-1} - 1)\varphi(g^c) = 0$, shows that B_c commutes with $(1-\mathcal{Y}^*)(q^c\mathcal{X}^{-1}-1)(1-\mathcal{Y})$, and hence with H. It follows that H commutes with $\widehat{\mathrm{B}}_c = \mathcal{F}_q^\beta \mathrm{B}_c \mathcal{F}_q^\beta$, and hence with $\mathrm{B}_c\widehat{\mathrm{B}}_c$. The eigenfunctions $\varphi_{q,c,n}^\beta$ of $\mathrm{B}_c\widehat{\mathrm{B}}_c$ are also eigenfunctions of H,

$$\mathrm{H}\varphi_{q,c,n}^\beta = \chi_n \cdot \varphi_{q,c,n}^\beta \quad \chi_n = \chi_{q,c,n}^\beta . \tag{10.6.46}$$

In the real limit Ⓡ: $q \to 1, n \to \infty$, such that $q^n/(1-q) \to \pi x^2$, the operators $\mathcal{X}, \mathcal{Y}$ converge to the operators on $\mathscr{H}_\eta^\beta$,

$$\frac{\mathcal{X}}{1-q} \underset{Ⓡ}{\longrightarrow} \pi x^2 \tag{10.6.47}$$

$$\frac{1-\mathcal{Y}}{1-q} \underset{Ⓡ}{\longrightarrow} -\frac{x}{2}\frac{\partial}{\partial x} \tag{10.6.48}$$

$$\frac{1-\mathcal{Y}^*}{1-q} \underset{Ⓡ}{\longrightarrow} \frac{\beta}{2} + \frac{x}{2}\frac{\partial}{\partial x}. \tag{10.6.49}$$

If we also include in the real limit Ⓡ the condition on $c \in \mathbb{Z}$,

$$c \to \infty, \quad \text{such that } \frac{q^c}{1-q} \to \pi c_\eta^2, \tag{10.6.50}$$

the projections B_c and $\widehat{\mathrm{B}}_c$ converge to their real counterparts B_{c_η} and $\widehat{\mathrm{B}}_{c_\eta}$, and so does the integral operator and its eigenfunctions:

$$\varphi_{q,c,n}^\beta \longrightarrow \varphi_{\eta,c,n}^\beta . \tag{10.6.51}$$

The limit of the differential operator H is given by

$$\begin{aligned} \mathrm{H}_{\eta,c}^\beta &= 4 \lim_{Ⓡ} \frac{1}{(1-q)^2} \mathrm{H}_{q,c}^{\beta/2} \\ &= 4\pi c_\eta^2 \left[\pi x^2 + \mathcal{F}_\eta^\beta \pi x^2 \mathcal{F}_\eta^\beta\right] + \left(\beta + x\frac{\partial}{\partial x}\right)\left(x\frac{\partial}{\partial x}\right) \\ &= (2\pi c_\eta x)^2 + \left(x^2 - c_\eta^2\right)\frac{\partial^2}{\partial x^2} + (1+\beta)x\frac{\partial}{\partial x} + c_\eta^2(1-\beta)\frac{1}{x}\frac{\partial}{\partial x} \\ &= (2\pi c_\eta x)^2 + x^{1-\beta}\frac{\partial}{\partial x}x^{\beta-1}(x^2 - c_\eta^2)\frac{\partial}{\partial x}. \end{aligned} \tag{10.6.52}$$

At $\beta = 1$, this is the operator (10.6.33).

Note that we can write

$$\mathrm{H}^{\beta}_{\eta,c} = 4\pi c^2_{\eta}\,\mathrm{H}^{\beta}_{\eta,\infty} + \mathrm{H}^{\beta}_{\eta,0}\,. \tag{10.6.53}$$

For $c_\eta \to 0$, $\mathrm{H}^{\beta}_{\eta,c}$ approaches

$$\mathrm{H}^{\beta}_{\eta,0} = \left(\beta + x\frac{\partial}{\partial x}\right)x\frac{\partial}{\partial x} = x^{1-\beta}\frac{\partial}{\partial x}x^{1+\beta}\frac{\partial}{\partial x}\,. \tag{10.6.54}$$

The more important limit is when $c_\eta \to \infty$; then $\mathrm{H}^{\beta}_{\eta,c}$ approaches

$$\begin{aligned}\mathrm{H}^{\beta}_{\eta,\infty} &= \pi x^2 + \mathcal{F}^{\beta}_{\eta}\pi x^2 \mathcal{F}^{\beta}_{\eta} \quad \text{and using the real limit of (10.6.43)}\\ &= \pi x^2 + \left(1 + \frac{x}{2}\frac{\partial}{\partial x}\right)\frac{1}{\pi x^2}\left(1 - \frac{\beta}{2} - \frac{x}{2}\frac{\partial}{\partial x}\right)\\ &= \pi x^2 - \frac{1}{4\pi}\left(\frac{\partial^2}{\partial x^2} + \frac{\beta - 1}{x}\frac{\partial}{\partial x}\right) = \pi x^2 - \frac{1}{4\pi}x^{1-\beta}\frac{\partial}{\partial x}x^{\beta-1}\frac{\partial}{\partial x}\\ &= 2\mathrm{D}^{+}_{\beta}\mathrm{D}^{-} + \frac{\beta}{2} = 2\mathrm{N}^{\beta}_{\eta} + \frac{\beta}{2}, \quad \text{cf. (10.6.22)}\,.\end{aligned} \tag{10.6.55}$$

The eigenfunctions $\varphi^{\beta}_{\eta,c,n}$ approximate (a scalar multiple of) the pure gamma basis $\varphi^{\beta}_{\eta,n}$,

$$\varphi^{\beta}_{\eta,c,n} \xrightarrow[c\to\infty]{} \varphi^{\beta}_{\eta,n}\,, \tag{10.6.56}$$

if we normalize the $\varphi^{\beta}_{\eta,c,n}$ by $(\varphi^{\beta}_{\eta,c,n}, \varphi^{\beta}_{\eta,n}) = \| \varphi^{\beta}_{\eta,n} \|^2$.

Let $M^c = (a^c_{m,n})$ denote the matrix of the inclusion $\mathscr{H}^{\beta}_{\eta,c} \hookrightarrow \mathscr{H}^{\beta}_{\eta}$ with respect to the basis $\{\varphi^{\beta}_{\eta,c,n}\}$, $\{\varphi^{\beta}_{\eta,n}\}$,

$$\varphi^{\beta}_{\eta,c,n} = \sum_m a^c_{m,n}\varphi^{\beta}_{\eta,m}\,, \tag{10.6.57}$$

and $a^c_{n,n} = 1$ by the normalization (10.6.56). From the recursion relation (10.6.16) of $\{\varphi^{\beta}_{\eta,m}\}$, we obtain

$$\begin{aligned}\mathrm{H}^{\beta}_{\eta,0}\varphi^{\beta}_{\eta,m} &= (m+1)(m+2)\varphi^{\beta}_{\eta,m+2} - \left[2m^2 + \beta\cdot m + \frac{\beta}{2}\left(1 + \frac{\beta}{2}\right)\right]\varphi^{\beta}_{\eta,m}\\ &\quad + \left(m - 1 + \frac{\beta}{2}\right)\left(m - 2 + \frac{\beta}{2}\right)\varphi^{\beta}_{\eta,m-2}\,.\end{aligned} \tag{10.6.58}$$

Putting (10.6.57) into the equation

$$\mathrm{H}^{\beta}_{\eta,c}\varphi^{\beta}_{\eta,c,n} = \chi_n\varphi^{\beta}_{\eta,c,n},\ \chi_n = \chi^{\beta}_{c,n}\,, \tag{10.6.59}$$

using (10.6.22), (10.6.53–10.6.58), and then comparing coefficients, we obtain the recursion relation of the $a^c_{m,n}$ in m (with n fixed):

$$\begin{aligned}&\left\{\chi_n - 4\pi c^2_{\eta}\left(2m + \frac{\beta}{2}\right) + \left[2m^2 + \beta\cdot m + \frac{\beta}{2}\left(1 + \frac{\beta}{2}\right)\right]\right\}\cdot a^c_{m,n}\\ &\quad = m(m-1)a^c_{m-2,n} + \left(m + \frac{\beta}{2}\right)\left(m + 1 + \frac{\beta}{2}\right)a^c_{m+2,n}\,.\end{aligned} \tag{10.6.60}$$

From this we deduce

$$a^c_{m,n} = \mathrm{O}\big(c_\eta^{-|m-n|}\big) \quad \text{as } c_\eta \to \infty\,, \tag{10.6.61}$$

and the asymptotic expansion

$$\begin{aligned}\chi^\beta_{c,n} \simeq 4\pi c_\eta^2\Big(2n+\frac{\beta}{2}\Big) - \Big[2n^2+\beta n+\frac{\beta}{2}\Big(1+\frac{\beta}{2}\Big)\Big] \\ -\frac{1}{8\pi c_\eta^2}\Big(2n+\frac{\beta}{2}\Big)\Big(2n^2+n\beta+1+\frac{\beta}{2}\Big)+\mathrm{O}\Big(\frac{1}{c_\eta^4}\Big)\,.\end{aligned} \tag{10.6.62}$$

$\varphi^\beta_{\eta,c,n}(x)$ is the restriction of an entire function satisfying the integral equation

$$\mathcal{F}^\beta_\eta \mathrm{B}_c \varphi^\beta_{\eta,c,n}(x) = \varepsilon_n \cdot \varphi^\beta_{\eta,c,n}(x) \quad \text{for all } x \in \mathbb{C},\ \varepsilon_n = \varepsilon^\beta_{c,n},\ \varepsilon_n^2 = \lambda_n\,. \tag{10.6.63}$$

The asymptotic expansion of the $\varphi^\beta_{\eta,c,n}(x)$ as $c_\eta \to \infty$ with x real and $|x| \to \infty$ was studied by Slepian (1964*a*) by solving the differential equation. The translation between our notation and Slepian's is given by

$$\varphi^\beta_{\eta,c,n}(x) = (2\pi)^{(1-\beta)/4}|x|^{(1-\beta)/2}\varphi^{\text{Slepian}}_{\beta/2-1,n}\Big(\frac{x}{c_\eta}\Big), \text{ with } c^{\text{Slepian}} = 2\pi c_\eta^2\,, \tag{10.6.64}$$

(for Slepian $N = \beta/2 - 1$ is an integer but the results hold for all $\beta \geq 1$).

We have for fixed $\beta \geq 1$ and fixed integer n (writing c for c_η)

$$\begin{aligned}\varphi^\beta_{\eta,c,n}(x) &\approx \varphi^\beta_{\eta,n}(x) = \mathrm{L}^{\beta/2-1}_n(2\pi x^2)\mathrm{e}^{-\pi x^2} \quad \text{for } 0 \leq |x| \leq (2\pi)^{-1/4}\cdot c^{1/2} \\
&\approx k\cdot c^{(\beta+1)/2}\cdot\frac{\mathrm{e}^{2\pi c\sqrt{c^2-x^2}}(c-\sqrt{c^2-x^2})^n}{(c^2-x^2)^{1/4}(c+\sqrt{c^2-x^2})^{n+\beta/2}} \quad \text{for } (2\pi)^{-1/4}c^{1/2} \leq |x| \leq c-\frac{1}{2\pi c} \\
&\approx k\cdot 2\pi c\cdot\begin{cases}\mathrm{I}_0(2\pi c\sqrt{c^2-x^2}) & \text{for } c-\dfrac{1}{2\pi c} \leq |x| \leq c \\ \mathrm{J}_0(2\pi c\sqrt{x^2-c^2}) & \text{for } c \leq |x| \leq c+\dfrac{1}{2\pi c}\end{cases} \\
&\approx k\cdot c\cdot 2\mathrm{Re}\frac{\mathrm{e}^{2\pi\mathrm{i}c\sqrt{x^2-c^2}}(c-\mathrm{i}\sqrt{x^2-c^2})^n}{\mathrm{e}^{\pi\mathrm{i}/4}(x^2-c^2)^{1/4}(c+\mathrm{i}\sqrt{x^2-c^2})^{n+\beta/2}} \quad \text{for } |x| \geq c+\frac{1}{2\pi c}\,,\end{aligned} \tag{10.6.65}$$

with

$$k = \frac{(-1)^n}{n!}(8\pi c^2)^n\mathrm{e}^{-2\pi c^2}\cdot 2^{\beta/2}\,.$$

The asymptotic behaviour of the eigenvalues $\varepsilon^\beta_{c,n}$ and $\lambda^\beta_{c,n}$ as $c \to \infty$ (with $\beta \geq 1$ and n fixed) is given by

$$\begin{aligned}\varepsilon^\beta_{c,n} &= (-1)^n\cdot\Big\{1-\frac{\pi(8\pi c^2)^{2n+\beta/2}\mathrm{e}^{-4\pi c^2}}{n!\,\Gamma(\beta/2+n)}\Big[1+\mathrm{O}\Big(\frac{1}{c^2}\Big)\Big]\Big\}\,, \\ \lambda^\beta_{c,n} &= (\varepsilon^\beta_{c,n})^2 = 1-\frac{2\pi(8\pi c^2)^{2n+\beta/2}\mathrm{e}^{-4\pi c^2}}{n!\,\Gamma(\beta/2+n)}\Big[1+\mathrm{O}\Big(\frac{1}{c^2}\Big)\Big]\,.\end{aligned} \tag{10.6.66}$$

When we let $c \to \infty$ and at the same time let $n \to \infty$ in such a way that

$$n = 2c^2 + \frac{\delta}{\pi} \log c + \mathrm{O}(1) \tag{10.6.67}$$

we have

$$\chi^{\beta}_{c,n} = (2\pi c^2)^2 + 4\pi c^2 \cdot \delta + \mathrm{O}(1) \tag{10.6.68}$$

and

$$\begin{aligned} \varepsilon^{\beta}_{c,n} &\approx \frac{(-1)^n}{\sqrt{2\pi}} \left| \Gamma\left(\frac{1+\mathrm{i}\delta}{2}\right) \right| \mathrm{e}^{(-\pi/4)\delta} = (-1)^n \cdot (1 + \mathrm{e}^{\pi\delta})^{-1/2} , \\ \lambda^{\beta}_{c,n} &\approx \frac{1}{2\pi} \left| \Gamma\left(\frac{1+\mathrm{i}\delta}{2}\right) \right|^2 \mathrm{e}^{-(\pi/2)\delta} = (1 + \mathrm{e}^{\pi\delta})^{-1} . \end{aligned} \tag{10.6.69}$$

This shows most clearly the exponential decay of the $\lambda^{\beta}_{c,n}$ as n varies in an interval of length $\mathrm{O}(\log c)$ about $2c^2$ (the decay being uniform with respect to β). The asymptotic expansion of the $\varphi^{\beta}_{\eta,c,n}(x)$ as c, $n \to \infty$ assuming (10.6.67) was determined by Slepian and reads

$$\begin{aligned} \varphi^{\beta}_{\eta,c,n}(x) &\approx |2\pi x^2|^{(1-\beta)/4} \cdot \left|\frac{x}{c}\right|^{1/2} \cdot \mathrm{J}_{\beta/2-1}(2\pi c x) \quad \text{for } 0 \le |x| \le \frac{1}{\sqrt{2\pi}} , \\ &\approx \operatorname{Re}\, (2\pi \mathrm{i} x^2)^{(1-\beta)/4} \cdot \frac{\mathrm{e}^{2\pi \mathrm{i} c x}}{\pi\sqrt{c^2 - x^2}} \left[\frac{c+x}{c-x}\right]^{\mathrm{i}\delta/2} \quad \text{for } \frac{1}{\sqrt{2\pi}} \le |x| \le c - \frac{1}{\sqrt{2\pi}} , \\ &\approx \frac{(-1)^n}{2} (2\pi c^2)^{(1-\beta)/4} |\zeta_\eta(1+\mathrm{i}\delta)| \mathrm{e}^{(\pi/4)\delta} \cdot \mathrm{e}^{2\pi \mathrm{i} c(c-x)} \cdot {}_1\phi_1\left(\frac{1-\mathrm{i}\delta}{2}; 1; 4\pi \mathrm{i} c(x-c)\right) \\ &\qquad\qquad \text{for } c - \frac{1}{\sqrt{2\pi}} \le |x| \le c + \frac{1}{\sqrt{2\pi}} , \\ &\approx \operatorname{Re}\, (2\pi \mathrm{i} x^2)^{(1-\beta)/4} \cdot \frac{\mathrm{e}^{2\pi \mathrm{i} c x}}{\pi\sqrt{x^2 - c^2}} \left[\frac{c+x}{c-x}\right]^{\mathrm{i}\delta/2} \cdot \frac{\mathrm{e}^{-(\pi \mathrm{i}/2)(1+\mathrm{i}\delta)}}{\sqrt{2}} \quad \text{for } |x| \ge c + \frac{1}{\sqrt{2\pi}} . \end{aligned} \tag{10.6.70}$$

10.7 THE PURE BETA BASIS

We can proceed in a similar way to §10.6 in the context of the beta measure of §10.5. We decompose the analytic space $\widehat{\mathscr{H}}^{\alpha,\beta}_{p}$ into its even and odd subspaces $\widehat{\mathscr{H}}^{\alpha,\beta}_{p,\pm}$. For $p \neq \eta$ we obtain a basis for $\widehat{\mathscr{H}}^{\alpha,\beta}_{p,+}$ (respectively $\widehat{\mathscr{H}}^{\alpha,\beta}_{p,-}$) by orthogonalizing $\{\mathrm{X}^n 1\}_{n\geq 0}$ (respectively $\{\mathrm{X}^n \mathrm{Y} 1\}_{n\geq 0}$); the resulting basis will be denoted $\left\{\hat{f}^{\alpha,\beta}_{p,+,n}(s)\right\}_{n\geq 0}$ $\left(\text{respectively } \left\{\hat{f}^{\alpha,\beta}_{p,-,n}(s)\right\}_{n\geq 1}\right)$. For the real prime we obtain a basis alternating in sign, by orthogonalizing $\{s^n 1\}_{n\geq 0}$; we will denote it by $\left\{\hat{f}^{\alpha,\beta}_{\eta,n}(s)\right\}_{n\geq 0}$. Thus

$$\hat{f}^{\alpha,\beta}_{p,\pm,n}(-s) = \pm \hat{f}^{\alpha,\beta}_{p,\pm,n}(s) \quad \text{for } p \neq \eta, \quad \hat{f}^{\alpha,\beta}_{\eta,n}(-s) = (-1)^n \hat{f}^{\alpha,\beta}_{\eta,n}(s) . \tag{10.7.1}$$

The p-adic basis is given by $\hat{f}^{\alpha,\beta}_{p,+,0}(s) = 1$ and for $n \geq 1$,

$$\hat{f}^{\alpha,\beta}_{p,\pm,n}(s) = \zeta_p((\alpha+\beta)/2) \cdot \Big[(1-p^{s-\alpha/2})(1-p^{s-\beta/2})p^{-ns} \pm (1-p^{-s-\alpha/2})(1-p^{-s-\beta/2})p^{ns}\Big]. \tag{10.7.2}$$

The corresponding basis for the space $\mathscr{H}^{\alpha,\beta}_p$ will be denoted $\{\varphi^{\alpha,\beta}_{p,\pm,n}\}$ and referred to as the *p-adic symmetric Jacobi basis* (recall that $\mathscr{H}^{\alpha,\beta}_p$ is the boundary space of the symmetric beta chain we encountered in §4.3). Thus

$$\hat{f}^{\alpha,\beta}_{p,\pm,n}(s) = \tau_p^{\alpha/2-s,\beta/2+s}\left(\varphi^{\alpha,\beta}_{p,\pm,n}\right), \tag{10.7.3}$$

and $\varphi^{\alpha,\beta}_{p,\pm,n}$ are given explicitly by $\varphi^{\alpha,\beta}_{p,+,0} = 1$, and for $n \geq 1$,

$$\varphi^{\alpha,\beta}_{p,\pm,n}(x) = (-1)^n\Big[p^{-\beta n/2}\varphi^{\beta}_{\mathbb{Z}_p,n}(x) \pm p^{-\alpha n/2}\varphi^{\alpha}_{\mathbb{Z}_p,n}(x^{-1})\Big] \tag{10.7.4}$$

with $(-1)^n\varphi^{\beta}_{\mathbb{Z}_p,n} = p^{n\beta}\phi_{p^n\mathbb{Z}_p} - p^{(n-1)\beta}\phi_{p^{n-1}\mathbb{Z}_p}$, the p-adic Laguerre basis (4.2.1).

For example, for $n = 1$, $\varphi^{\alpha,\beta}_{p,\pm,1}(x)$ is given by

$$(p^{\beta/2}-p^{-\beta/2})\phi_{p\mathbb{Z}_p}(x) - (p^{-\beta/2} \pm p^{\alpha/2})\phi_{\mathbb{Z}_p^*}(x) \pm (p^{\alpha/2}-p^{-\alpha/2})\phi_{p\mathbb{Z}_p}(x^{-1}). \tag{10.7.5}$$

For $N \geq n$, $\varphi^{\alpha,\beta}_{p,\pm,n}$ can be viewed as a function defined on $\mathrm{X}_{p(N)} = (\mathbb{Z}/p^N)^*\backslash\, \mathbb{P}^1(\mathbb{Z}/p^N)$, and $\{\varphi^{\alpha,\beta}_{p,\pm,n}\}_{n\leq N}$ given an orthogonal basis for the mod p^N approximating space $\mathscr{H}^{\alpha,\beta}_{p(N)}$.

For the real prime, the basis $\{\hat{f}^{\alpha,\beta}_{\eta,n}(s)\}$ are the *continuous symmetric Hahn polynomials*; they correspond to the Jacobi basis $\{\varphi^{\alpha,\beta}_{\eta,n}(x)\}$ of $\mathscr{H}^{\alpha,\beta}_\eta$ (cf. (8.2.20)); thus

$$\begin{aligned}\hat{f}^{\alpha,\beta}_{\eta,n}(s) &= \tau_\eta^{\alpha/2-s,\,\beta/2+s}\left(\varphi^{\alpha,\beta}_{\eta,n}\right)\\ &= \frac{\pi^n}{n!}\frac{\zeta_\eta(\alpha+2n)\zeta_\eta(\beta+2n)}{\zeta_\eta(\alpha/2-s)\zeta_\eta(\beta/2+s)}\sum_{i_0+j_0=n}\binom{n}{i_0}(-1)^{i_0}\frac{\zeta_\eta(\alpha/2-s+2i_0,\,\beta/2+s+2j_0)}{\zeta_\eta(\alpha+2i_0)\,\zeta_\eta(\beta+2j_0)}.\end{aligned} \tag{10.7.6}$$

We can translate the raising and lowering operators of the Jacobi base $\varphi^{\alpha,\beta}_{\eta,n}$ (8.2.17) into the analytic context. They are given by

$$\widehat{\mathrm{D}}^-:\ \widehat{\mathscr{H}}^{\alpha,\beta}_\eta \longrightarrow \widehat{\mathscr{H}}^{\alpha+2,\beta+2}_\eta,\quad \widehat{\mathrm{D}}^-\hat{f}(s) = \tfrac{1}{2}[\hat{f}(s+1)-\hat{f}(s-1)], \tag{10.7.7}$$

$$\begin{aligned}\widehat{\mathrm{D}}^+_{\alpha,\beta}:\ \widehat{\mathscr{H}}^{\alpha+2,\beta+2}_\eta \longrightarrow \widehat{\mathscr{H}}^{\alpha,\beta}_\eta,\ \mathrm{D}^+_{\alpha,\beta}\hat{f}(s) = \tfrac{1}{2}[&(\alpha/2+s)(\beta/2+s)\hat{f}(s+1)\\ &-(\alpha/2-s)(\beta/2-s)\hat{f}(s-1)],\end{aligned}$$

preserving the Heisenberg relations (8.2.18),

$$\widehat{\mathrm{D}}^-\widehat{\mathrm{D}}^+_{\alpha,\beta} - \widehat{\mathrm{D}}^+_{\alpha+2,\beta+2}\widehat{\mathrm{D}}^- = (\alpha+\beta)/2\cdot \mathrm{id}_{\widehat{\mathscr{H}}^{\alpha+2,\beta+2}_\eta}. \tag{10.7.8}$$

From the fact that $\hat{f}^{\alpha,\beta}_{\eta,n}(s) = 2/n(\alpha+\beta)\cdot\widehat{\mathrm{D}}^+_{\alpha,\beta}\hat{f}^{\alpha+2,\beta+2}_{\eta,n-1}(s)$ we can deduce (by induction on n) that $\hat{f}^{\alpha,\beta}_{\eta,n}(s)$ is a polynomial of degree n in s, all of its zeros are on the line $\mathrm{Re}(s) = 0$ (without recourse to any general facts of orthogonal polynomials), and similarly we can deduce all the properties of these polynomials (see Askey and Wilson (1985) for these).

10.8 THE ASKEY–WILSON POLYNOMIALS

We have seen that all our orthogonal bases came equipped with raising and lowering operators satisfying the Heisenberg relations up the ladder, from which their properties are deducible. The question emerges whether this structure exists for the most general (explicitly known) orthogonal polynomials of Askey and Wilson (1985). That the answer is affirmative is the subject of this section. The Askey–Wilson weight function has a factor $\zeta_{(q)}(2\mathrm{i}t)^{-1}\zeta_{(q)}(-2\mathrm{i}t)^{-1}$ which is missing in the weight function of our analytic space $\widehat{\mathscr{H}}^{\beta}_{q}$, and we shall start by reviewing the beta integrals of imaginary argument in order to explain the root of this extra factor, and how it can disappear in the real and p-adic limits, but not in the q-world. To make the comparison of this section with the literature more transparent, we shall use the more standard notation

$$\begin{aligned}(a)_n = (a;q)_n &= \prod_{0\le m<n}(1-q^m a) = \frac{\zeta_{(q)}(\alpha+n)}{\zeta_{(q)}(\alpha)}\,,\\ (a)_\infty = (a;q)_\infty &= \prod_{n\ge 0}(1-q^n a) = \frac{1}{\zeta_{(q)}(\alpha)} \qquad \text{for } a=q^\alpha\,,\end{aligned} \tag{10.8.1}$$

and write $\mathrm{e}^{\mathrm{i}\theta}$ for $q^{\mathrm{i}t}$, so that $(\mathrm{e}^{2\mathrm{i}\theta})_\infty = \zeta_{(q)}(2\mathrm{i}t)^{-1}$.

We have the following q-beta-integrals of imaginary arguments

$$\mathrm{I}^{(4)}_q: \quad \frac{1}{2\pi}\int_0^\pi \left|\frac{(\mathrm{e}^{2\mathrm{i}\theta})_\infty}{\prod_{1\le j\le 4}(a_j\mathrm{e}^{\mathrm{i}\theta})_\infty}\right|^2 \mathrm{d}\theta = \frac{(a_1a_2a_3a_4)_\infty}{(q)_\infty\prod_{i<j\le 4}(a_ia_j)_\infty}\,,$$

$$a_4 = 0\downarrow$$

$$\mathrm{I}^{(3)}_q: \quad \frac{1}{2\pi}\int_0^\pi \left|\frac{(\mathrm{e}^{2\mathrm{i}\theta})_\infty}{\prod_{1\le j\le 3}(a_j\mathrm{e}^{\mathrm{i}\theta})_\infty}\right|^2 \mathrm{d}\theta = \frac{1}{(q)_\infty\prod_{i<j\le 3}(a_ia_j)_\infty}\,,$$

$$a_3 = 0\downarrow$$

$$\mathrm{I}^{(2)}_q: \quad \frac{1}{2\pi}\int_0^\pi \left|\frac{(\mathrm{e}^{2\mathrm{i}\theta})_\infty}{(a_1\mathrm{e}^{\mathrm{i}\theta})_\infty(a_2\mathrm{e}^{\mathrm{i}\theta})_\infty}\right|^2 \mathrm{d}\theta = \frac{1}{(q)_\infty(a_1a_2)_\infty}\,.$$

The 'extra factor' $(\mathrm{e}^{2\mathrm{i}\theta})_\infty$ can be written as

$$\begin{aligned}(\mathrm{e}^{2\mathrm{i}\theta})_\infty &= (\mathrm{e}^{\mathrm{i}\theta};q^{1/2})_\infty\cdot(-\mathrm{e}^{\mathrm{i}\theta};q^{1/2})_\infty\\ &= (\mathrm{e}^{2\mathrm{i}\theta};q^2)_\infty\cdot(q\mathrm{e}^{2\mathrm{i}\theta};q^2)_\infty\\ &= (\mathrm{e}^{\mathrm{i}\theta})_\infty\cdot(-\mathrm{e}^{\mathrm{i}\theta})_\infty\cdot(q^{1/2}\mathrm{e}^{\mathrm{i}\theta})_\infty\cdot(-q^{1/2}\mathrm{e}^{\mathrm{i}\theta})_\infty\,.\end{aligned}$$

That is, it is the product of $(a\mathrm{e}^{\mathrm{i}\theta})_\infty$ where a varies over the *four* two-torsion points of the elliptic curve $E_q = \mathbb{C}^*/q^{\mathbb{Z}}$, and after applying a two-isogeny (either in the adic or the real direction) it can be written as a product of *two* such factors. Recall that the two-torsion points of an elliptic curve E_q are very special: dividing

E_q by the algebraic equivalence relation $\sim$ identifying a point with its inverse, $[\alpha]_{E_q} \sim [-\alpha]_{E_q}$, we obtain the projective line, $E_q/\!\sim\; \simeq \mathbb{P}^1(\mathbb{C})$, thus exhibiting E_q as a double cover of $\mathbb{P}^1(\mathbb{C})$, ramified precisely at the four two-torsion points where $[\alpha]_{E_q} = [-\alpha]_{E_q}$.

We can take the p-adic limits of the integrals I_q to obtain

$$\mathrm{I}_p^{(4)}\colon\ \frac{\log p}{2\pi}\int_0^{\frac{\pi}{\log p}} \left|\frac{1-p^{-2\mathrm{i}t}}{\prod_{1\le j\le 4}(1-p^{-\alpha_j-\mathrm{i}t})}\right|^2 \mathrm{d}t = \frac{1-p^{-\alpha_1-\alpha_2-\alpha_3-\alpha_4}}{\prod_{i<j\le 4}(1-p^{-\alpha_i-\alpha_j})}\,,$$

$$\alpha_4 \to \infty \downarrow$$

$$\mathrm{I}_p^{(3)}\colon\ \frac{\log p}{2\pi}\int_0^{\frac{\pi}{\log p}} \left|\frac{1-p^{-2\mathrm{i}t}}{\prod_{1\le j\le 3}(1-p^{-\alpha_j-\mathrm{i}t})}\right|^2 \mathrm{d}t = \frac{1}{\prod_{i<j\le 3}(1-p^{-\alpha_i-\alpha_j})}\,,$$

$$\alpha_3 \to \infty \downarrow$$

$$\mathrm{I}_p^{(2)}\colon\ \frac{\log p}{2\pi}\int_0^{\frac{\pi}{\log p}} \left|\frac{1-p^{-2\mathrm{i}t}}{(1-p^{-\alpha_1-\mathrm{i}t})(1-p^{-\alpha_2-\mathrm{i}t})}\right|^2 \mathrm{d}t = \frac{1}{(1-p^{-\alpha_1-\alpha_2})}\,.$$

Now we can get rid of the factor $(1-p^{-2\mathrm{i}t})$ by using *two* of our parameters, $\alpha_1 = 0$, $\alpha_2 = \pi\mathrm{i}/\log p$, obtaining from $\mathrm{I}_p^{(4)}$ and $\mathrm{I}_p^{(3)}$ the more familiar beta and gamma integrals

$$\widetilde{\mathrm{I}}_p^{(2)}\colon\ \frac{\log p}{2\pi}\int_0^{\frac{2\pi}{\log p}} \left|\frac{1-p^{-(\alpha+\beta)/2}}{(1-p^{-\alpha/2-\mathrm{i}t})(1-p^{-\beta/2-\mathrm{i}t})}\right|^2 \mathrm{d}t = \frac{1-p^{-(\alpha+\beta)}}{(1-p^{-\alpha})(1-p^{-\beta})} = \zeta_p(\alpha,\beta)\,,$$

$$\downarrow \alpha \to \infty$$

$$\widetilde{\mathrm{I}}_p^{(1)}\colon\ \frac{\log p}{2\pi}\int_0^{\frac{2\pi}{\log p}} \left|\frac{1}{1-p^{-\beta/2-\mathrm{i}t}}\right|^2 \mathrm{d}t = \frac{1}{1-p^{-\beta}} = \zeta_p(\beta)\,.$$

In the real limit of the integrals I_q we obtain

$$\mathrm{I}_\eta^{(4)}\colon\ \frac{1}{2\pi}\int_0^{\infty} \left|\frac{\prod_{1\le j\le 4}\Gamma(\alpha_j+\mathrm{i}t)}{\Gamma(2\mathrm{i}t)}\right|^2 \mathrm{d}t = \frac{\prod_{i<j\le 4}\Gamma(\alpha_i+\alpha_j)}{\Gamma(\alpha_1+\alpha_2+\alpha_3+\alpha_4)}\,,$$

$$\alpha_4 \to \infty \downarrow$$

$$\mathrm{I}_\eta^{(3)}\colon\ \frac{1}{2\pi}\int_0^{\infty} \left|\frac{\prod_{1\le j\le 3}\Gamma(\alpha_j+\mathrm{i}t)}{\Gamma(2\mathrm{i}t)}\right|^2 \mathrm{d}t = \prod_{i<j\le 3}\Gamma(\alpha_i+\alpha_j)\,.$$

Again we can get rid of the factor $1/\Gamma(2\mathrm{i}t)$ by using *two* of our parameters, $\alpha_1 = 0$, $\alpha_2 = 1/2$, obtaining the more familiar beta and gamma integrals

$$\widetilde{\mathrm{I}}_\eta^{(2)}\colon\ \frac{1}{4\pi}\int_0^{\infty} \left|\frac{\zeta_\eta(\alpha/2+\mathrm{i}t)\zeta_\eta(\beta/2+\mathrm{i}t)}{\zeta_\eta((\alpha+\beta)/2)}\right|^2 \mathrm{d}t = \frac{\zeta_\eta(\alpha)\zeta_\eta(\beta)}{\zeta_\eta(\alpha+\beta)} = \zeta_\eta(\alpha,\beta)\,,$$

$$\downarrow \alpha \to \infty$$

$$\widetilde{\mathrm{I}}_\eta^{(1)}\colon\ \frac{1}{4\pi}\int_0^{\infty} |\zeta_\eta(\beta/2+\mathrm{i}t)|^2 \mathrm{d}t = \zeta_\eta(\beta)\cdot 2^{\beta/2} = \zeta_\eta(\beta)\cdot \|\phi_{\mathbb{Z}_\eta}\|^{-2}_{\mathscr{H}^\beta_\eta}\,.$$

We can summarize the situation in the following diagram:

$$
\begin{array}{ccccccccc}
\widetilde{\mathrm{I}}_p^{(2)} & \longleftarrow & \mathrm{I}_p^{(4)} & \underset{\textcircled{p}}{\longleftarrow} & \mathrm{I}_q^{(4)} & \underset{\textcircled{\eta}}{\longrightarrow} & \mathrm{I}_\eta^{(4)} & \longrightarrow & \widetilde{\mathrm{I}}_\eta^{(2)} \\
\Big\downarrow{\scriptstyle \alpha\to\infty} & & \Big\downarrow{\scriptstyle \alpha_4\to\infty} & & \Big\downarrow{\scriptstyle a_4=0} & & \Big\downarrow{\scriptstyle \alpha_4\to\infty} & & \Big\downarrow{\scriptstyle \alpha\to\infty} \\
\widetilde{\mathrm{I}}_p^{(1)} & \longleftarrow & \mathrm{I}_p^{(3)} & \underset{\textcircled{p}}{\longleftarrow} & \mathrm{I}_q^{(3)} & \underset{\textcircled{\eta}}{\longrightarrow} & \mathrm{I}_\eta^{(3)} & \longrightarrow & \widetilde{\mathrm{I}}_\eta^{(1)} \\
 & & \Big\downarrow{\scriptstyle \alpha_3\to\infty} & & \Big\downarrow{\scriptstyle a_3=0} & & & & \\
 & & \mathrm{I}_p^{(2)} & \underset{\textcircled{p}}{\longleftarrow} & \mathrm{I}_q^{(2)} & & & &
\end{array}
\tag{10.8.2}
$$

After this introduction let us return to the Askey–Wilson weight function appearing in the integral $\mathrm{I}_q^{(4)}$, and writing $z = \mathrm{e}^{\mathrm{i}\theta}$, we denote it by

$$
\mathrm{W}_{(a_j)}(z) = \frac{(z^2)_\infty (z^{-2})_\infty}{\prod_{1\le j\le 4}(a_j z)_\infty (a_j z^{-1})_\infty}\,. \tag{10.8.3}
$$

Let $\widehat{\mathscr{H}}^+_{(a_j)}$ denote the even part of the associated Hilbert space, that is the even functions on $\mathbb{C}^{(1)} = \{z \in \mathbb{C} \mid |z| = 1\}$ square integrable with respect to $\mathrm{W}_{(a_j)}(z)$:

$$
\widehat{\mathscr{H}}^+_{(a_j)} = \left\{ f\colon \mathbb{C}^{(1)} \longrightarrow \mathcal{C} \;\middle|\; f(z) = f(z^{-1}),\ \frac{1}{2\pi\mathrm{i}} \oint_{|z|=1} |f(z)|^2\, \mathrm{W}_{(a_j)}(z) \frac{\mathrm{d}z}{z} < \infty \right\}. \tag{10.8.4}
$$

Consider the operator

$$
\widehat{\mathrm{D}}^-\colon \widehat{\mathscr{H}}^+_{(a_j)} \to \widehat{\mathscr{H}}^+_{(a_j q^{1/2})}, \quad \widehat{\mathrm{D}}^- f(z) = \frac{f(zq^{1/2}) - f(zq^{-1/2})}{(1-q)\cdot(z - z^{-1})}\,; \tag{10.8.5}
$$

its adjoint is given by

$$
\widehat{\mathrm{D}}^+_{(a_j)}\colon \widehat{\mathscr{H}}^+_{(a_j q^{1/2})} \to \widehat{\mathscr{H}}^+_{(a_j)}\,, \tag{10.8.6}
$$

$$
\widehat{\mathrm{D}}^+_{(a_j)} f(z) = \frac{1}{q^{1/2}(1-q)(z-z^{-1})} \left[z^{-2} \prod_j (1 - a_j z)\cdot f(zq^{1/2}) - z^2 \prod_j (1 - a_j z^{-1})\cdot f(zq^{-1/2}) \right].
$$

We have the ladder

$$
\widehat{\mathscr{H}}^+_{(a_j)} \;\underset{\widehat{\mathrm{D}}^+_{(a_j)}}{\overset{\widehat{\mathrm{D}}^-}{\rightleftarrows}}\; \widehat{\mathscr{H}}^+_{(a_j q^{1/2})} \;\underset{\widehat{\mathrm{D}}^+_{(a_j q^{1/2})}}{\overset{\widehat{\mathrm{D}}^-}{\rightleftarrows}}\; \widehat{\mathscr{H}}^+_{(a_j q)} \;\underset{\widehat{\mathrm{D}}^+_{(a_j q)}}{\overset{\widehat{\mathrm{D}}^-}{\rightleftarrows}}\; \cdots \tag{10.8.7}
$$

and a straightforward computation gives the Heisenberg relations up the ladder

$$
q\cdot \widehat{\mathrm{D}}^- \widehat{\mathrm{D}}^+_{(a_j)} - \widehat{\mathrm{D}}^+_{(a_j q^{1/2})} \widehat{\mathrm{D}}^- = \left(\frac{1 - a_1 a_2 a_3 a_4}{1-q} \right) \cdot \mathrm{id}_{\widehat{\mathscr{H}}^+_{(a_j q^{1/2})}}\,. \tag{10.8.8}
$$

It follows immediately from the Heisenberg relation (10.8.8) that

$$f_{q,n}^{(a_j)} = \frac{q^{n(n-1)/2}}{[n]_q!} (\widehat{\mathrm{D}}^+)^n \mathbb{1} \tag{10.8.9}$$

form an orthogonal basis for $\widehat{\mathscr{H}}^+_{(a_j)}$; these are the Askey–Wilson polynomials. We have

$$\begin{aligned} \widehat{\mathrm{D}}^+_{(a_j)} f_{q,n-1}^{(a_j q^{1/2})} &= [n]_q \cdot q^{1-n} \cdot f_{q,n}^{(a_j)}, \\ \widehat{\mathrm{D}}^- f_{q,n}^{(a_j)} &= \left(\frac{1 - q^{n-1} a_1 a_2 a_3 a_4}{1-q} \right) \cdot f_{q,n-1}^{(a_j q^{1/2})}, \\ \widehat{\mathrm{D}}^+_{(a_j)} \widehat{\mathrm{D}}^- f_{q,n}^{(a_j)} &= q^{1-n} \cdot [n]_q \cdot \left(\frac{1 - q^{n-1} a_1 a_2 a_3 a_4}{1-q} \right) \cdot f_{q,n}^{(a_j)}, \end{aligned} \tag{10.8.10}$$

$$f_{q,n}^{(a_j)}(z) = \frac{(-1)^n a_1^n}{(q)_n \cdot (q a_1 a_2)_n (q a_1 a_3)_n (q a_1 a_4)_n} \cdot {}_4\phi_3\Big(-n, n-1+\sum \alpha_j, \alpha_1 + \mathrm{i}t, \alpha_1 - \mathrm{i}t; \alpha_1+\alpha_2, \alpha_1+\alpha_3, \alpha_1+\alpha_4; 1 \Big) \tag{10.8.11}$$

(where $a_j = q^{\alpha_j}$, $z = q^{\mathrm{i}t}$). $f_{q,n}^{(a_j)}(z)$ is a polynomial of degree n in $(z+z^{-1})/2$, symmetric in $\{a_1, a_2, a_3, a_4\}$, etc.; all the basic properties of $f_{q,n}^{(a_j)}$ follow from the Heisenberg relations.

The Askey–Wilson polynomials give rise to a q-analogue of the polynomials $\hat{f}^{\alpha,\beta}_{\eta,n}$ and $\hat{f}^{\beta}_{\eta,n}$ (see Askey and Wilson 1985, p. 29). We take the parameters $\{a_1, a_2, a_3, a_4\}$ equal to $\{\mathrm{i}q^{\alpha/2}, -\mathrm{i}q^{\alpha/2}, \mathrm{i}q^{\beta/2}, -\mathrm{i}q^{\beta/2}\}$, and shift θ to $\theta - \pi/2$, changing $(z+z^{-1})/2 = (q^{\mathrm{i}t} + q^{-\mathrm{i}t})/2 = \cos\theta$ into $(z - z^{-1})/2\mathrm{i} = (q^{\mathrm{i}t} - q^{-\mathrm{i}t})/2\mathrm{i} = \sin\theta$. The resulting basis $\hat{f}^{\alpha,\beta}_{q,n}(s)$ are polynomials of degree n in $q^s - q^{-s}$, alternating in sign, $\hat{f}^{\alpha,\beta}_{q,n}(-s) = (-1)^n \hat{f}^{\alpha,\beta}_{q,n}(s)$. They are orthogonal with respect to the measure

$$\left| \frac{(-q^{2s})_\infty}{(q^{\alpha+2s}; q^2)_\infty (q^{\beta+2s}; q^2)_\infty} \right|^2 \mathrm{d}s \cdot \left(\frac{\log q^{-1}}{2\pi \mathrm{i}} \right).$$

They are given explicity by

$$\hat{f}^{\alpha,\beta}_{q,n}(s) = \frac{q^{n(n-1)/2}}{[n]_q!} \widehat{\mathrm{D}}^+_{\alpha,\beta} \widehat{\mathrm{D}}^+_{\alpha+1/2,\beta+1/2} \cdots \widehat{\mathrm{D}}^+_{\alpha+(n-1)/2,\beta+(n-1)/2} \mathbb{1} \tag{10.8.12}$$

where now

$$\widehat{\mathrm{D}}^+_{\alpha,\beta} \hat{f}(s) = \frac{q^{-2s}(1-q^{\alpha+2s})(1-q^{\beta+2s}) \hat{f}(s+1/2) - q^{2s}(1-q^{\alpha-2s})(1-q^{\beta-2s}) \hat{f}(s-1/2)}{(1-q)^2 \cdot (q^s + q^{-s})}. \tag{10.8.13}$$

The $\hat{f}^{\alpha,\beta}_{q,n}(s)$ satisfy the appropriate specializations of (10.8.8)–(10.8.11); their basic properties come out of the Heisenberg relations. In the real limit they converge to the continuous Hahn polynomials (10.7.6):

$$\lim_{q\to 1} \hat{f}^{\alpha/2,\beta/2}_{q^2,n}\left(\frac{s}{2}\right) = \hat{f}^{\alpha,\beta}_{\eta,n}(s)\,. \tag{10.8.14}$$

Taking $\alpha \to \infty$, we obtain the q-analogue of the Meixner–Pollaczeck polynomials (10.6.10). These polynomials $\hat{f}^{\beta}_{q,n}(s)$ are of degree n in $q^s - q^{-s}$, alternating in sign, and orthogonal with respect to the measure

$$\left|\frac{(-q^{2s})_\infty}{(q^{\beta+2s};q^2)_\infty}\right|^2 \mathrm{d}_q s\,.$$

They are given by

$$\hat{f}^{\beta}_{q,n}(s) = \frac{q^{n(n-1)/2}}{[n]_q!}\widehat{\mathrm{D}}^+_\beta\,\widehat{\mathrm{D}}^+_{\beta+1/2}\cdots\widehat{\mathrm{D}}^+_{\beta+(n-1)/2}1\,, \tag{10.8.15}$$

where

$$\widehat{\mathrm{D}}^+_\beta \hat{f}(s) = \frac{(q^{-2s}-q^\beta)\hat{f}(s+1/2)-(q^{2s}-q^\beta)\hat{f}(s-1/2)}{(1-q)^2\cdot(q^s+q^{-s})}\,. \tag{10.8.16}$$

In the real limit they give the Meixner–Pollaczeck polynomials

$$\lim_{q\to 1} \hat{f}^{\beta/2}_{q^2,n}\left(\frac{s}{2}\right) = \hat{f}^{\beta}_{\eta,n}(s)\,. \tag{10.8.17}$$

11
The quantum group SU(1,1)

11.1 THE QUANTUM ENVELOPING ALGEBRA U_q

The quantum enveloping algebra $\widehat{\mathrm{U}}_t = \widehat{\mathrm{U}}_t(s\ell(2))$ is defined as the complete $\mathbb{C}[[t]]$-algebra, generated by X_+, X_-, H, with relations

$$[\mathrm{H}, \mathrm{X}_\pm] = \pm 2\mathrm{X}_\pm \,, \tag{11.1.1}$$

$$[\mathrm{X}_+, \mathrm{X}_-] = \frac{\sinh(\mathrm{H}t/2)}{\sinh(t/2)} = \frac{q^{\mathrm{H}/2} - q^{-\mathrm{H}/2}}{q^{1/2} - q^{1/2}}, \quad q = \mathrm{e}^t \,. \tag{11.1.2}$$

We make it into a $*$-algebra by defining

$$\mathrm{H}^* = \mathrm{H}, \quad \mathrm{X}_\pm^* = -\mathrm{X}_\mp \,. \tag{11.1.3}$$

The algebra $\widehat{\mathrm{U}}_t$ with this $*$-structure is referred to as $\widehat{\mathrm{U}}_t(\mathrm{SU}(1,1))$. Note that while $\mathrm{SL}_2(\mathbb{R})$ and SU(1,1) are isomorphic groups and have the same enveloping algebra, their $*$-structures are different. We want to take $q \in (0,1)$, and so are obliged to take SU(1,1) (for $s\ell_2$, we would require $|q|=1$).

The 'algebraic' quantum enveloping algebra $\mathrm{U}_{\langle q \rangle}$ is the subalgebra of $\widehat{\mathrm{U}}_t$ generated by X_+, X_-, $q^{-\mathrm{H}/4}$, $q^{\mathrm{H}/4}$; it has the defining relation

$$q^{\pm\mathrm{H}/4} \cdot q^{\mp\mathrm{H}/4} = 1 \,, \tag{11.1.4}$$

$$q^{\mathrm{H}/4} \, \mathrm{X}_\pm q^{-\mathrm{H}/4} = q^{\pm 1/2} \cdot \mathrm{X}_\pm \,, \tag{11.1.5}$$

$$[\mathrm{X}_+, \mathrm{X}_-] = \frac{q^{\mathrm{H}/2} - q^{-\mathrm{H}/2}}{q^{1/2} - q^{-1/2}} \overset{\mathrm{def}}{=} \langle \mathrm{H} \rangle_{(q^{1/2})} \,. \tag{11.1.6}$$

We shall prefer to work with the algebra $\mathrm{U}_q \subseteq \mathrm{U}_{\langle q \rangle} \subseteq \widehat{\mathrm{U}}_t$ generated by $q^{\pm\mathrm{H}/2}$, and by

$$\mathrm{Y}_+ = \mathrm{X}_+ \cdot q^{(\mathrm{H}-1)/4}, \quad \mathrm{Y}_- = -q^{(\mathrm{H}-1)/4} \cdot \mathrm{X}_- \,. \tag{11.1.7}$$

It has the defining relations $q^{\pm\mathrm{H}/2} \cdot q^{\mp\mathrm{H}/2} = 1$, and

$$q^{\mathrm{H}/2}\,\mathrm{Y}_\pm q^{-\mathrm{H}/2} = q^{\pm 1}\cdot \mathrm{Y}_\pm\,, \tag{11.1.8}$$

$$\mathrm{Y}_-\mathrm{Y}_+ - q\mathrm{Y}_+\mathrm{Y}_- = \frac{1-q^{\mathrm{H}}}{1-q} \overset{\mathrm{def}}{=} [\mathrm{H}]_{(q)}\,. \tag{11.1.9}$$

Thus the difference between $\mathrm{U}_{\langle q\rangle}$ and U_q is the difference between the symmetric and non-symmertric q-numbers.

The $*$-structure corresponding to (11.1.3) is given by

$$(q^{\pm\mathrm{H}/2})^* = q^{\pm\mathrm{H}/2}, \quad (\mathrm{Y}_\pm)^* = \mathrm{Y}_\mp\,.$$

In the real limit $q\to 1$ (or $t\to 0$) the operators $[\mathrm{H}]_{(q)}$ or $\langle\mathrm{H}\rangle_{(q^{1/2})}$ converge to H, and we obtain the classical enveloping algebra U(SU(1,1)) generated by $\mathrm{Y}_\pm$ and H with relations

$$[\mathrm{H},\mathrm{Y}_\pm] = \pm 2\mathrm{Y}_\pm, \quad [\mathrm{Y}_-,\mathrm{Y}_+] = \mathrm{H}\,,$$

and $*$-structure:

$$\mathrm{H}^* = \mathrm{H}, \quad \mathrm{Y}_\pm^* = \mathrm{Y}_\mp\,.$$

Let U_q^+ denote the subalgebra of U_q generated by $\mathrm{Y}_+,\mathrm{Y}_-$ and $q^{\mathrm{H}/2}$ (without $q^{-\mathrm{H}/2}$); it has the defining relations (11.1.9), and (11.1.8) should be modified to

$$q^{\mathrm{H}/2}\mathrm{Y}_+ = q\cdot\mathrm{Y}_+q^{\mathrm{H}/2}, \quad q\cdot q^{\mathrm{H}/2}\mathrm{Y}_- = \mathrm{Y}_-q^{\mathrm{H}/2}\,. \tag{11.1.8}$$

The algebra U_q^+ converges in the p-adic limit: $q=p^{-d}$, H replaced by H/d and $d\to\infty$; the limit algebra U_p^+ is generated by $\mathrm{Y}_+,\mathrm{Y}_-$ and $p^{-\mathrm{H}/2}$ with defining relations

$$p^{-\mathrm{H}/2}\mathrm{Y}_+ = 0, \quad \mathrm{Y}_-p^{-\mathrm{H}/2} = 0\,, \tag{11.1.10}$$

$$\mathrm{Y}_-\mathrm{Y}_+ = 1-p^{-\mathrm{H}}\,. \tag{11.1.11}$$

Another subalgebra of U_q that will be useful is the subalgebra U_q^- generated by $\mathrm{Y}_+q^{-\mathrm{H}/2} = \mathrm{X}_+q^{-(\mathrm{H}+1)/4}$, $q^{-\mathrm{H}/2}\mathrm{Y}_- = -q^{-(\mathrm{H}+1)/4}\mathrm{X}_-$, and $q^{-\mathrm{H}/2}$ (without $q^{\mathrm{H}/2}$); its defining relations are similarly derived from (11.1.5), (11.1.6).

The arithmetic of the algebra U_q is governed by the following generalization of defining relations:

$$q^{n\mathrm{H}/2}\mathrm{Y}_\pm^m = q^{\pm n\cdot m}\mathrm{Y}_\pm^m q^{n\mathrm{H}/2} \tag{11.1.12}$$

$$\begin{aligned}\mathrm{Y}_-^n\mathrm{Y}_+^m = &\sum_{0\le k\le\min\{n,m\}} \begin{bmatrix} n\\ k\end{bmatrix}_q \begin{bmatrix} m\\ k\end{bmatrix}_q [k]_q!\,q^{(n-k)(m-k)}\mathrm{Y}_+^{m-k}\\ &[\mathrm{H}+m+n-k-1]_q\cdots[\mathrm{H}+m+n-2k]_q\mathrm{Y}_-^{n-k}\end{aligned} \tag{11.1.13}$$

where (11.1.12) follows from (11.1.8), and (11.1.13) follows by induction from (11.1.9).

11.2 HIGHEST WEIGHT REPRESENTATION

A vector v_0^β in a U_q-module is called a highest weight β vector if it satisfies

$$q^{\mathrm{H}/2} v_0^\beta = q^{\beta/2} \cdot v_0^\beta , \tag{11.2.1}$$

$$\mathrm{Y}_- v_0^\beta = 0 . \tag{11.2.2}$$

The highest weight β U_q-module is the Hilbert space $\mathcal{V}^\beta$ with the U_q unitary representation (i.e. the $*$-homomorphism $\mathrm{U}_q \to \mathrm{End}(\mathcal{V}^\beta)$) generated by a highest weight β vector $v_0^\beta \in \mathcal{V}^\beta$ (i.e. $\mathrm{U}_q \cdot v_0^\beta \subseteq \mathcal{V}^\beta$ is dense).

We set

$$v_n^\beta = \frac{q^{-n\beta/2}}{[n]_q!} \mathrm{Y}_+^n v_0^\beta , \tag{11.2.3}$$

and we obtain

$$q^{\mathrm{H}/2} v_n^\beta = q^{\beta/2+n} \cdot v_n^\beta , \tag{11.2.4}$$

$$\mathrm{Y}_+ v_n^\beta = [n+1]_q q^{\beta/2} \cdot v_{n+1}^\beta , \tag{11.2.5}$$

$$\mathrm{Y}_- v_n^\beta = [n-1+\beta]_q q^{-\beta/2} \cdot v_{n-1}^\beta . \tag{11.2.6}$$

Note that by (11.2.4) the v_n^β are orthogonal, being eigenvectors of $q^{\mathrm{H}/2}$ with different eigenvalues. By (11.2.5), (11.2.6) the span of v_n^β is closed under the U_q-action. Hence the v_n^β form an orthogonal basis for $\mathcal{V}^\beta$:

$$\mathcal{V}^\beta = \bigoplus_{n \geq 0} \mathbb{C} \cdot v_n^\beta . \tag{11.2.7}$$

The $*$-structure determines the norms of the v_n^β:

$$\begin{aligned}
\| v_n^\beta \|^2 &= \frac{q^{-\beta}}{[n]_q^2} \left(\mathrm{Y}_+ v_{n-1}^\beta , \mathrm{Y}_+ v_{n-1}^\beta \right) = \frac{q^{-\beta}}{[n]_q^2} \left(\mathrm{Y}_- \mathrm{Y}_+ v_{n-1}^\beta , v_{n-1}^\beta \right) \\
&= \frac{q^{-\beta}}{[n]_q^2} \left(\left(\frac{1-q^{\mathrm{H}}}{1-q} + q \mathrm{Y}_+ \mathrm{Y}_- \right) v_{n-1}^\beta , v_{n-1}^\beta \right) \\
&= \frac{q^{-\beta}}{[n]_q^2} \left([\beta + 2(n-1)]_q + q[\beta+n-2]_q [n-1]_q \right) \cdot \| v_{n-1}^\beta \|^2 \\
&= q^{-\beta} \cdot \frac{1-q^{\beta+n-1}}{1-q^n} \| v_{n-1}^\beta \|^2 ,
\end{aligned}$$

and normalizing by $\| v_0^\beta \| = 1$, we obtain

$$\| v_n^\beta \|^2 = q^{-\beta n} \frac{\zeta_{(q)}(1)}{\zeta_{(q)}(1+n)} \frac{\zeta_{(q)}(\beta+n)}{\zeta_{(q)}(\beta)} = \mathrm{C}_{(q)}^\beta (n) . \tag{11.2.8}$$

Conversely, for $\beta > 0$ we can define $\mathcal{V}^\beta$ by (11.2.7) with the norms (11.2.8), and we check that (11.2.4)–(11.2.6) defines a unitary representation of U_q. Thus, for each $\beta > 0$ there exists a highest weight β U_q-module, and it is unique up to isomorphism.

Define the operators

$$\mathrm{D}\colon \mathcal{V}^\beta \to \mathcal{V}^{\beta+1}, \quad \text{by} \quad \mathrm{D}v_n^\beta = \frac{q^{-\beta}}{1-q} \cdot v_{n-1}^{\beta+1}. \tag{11.2.9}$$

Their 'adjoints'

$$\mathrm{D}_\beta^+ = (1-q^\beta) \cdot \mathrm{D}^*\colon \mathcal{V}^{\beta+1} \to \mathcal{V}^\beta, \tag{11.2.10}$$

satisfy

$$\mathrm{D}_\beta^+ v_{n-1}^{\beta+1} = [n]_q \cdot q^{1-n} \cdot v_n^\beta. \tag{11.2.11}$$

From (11.2.9), (11.2.11) we deduce the Heisenberg relations up the ladder

$$\mathrm{D}\mathrm{D}_\beta^+ - \mathrm{D}_{\beta+1}^+ \mathrm{D} = \frac{q^{-\beta}}{1-q} \cdot \mathrm{id}_{\mathcal{V}^{\beta+1}}, \tag{11.2.12}$$

and the equation

$$v_n^\beta = \frac{q^{n(n-1)/2}}{[n]_q!} \text{`}(\mathrm{D}^+)^n\text{'} v_0^{\beta+n}. \tag{11.2.13}$$

Similarly we can define an action of the q-Heisenberg algebra (cf. (10.4.14), (10.4.15)) on the space $\mathcal{V}^\beta$ by setting

$$\mathrm{A}_+ v_n^\beta = [n+1]_q q^{-n} \cdot v_{n+1}^\beta, \tag{11.2.14}$$

$$\mathrm{A}_- v_n^\beta = v_{n-1}^\beta, \tag{11.2.15}$$

$$\mathrm{N} v_n^\beta = q^{-n} \cdot v_n^\beta. \tag{11.2.16}$$

Note that the actions of the generators of the subalgebra $\mathrm{U}_q^- \subseteq \mathrm{U}_q$ are expressed in terms of the q-Heisenberg algebra by

$$q^{-\mathrm{H}/2} = q^{-\beta/2} \cdot \mathrm{N}, \tag{11.2.17}$$

$$\mathrm{Y}_+ q^{-\mathrm{H}/2} = \mathrm{A}_+, \tag{11.2.18}$$

$$q^{-\mathrm{H}/2}\mathrm{Y}_- = -\frac{1-q^{-\beta}\mathrm{N}}{1-q} \mathrm{A}_-. \tag{11.2.19}$$

Comparing the norms (11.2.8) with the norms of the basic and Laguerre basis (cf. §10.4), we see that we have isomorphisms

$$\begin{array}{ccccccc} \mathcal{V}^\beta & \overset{\sim}{\longleftrightarrow} & \mathscr{H}^\beta_{\mathbb{Q}_q/\mathbb{Z}_q} & \underset{\sim}{\overset{\mathcal{F}_q^\beta}{\longleftrightarrow}} & \phi_{\mathbb{Z}_q} \cdot \mathscr{H}^\beta_{\mathbb{Z}_q} & \underset{\sim}{\overset{\phi_{\mathbb{Z}_q}^{\pm 1}}{\longleftrightarrow}} & \mathscr{H}^\beta_{\mathbb{Z}_q} \\ \Cup & & \Cup & & \Cup & & \Cup \\ v_n^\beta & \longleftrightarrow & \phi_{(q),n} & \longleftrightarrow & \phi_{\mathbb{Z}_q} \cdot \varphi^\beta_{\mathbb{Z}_{q,n}} & \longleftrightarrow & \varphi^\beta_{\mathbb{Z}_{q,n}} \end{array} \tag{11.2.20}$$

Thus the spaces $\mathscr{H}^\beta_{\mathbb{Q}_q/\mathbb{Z}_q}$, $\phi_{\mathbb{Z}_q} \cdot \mathscr{H}^\beta_{\mathbb{Z}_q}$ and $\mathscr{H}^\beta_{\mathbb{Z}_q}$ are all concrete realizations of the highest weight β U_q-module. The Heisenberg operators up the ladder (11.2.9), (11.2.10) correspond to the operators we had on these spaces. The action of the q-Heisenberg algebra (11.2.14)–(11.2.16) corresponds to the action we had on these spaces, cf. (10.4.12), (10.4.30), (10.4.35) respectively. The action of the subalgebra $\mathrm{U}_q^- \subseteq \mathrm{U}_q$ on these spaces is by 'differential' operators, and is given via (11.2.17)–(11.2.19) by the action of the q-Heisenberg algebra. Explicity, the action on $\mathscr{H}^\beta_{\mathbb{Z}_q}$ is given by

$$q^{-\mathrm{H}/2}\varphi(g^i) = q^{-i}[(q^{-\beta/2} + q^{\beta/2-1})\varphi(g^i) - (1-q^i)q^{-\beta/2}\varphi(g^{i-1}) \\ - q^{\beta/2-1}\varphi(g^{i+1})],$$

$$\mathrm{Y}_+ q^{-\mathrm{H}/2}\varphi(g^i) = \frac{q^{-i}}{1-q}[\varphi(g^i) - (1+q^{1-\beta})(1-q^i)\varphi(g^{i-1}) \\ + q^{1-\beta}(1-q^i)(1-q^{i-1})\varphi(g^{i-2})], \tag{11.2.21}$$

$$q^{-\mathrm{H}/2}\mathrm{Y}_-\varphi(g^i) = \frac{q^{-i}}{1-q}[\varphi(g^i) - (1+q^{\beta-1})\varphi(g^{i+1}) + q^{\beta-1}\varphi(g^{i+2})].$$

To get the full action of U_q on these spaces use (10.4.24) and (10.4.25) to express the operator $q^{\mathrm{H}/2} = q^{\beta/2}\mathrm{N}^{-1}$. The action of this operator is not by a finite difference operator. For example, on the space $\mathscr{H}^\beta_{\mathbb{Q}_q/\mathbb{Z}_q}$ it is given by

$$q^{\mathrm{H}/2}\varphi(g^i) = q^{\beta/2-1}\sum_{k\geq 0}(-1)^k q^{k(k-1)/2-(1+k)(i-1)}\varphi(g^{i-k}). \tag{11.2.22}$$

In the real limit we obtain an action of the classical enveloping algebra $\mathrm{U}(\mathrm{SU}(1,1))$. On the space $\mathscr{H}^\beta_{\mathbb{Q}_\eta/\mathbb{Z}_\eta}$ it is given by

$$\begin{aligned} \mathrm{Y}_+ &= \pi x^2, \\ \mathrm{Y}_- &= -\mathrm{e}^{\pi x^2}\frac{1}{4\pi}\left(\frac{\partial^2}{\partial x^2} + \frac{\beta-1}{x}\frac{\partial}{\partial x}\right)\mathrm{e}^{\pi x^2}, \\ \mathrm{H} &= \mathrm{e}^{-\pi x^2}(x\partial/\partial x + \beta/2)\mathrm{e}^{\pi x^2}. \end{aligned} \tag{11.2.23}$$

To transform this action to the spaces $\phi_{\mathbb{Z}_\eta}\cdot\mathscr{H}^\beta_{\mathbb{Z}_\eta}$ and $\mathscr{H}^\beta_{\mathbb{Z}_\eta}$, it is useful to notice the equation (the real limit of (10.4.29))

$$\begin{aligned} \mathcal{F}^\beta_\eta(-\pi x^2)\mathcal{F}^\beta_\eta &= \left(1 + \frac{x}{2}\frac{\partial}{\partial x}\right)\frac{1}{(-\pi x^2)}\left(1 - \frac{\beta}{2} - \frac{x}{2}\frac{\partial}{\partial x}\right) \\ &= \frac{1}{4\pi}\left(\frac{\partial^2}{\partial x^2} + \frac{\beta-1}{x}\frac{\partial}{\partial x}\right) = \frac{1}{4\pi}x^{1-\beta}\frac{\partial}{\partial x}\,x^{\beta-1}\frac{\partial}{\partial x} \end{aligned} \tag{11.2.24}$$

The action on the space $\mathscr{H}^\beta_{\mathbb{Z}\eta}$ is then given by

$$\begin{aligned} \mathrm{Y}_+ &= e^{\pi x^2}\frac{1}{4\pi}\left(\frac{\partial^2}{\partial x^2} + \frac{\beta-1}{x}\frac{\partial}{\partial x}\right)e^{-\pi x^2}, \\ \mathrm{Y}_- &= \frac{1}{4\pi}\left(\frac{\partial^2}{\partial x^2} + \frac{\beta-1}{x}\frac{\partial}{2x}\right), \\ \mathrm{H} &= x\frac{\partial}{\partial x} + \frac{\beta}{2} - \frac{1}{4\pi}\left(\frac{\partial^2}{\partial x^2} + \frac{\beta-1}{x}\frac{\partial}{\partial x}\right). \end{aligned} \tag{11.2.25}$$

In the p-adic limit we obtain an action of the algebra U_p^+. On the abstract space

$$\mathcal{V}_p^\beta = \bigoplus_{n\geq o} \mathbb{C} v_n^\beta \quad \| v_0^\beta \| = 1, \quad \| v_n^\beta \|^2 = (1-p^{-\beta}) \cdot p^{\beta n}, \quad n>0\,, \qquad (11.2.26)$$

it is given by

$$\begin{aligned} \mathrm{Y}_+ v_n^\beta &= p^{-\beta/2} v_{n+1}^\beta\,, \\ \mathrm{Y}_- v_n^\beta &= \begin{cases} p^{\beta/2} v_{n-1}^\beta & n \geq 2\,, \\ (1-p^{-\beta}) p^{\beta/2} v_0^\beta & n = 1\,, \\ 0 & n = 0\,, \end{cases} \\ p^{-\mathrm{H}/2} v_n^\beta &= \begin{cases} 0 & n \geq 1\,, \\ p^{-\beta/2} v_0^\beta & n = 0\,. \end{cases} \end{aligned} \qquad (11.2.27)$$

Thus $p^{-\mathrm{H}/2}$ is ($p^{-\beta/2}$ times) the orthogonal projection onto $\mathbb{C} \cdot v_0^\beta$, $\mathrm{Y}_+ - p^{-\mathrm{H}/2}$ is a rank-1 isometry, $\mathrm{Y}_- - p^{-\mathrm{H}/2}$ is the adjoint coisometry; they generate the Toeplitz algebra. Using the isomorphism

$$\mathcal{V}_p^\beta \overset{\sim}{\longleftrightarrow} \mathscr{H}^\beta_{\mathbb{Q}_p/\mathbb{Z}_p} \overset{\mathcal{F}_p^\beta}{\underset{\sim}{\longleftrightarrow}} \phi_{\mathbb{Z}_p} \cdot \mathscr{H}^\beta_{\mathbb{Z}_p} = \mathscr{H}^\beta_{\mathbb{Z}_p}\,, \qquad (11.2.28)$$

we can transform the action of U_p^+ to $\mathscr{H}^\beta_{\mathbb{Z}_p}$; it is given by

$$\begin{aligned} \mathrm{Y}_+ \varphi(p^i) &= p^{-\beta/2} (\varphi, \phi_{\mathbb{Z}_p}) - p^{\beta/2} \varphi(p^{i-1})\,, \\ \mathrm{Y}_- \varphi(p^i) &= p^{-\beta/2} (\varphi, \phi_{\mathbb{Z}_p}) - p^{-\beta/2} \varphi(p^{i+1})\,, \\ p^{-\mathrm{H}/2} \varphi(p^i) &= p^{-\beta/2} (\varphi, \phi_{\mathbb{Z}_p})\,. \end{aligned} \qquad (11.2.29)$$

The highest weight β U_q-module can be defined also for negative β, although the norms (11.2.8) are not positive. For β a negative integer, $\beta = -N$, we let $\mathcal{V}^{-N}$ denote the finite-dimensional U_q-module with basis $v_0^{-N}, \ldots, v_N^{-N}$, and with the U_q-action given by (11.2.4)–(11.2.6). Imitating the classical theory, one proves that U_q is semi-simple: every finite-dimensional U_q-module is a direct sum of irreducible U_q-modules, and every irreducible U_q-module contains a highest weight vector, and hence is isomorphic to some $\mathcal{V}^{-N}$, $N \geq 0$. In this respect the algebra U_q is nicer than the algebra $\mathrm{U}_{\langle q \rangle}$ based on symmetric q-numbers: for $\mathrm{U}_{\langle q \rangle}$ there are *two* non-isomorphic irreducible $\mathrm{U}_{\langle q \rangle}$-modules of dimension $N+1$, one of highest weight $-N$, and the other of highest weight $-N + 2\pi i/\log q$ (i.e. $q^{-\mathrm{H}/2} v_0 = -q^{N/2} \cdot v_0$). For a concrete realization of the U_q-module $\mathcal{V}^{-N}$ we can take $\mathbb{C}[V_{(q^{1/2})}]_{(N)}$, the homogeneous elements of degree N in the algebra generated by X, Y with $\mathrm{XY} = q^{1/2}\mathrm{YX}$. The isomorphism is given by

$$\begin{array}{ccc} \mathcal{V}^{-N} & \overset{\sim}{\longleftrightarrow} & \mathbb{C}[V_{(q^{1/2})}]_{(N)} \\ \Cup & & \Cup \\ v_m^{-N} & \longleftrightarrow & \begin{bmatrix} N \\ m \end{bmatrix}_q q^{m(m-1)/4} \mathrm{Y}^m \mathrm{X}^{N-m} \end{array} \qquad (11.2.30)$$

The action of U_q on $\mathbb{C}[V_{(q^{1/2})}]_{(N)}$ is given by

$$\begin{aligned} \mathrm{Y}_+ \cdot \mathrm{Y}^m\mathrm{X}^n &= [n]_q q^{-n/2}\mathrm{Y}^{m+1}\mathrm{X}^{n-1} \\ \mathrm{Y}_- \cdot \mathrm{Y}^m\mathrm{X}^n &= [m]_q q^{-(n+1)/2}\mathrm{Y}^{m-1}\mathrm{X}^{n+1}\,, \\ q^{\mathrm{H}/2} \cdot \mathrm{Y}^m\mathrm{X}^n &= q^{(m-n)/2}\mathrm{Y}^m\mathrm{X}^n\,. \end{aligned} \tag{11.2.31}$$

Thus the algebra $\mathbb{C}[V_{(q^{1/2})}] = \oplus_{N\geq 0}\mathbb{C}[V_{q^{(1/2)}}]_{(N)}$ is the direct sum of *all* finite-dimensional irreducible U_q-modules (and 'half' the finite-dimensional irreducible $\mathrm{U}_{\langle q\rangle}$-modules).

11.3 THE HOPF ALGEBRA STRUCTURE

The algebra $\widehat{\mathrm{U}}_t$ has the additional structure of a formal Hopf algebra. $\mathrm{U}_{\langle q\rangle}$ and U_q are sub-Hopf algebras, and U_q^+, U_q^- are sub-bi-algebras. We describe these structures below.

The *co-multiplication* map.

$$\Delta\colon \widehat{\mathrm{U}}_t \to \widehat{\mathrm{U}}_t \widehat{\otimes} \widehat{\mathrm{U}}_t \quad (\widehat{\otimes} \text{ the completed tensor product}) \tag{11.3.1}$$

is given on the generators by

$$\Delta(\mathrm{H}) = 1\otimes \mathrm{H} + \mathrm{H}\otimes 1, \quad \Delta(\mathrm{X}_\pm) = q^{-\mathrm{H}/4}\otimes \mathrm{X}_\pm + \mathrm{X}_\pm \otimes q^{\mathrm{H}/4}\,.$$

It induces

$$\Delta\colon \mathrm{U}_q \to \mathrm{U}_q\otimes \mathrm{U}_q, \quad \text{and} \quad \Delta\colon \mathrm{U}_q^\pm \longrightarrow \mathrm{U}_q^\pm \otimes \mathrm{U}_q^\pm\,, \tag{11.3.2}$$

given by

$$\Delta(q^{\pm\mathrm{H}/2}) = q^{\pm\mathrm{H}/2}\otimes q^{\pm\mathrm{H}/2}, \quad \Delta(\mathrm{Y}_\pm) = 1\otimes \mathrm{Y}_\pm + \mathrm{Y}_\pm\otimes q^{\mathrm{H}/2}\,.$$

Δ is a $*$-homomorphism (so $\Delta\circ * = (*\otimes *)\circ\Delta$) satisfying co-associativity

$$(\mathrm{id}\otimes\Delta)\circ\Delta = (\Delta\otimes\mathrm{id})\circ\Delta\,.$$

The *co-unit map* is

$$\varepsilon\colon \widehat{\mathrm{U}}_t \to \mathbb{C}, \quad \varepsilon(\mathrm{H}) = \varepsilon(\mathrm{X}_\pm) = 0\,, \tag{11.3.3}$$

inducing

$$\varepsilon\colon \mathrm{U}_q \to \mathbb{C} \quad \text{and} \quad \varepsilon\colon \mathrm{U}_q^\pm \longrightarrow \mathbb{C}, \quad \varepsilon(q^{\pm\mathrm{H}/2}) = 1, \quad \varepsilon(\mathrm{Y}_\pm) = 0\,. \tag{11.3.4}$$

ε is a $*$-homomorphism (so $\varepsilon\circ * = *\circ\varepsilon$, where $*$ on $\mathbb{C}$ is conjugation) satisfying the co-unit axioms

$$(\varepsilon\otimes\mathrm{id})\circ\Delta = \mathrm{id} = (\mathrm{id}\otimes\varepsilon)\circ\Delta\,.$$

The *antipode map* is

$$\mathrm{S}\colon \widehat{\mathrm{U}}_t \to \widehat{\mathrm{U}}_t, \quad \mathrm{S}(\mathrm{H}) = -\mathrm{H}, \quad \mathrm{S}(\mathrm{X}_\pm) = -q^{\pm 1/2}\mathrm{X}_\pm\,, \tag{11.3.5}$$

inducing

$$\mathrm{S}\colon \mathrm{U}_q \to \mathrm{U}_q \text{ with } \mathrm{S}(q^{\pm \mathrm{H}/2}) = q^{\mp \mathrm{H}/2}, \quad \mathrm{S}(\mathrm{Y}_\pm) = -\mathrm{Y}_\pm q^{\mathrm{H}/2}\,. \tag{11.3.6}$$

Note that S does not induce a mapping on U_q^+ or U_q^-, but interchanges them, $\mathrm{U}_q^+ \overset{\mathrm{S}}{\underset{\sim}{\longleftrightarrow}} \mathrm{U}_q^-$. S is an anti-$*$-homomorphism and co-homomorphism: denoting by τ the flip $\tau(u_1 \otimes u_2) = u_2 \otimes u_1$, we have

$$\mathrm{S} \circ m = m \circ (\mathrm{S} \otimes \mathrm{S}) \circ \tau$$

$$(m\colon \mathrm{U}_q \otimes \mathrm{U}_q \to \mathrm{U}_q \text{ the multiplcation, so } \mathrm{S}(u_1 \cdot u_2) = \mathrm{S}(u_2)\cdot \mathrm{S}(u_1))\,,$$

$$\Delta \circ \mathrm{S} = \tau \circ (\mathrm{S} \otimes \mathrm{S}) \circ \Delta\,,$$

$$\mathrm{S}1 = 1, \quad \varepsilon \circ \mathrm{S} = \varepsilon \quad (1\colon \mathbb{C} \to \mathrm{U}_q \text{ the unit map } \lambda \mapsto \lambda \cdot 1)\,,$$

$$\mathrm{S} * \mathrm{S} * = \mathrm{id}\,. \tag{11.3.7}$$

S satisfies the antipode axiom

$$m \circ (\mathrm{S} \otimes \mathrm{id}) \circ \Delta = m \circ (\mathrm{id} \otimes \mathrm{S}) \circ \Delta = 1 \circ \varepsilon\,.$$

Thus for two U_q-modules $\mathscr{H}_1$ and $\mathscr{H}_2$, $\mathscr{H}_1 \otimes \mathscr{H}_2$ is a $\mathrm{U}_q \otimes \mathrm{U}_q$-module, and via Δ, it becomes a U_q-module: for $v_i \in \mathscr{H}_i$ we have

$$\begin{aligned} q^{\pm \mathrm{H}/2}(v_1 \otimes v_2) &= q^{\pm \mathrm{H}/2} v_1 \otimes q^{\pm \mathrm{H}/2} v_2\,, \\ \mathrm{Y}_\pm(v_1 \otimes v_2) &= v_1 \otimes \mathrm{Y}_\pm v_2 + \mathrm{Y}_\pm v_1 \otimes q^{\mathrm{H}/2} v_2\,. \end{aligned} \tag{11.3.8}$$

For example, the action (11.2.31) of U_q on $\mathbb{C}[V_{(q^{1/2})}]$ is designed so that $\mathbb{C}[V_{(q^{1/2})}]$ is a U_q-module-algebra, in the sense that the multiplication

$$m_V\colon \mathbb{C}[V_{(q^{1/2})}] \otimes \mathbb{C}[V_{(q^{1/2})}] \longrightarrow \mathbb{C}[V_{(q^{1/2})}]$$

is a map of U_q-modules, or explicitly

$$m_V(\Delta(u)(f_1 \otimes f_2)) = u(f_1 \cdot f_2) \quad \text{for } u \in \mathrm{U}_q,\ f_i \in \mathbb{C}[V_{(q^{1/2})}]\,. \tag{11.3.9}$$

Taking the two highest weight U_q-modules, $\mathcal{V}^\alpha$ and $\mathcal{V}^\beta$, we can form their tensor product $\mathcal{V}^\alpha \otimes \mathcal{V}^\beta$; it is a Hilbert space with orthogonal basis $v_i^\alpha \otimes v_j^\beta$, with norms

$$\| v_i^\alpha \otimes v_j^\beta \|^2 = \mathrm{C}^\alpha_{(q)}(i) \cdot \mathrm{C}^\beta_{(q)}(j)\,. \tag{11.3.10}$$

We have the $*$-representation of U_q on $\mathcal{V}^\alpha \otimes \mathcal{V}^\beta$ given on the generators by

$$\begin{aligned} q^{\pm \mathrm{H}/2}(v_i^\alpha \otimes v_j^\beta) &= q^{\pm[(\alpha+\beta)/2+i+j]} \cdot v_i^\alpha \otimes v_j^\beta\,, \\ \mathrm{Y}_+(v_i^\alpha \otimes v_j^\beta) &= [j+1]_q \cdot q^{\beta/2} \cdot (v_i^\alpha \otimes v_{j+1}^\beta) + [i+1]_q \cdot q^{(\alpha+\beta)/2+j} \cdot (v_{i+1}^\alpha \otimes v_j^\beta)\,, \\ \mathrm{Y}_-(v_i^\alpha \otimes v_j^\beta) &= [j-1+\beta]_q \cdot q^{-\beta/2} \cdot (v_i^\alpha \otimes v_{j-1}^\beta) + [i-1+\alpha]_q \cdot q^{(\beta-\alpha)/2+j} \cdot (v_{i-1}^\alpha \otimes v_j^\beta)\,. \end{aligned} \tag{11.3.11}$$

The linear span of $v_i^\alpha \otimes v_j^\beta$ with $i+j=m$ is the subspace of vectors of weight $\alpha+\beta+2m$.

Considering the equation $\mathrm{Y}_- v^{(\alpha)\beta}_{(m)0} = 0$, where $v^{(\alpha)\beta}_{(m)0} = \sum_{i+j=m} c_{ij} \cdot v^\alpha_i \otimes v^\beta_j$, we obtain the recursion relation

$$c_{ij} = -c_{i-1,j+1} q^{-\beta+\alpha/2-j} \cdot \frac{1-q^{\beta+j}}{1-q^{\alpha+i-1}} \,. \tag{11.3.12}$$

We deduce that there is a unique, up to constant multiple, highest weight $\alpha+\beta+2m$ vector $v^{(\alpha)\beta}_{(m)0} \in \mathcal{V}^\alpha \otimes \mathcal{V}^\beta$, given by

$$\begin{aligned} v^{(\alpha)\beta}_{(m)0} &= \sum_{i+j=m} (-1)^i q^{j(j-1)/2+j\beta+i\alpha/2} \frac{\zeta_{(q)}(\alpha+m)}{\zeta_{(q)}(\alpha+i)} \frac{\zeta_{(q)}(\beta+m)}{\zeta_{(q)}(\beta+j)} \cdot v^\alpha_i \otimes v^\beta_j \\ &= \frac{\zeta_{(q)}(\alpha+\beta+2m)}{\zeta_{(q)}(\alpha+\beta+m)} \cdot q^{m\beta} \cdot \sum_{i+j=m} q^{i\alpha/2} \cdot \varphi^{(\alpha)\beta}_{q(m)m}(i,j) \cdot v^\alpha_i \otimes v^\beta_j \,. \end{aligned} \tag{11.3.13}$$

We get the decomposition of $\mathcal{V}^\alpha \otimes \mathcal{V}^\beta$ into a direct sum of highest weight U_q-modules

$$\mathcal{V}^\alpha \otimes \mathcal{V}^\beta \simeq \bigotimes_{m\geq 0} \mathcal{V}^{\alpha+\beta+2m} \,. \tag{11.3.14}$$

Noting that the action (11.3.11) of Y_+ is given by the sum of two q-commuting operators, we can apply the operator q-binom (6.3.18), and write

$$\begin{aligned} v^{(\alpha)\beta}_{(m)n} &\overset{\text{def}}{=} \frac{q^{-[(\alpha+\beta)/2+m]n}}{[n]_q!} \mathrm{Y}^n_+ v^{(\alpha)\beta}_{(m)0} \\ &= q^{-[(\alpha+\beta)/2+m]n} \sum_{i+j=n} \frac{1}{[i]_q! [j]_q!} \left(1 \otimes \mathrm{Y}^j_+\right) \left(\mathrm{Y}^i_+ \otimes q^{i\mathrm{H}/2}\right) v^{(\alpha)\beta}_{(m)0} \,. \end{aligned} \tag{11.3.15}$$

Substituting (11.3.13) into (11.3.15) we obtain

$$\begin{aligned} v^{(\alpha)\beta}_{(m)n} &= q^{-[(\alpha+\beta)/2+m]n} \sum_{\substack{i_1+j_1=n \\ i_0+j_0=m}} \frac{(-1)^{i_0} q^{j_0(j_0-1)/2+j_0\beta+(i_1-j_0)\beta/2+i_1 j_0}}{[i_1]_q! [j_1]_q! [i_0]_q! [j_0]_q!} \frac{\zeta_{(q)}(\alpha+m)}{\zeta_{(q)}(\alpha+i_0)} \\ &\qquad \frac{\zeta_{(q)}(\beta+m)}{\zeta_{(q)}(\beta+j_0)} \mathrm{Y}^{i_0+i_1}_+ \otimes \mathrm{Y}^{j_0+j_1}_+ (v^\alpha_0 \otimes v^\beta_0) \\ &= q^{-mn} \sum_{i+j=m+n} q^{(i-n)\alpha/2} v^\alpha_i \otimes v^\beta_j \,. \\ &\quad \left[\sum_{i_0+j_0=m} \begin{bmatrix} i \\ i_0 \end{bmatrix}_q \begin{bmatrix} j \\ j_0 \end{bmatrix}_q (-1)^{i_0} q^{j_0(j_0-1)/2+j_0\beta+j_0(i-i_0)} \frac{\zeta_{(q)}(\alpha+m)}{\zeta_{(q)}(\alpha+i_0)} \frac{\zeta_{(q)}(\beta+m)}{\zeta_{(q)}(\beta+j_0)} \right] \\ &= \frac{\zeta_{(q)}(\alpha+\beta+n+2m)}{\zeta_{(q)}(\alpha+\beta+n+m)} q^{m(\beta-n)} \sum_{i+j=m+n} q^{\alpha(i-n)/2} \varphi^{(\alpha)\beta}_{q(m+n),m}(i,j) \cdot v^\alpha_i \otimes v^\beta_j \end{aligned} \tag{11.3.16}$$

where $\varphi^{(\alpha)\beta}_{q(m+n),m}$ are the q-Hahn polynomials of (7.3.20).

We deduce that the q-Hahn polynomials are the matrix coefficients expressing the orthogonal basis $\{v^{(\alpha)\beta}_{(m)n}\}$ in terms of the orthogonal basis $v^\alpha_i \otimes v^\beta_j$.

Note that the norm of $v^{(\alpha)\beta}_{(m)0}$ is given by (11.3.13), (11.3.10), and (6.3.15),

$$\left\|v^{(\alpha)\beta}_{(m)0}\right\|^2 = \frac{\zeta_{(q)}(1)}{\zeta_{(q)}(1+m)} \frac{\zeta_{(q)}(\alpha+m)}{\zeta_{(q)}(\alpha)} \frac{\zeta_{(q)}(\beta+m)}{\zeta_{(q)}(\beta)} \frac{\zeta_{(q)}(\alpha+\beta+2m-1)}{\zeta_{(q)}(\alpha+\beta+m-1)} \tag{11.3.17}$$

and that

$$\left\|v^{(\alpha)\beta}_{(m)n}\right\|^2 = \left\|v^{(\alpha)\beta}_{(m)0}\right\|^2 \cdot \mathrm{C}^{\alpha+\beta+2m}_{(q)}(n)\,.$$

Thus the orthonormal bases for the weight $\alpha+\beta+2N$ subspace of $\mathcal{V}^\alpha \otimes \mathcal{V}^\beta$

$$\left\{\frac{v^\alpha_i \otimes v^\beta_j}{\mathrm{C}^\alpha_{(q)}(i)^{1/2}\mathrm{C}^\beta_{(q)}(j)^{1/2}}\right\}_{i+j=N} \tag{11.3.18}$$

and

$$\left\{\frac{v^{(\alpha)\beta}_{(m)n}}{\left\|v^{(\alpha)\beta}_{(m)0}\right\| \cdot \mathrm{C}^{\alpha+\beta+2m}_{(q)}(n)^{1/2}}\right\}_{m+n=N} \tag{11.3.19}$$

are interchanged by the orthogonal matrix

$$\mathrm{M} = \left\{\frac{\zeta_{(q)}(\alpha+\beta+N+m)}{\zeta_{(q)}(\alpha+\beta+N)} q^{m(\beta-n)+\alpha(i-n)/2} \varphi^{(\alpha)\beta}_{q(N),m}(i,j) \cdot \frac{\mathrm{C}^\alpha_{(q)}(i)^{1/2}\mathrm{C}^\beta_{(q)}(j)^{1/2}}{\left\|v^{(\alpha)\beta}_{(m)0}\right\| \mathrm{C}^{\alpha+\beta+2m}_{(q)}(n)^{1/2}}\right\},$$
$$i+j=m+n=N\,. \tag{11.3.20}$$

Noticing that

$$\frac{\zeta_{(q)}(\alpha+\beta+N+m)^2}{\zeta_{(q)}(\alpha+\beta+N)^2} q^{2m(\beta-n)+\alpha(i-n)} \frac{\mathrm{C}^\alpha_{(q)}(i)\mathrm{C}^\beta_{(q)}(j)}{\left\|v^{(\alpha)\beta}_{(m)0}\right\|^2 \cdot \mathrm{C}^{\alpha+\beta+2m}_{(q)}(n)} = \frac{\tau^{(\alpha)\beta}_{q(N)}(i,j)}{\left\|\varphi^{(\alpha)\beta}_{q(N),m}\right\|^2} \tag{11.3.21}$$

we have

$$\mathrm{M} = \left\{\frac{\varphi^{(\alpha)\beta}_{q(N),m}(i,j)}{\left\|\varphi^{(\alpha)\beta}_{q(N),m}\right\|} \cdot \tau^{(\alpha)\beta}_{q(N)}(i,j)^{1/2}\right\}, \quad i+j=N, \ \ 0 \le m \le N\,.$$

The equation $\mathrm{M}\cdot\mathrm{M}^t = \mathrm{I}_{(N+1)}$ (the identity $(N+1)\times(N+1)$ matrix) gives the orthogonality of $\varphi^{(\alpha)\beta}_{q(N),m}$ with respect to $\tau^{(\alpha)\beta}_{q(N)}$.

Similarly, the equation $\mathrm{M}^t \cdot \mathrm{M} = \mathrm{I}_{(N+1)}$ gives the 'dual' orthogonality

$$\sum_{0 \le m \le N} \varphi^{(\alpha)\beta}_{q(N),m}(i,j) \cdot \varphi^{(\alpha)\beta}_{q(N),m}(i',j') \cdot \frac{1}{\left\| \varphi^{(\alpha)\beta}_{q(N),m} \right\|^2} = \delta_{(i,j),(i',j')} \cdot \frac{1}{\tau^{(\alpha)\beta}_{q(N)}(i,j).} \qquad (11.3.22)$$

11.4 THE UNIVERSAL R-MATRIX

The Hopf algebras $\mathrm{U}_q \subseteq \mathrm{U}_{\langle q \rangle}$ are non-co-commutative, in the sense that

$$\Delta \neq \tau \circ \Delta, \quad \tau(x \otimes y) = y \otimes x \text{ the flip}. \qquad (11.4.1)$$

For $\widehat{\mathrm{U}}_t$ (respectively $\mathrm{U}_{\langle q \rangle}$) the opposite comultiplication $\tau \circ \Delta$ is obtained from Δ by replacing t by $-t$ (respectively q by q^{-1}), so

$$\widehat{\mathrm{U}}_t^{\text{co-op}} = \widehat{\mathrm{U}}_{-t}, \quad \mathrm{U}_{\langle q \rangle}^{\text{co-op}} = \mathrm{U}_{\langle q^{-1} \rangle}. \qquad (11.4.2)$$

The universal R-matrix is the (invertible) element of $\widehat{\mathrm{U}}_t \otimes \widehat{\mathrm{U}}_t$ given by

$$\begin{aligned} \mathrm{R} &= q^{\mathrm{H}\otimes\mathrm{H}/4} \mathrm{e}_{(q^{-1})}\big((1-q^{-1}) q^{\mathrm{H}/4}\mathrm{X}_+ \otimes q^{-\mathrm{H}/4}\mathrm{X}_-\big) \qquad (11.4.3) \\ &= q^{\mathrm{H}\otimes\mathrm{H}/4} \mathrm{e}_{(q^{-1})}\big((1-q)\mathrm{Y}_+ \otimes q^{-\mathrm{H}/2}\mathrm{Y}_-\big) \\ &= q^{\mathrm{H}\otimes\mathrm{H}/4} \mathrm{e}_{(q)}\big(-(1-q)\mathrm{Y}_+ \otimes q^{-\mathrm{H}/2}\mathrm{Y}_-\big)^{-1} \\ &= q^{\mathrm{H}\otimes\mathrm{H}/4} \sum_{n \ge 0} \frac{(1-q)^n}{[n]_q!} \mathrm{Y}_+^n \otimes q^{-n\mathrm{H}/2}\mathrm{Y}_-^n \end{aligned}$$

where $\mathrm{e}_{(q^{-1})}(y) = \mathrm{e}_{(q)}(-y)^{-1}$ is the q-exponential. Inner conjugation by R interchanges Δ and $\tau \circ \Delta$:

$$\tau \circ \Delta(u) = \mathrm{R} \cdot \Delta(u) \cdot \mathrm{R}^{-1} \quad \text{for all } u \in \widehat{\mathrm{U}}_t. \qquad (11.4.4)$$

Note that it suffices to verify (11.4.4) for u one of the generators $\mathrm{X}_\pm$ or H and this follows by direct calculation.

It follows that given any two $\widehat{\mathrm{U}}_t$-modules $\mathcal{V}_1$ and $\mathcal{V}_2$ we obtain an isomorphism of $\widehat{\mathrm{U}}_t$-modules

$$\mathrm{R}_{\mathcal{V}_1,\mathcal{V}_2} = \tau\mathrm{R}\colon \mathcal{V}_1 \otimes \mathcal{V}_2 \xrightarrow{\sim} \mathcal{V}_2 \otimes \mathcal{V}_1, \quad v_1 \otimes v_2 \mapsto \tau(\mathrm{R} \cdot (v_1 \otimes v_2)). \qquad (11.4.5)$$

In particular, for the highest weight modules we obtain an isomorphism

$$\mathrm{R}^{\alpha,\beta} = \tau\mathrm{R}\colon \mathcal{V}^\alpha \otimes \mathcal{V}^\beta \xrightarrow{\sim} \mathcal{V}^\beta \otimes \mathcal{V}^\alpha. \qquad (11.4.6)$$

Direct calculation gives the identity in $\widehat{\mathrm{U}}_t \,\widehat{\otimes}\, \widehat{\mathrm{U}}_t \,\widehat{\otimes}\, \widehat{\mathrm{U}}_t$:

$$(\Delta \otimes \mathrm{id})(\mathrm{R}) = \mathrm{R}_{13} \cdot \mathrm{R}_{23}, \quad (\mathrm{id} \otimes \Delta)(\mathrm{R}) = \mathrm{R}_{13}\mathrm{R}_{12} \qquad (11.4.7)$$

where R_{ij} is R placed in the (i,j)-th position in $\widehat{\mathrm{U}}_t \,\widehat{\otimes}\, \widehat{\mathrm{U}}_t \,\widehat{\otimes}\, \widehat{\mathrm{U}}_t$. It follows from (11.4.7) that R satisfies the *quantum Yang–Baxter equation*

$$\mathrm{R}_{12}\mathrm{R}_{13}\mathrm{R}_{23} = \mathrm{R}_{23}\mathrm{R}_{13}\mathrm{R}_{12}, \qquad (11.4.8)$$

so that we have the *Braid relation*

$$(\mathrm{id}_{\mathcal{V}^{\alpha_3}} \otimes \mathrm{R}^{\alpha_1,\alpha_2})(\mathrm{R}^{\alpha_1,\alpha_3} \otimes \mathrm{id}_{\mathcal{V}^{\alpha_2}})(\mathrm{id}_{\mathcal{V}^{\alpha_1}} \otimes \mathrm{R}^{\alpha_2,\alpha_3}) = (\mathrm{R}^{\alpha_2,\alpha_3} \otimes \mathrm{id}_{\mathcal{V}^{\alpha_1}})(\mathrm{id}_{\mathcal{V}^{\alpha_2}} \otimes \mathrm{R}^{\alpha_1,\alpha_3})(\mathrm{R}^{\alpha_1,\alpha_2} \otimes \mathrm{id}_{\mathcal{V}^{\alpha_3}})\,. \tag{11.4.9}$$

Schematically:

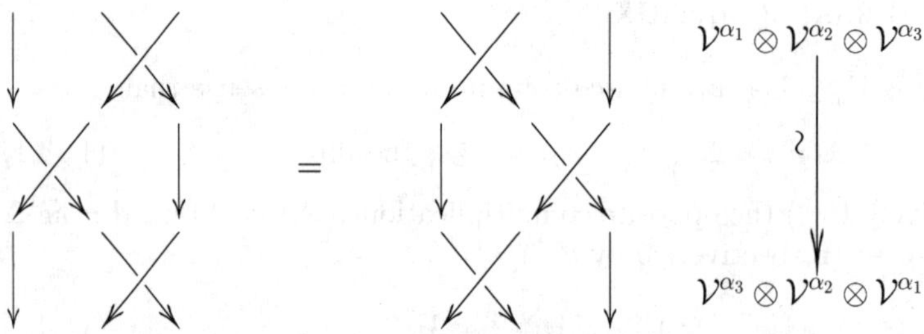

(11.4.10)

The Braid group B_n is the group with generators $\sigma_1, \ldots, \sigma_{n-1}$ and relations

$$\sigma_i\sigma_{i+1}\sigma_i = \sigma_{i+1}\sigma_i\sigma_{i+1} \tag{11.4.11}$$

$$\sigma_i\sigma_j = \sigma_j\sigma_i \quad \text{for } |i-j| \neq 1\,. \tag{11.4.12}$$

Adding the relations $\{\sigma_i^2\}$ we obtain the symmetric group Σ_n, so we have a homomorphism

$$\mathrm{B}_n \;\to\; \Sigma_n, \quad \sigma \mapsto \underline{\sigma}\,. \tag{11.4.13}$$

It follows from (11.4.9) that to every $\sigma \in \mathrm{B}_n$ we can associate a U_q-isomorphism

$$\mathrm{R}_\sigma^{\alpha_1,\ldots,\alpha_n}\colon \mathcal{V}^{\alpha_1} \otimes \cdots \otimes \mathcal{V}^{\alpha_n} \xrightarrow{\sim} \mathcal{V}^{\alpha_{\underline{\sigma}(1)}} \otimes \cdots \otimes \mathcal{V}^{\alpha_{\underline{\sigma}(n)}}\,. \tag{11.4.14}$$

We have

$$\mathrm{R}_{\tilde{\sigma}\sigma}^{\alpha_1,\ldots,\alpha_n} = \mathrm{R}_{\tilde{\sigma}}^{\alpha_{\underline{\sigma}(1)},\ldots,\alpha_{\underline{\sigma}(n)}} \circ \mathrm{R}_\sigma^{\alpha_1,\ldots,\alpha_n}\,, \tag{11.4.15}$$

and on the generators

$$\mathrm{R}_{\sigma_i}^{\alpha_1,\ldots\alpha_n} = \mathrm{id}_{\mathcal{V}^{\alpha_1}} \otimes \ldots \mathrm{R}^{\alpha_i,\alpha_{i+1}} \ldots \otimes \mathrm{id}_{\mathcal{V}^{\alpha_n}}\,. \tag{11.4.16}$$

In particular, B_n acts on $(\mathcal{V}^\alpha)^{\otimes n}$ by U_q-automorphisms.

The universal R-matrix also satisfies

$$(\mathrm{S} \otimes \mathrm{id})(\mathrm{R}) = \mathrm{R}^{-1}, \quad (\mathrm{id} \otimes \mathrm{S})(\mathrm{R}^{-1}) = \mathrm{R}\,, \tag{11.4.17}$$

$$\text{so } (\mathrm{S} \otimes \mathrm{S})(\mathrm{R}) = \mathrm{R}$$

and, moreover,

$$(* \otimes *)(\mathrm{R}) = \tau(\mathrm{R})\,, \qquad \text{so } (\mathrm{R}^{\alpha,\beta})^* = \mathrm{R}^{\beta,\alpha}\,. \tag{11.4.18}$$

We can calculate the explicit action of $\mathrm{R}^{\alpha,\beta}$ on the basis $\left\{v_i^\alpha \otimes v_j^\beta\right\}$ and $\left\{v_{(m)n}^{(\alpha)\beta}\right\}$ of $\mathcal{V}^\alpha \otimes \mathcal{V}^\beta$, using (11.4.3):

$$\mathrm{R}^{\alpha,\beta} v_{(m)n}^{(\alpha)\beta} = (-1)^m \cdot q^{m(m-1)/2+m(\alpha+\beta)/2+\alpha\beta/4} \cdot v_{(m)n}^{(\beta)\alpha} \tag{11.4.19}$$

$$\mathrm{R}^{\alpha,\beta} v_i^\alpha \otimes v_j^\beta = q^{(\alpha/2+i)(\beta/2+j)} \sum_{0\le k\le j} \begin{bmatrix} i+k \\ i \end{bmatrix}_q q^{-k(\beta/2+i)} \frac{\zeta_{(q)}(\beta+j)}{\zeta_{(q)}(\beta+j-k)} \cdot v_{j-k}^\beta \otimes v_{i+k}^\alpha \,. \tag{11.4.20}$$

Note that by putting the definitions of $v_{(m)n}^{(\alpha)\beta}$, (11.3.16) into (11.4.19) using (11.4.20) to compute the left-hand side, and comparing coefficients, we obtain

$$(-1)^m q^{m(m-1)/2+m\alpha-i_0 j_0} \varphi_{q(N),m}^{(\beta)\alpha}(i_0, j_0) = \sum_{i+j=N} \begin{bmatrix} j_0 \\ j_0 - i \end{bmatrix}_q q^{i\beta} \frac{\zeta_{(q)}(\beta+j)}{\zeta_{(q)}(\beta+i_0)} \varphi_{q(N),m}^{(\alpha)\beta}(i,j) \,. \tag{11.4.21}$$

Thus $\mathrm{R}^{\alpha,\beta}$ repairs the lost symmetry of (α,β) in $\varphi_{q(N),m}^{(\alpha)\beta}$ (i.e. the lost symmetry of the non-symmetric q-beta chain).

12
The Heisenberg group

12.1 THE HEISENBERG GROUP AND ITS FUNDAMENTAL REPRESENTATION

Associated with the symplectic phase space V is the Heisenberg group

$$\mathrm{Heis} = V \times \mathbb{A}^1 \tag{12.1.1}$$

with the multiplication rule:

$$\begin{gathered}(v_1, t_1) \cdot (v_2, t_2) = (v_1 + v_2, t_1 + t_2 + \frac{1}{2}\langle v_1, v_2\rangle)\,, \\ \langle v_1, v_2\rangle = x_1 y_2 - y_1 x_2 \quad \text{if } v_i = (x_i, y_i)\,;\end{gathered} \tag{12.1.2}$$

the inverse is given by

$$(v, t)^{-1} = (-v, -t)\,. \tag{12.1.3}$$

Heis is an algebraic group defined over $\mathbb{Z}[1/2]$. It is isomorphic to the upper triangular matrix group

$$\begin{pmatrix} 1 & * & * \\ 0 & 1 & * \\ 0 & 0 & 1 \end{pmatrix},$$

via

$$(x, y, t) \mapsto \begin{pmatrix} 1 & x & t + xy/2 \\ 0 & 1 & y \\ 0 & 0 & 1 \end{pmatrix}. \tag{12.1.4}$$

This matrix group is defined over $\mathbb{Z}$, and gives the extension of Heis to characteristic 2. We prefer to keep the symmetric form of Heis, which shows more clearly the symmetry between x and y, 'position' and 'momentum' (cf. Hörmander 1979); this will force us to make adjustments in characteristic 2.

We have the central extension

$$* \to \mathbb{A}^1 \to \mathrm{Heis} \to V \to * \tag{12.1.5}$$

with $\mathbb{A}^1 = \mathrm{centre(Heis)} = [\mathrm{Heis}, \mathrm{Heis}]$, and Heis is a two-step nilpotent group.

The group of automorphisms of Heis is isomorphic to the semi-direct product

$$\begin{aligned}\operatorname{Aut}(\mathrm{Heis}) &\cong \mathrm{GL}(V) \ltimes V \\ (g_1, v_1)\cdot(g_2, v_2) &= (g_1 g_2, v_1 g_2 + v_2), \quad g_i \in \mathrm{GL}(V),\ v_i \in V\,,\end{aligned} \tag{12.1.6}$$

the action of $\mathrm{GL}(V) \ltimes V$ on Heis being

$$(v,t)^{(g,v_0)} = (vg, t\cdot \det(g) + \langle vg, v_0\rangle)\,. \tag{12.1.7}$$

The subgroup of inner automorphisms is $V = \mathrm{Heis}/\mathbb{A}^1$. The action of an automorphism (g, v) on the centre $\mathbb{A}^1$ is via $\det(g)$, and in particular $\mathrm{Sp}(V) = \mathrm{SL}_2$ act as automorphisms inducing the identity on the centre.

We have the groups of rational points: $\mathrm{Heis}(\mathbb{Q})$, $\mathrm{Heis}(\mathbb{Q}_p)$ for $p \geq \eta$, $\mathrm{Heis}(\mathbb{Z}_p)$ for $p \neq 2$, η, and $\mathrm{Heis}(\mathbb{A}) = \prod_{p\geq\eta} \mathrm{Heis}(\mathbb{Q}_p)$: $\mathrm{Heis}(\mathbb{Z}_p)$; their centres will be identified with $\mathbb{Q}$, $\mathbb{Q}_p$, $\mathbb{Z}_p$ and $\mathbb{A}$ respectively. We can use the isomorphism (12.1.4) to define

$$\mathrm{Heis}(\mathbb{Z}_2) = \left\{(x,y,t) \in \mathrm{Heis}(\mathbb{Q}_2) \,\middle|\, x, y \in \mathbb{Z}_2, t - xy/2 \in \mathbb{Z}_2\right\}, \tag{12.1.8}$$

$$\begin{aligned}\mathrm{Heis}(\mathbb{Z}) &= \mathrm{Heis}(\mathbb{Q}) \cap \left(\mathrm{Heis}(\mathbb{R}) \times \prod_{p>\eta} \mathrm{Heis}(\mathbb{Z}_p)\right) \\ &= \{(x,y,t) \in \mathrm{Heis}(\mathbb{Q}) \,|\, x, y \in \mathbb{Z}, t - xy/2 \in \mathbb{Z}\}\,.\end{aligned} \tag{12.1.9}$$

Note that $\mathrm{Heis}(\mathbb{Z}_2)$ and $\mathrm{Heis}(\mathbb{Z})$ are subgroups of $\mathrm{Heis}(\mathbb{Q}_2)$ and $\mathrm{Heis}(\mathbb{Q})$, and if $(x, y, t) \in \mathrm{Heis}(\mathbb{Z}_2)$, (respectively $\mathrm{Heis}(\mathbb{Z})$), then $t \in \frac{1}{2}\mathbb{Z}_2$, (respectively $t \in \frac{1}{2}\,\mathbb{Z}$). We have the central extension

$$* \to \mathbb{Z} \to \mathrm{Heis}(\mathbb{Z}) \to V(\mathbb{Z}) \to * \tag{12.1.10}$$

and similarly for $\mathbb{Z}_2$. We also have the following normal subgroup of $\mathrm{Heis}(\mathbb{Z})$:

$$\mathrm{Heis}(\mathbb{Z};\mathrm{N}) = \left\{(x,y,t) \in \mathrm{Heis}(\mathbb{Z}) \,\middle|\, x, y \in \mathrm{N}\mathbb{Z}, t - xy/2 \in \mathrm{N}\mathbb{Z}\right\}, \tag{12.1.11}$$

and we let

$$\mathrm{Heis}(\mathbb{Z}/\mathrm{N}\mathbb{Z}) = \mathrm{Heis}(\mathbb{Z})/\mathrm{Heis}(\mathbb{Z};\mathrm{N})\,. \tag{12.1.12}$$

$\mathrm{Heis}(\mathbb{Z}/\mathrm{N}\mathbb{Z})$ is isomorphic to the $\mathbb{Z}/\mathrm{N}\mathbb{Z}$ rational points of the group

$$\begin{pmatrix} 1 & * & * \\ 0 & 1 & * \\ 0 & 0 & 1 \end{pmatrix},$$

and we have the central extension

$$* \to \mathbb{Z}/\mathrm{N}\mathbb{Z} \to \mathrm{Heis}(\mathbb{Z}/\mathrm{N}\mathbb{Z}) \to V(\mathbb{Z}/\mathrm{N}\mathbb{Z}) \to *\,. \tag{12.1.13}$$

The 'gist' of the Heisenberg group comes from the Stone–von Neumann theorem which states that given a non-trivial character ψ of the centre of Heis, there is a *unique*, up to equivalence, irreducible unitary representation $r = r_\psi$ of Heis whose restriction to the centre is given by ψ. The point is that this representation has different realizations, and the equivalence of these realizations gives rise to interesting dictionaries (e.g. between waves and particles). The centres $\mathbb{Q}_p$ of $\mathrm{Heis}(\mathbb{Q}_p)$, and $\mathbb{A}$ of $\mathrm{Heis}(\mathbb{A})$, are *self-dual* locally compact abelian groups, so that if we fix one character ψ, all other characters are given by $\psi^{(a)}(x) = \psi(a \cdot x)$ for some $a \in \mathbb{Q}_p$, respectively $a \in \mathbb{A}$. We shall concentrate our attention on the 'canonical' character ψ_p given by

$$\psi_p\colon \mathbb{Q}_p \twoheadrightarrow \mathbb{Q}_p/\mathbb{Z}_p \xleftarrow{\sim} \mathbb{Z}\left[\frac{1}{p}\right]\Big/\mathbb{Z} \hookrightarrow \mathbb{R}/\mathbb{Z} \xrightarrow[\sim]{e^{-2\pi i x}} \mathbb{C}^{(1)}, \quad p \neq \eta, \tag{12.1.14$_p$}$$

$$\psi_\eta\colon \mathbb{Q}_\eta = \mathbb{R} \twoheadrightarrow \mathbb{R}/\mathbb{Z} \xrightarrow[\sim]{e^{2\pi i x}} \mathbb{C}^{(1)}. \tag{12.1.14$_\eta$}$$

The global character $\psi_\mathbb{A}$ of $\mathbb{A}$ is given by the product of the local characters ψ_p

$$\psi_\mathbb{A}(x) = \prod_{p \geq \eta} \psi_p(x_p) \quad x = (x_p) \in \mathbb{A}. \tag{12.1.15}$$

Note that $\psi_\mathbb{A}$ is trivial on $\mathbb{Q} \subseteq \mathbb{A}$ (for $\mathbb{Q}$ this is the analogue of the fact that for a rational function the sum of its residues over all its poles is equal to zero), $\ker \psi_\mathbb{A} = \mathbb{Q} + \widehat{\mathbb{Z}}$, and $\psi_\mathbb{A}$ induces an isomorphism

$$\psi_\mathbb{A}\colon \mathbb{A}\Big/\mathbb{Q} + \widehat{\mathbb{Z}} \xrightarrow{\sim} \mathbb{C}^{(1)}. \tag{12.1.16}$$

The first realization of the basic representation $r = r_{\psi_p}$ of $\mathrm{Heis}(\mathbb{Q}_p)$ (respectively $r = r_{\psi_\mathbb{A}}$ of $\mathrm{Heis}(\mathbb{A})$) is the *Schrödinger model* on $\mathrm{L}_2(\mathbb{Q}_p)$ (respectively on $\mathrm{L}_2(\mathbb{A})$) given by

$$r(x, y, t)\varphi(z) = \psi(t) \cdot \bar{\psi}(xy/2) \cdot \psi(x \cdot z) \cdot \varphi(z - y). \tag{12.1.17}$$

The Fourier transform on $\mathrm{L}_2(\mathbb{Q}_p)$, $\mathrm{L}_2(\mathbb{A})$, is

$$\mathcal{F}\varphi(z) = \int \mathrm{d}y\, \varphi(y)\psi(yz). \tag{12.1.18}$$

There are two special distributions in the Schrödinger model that correspond to the polarization of V:

$$\delta_0(\varphi) = \varphi(0), \quad \delta_\infty(\varphi) = \delta_0(\mathcal{F}\varphi) = \int \varphi(z)\mathrm{d}z. \tag{12.1.19}$$

δ_0 (respectively δ_∞) is the unique, up to constant multiple, distribution invariant under all the operators $r(x, 0, 0)$ (respectively $r(0, y, 0)$):

$$\delta_0(r(x, 0, 0)\varphi) = \delta_0(\varphi), \quad \delta_\infty(r(0, y, 0)\varphi) = \delta_\infty(\varphi). \tag{12.1.20}$$

In terms of the Tate distribution $\tau^s(\varphi)$, we have

$$\delta_0 = \tau^0, \quad \delta_\infty = \tau^1 \tag{12.1.21}$$

(both over $\mathbb{Q}_p$ for $p \geq \eta$ and over $\mathbb{A}$, where $s = 0$, 1 correspond to the poles of $\zeta_\mathbb{A}(s)$).

More generally, for any line $\ell \in \mathbb{P}^1(\mathbb{Q}_p)$ (respectively $\ell \in \mathbb{P}^1(\mathbb{A})$), we can extend the central character ψ to the (maximal commutative) subgroup of Heis generated by ℓ and the centre,

$$\psi_\ell(x, y, t) = \psi(t) \quad \text{for } (x, y) \in \ell\,; \tag{12.1.22}$$

the induced representation of Heis is equivalent to r,

$$r_\psi \simeq \mathrm{Ind}_{\ell \times A^1}^{\mathrm{Heis}}(\psi_\ell)\,. \tag{12.1.23}$$

It follows that there is a unique (up to constant multiple) distribution δ_ℓ in any realization of r_ψ that is invariant under the action of ℓ:

$$\delta_\ell(r(x, y, 0)\varphi) = \delta_\ell(\varphi) \quad \text{for } (x, y) \in \ell\,. \tag{12.1.24}$$

The mapping $\ell \mapsto \delta_\ell$ gives an embedding

$$\mathbb{P}^1(\mathbb{Q}_p) \hookrightarrow \mathbb{P}^1(\mathcal{S}'(\mathbb{Q}_p)), \quad \text{respectively } \mathbb{P}^1(\mathbb{A}) \hookrightarrow \mathbb{P}^1(\mathcal{S}'(\mathbb{A}))\,. \tag{12.1.25}$$

12.2 TWISTED CONVOLUTION AND MULTIPLICATION

We can view r as a projective representation of $V(\mathbb{Q}_p)$, respectively $V(\mathbb{A})$. For example, in the Schrödinger model (12.1.17) we set $t = 0$. It satisfies

$$r(v_1)r(v_2) = \psi(\langle v_1, v_2\rangle/2)r(v_1 + v_2)\,. \tag{12.2.1}$$

Associated with an L_1-function f on $V(\mathbb{Q}_p)$, or $V(\mathbb{A})$, we have the (compact) operator

$$r(f) = \int \mathrm{d}v\, f(v)r(v)\,. \tag{12.2.2}$$

The composition of such operators corresponds to *twisted convolution*, denoted by $\natural$, of functions

$$\begin{aligned} r(f_1)r(f_2) &= r(f_1 \natural f_2) \\ f_1 \natural f_2(v) &= \int_V \mathrm{d}v_1\, f_1(v_1) f_2(v - v_1)\psi(\langle v_1, v\rangle/2)\,. \end{aligned} \tag{12.2.3}$$

The operator $r(f)$ is given by a kernel K_f, which is a partial Fourier transform of f:

$$\begin{aligned} r(f)\varphi(y) &= \int \mathrm{d}x\, \mathrm{K}_f(y, x)\varphi(x), \\ \mathrm{K}_f(y, x) &= (\mathcal{F} \otimes \mathrm{id}) f((x + y)/2, y - x) \\ &= \int \mathrm{d}z\, f(z, y - x)\psi((x + y)z/2)\,. \end{aligned} \tag{12.2.4}$$

From this we see that $f \mapsto r(f)$ induces a unitary isomorphism from the algebra of L_2-functions on V (with multiplication given by $\natural$) to the algebra of Hilbert–Schmidt operators (with multiplication given by composition of operators). Since the only operators that commute with all Hilbert–Schmidt operators are the scalars, it follows that r is indeed an irreducible representation (and a modification of this argument gives the uniqueness of r_ψ, the Stone–von Neumann theorem; cf. Weil 1964). We have

$$\begin{aligned} \|f_1 \natural f_2\|_{\mathrm{L}_1} &\le \|f_1\|_{\mathrm{L}_1} \cdot \|f_2\|_{\mathrm{L}_1}\,, \\ \|f_1 \natural f_2\|_{\mathrm{L}_2} &\le \|f_1\|_{\mathrm{L}_2} \cdot \|f_2\|_{\mathrm{L}_2}\,, \\ \|f_1 \natural f_2\|_{\mathrm{L}_\infty} &\le \|f_1\|_{\mathrm{L}_2} \cdot \|f_2\|_{\mathrm{L}_2}\,. \end{aligned} \tag{12.2.5}$$

The *symplectic Fourier transform* $\mathcal{F}_\mathrm{v}$ is the unitary operator on $\mathrm{L}_2(V)$ given by

$$\mathcal{F}_\mathrm{v} f(v) = \hat{f}(v) = \int \mathrm{d}v' f(v')\psi(\langle v', v\rangle)\,; \tag{12.2.6}$$

it satisfies $\mathcal{F}_\mathrm{v}^2 = \mathrm{id}$, and it commutes with the natural action of $\mathrm{Sp}(V)$. The Weyl quantization associates the operator $\hat{r}(f) \overset{\mathrm{def}}{=} r(\hat{f})$ with a function f on V; it is given explicitly by

$$\hat{r}(f)\varphi(z) = \iint \mathrm{d}x\ \mathrm{d}y\ f(x, (y+z)/2)\psi(x(z-y))\varphi(y)\,. \tag{12.2.7}$$

Now the composition of operators corresponds to *twisted multiplication*, denoted by $\#$, of functions

$$\begin{aligned} &\hat{r}(f_1)\hat{r}(f_2) = \hat{r}(f_1 \# f_2) \\ &f_1\#f_2(v) = \mathcal{F}_\mathrm{v}(\mathcal{F}_\mathrm{v} f_1 \natural \mathcal{F}_\mathrm{v} f_2)(v) \\ &\qquad = \iint \mathrm{d}v_1 \mathrm{d}v_2 f_1(v+v_1) f_2(v+v_2)\bar{\psi}(2\langle v_1, v_2\rangle) \\ &\qquad = [(\mathcal{F}_\mathrm{v} \otimes \mathcal{F}_\mathrm{v})\psi(\langle v_1, v_2\rangle/2)(\mathcal{F}_\mathrm{v} \otimes \mathcal{F}_\mathrm{v})] f_1 \otimes f_2(v, v)\,. \end{aligned} \tag{12.2.8}$$

The Weyl quantization has nice properties:

(i) the adjoint $\hat{r}(f)^* = \hat{r}(\bar{f})$, $\bar{f}$ the complex conjugate of f;

(ii) if $f(x,y) = f(y)$ is independent of x, then $\hat{r}(f)\varphi(y) = f(y)\cdot\varphi(y)$; (12.2.9)

(iii) if $f(x,y) = f(x)$ is independent of y, then $\hat{r}(f)\varphi(y) = f * \varphi(y)$.

An important tool in the analysis of the operators $\hat{r}(f_i)$ is the control one has over the twisted multiplication of their 'symbols' $f_1\#f_2$. This is the symbolic calculus of pseudo-differential operators; see Hörmander (1979) for the theory over the reals, and Haran (1993*b*) for the theory over the p-adic numbers. One can similarly develop a theory over the adeles $\mathbb{A}$.

12.3 MATRIX COEFFICIENTS

Given two vectors φ_1, φ_2 in the space of the representation r we have the associated matrix coefficients

$$c_{\varphi_1,\varphi_2}(v) = (\varphi_1, r(v)\varphi_2)\,. \tag{12.3.1}$$

In the Schrödinger realization this is given (after Fourier inversion) by

$$c_{\varphi_1,\varphi_2}(x,y) = \int \mathrm{d}z\,\bar{\psi}(xz)\varphi_1(z+y/2)\overline{\varphi_2(z-y/2)}\,. \tag{12.3.2}$$

A direct calculation gives

$$(c_{\varphi_1,\varphi_2}, c_{\varphi_3,\varphi_4})_{\mathrm{L}_2(V)} = (\varphi_1,\varphi_3)\cdot(\varphi_4,\varphi_2)\,. \tag{12.3.3}$$

The Schwartz inequality gives

$$\|c_{\varphi_1,\varphi_2}\|_{\mathrm{L}_\infty} \leq \|\varphi_1\|_{\mathrm{L}_2}\cdot\|\varphi_2\|_{\mathrm{L}_2} \tag{12.3.4}$$

showing that $c_{\varphi_1,\varphi_2}(v)$ is an L_2-function on V which is continuous and vanishing at infinity.

From (12.3.3) we obtain

$$(\varphi', r(c_{\varphi_1,\varphi_2})\varphi) = (c_{\varphi',\varphi}, c_{\varphi_1,\varphi_2}) = (\varphi', (\varphi,\varphi_2)\cdot\varphi_1)\,, \tag{12.3.5}$$

that is $r(c_{\varphi_1,\varphi_2})$ is the rank-1 operator

$$r(c_{\varphi_1,\varphi_2})\varphi = (\varphi,\varphi_2)\cdot\varphi_1\,, \tag{12.3.6}$$

and hence

$$c_{\varphi_1,\varphi_2} \natural\, c_{\varphi_3,\varphi_4} = (\varphi_3,\varphi_2)\cdot c_{\varphi_1,\varphi_4}\,. \tag{12.3.7}$$

The symplectic Fourier transform $\hat{c}_{\varphi_1,\varphi_2} = \mathcal{F}_{\mathrm{v}} c_{\varphi_1,\varphi_2}$ again satisfies

$$(\hat{c}_{\varphi_1,\varphi_2}, \hat{c}_{\varphi_3,\varphi_4})_{\mathrm{L}_2(V)} = (\varphi_1,\varphi_3)\cdot(\varphi_4,\varphi_2)\,, \tag{12.3.8}$$

$$\|\hat{c}_{\varphi_1,\varphi_2}\|_{\mathrm{L}_\infty} \leq \|\varphi_1\|_{\mathrm{L}_2}\cdot\|\varphi_2\|_{\mathrm{L}_2}\,, \tag{12.3.9}$$

$$\hat{c}_{\varphi_1,\varphi_2} \in \mathrm{L}_2(V)\cap \mathrm{C}_0(V)\,. \tag{12.3.10}$$

It is given in the Schrödinger realization by a formula similar to (12.3.2):

$$\hat{c}_{\varphi_1,\varphi_2}(x,y) = \int \mathrm{d}z\,\bar{\psi}\,(xz)\varphi_1(y+z/2)\overline{\varphi_2(y-z/2)}\,. \tag{12.3.11}$$

We also have

$$\hat{c}_{\varphi_1,\varphi_2} = \overline{\hat{c}_{\varphi_2,\varphi_1}}\,, \tag{12.3.12}$$

$$\hat{c}_{\varphi_1,\varphi_2}(y,-x) = \hat{c}_{\mathcal{F}\varphi_1\mathcal{F}\varphi_2}(x,y)\,. \tag{12.3.13}$$

The *Wigner transform* is defined by

$$\mathrm{W}\varphi(v) = \hat{c}_{\varphi,\varphi}(v) \in \mathrm{L}_2(V) \cap \mathrm{C}_0(V)\,, \tag{12.3.14}$$

and although it is not always positive, it is real valued by (12.3.12), and serves as a kind of common expectation value for the position and momentum; it satisfies

$$\int \mathrm{d}x\,\mathrm{W}\varphi(x,y) = |\varphi(y)|^2\,, \tag{12.3.15}$$

$$\int \mathrm{d}y\,\mathrm{W}\varphi(x,y) = |\mathcal{F}\varphi(x)|^2\,, \tag{12.3.16}$$

$$\mathrm{W}\varphi_1 = \mathrm{W}\varphi_2 \Longleftrightarrow \varphi_1 = \lambda\cdot\varphi_2 \text{ for some } \lambda \in \mathbb{C}^{(1)}\,. \tag{12.3.17}$$

Note that $\mathrm{W}\varphi(x,y) \neq 0$ implies that there exists a z such that $y \pm z \in \operatorname{supp}\varphi$, and hence

$$y \in \Delta(\operatorname{supp}\varphi) \stackrel{\text{def}}{=} \{(y_1+y_2)/2 \,|\, y_i \in \operatorname{supp}\varphi\}\,. \tag{12.3.18}$$

Thus the projection onto the y variable of the support of $\mathrm{W}\varphi$ is contained in $\Delta(\operatorname{supp}\varphi)$,

$$\operatorname{proj}_y(\operatorname{supp}\mathrm{W}\varphi) \subseteq \Delta(\operatorname{supp}\varphi)\,, \tag{12.3.19}$$

and similarly, using (12.3.13)), we have

$$\operatorname{proj}_x(\operatorname{supp}\mathrm{W}\varphi) \subseteq \Delta(\operatorname{supp}\mathcal{F}\varphi)\,. \tag{12.3.20}$$

We have the following basic inequality:

$$\|\mathrm{W}\varphi\|_{\mathrm{L}_\infty} \leq \|\varphi\|_{\mathrm{L}_2}^2 = \int\!\!\int \mathrm{d}x\,\mathrm{d}y\,\mathrm{W}\varphi(x,y) \leq \|\mathrm{W}\varphi\|_{\mathrm{L}_\infty}\cdot \operatorname{vol}(\operatorname{supp}\mathrm{W}\varphi)\,, \tag{12.3.21}$$

from which we obtain the *uncertainty principle*

$$\operatorname{vol}(\operatorname{supp}\mathrm{W}\varphi) \geq \frac{\|\varphi\|_{\mathrm{L}_2}^2}{\|\mathrm{W}\varphi\|_{\mathrm{L}_\infty}} \geq 1\,. \tag{12.3.22}$$

Thus the uncertainty principle is a *mathematical* phenomenon, rather than a *physical* one, inasmuch as it is valid over all $\mathbb{Q}_p$, $p \geq \eta$, and over $\mathbb{A}$.

For a *unit* vector ϕ, in the space of the representation r, the map

$$\mathrm{C}_\phi(\varphi) = c_{\varphi,\phi} = (\varphi, r(v)\phi) \tag{12.3.23}$$

gives by (12.3.3) an isometry of the space of the representation r onto a closed subspace $\mathrm{L}_2(V)_\phi$ of $\mathrm{L}_2(V)$. The adjoint map is given by

$$\mathrm{C}_\phi^* f = r(f)\phi, \quad f \in \mathrm{L}_2(V)\,. \tag{12.3.24}$$

The projection from $\mathrm{L}_2(V)$ onto $\mathrm{L}_2(V)_\phi$ is given by the $\natural$-idempotent $c_{\phi,\phi}$,

$$\mathrm{C}_\phi\mathrm{C}_\phi^* f(v) = (r(f)\phi, r(v)\phi) = f \natural c_{\phi,\phi}(v)\,, \tag{12.3.25}$$

$$c_{\phi,\phi} \natural c_{\phi,\phi} = c_{\phi,\phi}\,. \tag{12.3.26}$$

Since $C_\phi^* C_\phi$ is the identity on the space of the representation r, any element of this space has the *wave packet expansion* as a superposition of translates of ϕ:

$$\varphi = C_\phi^* C_\phi \varphi = \int_V dv\, (\varphi, r(v)\phi) \cdot r(v)\phi\,. \tag{12.3.27}$$

We thus have our second realization of the fundamental representation r on the space

$$L_2(V)_\phi = L_2(V) \natural c_{\phi,\phi} = \{f \in L_2(V) \mid f \natural c_{\phi,\phi} = f\} \tag{12.3.28}$$

(in fact, infinitely many such realizations, one for each ϕ). The advantage of this ϕ-realization is that $L_2(V)_\phi$ is a reproducing kernel Hilbert space, for the integral equation in (12.3.28) gives

$$f(v) = f \natural c_{\phi,\phi}(v) = (r(f)\phi, r(v)\phi) = (f, \bar{E}_v)_{L_2(V)}\,, \tag{12.3.29}$$

with the kernel

$$E_v(v_0) = \delta_{v_0} \natural c_{\phi,\phi}(v) = (r(v_0)\phi, r(v)\phi)\,. \tag{12.3.30}$$

Thus any operator A (respectively functional τ) on the space of the representation r, when transformed to the space $L_2(V)_\phi$, becomes represented by a kernel (respectively a function) on V:

$$A f(v) = \int_V dv_0\, A(v, v_0) f(v_0), \quad A(v, v_0) = (A\, r(v_0)\phi, r(v)\phi)\,, \tag{12.3.31}$$

respectively,

$$\tau(f) = \int_V dv\, \tau(v) f(v), \quad \tau(v) = \tau(r(v)\phi)\,. \tag{12.3.32}$$

The decomposition of the identity (12.3.27) gives mathematical meaning to the concept of 'localizing a function in phase space': if ϕ is concentrated near $(0,0)$ in phase space then $r(v)\phi$ is concentrated near v. Thus for a subset B of our phase space V (over $\mathbb{Q}_p$, $p \geq \eta$; or over $\mathbb{A}$), let 1_B denote the characteristic function of the set B, viewed as a projection on $L_2(V)$. The 'localization-in-B' operator acts in the space of the representation r as

$$P_B^\phi = C_\phi^* 1_B C_\phi\,. \tag{12.3.33}$$

Note that P_B^ϕ is a self-adjoint operator satisfying

$$0 \leq P_B^\phi \leq 1\,, \tag{12.3.34}$$

but in general P_B^ϕ is not a projection.

12.4 THE LOCAL LATTICE MODEL

When we specialize to the ϕ_p-realization of the fundamental representation with ϕ_p given by

$$\phi_p = \phi_{\mathbb{Z}_p} \quad \text{for } p \neq \eta, \quad \phi_\eta(x) = 2^{1/4}\phi_{\mathbb{Z}_\eta}(x) = 2^{1/4} e^{-\pi x^2}\,, \tag{12.4.1}$$

we obtain the $\natural$-idempotent $\Phi_{V(\mathbb{Z}_p)} = c_{\phi_p,\phi_p}$ explicitly given by

$$\Phi_{V(\mathbb{Z}_p)}(x,y) = \begin{cases} \phi_{\mathbb{Z}_p}(x)\phi_{\mathbb{Z}_p}(y) & p \neq 2,\ \eta\,, \\ \phi_{\mathbb{Z}_2}(x)\phi_{\mathbb{Z}_2}(y)\cdot(-1)^{x\cdot y} & p = 2\,, \\ \phi_{\mathbb{Z}_\eta}(x/\sqrt{2})\phi_{\mathbb{Z}_\eta}(y/\sqrt{2}) = \mathrm{e}^{-(\pi/2)(x^2+y^2)} & p = \eta\,. \end{cases} \tag{12.4.2}$$

The integral equation $f \natural \Phi_{V(\mathbb{Z}_p)} = f$, characterizing the elements of the *local lattice model* $\mathrm{L}_2(V(\mathbb{Q}_p))_{\phi_p}$, is equivalent for $p \neq \eta$ to the 'differential equation' $f \natural \delta_{v_0} = f$ for all $v_0 \in V(\mathbb{Z}_p)$, or explicitly

$$f(x+x_0, y+y_0) = f(x,y)\cdot\psi_p((xy_0 - yx_0)/2) \quad x_0, y_0 \in \mathbb{Z}_p, \quad p \neq 2,\ \eta \tag{12.4.3}$$

For the real prime, we think of $\Phi_{V(\mathbb{Z}_\eta)}$ as the 'characteristic function of $V(\mathbb{Z}_\eta)$' it is not a multiplicative idempotent, but only an idempotent with respect to twisted multiplication (cf. (2.2.3)). The integral equation $f \natural \Phi_{V(\mathbb{Z}_\eta)} = f$, characterizing the local lattice model (= the Bargmann–Fock model) $\mathrm{L}_2(V(\mathbb{Q}_\eta))_{\phi_\eta}$, is equivalent to the differential equation

$$\left.\frac{\partial}{\partial \bar{v}_0}\right|_{v_0=0} f \natural \delta_{v_0} = 0\,, \tag{12.4.4}$$

where we give $V(\mathbb{Q}_\eta) = \mathbb{R} \oplus \mathbb{R}$ the complex structure by identifying

$$v = (x,y) \text{ with } v = x + \mathrm{i}y \in \mathbb{C}, \quad \text{so } \frac{\partial}{\partial \bar{v}} = \frac{\partial}{\partial x} + \mathrm{i}\frac{\partial}{\partial y}\,. \tag{12.4.5}$$

Explicitly, (12.4.4) reads

$$\mathcal{L}_{V(\mathbb{Z}_\eta)} f = 0\,, \tag{12.4.6}$$

where the operator $\mathcal{L}_{V(\mathbb{Z}_\eta)}$ is given by

$$\mathcal{L}_{V(\mathbb{Z}_\eta)} = \frac{\partial}{\partial \bar{v}} + \pi v = \Phi_{V(\mathbb{Z}_\eta)} \frac{\partial}{\partial \bar{v}} \Phi^{-1}_{V(\mathbb{Z}_\eta)}\,, \tag{12.4.7}$$

or equivalently

$$f(v) = \mathrm{e}^{-(\pi/2)|v|^2}\cdot h(v)\,, \tag{12.4.8}$$

where $h(v)$ is an entire function of $v = x + \mathrm{i}y$.

The reproducing kernel of the space $\mathrm{L}_2(V(\mathbb{Q}_p))_{\phi_p}$, $p \geq \eta$ is given by

$$E_v(v_0) = (r(v_0)\phi_p, r(v)\phi_p)_{\mathrm{L}_2(\mathbb{Q}_p)} = \Phi_{V(\mathbb{Z}_p)}(v - v_0)\psi_p(\langle v_0, v\rangle/2) \tag{12.4.9}$$

(for $p = \eta$ this is equal to $\Phi_{V(\mathbb{Z}_\eta)}(v)\Phi_{V(\mathbb{Z}_\eta)}(v_0)\mathrm{e}^{\pi v\cdot\bar{v}_0}$, the Bargmann kernel).

The action of $\mathrm{Heis}(\mathbb{Q}_p)$ on the space $\mathrm{L}_2(V(\mathbb{Q}_p))_{\phi_p}$, $p \geq \eta$ is given by

$$\begin{aligned} r(v_0)f(v) &= \delta_{v_0} \natural f(v) = f(v - v_0)\psi(\langle v_0, v\rangle/2)\,, \\ r(f_0)f &= f_0 \natural f, \quad \hat{r}(f_0)f = f_0 \# f\,. \end{aligned} \tag{12.4.10}$$

For $p \neq \eta$, the action of $\mathrm{Heis}(\mathbb{Z}_p)$ on $\mathrm{L}_2(V(\mathbb{Q}_p))_{\phi_p}$ takes a simple form. Note the equation

$$\delta_{v_1} \natural f \natural \delta_{v_2}(v) = f(v - v_1 - v_2)\cdot\psi_p(\langle v_1, v\rangle/2 + \langle v, v_2\rangle/2 + \langle v_2, v_1\rangle/2)\,, \tag{12.4.11}$$

from which we have

$$\delta_{v_0} \natural f \natural \delta_{v_0}(v) = f(v - 2v_0), \quad \delta_{v_0} \natural f \natural \delta_{-v_0}(v) = f(v) \cdot \psi_p(\langle v_0, v\rangle). \tag{12.4.12}$$

From this we obtain for $f \in \mathrm{L}_2(V(\mathbb{Q}_p))_{\phi_p}, v_0 \in V(\mathbb{Z}_p), p \neq \eta, 2$,

$$\delta_{v_0} \natural f(v) = \delta_{v_0} \natural f \natural \delta_{\pm v_0}(v) = f(v - 2v_0) = f(v) \cdot \psi_p(\langle v_0, v\rangle). \tag{12.4.13}$$

For $p = 2$ we have only a projective representation of $V(\mathbb{Z}_2)$, and we should multiply the right-hand side of (12.4.13) by $(-1)^{x_0 \cdot y_0}$, $v_0 = (x_0, y_0)$.

The real analogue of (12.4.13) is given by the *creation operator*

$$\begin{aligned} a^* f(v) &= \frac{1}{2\sqrt{\pi}} \frac{\partial}{\partial \bar{v}_0}\bigg|_{v_0 = 0} \delta_{v_0} \natural f(v) = \frac{1}{2\sqrt{\pi}} \frac{\partial}{\partial \bar{v}_0}\bigg|_{v_0 = 0} \delta_{v_0} \natural f \natural \delta_{\pm v_0}(v) \\ &= -\frac{1}{\sqrt{\pi}} \frac{\partial}{\partial \bar{v}} f(v) = \sqrt{\pi} \cdot v \cdot f(v). \end{aligned} \tag{12.4.14}$$

Since we have only a projective representation, a^* is not self-adjoint, and its adjoint a is the *annihilation operator*

$$\begin{aligned} a f(v) &= -\frac{1}{2\sqrt{\pi}} \frac{\partial}{\partial v_0}\bigg|_{v_0 = 0} \delta_{v_0} \natural f(v) \\ &= \frac{1}{2\sqrt{\pi}} \left(\frac{\partial}{\partial v} + \pi \bar{v}\right) \cdot f(v) = \frac{1}{2\sqrt{\pi}} \Phi_{V(\mathbb{Z}_\eta)} \frac{\partial}{\partial v} \Phi_{V(\mathbb{Z}_\eta)}^{-1} f(v). \end{aligned} \tag{12.4.15}$$

We now have the (ordinary) *Heisenberg relation*

$$a a^* - a^* a = \mathrm{id}_{\mathrm{L}_2(V(\mathbb{Q}_\eta))_{\phi\eta}}. \tag{12.4.16}$$

The linear action of $\mathrm{Sp}(V(\mathbb{Q}_p)) = \mathrm{SL}_2(\mathbb{Q}_p)$ on $V(\mathbb{Q}_p)$ preserves twisted convolution

$$\mathrm{R}(g) f_1 \natural \mathrm{R}(g) f_2 = \mathrm{R}(g)(f_1 \natural f_2), \quad \mathrm{R}(g) f(v) = f(vg). \tag{12.4.17}$$

The subgroup $\mathrm{SL}_2(\mathbb{Q}_p)_{\Phi_{V(\mathbb{Z}_p)}}$ stabilizing $\Phi_{V(\mathbb{Z}_p)}$ acts linearly on the space $\mathrm{L}_2(V(\mathbb{Q}_p))_{\phi_p}$. For $p \neq \eta$, 2, we obtain the unitary action of $\mathrm{SL}_2(\mathbb{Z}_p)$

$$\begin{aligned} &\mathrm{R}\colon \mathrm{SL}_2(\mathbb{Z}_p) \to \mathrm{U}(\mathrm{L}_2(V(\mathbb{Q}_p))_{\phi_p}), \quad \mathrm{R}(g) f(v) = f(vg), \\ &\qquad \mathrm{R}(g^{-1}) r(v) \mathrm{R}(g) = r(vg), \quad v \in V(\mathbb{Q}_p), \ g \in \mathrm{SL}_2(\mathbb{Z}_p). \end{aligned} \tag{12.4.18}$$

For $p = 2$,

$$g = \begin{pmatrix} a & b \\ c & d \end{pmatrix} \in \mathrm{SL}_2(\mathbb{Z}_2)$$

stabilizes $\Phi_{V(\mathbb{Z}_2)}$ if and only if it satisfies the further condition

$$(-1)^{x \cdot y} = (-1)^{(ax+cy)\cdot(bx+dy)} \quad \text{for all } x, y \in \mathbb{Z}_2, \tag{12.4.19}$$

or equivalently,

$$g \equiv \begin{pmatrix} 1 & 0 \\ 0 & 1 \end{pmatrix} \quad \text{or} \quad \begin{pmatrix} 0 & 1 \\ 1 & 0 \end{pmatrix} \pmod 2. \tag{12.4.20}$$

For the real prime $p = \eta$, identifying $V(\mathbb{Q}_\eta) = \mathbb{C}$ as (12.4.5), the stabilizer of $\Phi_{V(\mathbb{Z}_\eta)}$ is '$\mathrm{SL}_2(\mathbb{Z}_\eta)$'$= \mathrm{SO}_2(\mathbb{R}) = \mathbb{C}^{(1)}$, and this acts linearly on $\mathrm{L}_2(V(\mathbb{Q}_\eta))_{\phi_\eta}$ in analogy to (12.4.18):

$$\begin{aligned} \mathrm{R}\colon \mathbb{C}^{(1)} \to \mathrm{U}(\mathrm{L}_2(V(\mathbb{Q}_\eta))_{\phi_\eta}), \quad \mathrm{R}(g)f(v) = f(vg)\,, \\ \mathrm{R}(g^{-1})r(v)\mathrm{R}(g) = r(vg), \quad v \in V(\mathbb{Q}_\eta), \quad g \in \mathbb{C}^{(1)}\,. \end{aligned} \tag{12.4.21}$$

The Fourier transform $\mathcal{F}$ on the local lattice model is given by

$$\mathcal{F}f(x,y) = f(-y,x)\,. \tag{12.4.22}$$

The distributions δ_0 and δ_∞ of (12.1.19) are given on the local lattice model by

$$\delta_0(f) = \int f(x,0)\mathrm{d}x, \quad \delta_\infty(f) = \int f(0,y)\mathrm{d}y\,. \tag{12.4.23}$$

Denoting the unitary action of the multiplicative group on the Schrödinger model by $\pi(a)\varphi(z) = |a|^{-1/2}\varphi(a^{-1}z)$, the vector $\pi(a)\phi_p$ corresponds in the local lattice model to the vector

$$\Phi_p^{(a)}(x,y) = (\pi(a)\phi_p, r(x,y)\phi_p)_{\mathrm{L}_2(\mathbb{Q}_p)}\,. \tag{12.4.24}$$

A direct calculation gives that $\Phi_p^{(a)}$ can be written for all $p \geq \eta$ as

$$\Phi_p^{(a)}(x,y) = \rho_0(a)^{1/2}\phi_p(\tilde{\rho}_0(a)x) \cdot \rho_\infty(a)^{1/2}\phi_p(\tilde{\rho}_\infty(a)y) \cdot \psi_p(xyz_p(a)/2), \tag{12.4.25}$$

where $\tilde{\rho}_0(a),\ \tilde{\rho}_\infty(a) \in \mathbb{Q}_p^*$ are any elements such that

$$|\tilde{\rho}_0(a)|_p = \rho_0(a), \quad |\tilde{\rho}_\infty(a)|_p = \rho_\infty(a)\,, \tag{12.4.26}$$

and where $z_p(a)$ is given by

$$\begin{aligned} z_\eta(a) &= \rho_\infty(a)^2 - \rho_0(a)^2 = \frac{1-a^2}{1+a^2}\,, \\ z_p(a) &= \text{`}\rho_\infty(a)^\infty - \rho_0(a)^\infty\text{'} = \begin{cases} -1 & |a| > 1 \\ 1 & |a| \leq 1 \end{cases} \quad p \neq \eta\,. \end{aligned} \tag{12.4.27}$$

$\Phi_p^{(a)}(x,y)\bar{\psi}_p(t)$ is a positive-definite function on the (unimodular) semi-direct product group $\mathbb{Q}_p^* \ltimes \mathrm{Heis}(\mathbb{Q}_p)$, bi-invariant under $\mathbb{Z}_p^* \ltimes \mathrm{Heis}(\mathbb{Z}_p)$; it generalizes both $\Phi_{V(\mathbb{Z}_p)}(x,y)\bar{\psi}_p(t)$ on $\mathrm{Heis}(\mathbb{Q}_p)$ and $\tilde{\rho}_p(a) = \rho_0(a)^{1/2}\rho_\infty(a)^{1/2}$ on $\mathbb{Q}_p^*$.

The operator $\pi(a)$, when transformed to the local lattice model, is given by the kernel (cf. (12.3.31))

$$\begin{aligned} \mathrm{K}_p^{(a)}((x,y),(x_0,y_0)) &= (\pi(a)r(x_0,y_0)\phi_p, r(x,y)\phi_p)_{\mathrm{L}_2(\mathbb{Q}_p)} \\ &= (r(x_0a^{-1}, y_0a)\pi(a)\phi_p, r(x,y)\phi_p)_{\mathrm{L}_2(\mathbb{Q}_p)} \\ &= (\pi(a)\phi_p, r(-x_0a^{-1}, -y_0a)r(x,y)\phi_p)_{\mathrm{L}_2(\mathbb{Q}_p)} \\ &= \Phi_p^{(a)}(x - x_0a^{-1}, y - y_0a) \cdot \psi_p((x_0a^{-1}y - y_0ax)/2)\,; \end{aligned} \tag{12.4.28}$$

equivalently, for $f \in L_2(V(\mathbb{Q}_p))_{\phi_p}$,

$$\pi(a)f = f^{(a)} \natural \Phi_p^{(a)}, \quad \text{with } f^{(a)}(x,y) = f(xa, ya^{-1}). \tag{12.4.29}$$

In the local lattice model, we have the unique (up to constant multiple) Schwartz function $\Phi_{V(\mathbb{Z}_p)}$, 'invariant under $V(\mathbb{Z}_p)$', $p \geq \eta$:

$$\delta_{V(\mathbb{Z}_p)}(f) = f(0,0) = (f, \Phi_{V(\mathbb{Z}_p)})_{L_2(V(\mathbb{Q}_p))}, \quad f \in L_2(V(\mathbb{Q}_p))_{\phi_p}. \tag{12.4.30}$$

In the Schrödinger model, it corresponds to the unit vector ϕ_p

$$\delta_{V(\mathbb{Z}_p)}(\varphi) = (\varphi, \phi_p)_{L_2(\mathbb{Q}_p)}, \quad \varphi \in L_2(\mathbb{Q}_p). \tag{12.4.31}$$

We have worked throughout with the 'standard' two-dimensional metric (1.3.1). We can change this metric, as long as we keep its property of being *self-dual* with respect to the symplectic form on V. The self-dual metrics on $V(\mathbb{Q}_p)$ are parametrized by the space

$$\widetilde{X}_p^+ = SL_2(\mathbb{Q}_p)_{\Phi_{V(\mathbb{Z}_p)}} \backslash SL_2(\mathbb{Q}_p). \tag{12.4.32}$$

For $p \neq 2,\ \eta$, $\widetilde{X}_p^+$ consists of *half* the $(p+1)$-regular tree $\widetilde{X}_p$:

$$\widetilde{X}_p^+ = \coprod_{N \geq 0} \widetilde{X}_{p(2N)}. \tag{12.4.33}$$

Its elements $g \in \widetilde{X}_p^+$ correspond to the $\mathbb{Z}_p$-lattices $\mathcal{O} = V(\mathbb{Z}_p)g \subseteq V(\mathbb{Q}_p)$ which are *self-dual* in the sense that $\mathcal{O} = \mathcal{O}^V$, where the dual lattice $\mathcal{O}^V$ is defined by

$$\mathcal{O}^V = \{v \in V(\mathbb{Q}_p) | \langle v, \mathcal{O}\rangle \subseteq \mathbb{Z}_p\}. \tag{12.4.34}$$

The central character ψ_p can be extended to the (maximal commutative) subgroup $\mathcal{O} \times \mathbb{Q}_p/\mathbb{Z}_p$ of $\mathrm{Heis}(\mathbb{Q}_p)/\mathbb{Z}_p$

$$\psi_{\mathcal{O}}(x,y,t) = \psi_p(t) \quad \text{for } (x,y) \in \mathcal{O}, \tag{12.4.35}$$

and the induced representation is equivalent to r_{ψ_p}

$$r_{\psi_p} \simeq \mathrm{Ind}_{\mathcal{O} \times \mathbb{Q}_p/\mathbb{Z}_p}^{\mathrm{Heis}(\mathbb{Q}_p)/\mathbb{Z}_p}(\psi_{\mathcal{O}}). \tag{12.4.36}$$

It follows that in any realization of r_{ψ_p} (and in particular in the Schrödinger model) there is a unique (up to constant multiple) Schwartz function $\phi_p^{\mathcal{O}}$ that is invariant under the action of $r(\mathcal{O})$. Alternatively, we have a realization of r_{ψ_p} on the subspaces $L_2(V(\mathbb{Q}_p))_{\phi_p^{\mathcal{O}}}$ of $L_2(V(\mathbb{Q}_p))$, where

$$L_2(V(\mathbb{Q}_p))_{\phi_p^{\mathcal{O}}} = L_2(V(\mathbb{Q}_p)) \natural \Phi_{\mathcal{O}} = \{f \in L_2(V(\mathbb{Q}_p)) | f \natural \Phi_{\mathcal{O}} = f\},$$

$$r(v_0)f(v) = \delta_{v_0} \natural f(v) = f(v - v_0)\psi(\langle v_0, v\rangle/2) \quad \text{for } f \in L_2(V(\mathbb{Q}_p))_{\phi_p^{\mathcal{O}}}. \tag{12.4.37}$$

Here $\Phi_{\mathcal{O}}$ is the $\natural$-idempotent given by the matrix coefficients of $\phi_p^{\mathcal{O}}$

$$\Phi_{\mathcal{O}}(v) = (\phi_p^{\mathcal{O}}, r(v)\phi_p^{\mathcal{O}}) = \Phi_{V(\mathbb{Z}_p)}(vg^{-1}), \quad \text{for } \mathcal{O} = V(\mathbb{Z}_p)g. \tag{12.4.38}$$

For $p \neq \eta$, 2, the integral equation $f \natural \Phi_{\mathcal{O}} = f$ is equivalent to the 'differential equation'

$$f(v + v_0) = f(v) \cdot \psi_p(\langle v, v_0 \rangle/2) \quad \text{for all } v_0 \in \mathcal{O},\ v \in V(\mathbb{Q}_p)\,. \tag{12.4.39}$$

For the real prime, $\widetilde{\mathrm{X}}_\eta^+ = \mathrm{SO}_2(\mathbb{R}) \backslash \mathrm{SL}_2(\mathbb{R})$ consists of self-dual metrics on the plane $V(\mathbb{R})$, and can be identified with the upper half-plane $\{\tau \in \mathbb{C}, \mathrm{Im}(\tau)\rangle 0\}$. The associated Schwartz function in the Schrödinger model $\mathrm{L}_2(\mathbb{R})$ is

$$\phi_\eta^{(\tau)}(z) = (2\,\mathrm{Im}(\tau))^{1/4} \mathrm{e}^{-\pi i \bar{\tau} z^2}\,. \tag{12.4.40}$$

Its matrix coefficient is the $\natural$-idempotent

$$\Phi_{\mathcal{O}_\eta^{(\tau)}}(x, y) = \left(\phi_\eta^{(\tau)}, r(x, y)\phi_\eta^{(\tau)}\right)_{\mathrm{L}_2(\mathbb{R})} = \mathrm{e}^{-[\pi/2\mathrm{Im}(\tau)]|x+\tau y|^2}\,. \tag{12.4.41}$$

The integral equation $f \natural \Phi_{\mathcal{O}_\eta^{(\tau)}} = f$ characterizing $\mathrm{L}_2(V(\mathbb{R}))_{\phi_\eta^{(\tau)}}$ is equivalent to the differential equation $\mathcal{L}_{\mathcal{O}_\eta^{(\tau)}} f = 0$, where

$$\mathcal{L}_{\mathcal{O}_\eta^{(\tau)}} = \left(\frac{\partial}{\partial x} - \frac{1}{\tau}\frac{\partial}{\partial y}\right) + \frac{\pi i}{\tau}(x + \tau y) = \Phi_{\mathcal{O}_\eta^{(\tau)}}\left(\frac{\partial}{\partial x} - \frac{1}{\tau}\frac{\partial}{\partial y}\right)(\Phi_{\mathcal{O}_\eta^{(\tau)}})^{-1}\,, \tag{12.4.42}$$

or equivalently

$$f(x, y) = \mathrm{e}^{-[\pi/2\mathrm{Im}(\tau)]|x+\tau y|^2} \cdot h(x + \tau y)\,, \tag{12.4.43}$$

where h is an entire function of $x + \tau y$.

The analogue of (12.4.30) in the $\mathcal{O}_\eta^{(\tau)}$-model gives

$$\delta_{\mathcal{O}_\eta^{(\tau)}}(f) = f(0,0) = \left(f, \Phi_{\mathcal{O}_\eta^{(\tau)}}\right)_{\mathrm{L}_2(V(\mathbb{R}))}, \quad \text{for } f \in \mathrm{L}_2(V(\mathbb{R}))_{\phi_\eta^{(\tau)}} \tag{12.4.44}$$

and it corresponds in the Schrödinger model to

$$\delta_{\mathcal{O}_\eta^{(\tau)}}(\varphi) = \left(\varphi, \phi_\eta^{(\tau)}\right)_{\mathrm{L}_2(\mathbb{R})} = (2\,\mathrm{Im}(\tau))^{1/4} \int \varphi(z) \mathrm{e}^{\pi i \tau z^2} \mathrm{d}z\,. \tag{12.4.45}$$

The mapping $\mathcal{O} \mapsto \phi_p^{\mathcal{O}}$ gives an embedding

$$\widetilde{\mathrm{X}}_p^+ \hookrightarrow \mathbb{P}^1(\mathcal{S}(\mathbb{Q}_p)), \quad p \geq \eta \quad \text{and} \quad \widetilde{\mathrm{X}}_{\mathbb{A}}^+ \hookrightarrow \mathbb{P}^1(\mathcal{S}(\mathbb{A}))\,. \tag{12.4.46}$$

12.5 SPECIAL BASIS

The Heisenberg operators a^* and a give us an orthogonal basis for $\mathrm{L}_2(V(\mathbb{Q}_\eta))_{\phi_\eta}$

$$\begin{gathered} \mathrm{L}_2(V(\mathbb{Q}_\eta))_{\phi_\eta} = \bigoplus_{n \geq 0} \mathbb{C} \cdot \Phi_{V(\mathbb{Z}_\eta),n}\,, \\ \Phi_{V(\mathbb{Z}_\eta),n}(v) = (1/n!)(a^*)^n \Phi_{V(\mathbb{Z}_\eta)}(v) = ((\sqrt{\pi} v)^n/n!\,)\mathrm{e}^{-(\pi/2)|v|^2}\,. \end{gathered} \tag{12.5.1}$$

The decomposition (12.5.1) is just the spectral decomposition of the action of the *commutative* group $\mathbb{C}^{(1)} = \mathrm{SL}_2(\mathbb{Z}_\eta)$:

$$\mathrm{R}(g)\Phi_{V(\mathbb{Z}_\eta),n} = g^n \cdot \Phi_{V(z_\eta),n}, \quad g \in \mathbb{C}^{(1)}. \tag{12.5.2}$$

We can translate the operators a^* and a, and the basis $\Phi_{V(\mathbb{Z}_\eta),n}$, to the Schrödinger model $\mathrm{L}_2(\mathbb{R})$:

$$\begin{aligned} a^* &= \phi_{\mathbb{Z}_\eta}^{-1} \frac{1}{2\mathrm{i}\sqrt{\pi}} \frac{\partial}{\partial z} \phi_{\mathbb{Z}_\eta} = \frac{1}{2\mathrm{i}\sqrt{\pi}} \frac{\partial}{\partial z} + \mathrm{i}\sqrt{\pi} z, \\ a &= \phi_{\mathbb{Z}_\eta} \frac{1}{2\mathrm{i}\sqrt{\pi}} \frac{\partial}{\partial z} \phi_{\mathbb{Z}_\eta}^{-1} = \frac{1}{2\mathrm{i}\sqrt{\pi}} \frac{\partial}{\partial z} - \mathrm{i}\sqrt{\pi} z, \end{aligned} \tag{12.5.3}$$

$$\begin{aligned} \tilde{\phi}_{\eta,n}(z) &= r(\Phi_{V(\mathbb{Z}_\eta),n})\phi_\eta(z) = 1/n!\,(a^*)^n \phi_\eta(z) \\ &= \frac{2^{1/4}}{n!(2\mathrm{i}\sqrt{\pi})^n} \mathrm{e}^{\pi z^2} \left(\frac{\partial}{\partial z}\right)^n \mathrm{e}^{-2\pi z^2} = \frac{1}{(2\mathrm{i}\sqrt{\pi})^n} \cdot \mathrm{H}_n(z) \cdot \phi_\eta(z), \end{aligned} \tag{12.5.4}$$

where $\mathrm{H}_n(z)$ are the Hermite polynomials.

We have immediately from the Heisenberg relations (12.4.16)

$$\mathrm{L}_2(\mathbb{Q}_\eta) = \bigoplus_{n \geq 0} \mathbb{C} \cdot \tilde{\phi}_{\eta,n}, \tag{12.5.5}$$

$$\begin{aligned} a^* \tilde{\phi}_{\eta,n-1} &= n \cdot \tilde{\phi}_{\eta,n}, \\ a\tilde{\phi}_{\eta,n} &= \tilde{\phi}_{\eta,n-1}, \\ a^* a \tilde{\phi}_{\eta,n} &= n \cdot \tilde{\phi}_{\eta,n}, \end{aligned} \tag{12.5.6}$$

$$\left\|\tilde{\phi}_{\eta,n}\right\|^2 = 1/n!. \tag{12.5.7}$$

The polynomials $\mathrm{H}_n(z)$ are just the polynomials $\varphi_{\eta,n}^{\infty,\infty}(\sqrt{2}z)$ we obtained in the limit $\alpha = \beta \to \infty$ of the real beta chain (cf. (8.3.39)). A more appropriate interpretation is via the $\beta = 1, 3$ Laguerre basis we had in (10.6.12), using the one-dimensional case of (9.2.69):

$$\mathrm{L}_2(\mathbb{Q}_\eta) = \mathrm{L}_2(\mathbb{R})^{\pm 1} \oplus \mathrm{L}_2(\mathbb{R})^{\mathrm{odd}} = \mathscr{H}_\eta^1 \oplus \sqrt{2\pi} x \cdot \mathscr{H}_\eta^3. \tag{12.5.8}$$

We have

$$\begin{aligned} \tilde{\phi}_{\eta,2n}(z) &= (-1)^n \frac{2^n \cdot n!}{(2n)!} \varphi_{\mathbb{Z}_\eta,n}^1(\sqrt{2}z) \cdot \phi_\eta(z) = (-1)^n \frac{2^n \cdot n!}{(2n)!}\, \varphi_{\eta,n}^1(z), \\ \tilde{\phi}_{\eta,2n+1}(z) &= (-1)^n \frac{2^n \cdot n!}{(2n+1)!} 2\mathrm{i}\sqrt{\pi} z \varphi_{\mathbb{Z}_\eta,n}^3(\sqrt{2}z) \cdot \phi_\eta(z) = (-1)^n \frac{2^n \cdot n!}{(2n+1)!} 2\mathrm{i}\sqrt{\pi} z \varphi_{\eta,n}^3(z). \end{aligned} \tag{12.5.9}$$

It is remarkable that the Heisenberg relations up the ladder (10.6.20) for $\varphi^{\beta}_{\eta,n}(z)$ reduce via the isomorphism (12.5.8) to the ordinary Heisenberg relations (12.4.16) for the $\tilde{\phi}_{\eta,n}(z)$.

The Hermite basis $\{\tilde{\phi}_{\eta,n}\}$ are 'packed together' inside discs $\mathrm{B}_R = \{(x,y) \in \mathbb{R}\times\mathbb{R} \,|\, |x,y|_\eta = \sqrt{x^2+y^2} \le R\}$, in the sense that the first $\pi R^2 = \mathrm{vol}(\mathrm{B}_R)$ of them are 'localized' in B_R. They are eigenfunctions of the localization operator $P^{\phi}_{\mathrm{B}_R} = \mathrm{C}^*_\phi 1_{\mathrm{B}_R} \mathrm{C}_\phi$, with eigenvalues

$$\begin{aligned}
\lambda_{R,n} &= \|\tilde{\phi}_{\eta,n}\|^{-2}\cdot\left(\mathrm{C}^*_\phi 1_{\mathrm{B}_R}\mathrm{C}_\phi\tilde{\phi}_{\eta,n}, \tilde{\phi}_{\eta,n}\right) \\
&= \frac{\pi^n}{n!}\iint_{\mathrm{B}_R} \mathrm{d}v\, |v|^{2n}\mathrm{e}^{-\pi|v|^2} \\
&= \frac{\pi^n}{n!}\cdot 2\pi\int_0^R \mathrm{d}^*a\cdot a^{2(1+n)}\mathrm{e}^{-\pi a^2} \\
&= \frac{1}{n!}\int_0^{\pi R^2} \mathrm{d}a\cdot a^n\cdot\mathrm{e}^{-a} = \frac{\gamma(1+n,\pi R^2)}{\Gamma(1+n)}\,,
\end{aligned}$$

the incomplete gamma function. For fixed R, the eigenvalues $\lambda_{R,n}$ decrease monotonically; they are close to 1 for $n \le \pi R^2$, and then they plunge to near zero for $n \ge \pi R^2 + \mathrm{O}(R)$. Thus the width of the plunge region $\mathrm{O}(R)$ is wider than the logarithmic width of the prolate spheroidal wave functions of §10.6, but the $\tilde{\phi}_{\eta,n}$ have the advantage of being independent of R. When both $n\to\infty$ and $R\to\infty$ in such a way that $\delta = (\pi R^2 - n)/\sqrt{2\pi n}$ remains bounded, we have Tricomi's estimate

$$\lambda_{R,n} = \lambda(\delta) + \mathrm{O}(1/R)\,,$$

with

$$\lambda(\delta) = \frac{1}{2} + \frac{1}{\sqrt{\pi}}\mathrm{Erf}(\sqrt{\pi}\delta) = \int_{-\infty}^{\delta} \mathrm{e}^{-\pi x^2}\mathrm{d}x\,,$$

cf. Erdélyi *et al.* (1953), p. 141.

We can also replace the characteristic function 1_{B_R} by a Gaussian $\mathrm{e}^{-(\pi/R^2)|v|^2}$. The Hermite functions are eigenfunctions of the operator $\mathrm{C}^*_\phi\mathrm{e}^{-(\pi/R^2)|v|^2}\mathrm{C}_\phi$, with eigenvalues

$$\begin{aligned}
\tilde{\lambda}_{R,n} &= \|\tilde{\phi}_{\eta,n}\|^{-2}\cdot\left(\mathrm{C}^*_\phi\mathrm{e}^{-(\pi/R^2)|v|^2}\mathrm{C}_\phi\tilde{\phi}_{\eta,n}, \tilde{\phi}_{\eta,n}\right) \\
&= \frac{\pi^n}{n!}\iint \mathrm{d}v\, |v|^{2n}\mathrm{e}^{-\pi|v|^2(1+R^{-2})} \\
&= \frac{\pi^n}{n!}2\pi\int_0^\infty d^*a\cdot a^{2(1+n)}\cdot\mathrm{e}^{-\pi a^2}\cdot(1+R^{-2})^{-(1+n)} \\
&= (1+R^{-2})^{-(1+n)}\,.
\end{aligned}$$

We can ask: is there a p-adic analogue of the basis $\tilde{\phi}_{\eta,n}$?

The first answer is the orthonormal basis for $\mathrm{L}_2(V(\mathbb{Q}_p))_{\phi_p}$ consisting of the $\mathrm{Heis}(\mathbb{Z}_p)$-eigenfunctions $\{\bar{E}_v \,|\, v \in V(\mathbb{Q}_p)/V(\mathbb{Z}_p)\}$, or in the Schrödinger model by $\{r(x,y)\phi_p \,|\, x,y \in \mathbb{Q}_p/\mathbb{Z}_p\}$. But there are other answers which we shall soon meet.

First of all note that for $p \neq \eta$ the group $\mathrm{SL}_2(\mathbb{Z}_p)$ is not commutative, and the decomposition of $\mathrm{L}_2(V(\mathbb{Q}_p))_{\phi_p}$ under the $\mathrm{SL}_2(\mathbb{Z}_p)$-action is not into one-dimensional subspaces (as we had in (12.5.1), (12.5.2)). In fact we now have the decomposition ($p \neq \eta$, 2):

$$\mathrm{L}_2(V(\mathbb{Q}_p))_{\phi_p} = \mathcal{V}_0 \oplus \bigoplus_{n \geq 1} (\mathcal{V}_{n,+} \oplus \mathcal{V}_{n,-}) \tag{12.5.10}$$

where $\mathcal{V}_0 = \mathbb{C} \cdot \Phi_{V(\mathbb{Z}_p)} = (\mathrm{L}_2(V(\mathbb{Q}_p))_{\phi_p})^{\mathrm{SL}_2(\mathbb{Z}_p)}$ is the one-dimensional trivial representation,

$$\mathcal{V}_{n,\pm} = \left\{ f \in \mathrm{L}_2(V(\mathbb{Q}_p))_{\phi_p} \,\middle|\, |v| \cdot f(v) = p^n \cdot f(v), f(-v) = \pm f(v) \right\}, \quad n \geq 1 . \tag{12.5.11}$$

It is clear that since $\mathrm{SL}_2(\mathbb{Z}_p)$ preserves the absolute value $|v|$, its linear action on $\mathrm{L}_2(V(\mathbb{Q}_p))_{\phi_p}$ preserves the functions supported on $|v| = p^n$. Similarly, since ± 1 is in the centre of $\mathrm{SL}_2(\mathbb{Z}_p)$, the even and odd functions are preserved. Moreover, an application of Mackey's irreducibility criterion (cf. Serre 1977) shows that $\mathcal{V}_{n,\pm}$ is an irreducible $\mathrm{SL}_2(\mathbb{Z}/p^{2n}\mathbb{Z})$-representation. It is induced from the one-dimensional character $\psi_{n,\pm}$ of the subgroup

$$\mathrm{H}_{p,n} = \left\{ g \in \mathrm{SL}_2(\mathbb{Z}/p^{2n}\mathbb{Z}) \,\middle|\, g \equiv \begin{pmatrix} \varepsilon & 0 \\ * & \varepsilon \end{pmatrix} (\mathrm{mod}\ p^n),\ \varepsilon = \pm 1 \right\}, \tag{12.5.12}$$

where

$$\psi_{n,+}\begin{pmatrix} \varepsilon + p^n a & p^n b \\ c & \varepsilon + p^n d \end{pmatrix} = \psi_p\Big(\frac{\varepsilon}{2} b p^{-n}\Big), \quad \psi_{n,-}\begin{pmatrix} \varepsilon + p^n a & p^n b \\ c & \varepsilon + p^n d \end{pmatrix} = \varepsilon \cdot \psi_p\Big(\frac{\varepsilon}{2} b p^{-n}\Big);$$

$$\mathcal{V}_{n,\pm} = \mathrm{Ind}_{\mathrm{H}_{p,n}}^{\mathrm{SL}_2(\mathbb{Z}/p^{2n})} (\psi_{n,\pm}) .$$

Thus (12.5.10) gives the decomposition of $\mathrm{L}_2(V(\mathbb{Q}_p))_{\phi_p}$ into $\mathrm{SL}_2(\mathbb{Z}_p)$-irreducible representations. The description (12.5.11) implies that the commutant of $\mathrm{SL}_2(\mathbb{Z}_p)$ is generated by the commutative compact group

$$\{\pm \mathrm{I}\} \times \mathrm{K}^{\mathrm{i}\mathbb{R}/\mathbb{Z}(2\pi\mathrm{i}/\log p)} \tag{12.5.13}$$

where

$$\mathrm{K}^s f(v) = |1, v|_p^s \cdot f(v) . \tag{12.5.14}$$

One way of obtaining a decomposition of $\mathrm{L}_2(V(\mathbb{Q}_p))_{\phi_p}$ into one-dimensional subspaces, analogous to (12.5.1), is to use the decomposition with respect to certain commutative subgroups $\mathcal{U}_p \subseteq \mathrm{SL}_2(\mathbb{Z}_p)$, given by the kernel of the norm from a quadratic extension $\mathbb{Q}_p(\sqrt{\varepsilon})$ of $\mathbb{Q}_p$; see Haran (1993*b*). For the unramified extension (and $p \neq 2$; the ramified extensions and $p = 2$ are similar, and will be left to the reader), we take $\varepsilon \in \mathbb{Z}_p^* \backslash (\mathbb{Z}_p^*)^2$, and identify $V(\mathbb{Q}_p)$ with $\mathbb{Q}_p(\sqrt{\varepsilon})$, and $V(\mathbb{Z}_p)$ with $\mathbb{Z}_p[\sqrt{\varepsilon}]$, via (cf. (12.4.5))

$$(x, y) \leftrightarrow x + \sqrt{\varepsilon} y . \tag{12.5.15}$$

We have

$$\mathcal{U}_p = \ker\{\mathrm{N}\colon \mathbb{Q}_p(\sqrt{\varepsilon})^* \to \mathbb{Q}_p^*\} \simeq \mu_{p+1} \times \mathrm{e}^{\sqrt{\varepsilon}p\mathbb{Z}_p}$$
$$= \left\{x + \sqrt{\varepsilon}\, y \in \mathbb{Z}_p[\sqrt{\varepsilon}]^* \,\middle|\, x^2 - \varepsilon y^2 = 1\right\}, \tag{12.5.16}$$

which we view as a subgroup of $\mathrm{SL}_2(\mathbb{Z}_p)$ via

$$\mathcal{U}_p \hookrightarrow \mathrm{SL}_2(\mathbb{Z}_p), \quad x + \sqrt{\varepsilon}\, y \mapsto \begin{pmatrix} x & \varepsilon y \\ y & x \end{pmatrix}. \tag{12.5.17}$$

The character group of $\mathcal{U}_p$ is given by

$$\mathrm{Z} = (\mathcal{U}_p)^V \cong (\mu_{p+1})^V \times \mathbb{Q}_p/\mathbb{Z}_p\,, \tag{12.5.18}$$

and we can write a typical character $\chi \in Z$ as

$$\chi(\zeta \cdot \mathrm{e}^{\sqrt{\varepsilon}pw}) = \chi_t(\zeta) \cdot \bar{\psi}_p(q_\chi pw/2), \quad \chi_t \in (\mu_{p+1})^V, \; q_\chi \cdot p \in \mathbb{Q}_p/\mathbb{Z}_p\,. \tag{12.5.19}$$

The decomposition of $\mathrm{L}_2(V(\mathbb{Q}_p))_{\phi_p}$ with respect to the linear action of $\mathcal{U}_p$ has multiplicity one, and is given by

$$\mathrm{L}_2(V(\mathbb{Q}_p))_{\phi_p} = \bigoplus_{\chi \in Z^+} \mathbb{C} \cdot \Phi_\chi, \quad \mathrm{R}(u)\Phi_\chi = \chi(u) \cdot \Phi_\chi\,. \tag{12.5.20}$$

The sum in (12.5.20) is taken over

$$\mathrm{Z}^+ = \{\chi^0, \text{the trivial character}\} \cup \left\{\chi \in \mathrm{Z} \,\middle|\, |q_\chi| = p^{2n},\; n \geq 1\right\}; \tag{12.5.21}$$

that is, over 'half' the characters in Z (in analogy with the real case, (12.5.1), where $\mathbb{N}$ are 'half' the characters $\mathbb{Z}$ of the circle $\mathcal{U}_\eta = \mathbb{C}^{(1)}$). The associated orthogonal basis of the Schrödinger model $\mathrm{L}_2(\mathbb{Q}_p)$ is given by

$$\tilde{\phi}_\chi = r(\Phi_\chi)\phi_p = \int_{\mathcal{U}_p} \mathrm{d}^*u\, \bar{\chi}(u) r(u \cdot z_\chi)\phi_p\,, \tag{12.5.22}$$

where z_χ is chosen so that $q_\chi = \mathrm{N}z_\chi$, and d^*u is the Haar measure on $\mathcal{U}_p$.

One drawback of the use of $\mathcal{U}_p$ is that it does not contain the units $\mathbb{Z}_p^*$ so that the basis $\tilde{\phi}_\chi$ of (12.5.22) does not have any nice covariance properties with respect to the $\mathbb{Z}_p^*$-action. An alternative approach is obtained by enriching the group (12.5.13) with the commutative compact group

$$\widetilde{\mathrm{Z}} = \mathbb{Z}_p^* \times \mathrm{I}^{\mathrm{i}\mathbb{R}/(2\pi\mathrm{i}/\log p)\mathbb{Z}} \times \mathrm{J}^{\mathrm{i}\mathbb{R}/(2\pi\mathrm{i}/\log p)\mathbb{Z}}\,, \tag{12.5.23}$$

where

$$\mathrm{I}^s f(v) = |1, y|_p^s \cdot f(v), \quad \mathrm{J}^s f(v) = |1, x|_p^s \cdot f(v), \quad v = (x, y)\,, \tag{12.5.24}$$

and $u \in \mathbb{Z}_p^*$ is identified with

$$\begin{pmatrix} u & 0 \\ 0 & u^{-1} \end{pmatrix} \in \mathrm{SL}_2(\mathbb{Z}_p)\,,$$

so that

$$\mathrm{R}(u)\cdot f(x,y)=f(xu,yu^{-1})\,. \tag{12.5.25}$$

A straightforward verification shows that the decomposition of $\mathrm{L}_2(V(\mathbb{Q}_p))_{\phi_p}$ with respect to the action of $\widetilde{\mathrm{Z}}$ has multiplicity one, and is given by

$$\begin{aligned}\mathbb{C}\Phi_{i,j,\chi} &= \bigoplus_{\substack{i,j\geq 0\\ \chi\in(\mathbb{Z}_p^*)^v}} \mathrm{L}_2(V(\mathbb{Q}_p))_{\phi_p}\,,\\ \mathrm{R}(u)\mathrm{I}^{s_1}\mathrm{J}^{s_2}\Phi_{i,j,\chi} &= \chi(u)p^{is_1+js_2}\cdot\Phi_{i,j,\chi}\,,\end{aligned} \tag{12.5.26}$$

where in (12.5.26) we impose the further condition on (χ,i,j):

$$\begin{aligned}&\min\{i,j\}>0 \Rightarrow \mathrm{cond}(\chi)=p^{i+j}\\ &\min\{i,j\}=0 \Rightarrow \mathrm{cond}(\chi)\leq p^{i+j}\\ &\ \mathrm{cond}(\chi)=\text{the conductor of the character } \chi\,.\end{aligned} \tag{12.5.27}$$

Indeed, note that for the orthonormal basis $\{overlineE_v \,|\, v\in V(\mathbb{Q}_p)/V(\mathbb{Z}_p)\}$ and for $u\in\mathbb{Z}_p^*$ we have

$$\mathrm{R}(u)\bar{E}_{(x,y)}=\bar{E}_{(xu^{-1},yu)}\,. \tag{12.5.28}$$

When $|1,x|_p=p^i, |1,y|_p=p^j$, we have $(xu^{-1},yu)\equiv(x,y)\mathrm{mod}\,V(\mathbb{Z}_p)$ if and only if $u\equiv 1 \bmod p^{\max(i,j)}$, and in this case

$$\mathrm{R}(u)\bar{E}_{(x,y)}=\psi_p((u^{-1}-u)xy/2)\cdot\bar{E}_{(x,y)}=\bar{\psi}_p((u-1)xy)\cdot\overline{E}_{(x,y)} \tag{12.5.29}$$

(for $p\neq\eta,2$; we leave the modifications needed when $p=2$ to the reader). Identifying $\mathbb{Z}_p^*$ with $\mu_{p-1}\times \mathrm{e}^{p\mathbb{Z}_p}$, we write a character $\chi\in(\mathbb{Z}_p^*)^{\mathrm{v}}$ as

$$\chi(\zeta\cdot \mathrm{e}^{pw})=\chi_t(\zeta)\cdot\bar{\psi}_p(pwq_\chi),\quad \zeta\in\mu_{p-1},\ w\in\mathbb{Z}_p\,, \tag{12.5.30}$$

so that $\mathrm{cond}(\chi)=p^n$ when $|q_\chi|_p=p^n$.

Writing $q_\chi=x_\chi\cdot y_\chi$, with $|1,x_\chi|_p=p^i$, $|1,y_\chi|_p=p^j$ (so that condition (12.5.27) is imposed on (χ,i,j)), we have

$$\Phi_{i,j,\chi}=\int_{(\mathbb{Z}/p^{\max(i,j)})^*}\mathrm{d}^*u\,\bar{\chi}(u)\bar{E}_{(x_\chi u^{-1},y_\chi u)}\,. \tag{12.5.31}$$

Translated to the Schrödinger model $\mathrm{L}_2(\mathbb{Q}_p)$, our basis is

$$\tilde{\phi}_{i,j,\chi}(z)=r(\Phi_{i,j,\chi})\phi_p(z)=\bar{\psi}_p(q_\chi/2)\int_{(\mathbb{Z}/p^{\max(i,j)})^*}d^*u\ \bar{\chi}(u)\psi_p(x_\chi u^{-1}z)\phi_{\mathbb{Z}_p}(z-y_\chi\cdot u)\,. \tag{12.5.32}$$

It satisfies

$$\tilde{\phi}_{i,j,\chi}(u^{-1}z)=\chi(u)\cdot\tilde{\phi}_{i,j,\chi}\,, \tag{12.5.33}$$

$$|1,z|^s\cdot\tilde{\phi}_{i,j,\chi}=p^{js}\cdot\tilde{\phi}_{i,j,\chi},\quad \mathcal{F}^{-1}|1,z|^s\mathcal{F}\tilde{\phi}_{i,j,\chi}=p^{is}\cdot\tilde{\phi}_{i,j,\chi}\,. \tag{12.5.34}$$

Thus in the Schrödinger model the basis $\tilde{\phi}_{i,j,\chi}$ is quite trivial: for $j > 0, \tilde{\phi}_{i,j,\chi}(z)$ is supported on $p^{-j}\mathbb{Z}_p^*$ where it is given by a constant multiple of $\chi(p^j z)$; for $j = 0$, $i > 0$, $\tilde{\phi}_{i,j,\chi}$ is a constant multiple of the Fourier transform of $\tilde{\phi}_{0,i,\chi^{-1}}, \tilde{\phi}_{i,j,\chi} = \text{cons} \cdot \mathcal{F}\tilde{\phi}_{j,i,\chi^{-1}}$; and for $i = j = 0$, χ is the trivial character χ^0, and $\tilde{\phi}_{0,0,\chi^0} = \phi_{\mathbb{Z}_p}$. The point is that the decomposition of $\mathrm{L}_2(\mathbb{Q}_p)$ into the one-dimensional spaces $\mathbb{C} \cdot \tilde{\phi}_{i,j,\chi}$ is finer than the decomposition into the irreducible subrepresentations of the $\mathrm{SL}_2(\mathbb{Z}_p)$-action (cf. (12.5.10)): the subrepresentation $\mathcal{V}_{n,\pm}$ is the sum of $\mathbb{C} \cdot \tilde{\phi}_{i,j,\chi}$ with $\max\{i, j\} = n, \chi(-1) = \pm 1$. The basis $\tilde{\phi}_{i,j,\chi}$ thus gives the simultaneous $\mathbb{Z}_p^* \subseteq \mathrm{SL}_2(\mathbb{Z}_p)$ decomposition.

The Tate distribution τ_p^s when viewed as a functional on $\mathrm{L}_2(V(\mathbb{Q}_p))_{\phi_p}$, is given by the function (cf. (12.3.32))

$$\zeta_p(s, x, y) = \tau_p^s(r(x, y)\phi_p)\,. \tag{12.5.35}$$

For $p \neq \eta$, 2, it is given explicitly by

$$\zeta_p(s, x, y) = \begin{cases} 1 & |x|, |y| \leq 1\,, \\ \dfrac{\zeta_p(1)}{\zeta_p(s)} |y|^{s-1} \psi_p(xy/2) & |x| \leq 1 < |y|\,, \\ \dfrac{\zeta_p(1)}{\zeta_p(1-s)} |x|^{-s} \bar{\psi}_p(xy/2) & |y| \leq 1 < |x|\,, \\ 0 & 1 < |x|, |y|\,. \end{cases} \tag{12.5.36}$$

Alternatively, it can be expanded by using the pure basis as

$$\begin{aligned} \zeta_p(s, x, y) &= \Phi_{V(\mathbb{Z}_p)}(x, y) + \zeta_p(1)/2 \sum_{\substack{n\geq 1 \\ +,-}} p^{-n} \Phi_{p,\pm,n}(x, y) \cdot \hat{f}^1_{p,\pm,n}(s) \\ &= \Phi_{V(\mathbb{Z}_p)}(x, y) + \zeta_p(1)/2 \sum_{\substack{n\geq 1 \\ +,-}} p^{-n} \Phi_{p,\pm,n}(x, y) \cdot (-1)^n \left[\frac{p^{ns}}{\zeta_p(s)} \pm \frac{p^{n(1-s)}}{\zeta_p(1-s)} \right]. \end{aligned} \tag{12.5.37}$$

For the real prime it is given by

$$\begin{aligned} \zeta_\eta(s, x, y) &= 2^{1/4} \cdot \mathrm{e}^{-(\pi/2)(x^2+y^2)} \mathrm{e}^{(\pi/2)(x-\mathrm{i}y)^2} {}_1\mathrm{F}_1\Big(s/2; 1/2; -\pi(x - \mathrm{i}y)^2\Big) \\ &= 2^{1/4} \cdot \mathrm{e}^{-(\pi/2)(x^2+y^2)} \mathrm{e}^{(\pi/2)(x-\mathrm{i}y)^2} \cdot \sum_{n\geq 0} \frac{\big(2\pi(x - \mathrm{i}y)\big)^{2n}}{(2n)!} (-1)^n \frac{\zeta_\eta(s + 2n)}{\zeta_\eta(s)}\,. \end{aligned} \tag{12.5.38}$$

Alternatively, it can be expanded by using the pure basis as

$$\zeta_\eta(s, x, y) = 2^{1/4} \cdot \mathrm{e}^{-(\pi/2)(x^2+y^2)} \cdot \sum_{n\geq 0} \frac{(2n)!}{n!} (\pi/2)^n (x - \mathrm{i}y)^{2n} (-1)^n \hat{f}^1_{\eta,n}(s)\,. \tag{12.5.39}$$

The functions $\zeta_p(s,x,y)$, $p \geq \eta$, are entire function of s and satisfy the 'differential' equation (12.4.3), respectively (12.4.4) for $p = \eta$. As one can see directly from (12.5.36), respectively (12.5.38) for $p = \eta$, we have Tate's local function equation

$$\zeta_p(1-s,-y,x) = \zeta_p(s,x,y)\,. \tag{12.5.40}$$

Similarly, the global function

$$\zeta_{\mathbb{A}}(s,x,y) = \prod_{p\geq\eta} \zeta_p(s,x_p,y_p), \quad x,y \in \mathbb{A}\,, \tag{12.5.41}$$

is an entire function of s satisfying the differential equations (12.4.3), (12.4.4), and the functional equation

$$\zeta_{\mathbb{A}}(1-s,-y,x) = \zeta_{\mathbb{A}}(s,x,y)$$

It can be expanded using (12.5.37) and (12.5.39) in the form

$$\zeta_{\mathbb{A}}(s,x,y) = \sum_{\substack{n_\eta\geq 0\\ m_+,m_-\geq 1\\ (m_+,m_-)=1}} \frac{1}{\|\Phi_{m_+,m_-,n_\eta}\|^2}\,\overline{\Phi_{m_+,m_-,n_\eta}(x,y)}\cdot \hat{f}_{m_+,m_-,n_\eta}(s)\,. \tag{12.5.42}$$

Here $\{\Phi_{m_+,m_-,n_\eta}\}$ is the orthogonal basis of $\mathrm{L}_2(V(\mathbb{A}))_{\phi_{\mathbb{A}}}^{\widehat{\mathbb{Z}^*\mathbb{Z}^*_\eta}}$ given by

$$\begin{aligned}\Phi_{m_+,m_-,n_\eta}(v) = &\prod_{p\,\nmid\, m_+m_-\eta} \Phi_{V(\mathbb{Z}_p)}(v_p)\cdot \prod_{p^{n_p}\|m_+} \Phi_{p,+,n_p}(v_p)\cdot\\ &\prod_{p^{n_p}\|m_-} \Phi_{p,-,n_p}(v_p)\cdot \Phi_{V(\mathbb{Z}_\eta),2n_\eta}(v_\eta)\end{aligned} \tag{12.5.43}$$

and

$$\hat{f}_{m_+,m_-,n_\eta}(s) = \tau^s_{\mathbb{A}}(r(\Phi_{m_+,m_-,n_\eta})\phi_{\mathbb{A}}) \in \mathbb{C}[s,q^s;q\in\mathbb{Q}^+] \tag{12.5.44}$$

has all its zeros on the line $1/2 + \mathrm{i}\mathbb{R}$

12.6 THE GLOBAL LATTICE MODEL

In this section we work over the adeles $\mathbb{A}$. The main property we use is that the rationals $\mathbb{Q}$ embed as a discrete co-compact subgroup of $\mathbb{A}$. The global lattice model can be viewed as the matrix coefficients of the distribution $\delta_{V(\mathbb{Q})}$ given in the Schrödinger model by

$$\delta_{V(\mathbb{Q})}(\varphi) = \sum_{q\in\mathbb{Q}} \varphi(q), \quad \varphi \in \mathcal{S}(\mathbb{A}) \subseteq \mathrm{L}_2(\mathbb{A})\,. \tag{12.6.1}$$

The associated realization of the fundamental representation of $\mathrm{Heis}(\mathbb{A})$ is on the space $\mathrm{L}_2(V(\mathbb{A}))_{\delta_{V(\mathbb{Q})}}$ of function F on $V(\mathbb{A})$ satisfying

$$\mathrm{F}(v+v_0) = \mathrm{F}(v)\cdot\psi_{\mathbb{A}}(\langle v,v_0\rangle/2) \quad \text{for all } v\in V(\mathbb{A}), v_0\in V(\mathbb{Q}) \tag{12.6.2}$$

and $\|\mathrm{F}\|^2 = (\mathrm{F},\mathrm{F}) < \infty$, where the Hilbert space inner product is

$$(\mathrm{F}_1,\mathrm{F}_2)_{\mathrm{L}_2(V(\mathbb{A}))_{\delta_{V(\mathbb{Q})}}} = \int_{V(\mathbb{A})/V(\mathbb{Q})} F_1(v)\overline{\mathrm{F}_2(v)}\,\mathrm{d}v\,. \tag{12.6.3}$$

The isomorphism with the Schrödinger model is given by

$$\mathrm{L}_2(V(\mathbb{A}))_{\delta_{V(\mathbb{Q})}} \stackrel{\sim}{\longleftrightarrow} \mathrm{L}_2(\mathbb{A})$$

$$\mathrm{F}(x,y) = (\varphi, r(x,y)\delta_{V(\mathbb{Q})}) \longleftarrow\!\!\!| \; \varphi \tag{12.6.4}$$

$$= \bar{\psi}_{\mathbb{A}}(xy/2) \sum_{q\in\mathbb{Q}} \bar{\psi}_{\mathbb{A}}(qx)\varphi(q+y)\mathrm{F} \longmapsto \varphi(y) = \int_{\mathbb{A}/\mathbb{Q}} \mathrm{d}x\, \mathrm{F}(x,y)\psi_{\mathbb{A}}(xy/2)\,.$$

The distribution $\delta_{V(\mathbb{Q})}$ in the global lattice model becomes simply

$$\delta_{V(\mathbb{Q})}(\mathrm{F}) = \mathrm{F}(0,0)\,. \tag{12.6.5}$$

The action of $\mathrm{Heis}(\mathbb{A})$ on $\mathrm{L}_2(V(\mathbb{A}))_{\delta_{V(\mathbb{Q})}}$ is given by

$$r(v_0)\mathrm{F}(v) = \delta_{v_0} \natural\, \mathrm{F}(v) = \mathrm{F}(v-v_0)\psi_{\mathbb{A}}(\langle v_0, v\rangle/2), \quad v_0,\ v \in V(\mathbb{A})\,. \tag{12.6.6}$$

In particular, we have a true representation of the commutative group $V(\mathbb{Q})$:

$$r(v_0)\mathrm{F}(v) = \mathrm{F}(v-2v_0) = \mathrm{F}(v)\psi_{\mathbb{A}}(\langle v_0, v\rangle) \quad \text{for } v \in V(\mathbb{A}),\ v_0 \in V(\mathbb{Q})\,. \tag{12.6.7}$$

The group $\mathrm{SL}_2(\mathbb{Q})$ acts linearly on $\mathrm{L}_2(V(\mathbb{A}))_{\delta_{V(\mathbb{Q})}}$:

$$\mathrm{R}\colon \mathrm{SL}_2(\mathbb{Q}) \longrightarrow \mathrm{U}\Big(\mathrm{L}_2(V(\mathbb{A}))_{\delta_{V(\mathbb{Q})}}\Big)$$

$$\mathrm{R}(g)\mathrm{F}(v) = \mathrm{F}(vg)\,, \tag{12.6.8}$$

$$\mathrm{R}(g^{-1})r(v)\mathrm{R}(g) = r(vg)\,.$$

The Fourier transform $\mathcal{F}$ is the same as the operator

$$\mathrm{R}\begin{pmatrix} 0 & 1 \\ -1 & 0 \end{pmatrix}$$

that is

$$\mathcal{F}\mathrm{F}(x,y) = \mathrm{F}(-y,x)\,. \tag{12.6.9}$$

The equation $\delta_{V(\mathbb{Q})}(\mathcal{F}\mathrm{F}) = \delta_{V(\mathbb{Q})}(\mathrm{F})$, cf. (12.6.5) and (12.6.9), is equivalent to the *Poisson summation formula*

$$\sum_{q\in\mathbb{Q}} \mathcal{F}\varphi(q) = \sum_{q\in\mathbb{Q}} \varphi(q), \quad \varphi \in \mathcal{S}(\mathbb{A})\,. \tag{12.6.10}$$

The distributions δ_0 and δ_∞ of (12.1.19) are now given by

$$\delta_0(\mathrm{F}) = \int_{\mathbb{A}/\mathbb{Q}} \mathrm{d}x\, \mathrm{F}(x,0), \quad \delta_\infty(\mathrm{F}) = \int_{\mathbb{A}/\mathbb{Q}} \mathrm{d}y\, \mathrm{F}(0,y)\,. \tag{12.6.11}$$

In fact, for any $q = (q_0 : q_1) \in \mathbb{P}^1(\mathbb{Q}) \subseteq \mathbb{P}^1(\mathbb{A})$, the distribution δ_q (cf. (12.1.24)) is given by

$$\delta_q(f) = \int_{\mathbb{A}/\mathbb{Q}} \mathrm{d}x \, \mathrm{F}(xq_0, xq_1)\,. \tag{12.6.12}$$

The function

$$\phi_{\mathbb{A}} = \prod_{p \geq \eta} \phi_p \in \mathrm{L}_2(\mathbb{A})$$

corresponds to the 'theta function' in $\mathrm{L}_2(V(\mathbb{A}))_{\delta_{V(\mathbb{Q})}}$

$$\mathrm{F}_{\phi_{\mathbb{A}}}(x, y) = \bar{\psi}_{\mathbb{A}}(xy/2) \sum_{q \in \mathbb{Q}} \bar{\psi}_{\mathbb{A}}(qx)\phi_{\mathbb{A}}(q + y)\,. \tag{12.6.13}$$

It is the unique (up to constant multiple) function satisfying

$$\mathrm{F}_{\phi_{\mathbb{A}}} \natural\, \delta_{v_0} = F_{\phi_{\mathbb{A}}} \quad \text{for } v_0 \in V(\mathbb{Q})\,, \tag{12.6.14}$$

$$\Phi_{V(\mathbb{Z}_{\mathbb{A}})} \natural\, \mathrm{F}_{\phi_{\mathbb{A}}} = \mathrm{F}_{\phi_{\mathbb{A}}}\,, \tag{12.6.15}$$

where

$$\Phi_{V(\mathbb{Z}_{\mathbb{A}})}(x, y) = \prod_{p \geq \eta} \Phi_{V(\mathbb{Z}_p)}(x_p, y_p) \tag{12.6.16}$$

is the $\natural$-idempotent characterizing the local lattice model. Alternatively, the integral equation (12.6.15) can be replaced by the collection of 'differential' equation (12.4.3) and (12.4.4) for $\overline{\mathrm{F}_{\phi_{\mathbb{A}}}}$. The isomorphism between the global lattice model and the local lattice model is given by twisted convolution with $\mathrm{F}_{\phi_{\mathbb{A}}}$:

$$\begin{array}{ccc} \mathrm{L}_2(V(\mathbb{A}))_{\delta_{V(\mathbb{Q})}} & \xleftrightarrow{\ \sim\ } & \mathrm{L}_2(V(\mathbb{A}))_{\phi_{\mathbb{A}}} \\ f \natural\, \mathrm{F}_{\phi_{\mathbb{A}}} & \longleftarrow\!\!\dashv & f \\ \mathrm{F} & \longmapsto & \mathrm{F} \underset{V(\mathbb{Q})}{\natural} \overline{\mathrm{F}}_{\phi_{\mathbb{A}}}(v) \end{array} \tag{12.6.17}$$

$$\stackrel{\text{def}}{=} \int_{V(\mathbb{A})/V(\mathbb{Q})} \mathrm{d}v_0 \, \mathrm{F}(v_0)\overline{\mathrm{F}}_{\phi_{\mathbb{A}}}(v - v_0)\psi_{\mathbb{A}}(\langle v_0, v\rangle/2)\,.$$

The global lattice model induces local models using the embeddings, and adjoint projections, in the following diagram:

$$\begin{array}{ccccc} & \varphi \longmapsto \varphi \otimes \phi_{\mathbb{A}^{(p)}} & & & \\ \mathrm{L}_2(\mathbb{Q}_p) & \rightleftarrows & \mathrm{L}_2(\mathbb{A}) & & \\ & r(\Phi_{V(\mathbb{Z}_{\mathbb{A}^{(p)}})}) & & \searrow^{\sim} & \\ \updownarrow\wr & & \updownarrow\wr & & \mathrm{L}_2(V(\mathbb{A}))_{\delta_{V(\mathbb{Q})}} \\ & f \longmapsto f \otimes \Phi_{V(\mathbb{Z}_{\mathbb{A}}^{(p)})} & & \swarrow^{\sim} & \\ \mathrm{L}_2(V(\mathbb{Q}_p))_{\phi_p} & \rightleftarrows & \mathrm{L}_2(V(\mathbb{A}))_{\phi_{\mathbb{A}}} & & \\ & r(\Phi_{V(\mathbb{Z}_{\mathbb{A}^{(p)}})}) & & & \end{array}$$

(12.6.18)

(12.6.19)

where

$$\phi_{\mathbb{A}^{(p)}}(x) = \prod_{p' \neq p} \phi_{p'}(x_{p'}), \qquad x \in \mathbb{A}^{(p)} = \mathbb{A}/\mathbb{Q}_p\,, \tag{12.6.20}$$

$$\Phi_{V(\mathbb{Z}_{\mathbb{A}^{(p)}})}(v) = \prod_{p' \neq p} \Phi_{V(\mathbb{Z}_{p'})}(v_{p'}), \qquad v \in V\left(\mathbb{A}^{(p)}\right) = V(\mathbb{A})/V(\mathbb{Q}_p)\,. \tag{12.6.21}$$

Thus for $p \geq \eta$ we obtain a realization of the fundamental representation of $\mathrm{Heis}(\mathbb{Q}_p)$ on the space

$$\mathscr{H}^{(p)}_{V(\mathbb{Q})} = \Phi_{V(\mathbb{Z}_{\mathbb{A}^{(p)}})} \natural\ \mathrm{L}_2(V(\mathbb{A}))_{\delta_{V(\mathbb{Q})}}\,. \tag{12.6.22}$$

For the real prime $p = \eta$, a function $\mathrm{F} \in \mathscr{H}^{(\eta)}_{V(\mathbb{Q})}$ is determined by its restriction to $V(\mathbb{R})$, since

$$V(\widehat{\mathbb{Z}})\backslash V(\mathbb{A})/V(\mathbb{Q}) \simeq V(\mathbb{R})/V(\mathbb{Z})\,. \tag{12.6.23}$$

F must satisfy the equation

$$\mathrm{F}(x+n, y+m) = \mathrm{F}(x,y)\cdot(-1)^{n\cdot m}\cdot \mathrm{e}^{\pi\mathrm{i}(xm-yn)} \quad x,y \in \mathbb{R}, n, m, \in \mathbb{Z}\,, \tag{12.6.24}$$

and $\|\mathrm{F}\|^2 = (\mathrm{F},\mathrm{F})_{\mathscr{H}^{(\eta)}_{V(\mathbb{Q})}} < \infty$, where the inner product is given by

$$(\mathrm{F}_1, \mathrm{F}_2)_{\mathscr{H}^{(\eta)}_{V(\mathbb{Q})}} = \int_{V(\mathbb{R})/V(\mathbb{Z})} \mathrm{d}v\ \mathrm{F}_1(v)\overline{\mathrm{F}_2(v)}\,. \tag{12.6.25}$$

This is the well-known Weil–Berezin model consisting of L_2-sections of the line bundle $\mathbb{C}_{\psi_\eta}$ associated with ψ_η over the torus $V(\mathbb{R})/V(\mathbb{Z})$. The isomorphism with the Schrödinger model is given by

$$\begin{aligned} \mathrm{L}_2(\mathbb{R}) &\xleftrightarrow{\ \sim\ } \mathscr{H}^{(\eta)}_{V(\mathbb{Q})} \\ \varphi &\longmapsto \mathrm{F}(x,y) = (\varphi, r(x,y)\delta_{V(\mathbb{Z})}) \\ &\qquad\qquad = \mathrm{e}^{-\pi\mathrm{i}xy}\sum_{n\in\mathbb{Z}} \mathrm{e}^{-2\pi\mathrm{i}nx}\varphi(n+y) \\ \varphi(y) = \int_{\mathbb{R}/\mathbb{Z}} \mathrm{d}x\ \mathrm{F}(x,y)\mathrm{e}^{\pi\mathrm{i}xy} &\longleftarrow\!\!\mid \mathrm{F}\,. \end{aligned} \tag{12.6.26}$$

We can now translate all the notions we had in $\mathrm{L}_2(\mathbb{R})$ or $\mathrm{L}_2(V(\mathbb{R}))_{\phi_\eta}$ to the space $\mathscr{H}^{(\eta)}_{V(\mathbb{Q})}$.

For example, the position (respectively momentum) operators given in the Schrödinger model by

$$\mathrm{P}\varphi(z) = z\cdot\varphi(z), \quad \text{respectively } \mathrm{Q}\,\varphi(z) = \frac{1}{2\pi\mathrm{i}}\frac{\partial}{\partial z}\varphi(z)\,, \tag{12.6.27}$$

are given in the space $\mathscr{H}^{(\eta)}_{V(\mathbb{Q})}$ by

$$\mathrm{P}\,\mathrm{F}(x,y) = \frac{1}{2}\left(y - \frac{1}{\pi\mathrm{i}}\frac{\partial}{\partial x}\right)\mathrm{F}(x,y), \text{ respectively } \mathrm{Q}\,\mathrm{F}(x,y) = \frac{1}{2}\left(x + \frac{1}{\pi\mathrm{i}}\frac{\partial}{\partial y}\right)\mathrm{F}(x,y)\,. \tag{12.6.28}$$

The Heisenberg operators (12.5.3) correspond to the operators on $\mathscr{H}^{(\eta)}_{V(\mathbb{Q})}$ given by

$$\begin{aligned} a^*\mathrm{F}(x,y) &= \sqrt{\pi}(\mathrm{Q}+\mathrm{iP}) = -\frac{1}{2\sqrt{\pi}}\left[\left(\frac{\partial}{\partial x}+\mathrm{i}\frac{\partial}{\partial y}\right)-\pi(x+\mathrm{i}y)\right] \\ &= -\frac{1}{2\sqrt{\pi}}\,\Phi^{-1}_{V(\mathbb{Z}_\eta)}\frac{\partial}{\partial v}\,\Phi_{V(\mathbb{Z}_\eta)}, \\ a\,\mathrm{F}(x,y) &= \sqrt{\pi}(\mathrm{Q}-\mathrm{iP}) = \frac{1}{2\sqrt{\pi}}\left[\left(\frac{\partial}{\partial x}-\mathrm{i}\frac{\partial}{\partial y}\right)+\pi(x-iy)\right] \\ &= \frac{1}{2\sqrt{\pi}}\,\Phi_{V(\mathbb{Z}_\eta)}\frac{\partial}{\partial v}\,\Phi^{-1}_{V(\mathbb{Z}_\eta)}\,. \end{aligned} \tag{12.6.29}$$

For $p \neq \eta$ a function $\mathrm{F} \in \mathscr{H}^{(p)}_{V(\mathbb{Q})}$ is determined by its restriction to $V(\mathbb{Z}_p) \times V(\mathbb{R})$ (and even by its restriction to $V(\mathbb{R})$, but it will be convenient to keep the $V(\mathbb{Z}_p)$ variables). F satisfies

$$\overline{\mathcal{L}_{V(\mathbb{Z}_\eta)}}\mathrm{F} = 0\,, \tag{12.6.30}$$

where $\overline{\mathcal{L}_{V(\mathbb{Z}_\eta)}} = 2\sqrt{\pi}a$ is the annihilation operator

$$\overline{\mathcal{L}_{V(\mathbb{Z}_\eta)}} = \Phi_{V(\mathbb{Z}_\eta)}\frac{\partial}{\partial v}\Phi^{-1}_{V(\mathbb{Z}_\eta)} = \frac{\partial}{\partial x_\eta} - \mathrm{i}\frac{\partial}{\partial y_\eta} + \pi(x_\eta - \mathrm{i}y_\eta) = 2\pi(\mathrm{Q}-\mathrm{iP})\,. \tag{12.6.31}$$

For $x_p, y_p \in \mathbb{Z}_p$, $x_\eta, y_\eta \in \mathbb{R}$, $n, m \in \mathbb{Z}$, $p \neq 2$, F also satisfies

$$\mathrm{F}(x_p+n, y_p+m, x_\eta+n, y_\eta+m) = (-1)^{n\cdot m}\mathrm{e}^{\pi\mathrm{i}(x_\eta m - y_\eta n)}\mathrm{F}(x_p, y_p, x_\eta, y_\eta) \tag{12.6.32}$$

(for p=2, $(-1)^{n\cdot m}$ in (12.6.32) should be replaced by $(-1)^{x_2 m - y_2 n}$). Also $(\mathrm{F},\mathrm{F})_{\mathscr{H}^{(p)}_{V(\mathbb{Q})}} < \infty$, where the inner product is

$$(\mathrm{F}_1, \mathrm{F}_2)_{\mathscr{H}^{(p)}_{V(\mathbb{Q})}} = \int_{V(\mathbb{Z}_p)\times V(\mathbb{R})/V(\mathbb{Z})} dv\;\mathrm{F}_1(v)\overline{\mathrm{F}_2(v)}\,. \tag{12.6.33}$$

Thus roughly $\mathscr{H}^{(p)}_{V(\mathbb{Q})}$ consists of p-almost periodic L_2-holomorphic sections, or L_2-holomorphic sections defined over p-power coverings of the elliptic curve $\mathbb{C}/\mathbb{Z}+\mathbb{Z}\mathrm{i}$. The isomorphism with the Schrödinger model is given by

$$\begin{aligned} \mathrm{L}_2(\mathbb{Q}_p) &\longleftrightarrow \mathscr{H}^{(p)}_{V(\mathbb{Q})} \\ \varphi &\longmapsto (\phi_{\mathbb{A}^{(p)}} \otimes \varphi, r(x,y)\delta_{V(\mathbb{Q})}) = \mathrm{F}(x,y) \end{aligned} \tag{12.6.34}$$

$$= \psi_p(x_p y_p/2)\mathrm{e}^{-\pi/2(x_\eta^2+y_\eta^2)}\mathrm{e}^{\pi/2(x_\eta - \mathrm{i}y_\eta)^2}2^{1/4}\sum_{q\in\mathbb{Z}[1/p]}\varphi(q+y_p)\bar{\psi}_p(x_p q)\mathrm{e}^{-\pi q^2}\mathrm{e}^{-2\pi\mathrm{i}(x_\eta - \mathrm{i}y_\eta)q}\,.$$

For the inverse isomorphism, we write $z \in \mathbb{Q}_p$ as $z = q + y_p$ with $q \in \mathbb{Z}[1/p]$, $y_p \in \mathbb{Z}_p$. Then

$$\varphi(z) = \iint_{\mathbb{Z}_p\times\mathbb{R}/\mathbb{Z}} dx_p\,dx_\eta\,\mathrm{F}(x_p, y_p, x_\eta, -q)\cdot\mathrm{e}^{\pi\mathrm{i}x_\eta\cdot q}\cdot\psi_p(x_p\cdot q)\cdot\mathrm{e}^{\pi\mathrm{i}x_\eta\cdot q}\cdot\psi_p(x_p y_p/2)\,. \tag{12.6.35}$$

The distribution $\delta^{(p)}_{V(\mathbb{Q})}(\mathrm{F}) = \mathrm{F}(0,0)$ on $\mathscr{H}^{(p)}_{V(\mathbb{Q})}$ corresponds in the local Schrödinger model to the distribution

$$\begin{aligned} \delta^{(p)}_{V(\mathbb{Q})}(\varphi) &= \sum_{n\in\mathbb{Z}} \varphi(n) \quad \text{for } p=\eta\,, \\ \delta^{(p)}_{V(\mathbb{Q})}(\varphi) &= 2^{1/4} \sum_{q\in\mathbb{Z}[1/p]} \varphi(q)\cdot \mathrm{e}^{-\pi q^2} \quad \text{for } p\neq\eta\,. \end{aligned} \tag{12.6.36}$$

Note that for $\mathrm{Re}(s) > 1/2$

$$\begin{aligned} &\int_{\mathbb{R}^*} \mathrm{d}^*a\, |a|^s_\eta \Big(\delta^{(\eta)}_{V(\mathbb{Q})} - \delta_0\Big)(\pi(a^{-1})\phi_\eta) \\ &= \int \mathrm{d}^*a\, |a|^{s+1/2}\, 2^{1/4} \sum_{n\neq 0} \mathrm{e}^{-\pi a^2 n^2} = 2^{1/4}\zeta_{\mathbb{A}}(1/2+s)\,. \end{aligned} \tag{12.6.37}$$

Similarly we can define the p-zeta function for $\mathrm{Re}(s) > 1/2$ by

$$\begin{aligned} \zeta_{\mathbb{A}^{(p)}}(1/2+s) &= \int_{\mathbb{Q}_p^*} \mathrm{d}^*a\, |a|^s_p \Big(2^{-1/4}\delta^{(p)}_{V(\mathbb{Q})} - \delta_0\Big)(\pi(a^{-1})\phi_p) \\ &= \sum_{n\in\mathbb{Z}} p^{n(s+1/2)} \sum_{m\neq 0} \mathrm{e}^{-\pi m^2 p^{2n}}\,. \end{aligned} \tag{12.6.38}$$

Imitating Tate's thesis trick (cf. (13.1.19) below), we see that

$$\begin{aligned} \zeta_{\mathbb{A}^{(p)}}(s) = &\left(\sum_{m\in\mathbb{Z}} \mathrm{e}^{-\pi m^2}\right) + \sum_{n\geq 1} [p^{ns} + p^{n(1-s)}]\cdot \left(\sum_{m\neq 0} \mathrm{e}^{-\pi m^2 p^{2n}}\right) \\ &- \frac{1}{1-p^{-s}} - \frac{1}{1-p^{-(1-s)}}\,, \end{aligned} \tag{12.6.39}$$

showing that $\zeta_{\mathbb{A}^{(p)}}(s)$ has meromorphic continuation to all s, is $(2\pi\mathrm{i}/\log p)\mathbb{Z}$-periodic, has the unique $(\mathrm{mod}(2\pi\mathrm{i}/\log p)\mathbb{Z})$ simple poles at $s = 0, 1$, and satisfies the functional equation $\zeta_{\mathbb{A}^{(p)}}(s) = \zeta_{\mathbb{A}^{(p)}}(1-s)$. What are the locations of its zeros?

We have worked in this section with the 'standard' lattice $V(\mathbb{Q}) \subseteq V(\mathbb{A})$. We can change this lattice to any lattice $\mathcal{L} \subseteq V(\mathbb{A})$, where $\mathcal{L}$ is a discrete co-compact two-dimensional $\mathbb{Q}$-vector space such that $\mathcal{L}$ is self-dual: $\mathcal{L} = \mathcal{L}^V$; the dual lattice $\mathcal{L}^V$ is defined by

$$\mathcal{L}^V = \big\{v \in V(\mathbb{A}) \,\big|\, \langle v, \mathcal{L}\rangle \subseteq \mathbb{Q}\big\}\,. \tag{12.6.40}$$

The space of self-dual lattices $\mathcal{L} \subseteq V(\mathbb{A})$ is the homogeneous space $\mathrm{X} = \mathrm{SL}_2(\mathbb{Q})\backslash\mathrm{SL}_2(\mathbb{A})$. For $\mathcal{L} \in \mathrm{X}$ we can extend the central character $\psi_{\mathbb{A}}$ to the (maximal commutative) subgroup $\mathcal{L} \times \mathbb{A}/\mathbb{Q}$ of $\mathrm{Heis}(\mathbb{A})/\mathbb{Q}$ by

$$\psi_{\mathcal{L}}(x,y,t) = \psi_{\mathbb{A}}(t) \quad \text{for } (x,y) \in \mathcal{L}\,, \tag{12.6.41}$$

and the induced representation is equivalent to $r_{\psi_{\mathbb{A}}}$:

$$r_{\psi_{\mathbb{A}}} = \mathrm{Ind}_{\mathcal{L}\times\mathbb{A}/\mathbb{Q}}^{\mathrm{Heis}(\mathbb{A})/\mathbb{Q}}(\psi_{\mathcal{L}})\,. \tag{12.6.42}$$

It is realized in the space $\mathrm{L}_2(V(\mathbb{A}))_{\delta_{\mathcal{L}}}$ of functions F on $V(\mathbb{A})$ satisfying

$$\mathrm{F}(v+v_0) = \mathrm{F}(v)\cdot\psi_{\mathbb{A}}(\langle v, v_0\rangle/2) \quad \text{for } v\in V(\mathbb{A}),\ v_0\in\mathcal{L}\,, \tag{12.6.43}$$

and $(\mathrm{F},\mathrm{F}) < \infty$, where

$$(\mathrm{F}_1,\mathrm{F}_2) = \int_{V(\mathbb{A})/\mathcal{L}} dv\,\mathrm{F}_1(v)\overline{\mathrm{F}_2(v)}\,. \tag{12.6.44}$$

The distribution $\delta_{\mathcal{L}}(\mathrm{F}) = \mathrm{F}(0,0)$ is the unique (up to constant multiple) distribution that is invariant under the action of $r_\psi(\mathcal{L})$.

The mapping $\mathcal{L}\mapsto\delta_{\mathcal{L}}$ gives an embedding

$$\mathrm{X}\hookrightarrow\mathbb{P}^1(\mathcal{S}'(\mathbb{A}))\,. \tag{12.6.45}$$

Any $\mathcal{L}\in\mathrm{X}$ induces local models of the fundamental representation of $\mathrm{Heis}(\mathbb{Q}_p)$, $p\geq\eta$:

$$\mathscr{H}_{\mathcal{L}}^{(p)} = \Phi_{V(\mathbb{Z}_{\mathbb{A}^{(p)}})} \natural\ \mathrm{L}_2(V(\mathbb{A}))_{\delta_{\mathcal{L}}}\,. \tag{12.6.46}$$

12.7 AUTOMORPHIC FORMS ON THE HEISENBERG GROUP

The Haar measure on $\mathrm{Heis}(\mathbb{A})$ induces a probability measure on the symmetric space $\mathrm{Heis}(\mathbb{Q})\backslash\mathrm{Heis}(\mathbb{A})$, and we have a unitary representation of $\mathrm{Heis}(\mathbb{A})$ by translations on the associated Hilbert space:

$$\begin{aligned}&\tilde{r}\colon \mathrm{Heis}(\mathbb{A}) \longrightarrow \mathrm{U}(\mathrm{L}_2(\mathrm{Heis}(\mathbb{Q})\backslash\mathrm{Heis}(\mathbb{A})))\\ &\tilde{r}(h_0)f(h) = f(hh_0), \quad h, h_0\in\mathrm{Heis}(\mathbb{A})\,.\end{aligned} \tag{12.7.1}$$

We can decompose this representation with respect to the central characters $\left\{\psi_{\mathbb{A}}^{(q)}, q\in\mathbb{Q}\right\}$ of $\mathbb{Q}\backslash\mathbb{A}$, and write $\tilde{r} = \oplus_{q\in\mathbb{Q}}\tilde{r}^{(q)}$,

$$\mathrm{L}_2(\mathrm{Heis}(\mathbb{Q})\backslash\mathrm{Heis}(\mathbb{A})) = \bigoplus_{q\in\mathbb{Q}} \mathrm{L}_2(V(\mathbb{Q})\backslash V(\mathbb{A}); \psi_{\mathbb{A}}^{(q)})\,. \tag{12.7.2}$$

For $q\neq 0$, the representation $\tilde{r}^{(q)}$ is just the global lattice realization of the irreducible fundamental representation $r_{\psi_{\mathbb{A}}^{(q)}}$ with central character $\psi_{\mathbb{A}}^{(q)}$. For $q=0$

$$\mathrm{L}_2(V(\mathbb{Q})\backslash V(\mathbb{A}); \psi_{\mathbb{A}}^{(0)}) = \mathrm{L}_2(V(\mathbb{Q})\backslash V(\mathbb{A})) = \bigoplus_{v_0\in V(\mathbb{Q})} \mathbb{C}\cdot\psi_{\mathbb{A}}(\langle v_0, v\rangle) \tag{12.7.3}$$

decomposes as the sum of one-dimensional representations of the compact commutative group $V(\mathbb{Q})\backslash V(\mathbb{A})$.

We can 'cut down the level' and look at the $\mathrm{Heis}(\widehat{\mathbb{Z}}) = \prod_{p\neq\eta} \mathrm{Heis}(\mathbb{Z}_p)$-invariant vectors in our representation

$$\begin{aligned}\mathrm{L}_2(\mathrm{Heis}(\mathbb{Q})\backslash\mathrm{Heis}(\mathbb{A}))^{\mathrm{Heis}(\widehat{\mathbb{Z}})} &= \mathrm{L}_2(\mathrm{Heis}(\mathbb{Q})\backslash\mathrm{Heis}(\mathbb{A})/\mathrm{Heis}(\widehat{\mathbb{Z}}))\\ &= \mathrm{L}_2(\mathrm{Heis}(\mathbb{Z})\backslash\mathrm{Heis}(\mathbb{R})). \qquad (12.7.4)\end{aligned}$$

The group $\mathrm{Heis}(\mathbb{R})$ and the 'Hecke algebra'

$$\mathcal{S}(\mathrm{Heis}(\widehat{\mathbb{Z}})\backslash\mathrm{Heis}(\mathbb{A}_{\mathrm{fin}})/\mathrm{Heis}(\widehat{\mathbb{Z}})) = \mathcal{S}(\mathrm{Heis}(\mathbb{Z})\backslash\mathrm{Heis}(\mathbb{Q})/\mathrm{Heis}(\mathbb{Z})), \qquad (12.7.5)$$

act on the space (12.7.4). We can decompose this representation with respect to the central characters $\{\psi^{(n)} \mid n \in \mathbb{Z}\}$ of $\mathbb{Q}\backslash\mathbb{A}/\widehat{\mathbb{Z}} = \mathbb{Z}\backslash\mathbb{R}$, and write

$$\mathrm{L}_2(\mathrm{Heis}(\mathbb{Z})\backslash\mathrm{Heis}(\mathbb{R})) = \bigoplus_{n\in\mathbb{Z}} \mathrm{L}_2(V(\mathbb{Z})\backslash V(\mathbb{R}); \psi^{(n)}). \qquad (12.7.6)$$

For $n \neq 0$, the representation $r^{(n)}$ of $\mathrm{Heis}(\mathbb{R})$ on $\mathrm{L}_2(V(\mathbb{Z})\backslash V(\mathbb{R}); \psi^{(n)})$ is the sum of $|n|$ copies of the irreducible fundamental representation $r_{\psi_\eta^{(n)}}$ with central character $\psi_\eta^{(n)}$.

For $n = 0$

$$\mathrm{L}_2(V(\mathbb{Z})\backslash V(\mathbb{R}); \psi^{(0)}) = \mathrm{L}_2(V(\mathbb{Z})\backslash V(\mathbb{R})) = \bigoplus_{v_0\in V(\mathbb{Z})} \mathbb{C}\cdot\psi_\eta(\langle v_0, v\rangle) \qquad (12.7.7)$$

decomposes as the sum of the one-dimensional representations of the compact commutative torus $V(\mathbb{Z})\backslash V(\mathbb{R})$.

We can 'cut down the level' further, and look at the '$\mathrm{Heis}^{(\tau)}(\mathbb{Z}_\eta)$-invariant vectors'

$$\begin{aligned}\widetilde{S}^{(\tau)} &= \mathrm{L}_2(\mathrm{Heis}(\mathbb{Z})\backslash\mathrm{Heis}(\mathbb{R}))^{\mathrm{Heis}^{(\tau)}(\mathbb{Z}_\eta)} \qquad (12.7.8)\\ &= \text{`}\mathrm{L}_2(\mathrm{Heis}(\mathbb{Q})\backslash\mathrm{Heis}(\mathbb{A})/\mathrm{Heis}(\widehat{\mathbb{Z}})\cdot\mathrm{Heis}^{(\tau)}(\mathbb{Z}_\eta))\text{'}\end{aligned}$$

where a function f is right $\mathrm{Heis}^{(\tau)}(\mathbb{Z}_\eta)$-invariant if (cf. (12.4.42))

$$\mathcal{L}^{(\tau)} f = \left(\frac{\partial}{\partial x_0} - \frac{1}{\tau}\frac{\partial}{\partial y_0}\right)\bigg|_{x_0=y_0=0} f((x,y,t)\cdot(x_0,y_0,0)) = 0\,. \qquad (12.7.9)$$

We have

$$\mathcal{L}^{(\tau)} = \left(\frac{\partial}{\partial x} - \frac{1}{\tau}\frac{\partial}{\partial y}\right) - \frac{1}{2\tau}(x+\tau y)\frac{\partial}{\partial t} = \mathcal{L}_x - \frac{1}{\tau}\mathcal{L}_y\,, \qquad (12.7.10)$$

where $\mathcal{L}_x$ and $\mathcal{L}_y$ are given by

$$\mathcal{L}_x = \frac{\partial}{\partial x} - \frac{y}{2}\frac{\partial}{\partial t}, \quad \mathcal{L}_y = \frac{\partial}{\partial y} + \frac{x}{2}\frac{\partial}{\partial t}; \qquad (12.7.11)$$

$\mathcal{L}_x$ and $\mathcal{L}_y$ are the left-variant vector fields on $\mathrm{Heis}(\mathbb{R})$.

Decomposing $\widetilde{S}^{(\tau)}$ with respect to the central characters we have, as in (12.7.6),

$$\widetilde{S}^{(\tau)} = \bigoplus_{n\in\mathbb{Z}} S_n^{(\tau)}, \quad S_n^{(\tau)} = \text{`L}_2\Big(V(\mathbb{Q})\backslash V(\mathbb{A})/V(\widehat{\mathbb{Z}})V^{(\tau)}(\mathbb{Z}_\eta); \bar{\psi}_{\mathbb{A}}^{(n)}\Big)\text{'}. \tag{12.7.12}$$

For $f \in \widetilde{S}^{(\tau)}$ we have

$$\begin{aligned} f \in S_n^{(\tau)} &\iff f(x,y,t) = f(x,y,0)\cdot \mathrm{e}^{-2\pi i n t} \quad \text{for all } (x,y,t) \in \text{ Heis}(\mathbb{R}) \\ &\iff (\partial/\partial t) f = -2\pi i n \cdot f\,, \end{aligned} \tag{12.7.13}$$

and in this case (12.7.9) becomes

$$\mathcal{L}_n^{(\tau)} f = 0\,, \tag{12.7.14}$$

with

$$\mathcal{L}_n^{(\tau)} = \left(\frac{\partial}{\partial x} - \frac{1}{\tau}\frac{\partial}{\partial y}\right) + \frac{\pi i n}{\tau}(x+\tau y) = \mathrm{e}^{-[\pi n/2\mathrm{Im}(\tau)]|x+\tau y|^2}\left(\frac{\partial}{\partial x} - \frac{1}{\tau}\frac{\partial}{\partial y}\right)\mathrm{e}^{[\pi n/2\mathrm{Im}(\tau)]|x+\tau y|^2},$$

or equivalently,

$$f(x,y) = \mathrm{e}^{-[\pi n/2\mathrm{Im}(\tau)]|x+\tau y|^2}\cdot h(x+\tau y) \quad \text{with } h \text{ entire}. \tag{12.7.15}$$

The left Heis($\mathbb{Z}$)-invariance of f, for f satisfying (12.7.13), can be written as

$$f(x+k, y+\ell) = f(x,y)\cdot \mathrm{e}^{-\pi i n(\ell x - k y)}\cdot(-1)^{nk\ell}, \quad k,\ell\in\mathbb{Z}, \quad x,y\in\mathbb{R}. \tag{12.7.16}$$

Note that by (12.7.16) $|f|^2$ is $V(\mathbb{Z})$-periodic, and hence by (12.7.15) $f \neq 0$ only for $n \geq 0$. Thus $\widetilde{\mathrm{S}}_n^{(\tau)} = \{0\}$ for negative $n < 0$. Equation (12.7.15) can also be written as

$$f(x,y) = \mathrm{e}^{[\pi n/2\mathrm{Im}(\tau)][(x+\tau y)^2 - |x+\tau y|^2]}\theta_n(x+\tau y) = \mathrm{e}^{\pi i n(x+\tau y)\cdot y}\cdot\theta_n(x+\tau y) \text{ with } \theta_n \text{ entire}. \tag{12.7.17}$$

Then (12.7.16) translates into the equation for $\theta_n(z)$:

$$\theta_n(z+k+\tau\ell) = \theta_n(z)\cdot \mathrm{e}^{-\pi i n(2\ell z+\tau\ell^2)}, \tag{12.7.18}$$

or equivalently,

$$\theta_n(z+1) = \theta_n(z)\,, \tag{12.7.19}$$

$$\theta_n(z+\tau) = \theta_n(z)\cdot \mathrm{e}^{-\pi i n(2z+\tau)}, \tag{12.7.20}$$

These equations characterize the space $\Theta_n(\tau)$ of classical Jocobi theta functions of period τ and weight n. Equation (12.7.19) is equivalent to the existence of a Fourier expansion

$$\theta_n(z) = \sum_{m\in\mathbb{Z}} a_m \cdot \mathrm{e}^{2\pi i m z}, \tag{12.7.21}$$

and then (12.7.20) is equivalent to the recursive relation

$$a_{m+n} = a_m \cdot \mathrm{e}^{\pi\mathrm{i}\tau(2m+n)} . \tag{12.7.22}$$

It follows that for $n \geq 1$

$$\dim \mathrm{S}_n^{(\tau)} = \dim \Theta_n(\tau) = n . \tag{12.7.23}$$

A basis for $\Theta_1(\tau)$ is given by the classical theta function

$$\theta(z,\tau) = \sum_{m\in\mathbb{Z}} \mathrm{e}^{\pi\mathrm{i}\tau m^2 + 2\pi\mathrm{i}zm} . \tag{12.7.24}$$

Note that we have

$$\begin{aligned}\theta\left(z - \frac{1+\tau}{2}, \tau\right) &= \sum_{m\in\mathbb{Z}} (-1)^m \mathrm{e}^{\pi\mathrm{i}\tau m(m-1) + 2\pi\mathrm{i}zm} \\ &= \sum_{m\in\mathbb{Z}} (-1)^m q^{m(m-1)/2} q^{\beta m} \quad \text{with } q = \mathrm{e}^{2\pi\mathrm{i}\tau},\ q^\beta = \mathrm{e}^{2\pi\mathrm{i}z}, \\ &= \frac{1}{\zeta_{(q)}(1)\zeta_{(q)}(\beta)\zeta_{(q)}(1-\beta)}, \quad \text{by (6.3.30)} .\end{aligned} \tag{12.7.25}$$

The coresponding basis for $\mathrm{S}_1^{(\tau)}$ is

$$\begin{aligned} f_{(1)0}^{(\tau)}(x,y) &= (2\,\mathrm{Im}(\tau))^{1/4} \cdot \mathrm{e}^{\pi\mathrm{i}(x+\tau y)\cdot y} \cdot \theta(x+\tau y, \tau) \\ &= \mathrm{e}^{\pi\mathrm{i}xy} \sum_{m\in\mathbb{Z}} \mathrm{e}^{2\pi\mathrm{i}xm} \overline{\phi^{(\tau)}}(m+y) = \left(\delta_{V(\mathbb{Q})}, r_\psi(x,y)\phi^{(\tau)}\right) . \end{aligned} \tag{12.7.26}$$

A basis for $\mathrm{S}_n^{(\tau)}$ is given by $\left\{ f_{(n)a}^{(\tau)} \,\middle|\, a \in \mathbb{Z}/n\mathbb{Z} \right\}$ with

$$\begin{aligned} f_{(n)a}^{(\tau)}(x,y) &= (2n\,\mathrm{Im}(\tau))^{1/4} \cdot \mathrm{e}^{\pi\mathrm{i}n(x+\tau y)y} \cdot \\ &\qquad \sum_{m\in Z} \mathrm{e}^{\pi\mathrm{i}n\tau(m+a/n)^2 + 2\pi\mathrm{i}n(m+a/n)(x+\tau y)} \\ &= \mathrm{e}^{\pi\mathrm{i}nxy} \sum_{m\in Z} \mathrm{e}^{2\pi\mathrm{i}nx(m+a/n)} \overline{\phi^{(n\tau)}}(m+a/n+y) \\ &= \left(\delta_{V(\mathbb{Q})}, r_{\psi^{(n)}}(0,a/n) r_{\psi^{(n)}}(x,y)\phi^{(n\tau)}\right) . \end{aligned} \tag{12.7.27}$$

The sum in (12.7.12) is the ℓ_2 sum. We can form the algebraic sum

$$\mathrm{S}^{(\tau)} = \bigoplus_{n\geq 0} \mathrm{S}_n^{(\tau)} \tag{12.7.28}$$

which is a graded algebra with respect to ordinary multiplication:

$$f_i \in \mathrm{S}_{n_i}^{(\tau)} \Longrightarrow f_1 \cdot f_2 \in \mathrm{S}_{n_1+n_2}^{(\tau)} . \tag{12.7.29}$$

A straightforward computation shows that the multiplication of the basis elements (12.7.27) is given by

$$f^{(\tau)}_{(n_1)a_1}(x,y)\cdot f^{(\tau)}_{(n_2)a_2}(x,y)$$
$$= \frac{1}{(n_1+n_2)^{1/2}} \sum_{c\in\mathbb{Z}/(n_1+n_2)} f^{(\tau)}_{((n_1+n_2)n_1n_2),n_2a_1-n_1a_2+cn_1n_2}(0,0)$$
$$\cdot f^{(\tau)}_{(n_1+n_2),a_1+a_2+cn_1}(x,y)\,. \qquad (12.7.30)$$

The projective variety $E^{(\tau)} = \mathrm{Proj}\ \mathrm{S}^{(\tau)}$ is just the elliptic curve with period lattice $\mathbb{Z}+\mathbb{Z}\tau$, $E^{(\tau)}(\mathbb{C}) \cong \mathbb{C}/\mathbb{Z}+\mathbb{Z}\cdot\tau$. For a weight that is a square n^2, another basis for $\mathrm{S}^{(\tau)}_{n^2}$ is given by $\left\{g^{(\tau)}_{(n^2)a,b} \,\middle|\, a,b\in\mathbb{Z}/n\mathbb{Z}\right\}$ with

$$g^{(\tau)}_{(n^2)a,b}(x,y) = \left(\delta_{V(\mathbb{Q})}, r_\psi(a/n,b/n)r_\psi(nx,ny)\phi^{(\tau)}\right). \qquad (12.7.31)$$

For $n \geq 2$ we obtain an embedding into projective space

$$E^{(\tau)}(\mathbb{C}) \hookrightarrow \mathbb{P}^{n^2-1}(\mathbb{C}) \qquad (12.7.32)$$

induced by

$$x+\tau y \longmapsto \left(g^{(\tau)}_{(n^2)a,b}(x,y)\right)_{a,b\in\mathbb{Z}/n}. \qquad (12.7.33)$$

The above description of the Heisenberg group is just the beginning. The group SL_2, acting on Heis by automorphisms inducing the identity on the centre, gives rise to the metaplectic group $\widehat{\mathrm{SL}}_2$, that is the double cover of SL_2, and the Weil representation:

$$\mathrm{R}\colon \widehat{\mathrm{SL}}_2(\mathbb{Q}_p) \to \mathrm{U}(\mathrm{L}_2(\mathbb{Q}_p)), \quad p\geq\eta\,, \qquad (12.7.34)$$

so that

$$\mathrm{R}(g^{-1})r(v)\mathrm{R}(g) = r(vg),\ g\in\widehat{\mathrm{SL}}_2(\mathbb{Q}_p),\ v\in\mathrm{Heis}(\mathbb{Q}_p)\,,$$

and similarly over the adeles $\mathbb{A}$. The local (respectively global) lattice model shows that $\widehat{\mathrm{SL}}_2$ splits over $\mathrm{SL}_2(\mathbb{Z}_p)$(12.4.18) (respectively over $\mathrm{SL}_2(\mathbb{Q})$, (12.6.8)).

The semi-direct product, the Jacobi group $\mathrm{J} = \widehat{\mathrm{SL}}_2 \ltimes \mathrm{Heis}$, thus has an important role. Automorphic forms on the Jacobi group J 'live' on the space

$$\mathbb{J} = \mathrm{J}(\mathbb{Q})\backslash\mathrm{J}(\mathbb{A})\Big/\prod_{p\geq\eta}\mathrm{J}(\mathbb{Z}_p)\,. \qquad (12.7.35)$$

This space is a fibration over the modular curve (cf. (13.3.17) below)

$$\mathrm{X}_0 = \mathrm{SL}_2(\mathbb{Q})\backslash\mathrm{SL}_2(\mathbb{A})\Big/\prod_{p\geq\eta}\mathrm{SL}_2(\mathbb{Z}_p) = \mathrm{SL}_2(\mathbb{Z})\backslash\mathrm{SL}_2(\mathbb{R})/\mathrm{SO}_2(\mathbb{R}) \qquad (12.7.36)$$

with fiber the 'space' appearing in (12.7.8)

$$\mathrm{Heis}(\mathbb{Q})\backslash\mathrm{Heis}(\mathbb{A}) \Big/ \prod_{p\neq\eta} \mathrm{Heis}(\mathbb{Z}_p) \times \text{`}\mathrm{Heis}\left(\mathbb{Z}_\eta^{(\tau)}\right)\text{'}; \tag{12.7.37}$$

that is, it is the universal elliptic curve over X_0, X_0 being the moduli space for elliptic curves. Points on $\mathbb{J}$ are described by their 'modular coordinates' $\tau \in \mathrm{SL}_2(\mathbb{Z})\backslash\{\tau \in \mathbb{C}|\mathrm{Im}(\tau) > 0\}$, or $q = \mathrm{e}^{2\pi i\tau}$, corresponding to the elliptic curve $E_q = \mathbb{C}^*/q^{\mathbb{Z}} = \mathbb{C}/\mathbb{Z} + \mathbb{Z}\tau$, together with their 'Heisenberg coordinates' corresponding to a point on the elliptic curve, $[\beta]_{E_q} \in E_q$.

We see that beginning with the *additive* structure of $\mathbb{Q}_p$, $p \geq \eta$, and $\mathbb{A}$, as a self-dual locally compact abelian group, leads to the Heisenberg group and the Jacobi group, hence to elliptic curves and the theta function, and finally to the space (12.7.35) parametrized by our old parameters q, β—which initially arose from the *multiplicative* structure. The theory of $\widehat{\mathrm{SL}}_2$ (see Gelbart 1976) and of $\mathbb{J}$ (see Eichler and Zagier 1985) is too long to recall here; some of the theory of SL_2 will be presented in the next chapter.

13
The Riemann zeta function

13.1 THE RIEMANN ZETA FUNCTION AND THE THETA FUNCTION

The theory we presented in Chapter 10 works globally over the adeles by simply taking the product of the local theories for $p \geq \eta$. We have the Hilbert space

$$\mathscr{H}^{\beta}_{\mathbb{A}} = \mathrm{L}_2\big(\mathbb{A}, \tau^{\beta}_{\mathbb{A}}\big)^{\mathbb{Z}^*_{\mathbb{A}}} = \bigotimes_{p \geq \eta} \mathscr{H}^{\beta}_{p} \quad (\otimes \text{ with respect to } \phi_{\mathbb{Z}_p}), \tag{13.1.1}$$

of functions invariant under the units $\mathbb{Z}^*_{\mathbb{A}} = \widehat{\mathbb{Z}}^* \times \{\pm 1\}$.

We also have the unitary action

$$\pi^{\beta}_{\mathbb{A}}\colon \mathbb{A}^*/\mathbb{Z}^*_{\mathbb{A}} = \mathbb{Q}^+ \times \mathbb{R}^+ \to \mathrm{U}\big(\mathscr{H}^{\beta}_{\mathbb{A}}\big), \tag{13.1.2}$$

and $\phi_{\mathbb{Z}_{\mathbb{A}}} = \otimes_{p \geq \eta} \phi_{\mathbb{Z}_p}$ is a cyclic vector. It gives rise to the positive-definite function on $\mathbb{A}^*/\mathbb{Z}^*_{\mathbb{A}}$

$$\begin{aligned} \tilde{\rho}^{\beta}_{\mathbb{A}}(a) &= \big(\pi^{\beta}_{\mathbb{A}}(a)\phi_{\mathbb{Z}_{\mathbb{A}}}, \phi_{\mathbb{Z}_{\mathbb{A}}}\big)_{\mathscr{H}^{\beta}_{\mathbb{A}}} = \prod_{p \geq \eta} \tilde{\rho}^{\beta}_{p}(a_p) \quad \text{for } a = (a_p), \\ &= |a_\eta m^2 + a_\eta^{-1} n^2|^{-\beta/2} \quad \text{for } a = a_\eta m/n, \end{aligned} \tag{13.1.3}$$

with a_η real and m, n relatively prime integers $(m, n) = 1$. We obtain the 'analytic' realization of the representation $\pi^{\beta}_{\mathbb{A}}$

$$\mathscr{H}^{\beta}_{\mathbb{A}} \xrightarrow[\sim]{\tau^{\beta/2+\underline{s}}_{\mathbb{A}}} \widehat{\mathscr{H}}^{\beta}_{\mathbb{A}} = \mathrm{L}_2\big(\mathbb{T}, \hat{\tau}^{\beta}_{\mathbb{A}}(\mathrm{d}\underline{s})\big) \tag{13.1.4}$$

where $\mathbb{T}$ is the infinite-dimensional torus

$$\mathbb{T} = (\mathbb{A}^*/\mathbb{Z}^*_{\mathbb{A}})^{\mathrm{v}} = \prod_{p \geq \eta} (\mathbb{Q}^*_p/\mathbb{Z}^*_p)^{\mathrm{v}} = \mathrm{i}\mathbb{R} \times \prod_{p \neq \eta} \mathrm{i}\mathbb{R} \Big/ \frac{2\pi \mathrm{i}}{\log p}\mathbb{Z}. \tag{13.1.5}$$

The global dual measure is the product of the local ones

$$\hat{\tau}^{\beta}_{\mathbb{A}}(\mathrm{d}\underline{s}) = \prod_{p \geq \eta} \zeta_p(\beta/2 + s_p, \beta/2 - s_p)\mathrm{d}_p s_p, \qquad \underline{s} = (s_p) \in \mathbb{T}. \tag{13.1.6}$$

The operator $\hat{\mathcal{F}}^{\beta}_{\mathbb{A}} \hat{f}(\underline{s}) = \hat{f}(-\underline{s})$ on $\widehat{\mathscr{H}}^{\beta}_{\mathbb{A}}$ corresponds to the Fourier–Bessel transform $\mathcal{F}^{\beta}_{\mathbb{A}}$ on $\mathscr{H}^{\beta}_{\mathbb{A}}$; it is the product of the local operators, $\mathcal{F}^{\beta}_{\mathbb{A}} = \otimes_{p \geq \eta} \mathcal{F}^{\beta}_{p}$. The kernel $\mathcal{F}^{\beta}_{\mathbb{A}}(xy) = \prod_{p \geq \eta} \mathcal{F}^{\beta}_{p}(x_p y_p)$ can be expanded using the basis (cf. (12.5.43)) made from the local pure basis.

Similarly we have the global operations of β-(co)-convolution, $\varphi_1 *_\beta \varphi_2$ and $\Delta^\beta\varphi$. For $\beta = 1$ the measure $\tau^1_{\mathbb{A}}$ is the additive Haar measure on $\mathbb{A}$, projected down to $\mathbb{A}/\mathbb{Z}^*_{\mathbb{A}}$.

For $\beta > 1$ the measure $\tau^\beta_{\mathbb{A}}$ is supported on $\mathbb{A}^*/\mathbb{Z}^*_{\mathbb{A}} = \mathbb{Q}^+ \times \mathbb{R}^+$, and is $\mathbb{Q}^+$-invariant. We can consider the projection homomorphism

$$\mathbb{A}^*/\mathbb{Z}^*_{\mathbb{A}} = \mathbb{Q}^+ \times \mathbb{R}^+ \twoheadrightarrow \mathbb{A}^*/\mathbb{A}^{(1)} = \mathbb{R}^+ , \tag{13.1.7}$$

which is dual to the diagonal embedding

$$(\mathbb{A}^*/\mathbb{Z}^*_{\mathbb{A}})^{\vee} = \mathbb{T} \hookleftarrow \mathrm{i}\mathbb{R} = (\mathbb{R}^+)^{\vee} . \tag{13.1.8}$$

Summing the positive-definite function $\tilde{\rho}^\beta_{\mathbb{A}}$ along the fibers of (13.1.7) we obtain for $\mathrm{Re}(\beta) > 2$ the positive-definite function on $\mathbb{R}^+$:

$$\begin{aligned} \tilde{\rho}^\beta_{\mathbb{A}/\mathbb{Q}^*}(a) &= \sum_{q\in\mathbb{Q}^*} \tilde{\rho}^\beta_{\mathbb{A}}(q\cdot a) = 2\cdot \sum_{\substack{m,n\geq 1\\(m,n)=1}} |am^2 + a^{-1}n^2|^{-\beta/2} \\ &= \frac{a^{-\beta/2}}{\zeta_{\mathbb{A}}(\beta)} \int_{\mathbb{A}^*} \mathrm{d}^*x|x|^\beta \phi_{\mathbb{Z}_{\mathbb{A}}}(x) \sum_{q\in\mathbb{Q}^*} \phi_{\mathbb{Z}_{\mathbb{A}}}(a^{-1}qx) \\ &= \frac{a^{-\beta/2}}{\zeta_{\mathbb{A}}(\beta)} \int_{\mathbb{R}^+} \mathrm{d}^*x\cdot x^\beta \sum_{q_1\in\mathbb{Q}^*} \phi_{\mathbb{Z}_{\mathbb{A}}}(q_1x) \sum_{q_2\in\mathbb{Q}^*} \phi_{\mathbb{Z}_{\mathbb{A}}}(a^{-1}q_2x) . \end{aligned} \tag{13.1.9}$$

It correspond to the dual measure

$$\int_{\mathbb{R}^+} \mathrm{d}^*a\cdot a^s \cdot \tilde{\rho}^\beta_{\mathbb{A}/\mathbb{Q}^*}(a) = \zeta_{\mathbb{A}}(\beta/2+s, \beta/2-s), \quad s\in \mathrm{i}\mathbb{R} . \tag{13.1.10}$$

This gives rise to the global space, 'with interaction', which in the analytic realization is given by

$$\widehat{\mathscr{H}}^\beta_{\mathbb{A}/\mathbb{Q}^*} = \mathrm{L}_2(\mathrm{i}\mathbb{R}, \zeta_{\mathbb{A}}(\beta/2+s, \beta/2-s)\mathrm{d}s/2\pi\mathrm{i}) . \tag{13.1.11}$$

The point of the above discussion is that the global zeta function $\zeta_{\mathbb{A}}(s)$, which is given by the product of the local factors $\zeta_p(s)$ for $\mathrm{Re}(s) > 1$, is really the restriction along the diagonal embedding

$$\mathbb{T}_{\mathbb{C}} = \mathrm{Hom}(\mathbb{A}^*/\mathbb{Z}^*_{\mathbb{A}}, \mathbb{C}^*) = \mathbb{C} \times \prod_{p\neq\eta} \mathbb{C}\Big/\frac{2\pi\mathrm{i}}{\log p}\mathbb{Z} \hookleftarrow \mathbb{C} = \mathrm{Hom}(\mathbb{A}^*/\mathbb{Z}^*_{\mathbb{A}}\mathbb{Q}^*, \mathbb{C}^*) \tag{13.1.12}$$

$$\zeta_{\mathbb{A}}(s) = \prod_{p\geq\eta} \zeta_p(s) = \prod_{p\geq\eta} \zeta_p(s_p)\big|_{s_p=s} \tag{13.1.13}$$

and as such has to do with the (non-commutative) space $\mathbb{A}/\mathbb{A}^{(1)} = \mathbb{A}/\mathbb{Z}^*_{\mathbb{A}}\mathbb{Q}^*$.

Let us next review the functional equation of $\zeta_{\mathbb{A}}(s)$ from the point of view of the Heisenberg group. For $\mathrm{Re}(s) > 1/2$ we write

$$\zeta_{\mathbb{A}}(1/2+s) = \int_{\mathbb{A}^*} \mathrm{d}^*a \cdot |a|^{1/2+s} \cdot \phi_{\mathbb{Z}_{\mathbb{A}}}(a) = \int_{\mathbb{R}^+} \mathrm{d}^*a \cdot a^{1/2+s} \cdot \sum_{q \in \mathbb{Q}^*} \phi_{\mathbb{Z}_{\mathbb{A}}}(qa)$$
$$= \int_{\mathbb{R}^+} \mathrm{d}^*a \cdot a^s \cdot [\theta(a) - a^{1/2}] \qquad (13.1.14)$$

where

$$\theta(a) = a^{1/2} \cdot \sum_{q \in \mathbb{Q}} \phi_{\mathbb{Z}_{\mathbb{A}}}(qa) = \left(\delta_{V(\mathbb{Q})}, \pi^1_{\mathbb{A}}(a^{-1})\phi_{\mathbb{Z}_{\mathbb{A}}}\right). \qquad (13.1.15)$$

θ compares the global lattice model distribution $\delta_{V(\mathbb{Q})}$ with the local lattice ones. Denoting the Fourier transform by $\mathcal{F} = \mathcal{F}^1_{\mathbb{A}}$, we have:

$$\begin{aligned} \text{from the global lattice model:} \quad & \mathcal{F}\delta_{V(\mathbb{Q})} = \delta_{V(\mathbb{Q})}\,, \\ \text{from the local lattice model:} \quad & \mathcal{F}\phi_{\mathbb{Z}_{\mathbb{A}}} = \phi_{\mathbb{Z}_{\mathbb{A}}}\,, \\ \text{from the Schrödinger model:} \quad & \mathcal{F}\pi^1_{\mathbb{A}}(a^{-1}) = \pi^1_{\mathbb{A}}(a)\mathcal{F}\,. \end{aligned} \qquad (13.1.16)$$

Putting these together we have the functional equation of θ (Tate's Riemann–Roch for spec $\mathbb{Z}$):

$$\theta(a) = \theta(a^{-1})\,. \qquad (13.1.17)$$

Note that

$$\begin{aligned} \theta(a) &= a^{1/2} + \mathrm{O}(a^{-N}) \quad \text{as } a \to \infty, \text{ for all } N, \text{ and by (13.1.17)} \\ &= a^{-1/2} + \mathrm{O}(a^{N}) \quad \text{as } a \to 0, \text{ for all } N\,. \end{aligned} \qquad (13.1.18)$$

Now since $\zeta_{\mathbb{A}}(1/2+s)$ is the Mellin transform of $\theta(a)$ (13.1.14), the functional equation of θ is equivalent to the functional equation of $\zeta_{\mathbb{A}}$. We have

$$\zeta_{\mathbb{A}}(1/2+s) = \int_1^\infty \mathrm{d}^*a \cdot a^s \cdot [\theta(a) - a^{1/2}] + \int_0^1 \mathrm{d}^*a \cdot a^s \cdot \theta(a) - \frac{1}{1/2+s}$$
$$= \int_1^\infty \mathrm{d}^*a \cdot (a^s + a^{-s}) \cdot [\theta(a) - a^{1/2}] - \left[\frac{1}{1/2+s} + \frac{1}{1/2-s}\right]. \qquad (13.1.19)$$

This shows that $\zeta_{\mathbb{A}}(1/2+s)$ has meromorphic continuation to all $s \in \mathbb{C}$, with simple poles only at $s = \pm 1/2$ and residues ± 1, and satisfies the functional equation

$$\zeta_{\mathbb{A}}(1/2+s) = \zeta_{\mathbb{A}}(1/2-s)\,. \qquad (13.1.20)$$

Note that (13.1.19) can be rewritten as

$$(1/4 - s^2) \cdot \zeta_{\mathbb{A}}(1/2+s) = (1/4 - s^2) \cdot \int_1^\infty \mathrm{d}^*a \cdot (a^s + a^{-s}) \cdot [\theta(a) - a^{1/2}] - 1$$
$$= \int_1^\infty \mathrm{d}^*a \cdot (1/4 - (a\partial/\partial a)^2)(a^s + a^{-s}) \cdot [\theta(a) - a^{1/2}] - 1\,.$$

Using integration by parts we obtain

$$(1/4 - s^2) \cdot \zeta_{\mathbb{A}}(1/2 + s) = \int_1^\infty \mathrm{d}^* a \cdot (a^s + a^{-s}) \cdot (1/4 - (a\partial/\partial a)^2)\theta(a) \qquad (13.1.21)$$

$$= \int_0^\infty \mathrm{d}^* a \cdot a^s \cdot (1/4 - (a\partial/\partial a)^2)\theta(a) \,.$$

Thus the entire function $(1/4 - s^2)\zeta_{\mathbb{A}}(1/2 + s)$ is the Mellin transform of the Schwartz function on $\mathbb{R}^+$ given by $(1/4 - (a\partial/\partial a)^2)\theta(a)$. This leads to many equivalent formulations of the Riemann hypothesis, cf. Jansen's formulations in Polya (1974).

13.2 THE EXPLICIT SUMS

The explicit sums are the dictionary which gives the precise relation between the zeros of the global zeta function $\zeta_{\mathbb{A}}(s)$ and the distribution of the prime powers within $\mathbb{R}^+$. For a function $f \in \mathcal{S}(\mathbb{R}^+)$ we have its Mellin transform, the Fourier transform on the multiplicative group $\mathbb{R}^+$:

$$\hat{f}(s) = \int_0^\infty f(x) x^s \, \frac{\mathrm{d}x}{x} \,. \qquad (13.2.1)$$

This is an analytic function of $s \in \mathbb{C}$, satisfying $|\hat{f}(s)| = \mathrm{O}\big(|\mathrm{Im}(s)|^{-N}\big)$ uniformly in any vertical strip $a \leq \mathrm{Re}(s) \leq b$ for all N. The inverse Mellin transform reads

$$f(x) = \frac{1}{2\pi \mathrm{i}} \int_{\sigma - \mathrm{i}\infty}^{\sigma + \mathrm{i}\infty} \hat{f}(s) x^{-s} \, \mathrm{d}s, \quad \text{for any } \sigma \in \mathbb{R} \,. \qquad (13.2.2)$$

The Mellin transform takes multiplicative convolution to multiplication

$$\widehat{f_1 \circ f_2}(s) = \hat{f}_1(s) \cdot \hat{f}_2(s), \quad f_1 \circ f_2(x) = \int_0^\infty f_1(xy^{-1}) f_2(y) \frac{\mathrm{d}y}{y} \,, \qquad (13.2.3)$$

and also takes the adjoint

$$f^\circ(x) = \overline{f(x^{-1})} \quad \text{to} \quad \hat{f}^\circ(s) = \overline{\hat{f}(-\bar{s})} \,.$$

The idea of the explicit sums can be summarized by the following equality:

$$\sum_{\zeta_{\mathbb{A}}(1/2+s)=0} \hat{f}(s) - \hat{f}(1/2) - \hat{f}(-1/2)$$

$$= \frac{1}{2\pi \mathrm{i}} \oint \hat{f}(s) \mathrm{d} \log \zeta_{\mathbb{A}}(1/2 + s), \text{ by the residue theorem}$$

$$= \frac{1}{2\pi \mathrm{i}} \int_{c - \mathrm{i}\infty}^{c + \mathrm{i}\infty} + \int_{-c + \mathrm{i}\infty}^{-c - \mathrm{i}\infty} \hat{f}(s) \mathrm{d} \log \zeta_{\mathbb{A}}(1/2 + s), \quad \text{for any } c > 1/2$$

$$= \frac{1}{2\pi \mathrm{i}} \int_{c - \mathrm{i}\infty}^{c + \mathrm{i}\infty} (\hat{f}(s) - \hat{f}(-s)) \mathrm{d} \log \zeta_{\mathbb{A}}(1/2 + s) \,,$$

by the functional equation

$$= \sum_{p \geq \eta} \frac{1}{2\pi \mathrm{i}} \int_{c-\mathrm{i}\infty}^{c+\mathrm{i}\infty} (\hat{f}(s) - \hat{f}(-s)) \mathrm{d} \log \zeta_p(1/2 + s), \text{ by the Euler product}$$

$$= \sum_{p \geq \eta} \frac{1}{2\pi \mathrm{i}} \int_{c-\mathrm{i}\infty}^{c+\mathrm{i}\infty} \hat{f}(s) \mathrm{d} \log \frac{\zeta_p(1/2 + s)}{\zeta_p(1/2 - s)}, \quad \text{for any } c \in (-1/2, 1/2)$$

$$= - \sum_{p \geq \eta} \mathcal{W}_p(f) \tag{13.2.4}$$

where the *Weil distribution* $\mathcal{W}_p(f)$ is the inverse Mellin transform of $\mathrm{d} \log[\zeta_p(1/2 - s)/\zeta_p(1/2 + s)]$ (note the minus sign at the end of (13.2.4)).

Since we have enough functions to localize the zeros of $\zeta_{\mathbb{A}}(s)$, the

$$\text{Riemann hypothesis:} \quad \zeta_{\mathbb{A}}(1/2 + s) = 0 \Rightarrow s = -\bar{s}\,,$$

is equivalent to the

$$\sum_{\zeta_{\mathbb{A}}(1/2+s)=0} \hat{f}(s) \cdot \overline{\hat{f}(-\bar{s})} \geq 0 \quad \text{for all } f\,. \tag{13.2.5}$$

In view of the explicit sum formula (13.2.4) this is in turn equivalent to the

$$\text{fundamental inequality:} \quad \hat{f}(1/2) \cdot \overline{\hat{f}(-1/2)} + \overline{\hat{f}(1/2)} \cdot \hat{f}(-1/2) \geq \sum_{p \geq \eta} \mathcal{W}_p(f \circ f^{\circ})\,. \tag{13.2.6}$$

Note that this inequality will fail to be true if we ignore the factor $\mathcal{W}_\eta(f \circ f^{\circ})$ on the right-hand side, since $\zeta_{\mathbb{A}}(s)/\zeta_\eta(s)$ has the 'trivial' zeros at the poles of $\zeta_\eta(s)$, and similarly for any other factor $\mathcal{W}_p(f \circ f^{\circ})$.

For finite primes $p \neq \eta$ we have

$$\mathcal{W}_p(f) = \frac{1}{2\pi \mathrm{i}} \int_{-\mathrm{i}\infty}^{\mathrm{i}\infty} \hat{f}(s) \mathrm{d} \log \frac{\zeta_p(1/2 - s)}{\zeta_p(1/2 + s)}$$

$$= \frac{1}{2\pi \mathrm{i}} \int_{-\mathrm{i}\infty}^{\mathrm{i}\infty} \hat{f}(s) \log p \cdot \left[\sum_{n \geq 1} p^{-n(1/2+s)} + \sum_{n \geq 1} p^{-n(1/2-s)} \right] \mathrm{d}s$$

$$= \log p \cdot \sum_{\substack{n \in \mathbb{Z} \\ n \neq 0}} p^{-|n|/2} \frac{1}{2\pi \mathrm{i}} \int_{-\mathrm{i}\infty}^{\mathrm{i}\infty} \hat{f}(s) p^{-ns} \mathrm{d}s$$

$$= \log p \cdot \sum_{n \neq 0} p^{-|n|/2} f(p^n)\,. \tag{13.2.7}$$

Ignoring the contribution of the poles of $\zeta_{\mathbb{A}}(s)$ and the contribution of the real prime $\mathcal{W}_\eta(f)$, we can interpret the explicit sums as saying that the sum of the Mellin transform $\hat{f}(s)$ over the zeros of $\zeta_{\mathbb{A}}(s)$ is 'equal' to the weighted sum of f over the prime powers. Our first task is to understand $\mathcal{W}_\eta(f)$ on an equal footings with $\mathcal{W}_p(f), p \neq \eta$. For this we follow our general philosophy of using the q-world to interpolate between the finite p and the real η, and again it will be useful to approximate the gamma function by the beta function.

We start by defining the *q-beta–Weil distribution*:

$$\mathcal{W}^{(N)}_{(q|d)}(f) = \frac{1}{2\pi \mathrm{i}} \int_{-\mathrm{i}\infty}^{\mathrm{i}\infty} \hat{f}(s) \mathrm{d} \log \frac{\zeta_{(q^d)}((1/2 - s)/d, N/d)}{\zeta_{(q^d)}((1/2 + s)/d, N/d)}$$
$$= (\log q^{-1}) \cdot \sum_{n \neq 0} \frac{1 - q^{N|n|}}{1 - q^{d|n|}} \cdot q^{|n|/2} \cdot f(q^n) . \qquad (13.2.8)$$

This behaves nicely in both the p-adic and real limits:

$$\mathcal{W}^{(N)}_p(f) = \lim_{d\to\infty} \mathcal{W}^{(N)}_{(p^{-1}|d)}(f) = \frac{1}{2\pi \mathrm{i}} \int_{-\mathrm{i}\infty}^{\mathrm{i}\infty} \hat{f}(s) \mathrm{d} \log \frac{\zeta_p(1/2 - s, N)}{\zeta_p(1/2 + s, N)}$$
$$= \log p \cdot \sum_{n \neq 0} (1 - p^{-N|n|}) \cdot p^{-|n|/2} \cdot f(p^n) . \qquad (13.2.9)_p$$

$$\mathcal{W}^{(N)}_\eta(f) = \lim_{d\to\infty} \mathcal{W}^{(N)}_{(q^{1/d}|2)}(f) = \frac{1}{2\pi \mathrm{i}} \int_{-\mathrm{i}\infty}^{\mathrm{i}\infty} \hat{f}(s) \mathrm{d} \log \frac{\zeta_\eta(1/2 - s, N)}{\zeta_\eta(1/2 + s, N)}$$
$$= \lim_{q\to 1} (\log q^{-1}) \cdot \sum_{n \neq 0} \frac{1 - q^{N|n|}}{1 - q^{2|n|}} \cdot q^{|n|/2} \cdot f(q^n)$$
$$= \int_0^\infty \frac{1 - \min\{x, x^{-1}\}^N}{1 - \min\{x, x^{-1}\}^2} \cdot \min\{x, x^{-1}\}^{1/2} \cdot f(x) \frac{\mathrm{d}x}{x} . \qquad (13.2.9)_\eta$$

Define the *q-gamma–Weil distribution* by

$$\mathcal{W}_{(q|d)}(f) = \lim_{N\to\infty} \mathcal{W}^{(N)}_{(q|d)}(f) = \frac{1}{2\pi \mathrm{i}} \int_{-\mathrm{i}\infty}^{\mathrm{i}\infty} \hat{f}(s) \mathrm{d} \log \frac{\zeta_{(q^d)}((1/2 - s)/d)}{\zeta_{(q^d)}((1/2 + s)/d)}$$
$$= (\log q^{-1}) \cdot \sum_{n \neq 0} \frac{q^{|n|/2}}{1 - q^{d|n|}} \cdot f(q^n) . \qquad (13.2.10)$$

For $p \neq \eta$ we have the commutative limit relations

$$\begin{array}{ccc} \mathcal{W}^{(N)}_{(p^{-1}|d)}(f) & \xrightarrow[d\to\infty]{} & \mathcal{W}^{(N)}_p(f) \\ {\scriptstyle N\to\infty}\big\downarrow & & \big\downarrow{\scriptstyle N\to\infty} \\ \mathcal{W}_{(p^{-1}|d)}(f) & \xrightarrow[d\to\infty]{} & \mathcal{W}_p(f) \end{array} \qquad (13.2.11)_p$$

For the real prime we have, as usual, the more delicate limit relations: the bottom limit of $(13.2.11)_p$ has the real analogue (cf. (6.1.12))

$$\mathcal{W}_\eta(f) = \lim_{d\to\infty} \mathcal{W}_{(q^{1/d}|2)}(f) + \log[(1 - q^{2/d})^2 \cdot \zeta_{(q^{2/d})}(1/2, 1/2)] \cdot f(1)$$
$$= \lim_{d\to\infty} \mathcal{W}_{(q^{1/d}|2)}(f) + \log[(1 - q^{1/d})2\pi] \cdot f(1) ; \qquad (13.2.11a)$$

the right limit of $(13.2.11)_p$ has the real analogue (cf. (3.2.5))

$$\mathcal{W}_\eta(f) = \lim_{N\to\infty} \mathcal{W}^{(N)}_\eta(f) - \log(N/2\pi) \cdot f(1) . \qquad (13.2.11b)$$

We can now give a closed expression for the real Weil distribution $\mathcal{W}_\eta(f)$. Recalling Euler's constant

$$\gamma = \lim_{N\to\infty} \sum_{1\le n\le N} 1/n - \log N = \lim_{N\to\infty} \int_0^\infty \frac{1-m(x)^{2N}}{1-m(x)^2} \cdot m(x)^2 \frac{\mathrm{d}x}{x} - \log N\,, \tag{13.2.12}$$

with $m(x) = \min\{x, x^{-1}\}$, we replace N by $2N$ in (13.2.11a), and use $(13.2.9)_\eta$ to obtain

$$\begin{aligned}\mathcal{W}_\eta(f) &= \lim_{N\to\infty} \int_0^\infty \frac{1-m(x)^{2N}}{1-m(x)^2} \cdot \left(m(x)^{1/2} f(x) - m(x)^2 f(1)\right) \frac{\mathrm{d}x}{x} + (\gamma + \log \pi) f(1) \\ &= \int_0^\infty \frac{m(x)^{1/2} f(x) - m(x)^2 f(1)}{1-m(x)^2} \frac{\mathrm{d}x}{x} + (\gamma + \log\pi)\cdot f(1)\,. \end{aligned} \tag{13.2.13}$$

We can similarly use

$$\begin{aligned}\frac{\partial}{\partial\beta} \log \Gamma(\beta) &= \lim_{N\to\infty} \frac{\partial}{\partial\beta} \log \frac{\Gamma(\beta)\Gamma(N)}{\Gamma(\beta+N)} N^\beta \\ &= \lim_{N\to\infty} \left\{ \log N - \left[\frac{1}{\beta} + \cdots + \frac{1}{\beta+N-1}\right]\right\} \\ &= \lim_{N\to\infty} \left\{ \log N - \int_0^\infty \frac{1-m(x)^{2N}}{1-m(x)^2} \cdot m(x)^{2\beta} \frac{\mathrm{d}x}{x}\right\}\end{aligned}$$

to obtain

$$\begin{aligned}\mathcal{W}_\eta(f) &= \lim_{N\to\infty} \int_0^\infty \frac{1-m(x)^{2N}}{1-m(x)^2} \cdot \left(m(x)^{1/2} f(x) - m(x)^{2\beta} f(1)\right) \frac{\mathrm{d}x}{x} \\ &\qquad\qquad + \left(\log\pi - \frac{\partial}{\partial\beta}\log\Gamma(\beta)\right) f(1) \\ &= \int_0^\infty \frac{m(x)^{1/2} f(x) - m(x)^{2\beta} f(1)}{1-m(x)^2} \frac{\mathrm{d}x}{x} + \left(\log\pi - \frac{\partial}{\partial\beta}\log\Gamma(\beta)\right)\cdot f(1)\,.\end{aligned} \tag{13.2.14}$$

At $\beta = 1$ this is (13.2.13); at $\beta = 1/4$ it reads

$$\mathcal{W}_\eta(f) = \int_0^\infty \frac{f(x) - f(1)}{1-m(x)^2} \cdot m(x)^{1/2} \frac{\mathrm{d}x}{x} + (\gamma + \pi/2 + \log 8\pi)\cdot f(1)\,. \tag{13.2.15}$$

We can take the factor $\pi^{-s/2}$ out of $\zeta_\eta(s)$ and write the explicit sum as

$$\hat{f}(1/2) + \hat{f}(-1/2) - \sum_{\zeta_{\mathbb{A}}(\frac{1}{2}+\alpha)=0} \hat{f}(\alpha) = (\log\pi) f(1) + \sum_{p\ge\eta} \widetilde{\mathcal{W}}_p(f) \tag{13.2.16}$$

where $\widetilde{\mathcal{W}}_p(f) = \mathcal{W}_p(f)$ for $p \ne \eta$, and

$$\widetilde{\mathcal{W}}_\eta(f) = \mathcal{W}_\eta(f) - (\log\pi) f(1) = \frac{1}{2\pi\mathrm{i}} \int_{-\mathrm{i}\infty}^{\mathrm{i}\infty} \hat{f}(s) \mathrm{d} \log \frac{\Gamma(1/4 - s/2)}{\Gamma(1/4+s/2)}\,. \tag{13.2.17}$$

We then have the commutative diagram of limit relations:

$$
\begin{array}{ccccc}
\widetilde{\mathcal{W}}_p^{(N)}(f) & \xleftarrow[\ \textcircled{p}\ ]{} & \widetilde{\mathcal{W}}_{(q|d)}^{(N)}(f) & \xrightarrow[\ \textcircled{\eta}\ ]{} & \widetilde{\mathcal{W}}_\eta^{(N)}(f) \\
\Big\downarrow N \to \infty & & \Big\downarrow N \to \infty & & \Big\downarrow N \to \infty \\
\widetilde{\mathcal{W}}_p(f) & \xleftarrow[\ \textcircled{p}\ ]{} & \widetilde{\mathcal{W}}_{(q|d)}(f) & \xrightarrow[\ \textcircled{\eta}\ ]{} & \widetilde{\mathcal{W}}_\eta(f)
\end{array}
\tag{13.2.18}
$$

where

$$
\begin{aligned}
\widetilde{\mathcal{W}}_{(q|d)}^{(N)}(f) &= \mathcal{W}_{(q|d)}^{(N)}(f) - f(1) \cdot \log \frac{[N]_q}{[d]_q}\,, \\
\widetilde{\mathcal{W}}_p^{(N)}(f) &= \widetilde{\mathcal{W}}_{(p^{-1}|\infty)}^{(N)}(f) = \mathcal{W}_p^{(N)}(f) - f(1) \cdot \log(1 - p^{-N})\,, \\
\widetilde{\mathcal{W}}_\eta^{(N)}(f) &= \widetilde{\mathcal{W}}_{(1|2)}^{(N)}(f) = \mathcal{W}_\eta^{(N)}(f) - f(1) \cdot \log(N/2)\,, \\
\widetilde{\mathcal{W}}_{(q|d)}(f) &= \mathcal{W}_{(q|d)}(f) + f(1) \cdot \log(1 - q^d)\,.
\end{aligned}
\tag{13.2.19}
$$

The diagram (13.2.18) leads to many approximations to Weil's positivity (13.2.6). For example, taking f with $\hat{f}(\pm 1/2) = 0$, and approximating $\widetilde{\mathcal{W}}_\eta(f \circ f^\circ)$ by $\widetilde{\mathcal{W}}_{(q|2)}(f \circ f^\circ)$ and $f \circ f^\circ(p^n)$ by $(1-q)\sum_m f(q^m) \cdot \bar{f}(q^m p^{-n})$, we obtain

$$
\begin{aligned}
\log\left(\frac{(1-q)^{-1}}{2\pi}\right) \cdot \sum_m |f(q^m)|^2 \geq (\log q^{-1}) &\sum_{\substack{m_i \in \mathbb{Z} \\ m_1 \neq m_2}} \frac{q^{|m_1-m_2|\frac{1}{2}}}{1 - q^{|m_1-m_2|2}} f(q^{m_1}) \bar{f}(q^{m_2}) \\
&+ \sum_{p \neq \eta} (\log p) \sum_{\substack{m_i \in \mathbb{Z} \\ m_1 \neq m_2}} p^{-|m_1-m_2|\frac{1}{2}} f(q^{m_1}) \cdot \bar{f}(q^{m_1} \cdot p^{m_1-m_2})\,.
\end{aligned}
\tag{13.2.20}
$$

The validity of (13.2.20) for a sequence of $q \uparrow 1$ implies Weil's positivity, and hence the Riemann hypothesis. There are many variations on this theme: for example, approximate $\mathcal{W}_p(f \circ f^\circ)$ by $\widetilde{\mathcal{W}}_{(p^{-1}|d)}(f \circ f^\circ)$ and exponentiate to get

$$
\begin{aligned}
&\prod_{m \in \mathbb{Z}} \left(\frac{\zeta_{fin}(d)(1-q)^{-1}}{2\pi}\right)^{|f(q^m)|^2} \\
&\quad \geq \prod_{\substack{m_i \in \mathbb{Z} \\ m_1 \neq m_2}} (q^{-1})^{f(q^{m_1})\bar{f}(q^{m_2})q^{|m_1-m_2|\frac{1}{2}}/1-q^{|m_1-m_2|2}} \prod_{p \neq \eta} p^{f(q^{m_1})\bar{f}(q^{m_1}p^{m_1-m_2})p^{-|m_1-m_2|\frac{1}{2}}/1-p^{-|m_1-m_2|d}}\,.
\end{aligned}
\tag{13.2.21}
$$

The validity of (13.2.21) with $q \uparrow 1$, and $d \to \infty$, will again imply the Riemann hypothesis.

13.3 THE EISENSTEIN SERIES CONNECTIONS

We define the K-Bessel function as the multiplicative convolution of the functions $\phi_{\mathbb{Z}_p}(x)|x|^{\alpha}$ and $\phi_{\mathbb{Z}_p}(x)|x|^{-\alpha}$, $p \geq \eta$:

$$\mathrm{K}_p^{\alpha}(x) = (|\cdot|^{\alpha}\phi_{\mathbb{Z}_p}) \circ (|\cdot|^{-\alpha}\phi_{\mathbb{Z}_p}) = \int \mathrm{d}^*y\, \phi_{\mathbb{Z}_p}(y)\phi_{\mathbb{Z}_p}(x/y)|x/y^2|_p^{\alpha}. \tag{13.3.1}$$

A straightforward computation gives for $p \neq \eta$

$$\mathrm{K}_p^{\alpha}(p^n) = \frac{p^{(n+1)\alpha} - p^{-(n+1)\alpha}}{p^{\alpha} - p^{-\alpha}} = \langle n+1 \rangle_{(p^{\alpha})}, \quad n \geq 0,$$

or equivalently

$$\mathrm{K}_p^{\alpha}(x) = \zeta_p(2\alpha)\phi_{\mathbb{Z}_p}(x) \cdot \left(|x|_p^{-\alpha} - |p^2x|_p^{\alpha}\right). \tag{13.3.2}$$

Note that for a natural number n we have

$$\mathrm{K}_{\mathbb{A}_{\mathrm{fin}}}^{\alpha}(n) = \prod_{p \neq \eta} \mathrm{K}_p^{\alpha}(n) = n^{\alpha} \cdot \sigma_{-2\alpha}(n) = \sum_{d_1 d_2 = n} (d_1/d_2)^{\alpha}. \tag{13.3.3}$$

For the real prime we obtain $\mathrm{K}_{\eta}^{\alpha}(x) = 2\mathrm{K}_{\alpha}(2\pi x)$, where K_{α} is the classical Bessel function. Note that $\mathrm{K}_p^{\alpha}(x)$ is invariant under $\alpha \leftrightarrow -\alpha$, is 'fast decreasing' as $x \to \infty$, and satisfies $\mathrm{K}_p^{\alpha}(x) \approx \zeta_p(2\alpha) \cdot |x|_p^{-\alpha}$ as $x \to 0$, $p \geq \eta$.

From the gamma integral (2.2.1) and (13.3.1), we obtain the Mellin transform

$$\int \mathrm{K}_p^{\alpha}(x)|x|_p^s\, \mathrm{d}^*x = \zeta_p(s-\alpha)\zeta_p(s+\alpha), \quad p \geq \eta. \tag{13.3.4}$$

Similarly, using (3.2.7) and the beta integral (3.2.8), we obtain (Ramanujan's)

$$\begin{aligned}&\int \mathrm{K}_p^{\alpha}(x)\mathrm{K}_p^{\beta}(x)|x|_p^s\, \mathrm{d}^*x \\ &= \frac{\zeta_p(s+\alpha+\beta)\zeta_p(s-\alpha-\beta)\zeta_p(s+\alpha-\beta)\zeta_p(s-\alpha+\beta)}{\zeta_p(2s)}, \quad p \geq \eta.\end{aligned} \tag{13.3.5}$$

The K-Bessel function satisfies the 'differential equation'

$$x\frac{\partial}{\partial x}\mathrm{K}_{\eta}^{\alpha}(x) = \alpha \cdot \mathrm{K}_{\eta}^{\alpha}(x) - 2\pi x \cdot \mathrm{K}_{\eta}^{\alpha+1}(x) = -\alpha \cdot \mathrm{K}_{\eta}^{\alpha}(x) - 2\pi x \cdot \mathrm{K}_{\eta}^{\alpha-1}(x), \quad (13.3.6)_{\eta}$$

$$\begin{aligned}\mathrm{K}_p^{\alpha}(p^{n+1}) + \mathrm{K}_p^{\alpha}(p^{n-1}) &= (p^{\alpha} + p^{-\alpha}) \cdot \mathrm{K}_p^{\alpha}(p^n), \\ \mathrm{K}_p^{\alpha}(p^{n+1}) - \mathrm{K}_p^{\alpha}(p^{n-1}) &= p^{(n+1)\alpha} + p^{-(n+1)\alpha}, \quad n \geq 0.\end{aligned} \quad (13.3.6)_p$$

From these we obtain the second-order equations

$$\left(x\frac{\partial}{\partial x}\right)^2 \mathrm{K}_{\eta}^{\alpha}(x) = [\alpha^2 + (2\pi x)^2] \cdot \mathrm{K}_{\eta}^{\alpha}(x) \quad (13.3.7)_{\eta}$$

$$\mathrm{K}_p^{\alpha}(p^{n+2}) - 2\mathrm{K}_p^{\alpha}(p^n) + \mathrm{K}_p^{\alpha}(p^{n-2}) = (p^{\alpha} - p^{-\alpha})^2 \cdot \mathrm{K}_p^{\alpha}(p^n). \quad (13.3.7)_p$$

Note that $\mathrm{K}_p^{\alpha}(x)$ has for fixed x all its zeros at $\alpha \in \mathrm{i}\mathbb{R}$; for $p \neq \eta$ this is immediate:

$$\mathrm{K}_p^{\alpha}(p^n) = 0 \Longleftrightarrow p^{(n+1)\alpha} = p^{-(n+1)\alpha}, \quad p^{\alpha} \neq p^{-\alpha} \Longleftrightarrow \alpha \in \frac{\pi \mathrm{i}}{\log p}\left(\frac{1}{n+1}\mathbb{Z}\backslash\mathbb{Z}\right). \tag{13.3.8}$$

For $p=\eta$ this follows from the functional equation

$$x\frac{\partial}{\partial x}\left[\mathrm{K}^{\bar\alpha}_\eta(x)\cdot x\frac{\partial}{\partial x}\mathrm{K}^\alpha_\eta(x)\right]=\left|x\frac{\partial}{\partial x}\mathrm{K}^\alpha_\eta(x)\right|^2+[\alpha^2+(2\pi x)^2]\cdot|\mathrm{K}^\alpha_\eta(x)|^2\,; \qquad (13.3.9)$$

if $\mathrm{K}^\alpha_\eta(x_0)=0$ then the integral of (13.3.9) from x_0 to ∞ vanishes, and taking imaginary parts we obtain

$$2\,\mathrm{Im}(\alpha)\;\mathrm{Re}(\alpha)\int_{x_0}^\infty \mathrm{d}^*x\,|\mathrm{K}^\alpha_\eta(x)|^2=0\,,$$

from which we see that $\mathrm{Re}(\alpha)=0$.

The Fourier transform $\mathcal{K}^\alpha_p=\mathcal{F}\rho^\alpha_{\infty p}$ gives a convolution semi-group of probability measures on $\mathbb{Q}_p$, $p\geq\eta$, $\alpha>0$, the 'Bessel semi-group' and similarly globally $\mathcal{K}^\alpha_{\mathbb{A}}=\mathcal{F}\rho^\alpha_{\infty\mathbb{A}}$, $\rho^\alpha_{\infty\mathbb{A}}(x)=\prod_{p\geq\eta}\rho^\alpha_{\infty p}(x_p)$, gives a semi-group of probability measures on $\mathbb{A}$.

Using the functional equation

$$\mathcal{F}\frac{|x|^s\mathrm{d}^*x}{\zeta(s)}=\frac{|x|^{1-s}\mathrm{d}^*x}{\zeta(1-s)}\,,$$

we obtain from the beta integral

$$\int\mathcal{K}^\alpha_p(x)|x|^s_p\,\mathrm{d}^*x=\frac{\zeta_p(s)\zeta_p(\alpha+s-1)}{\zeta_p(\alpha)} \qquad (13.3.10)$$

($p\geq\eta$, and over $\mathbb{A}$ for $\mathrm{Re}(s)\gg0$). Comparing this with (13.3.4), we see that $\mathcal{K}^\alpha$ is given in terms of the K-Bessel function by

$$\mathcal{K}^\alpha_p(x)=\frac{|x|_p^{(\alpha-1)/2}}{\zeta_p(\alpha)}\mathrm{K}_p^{(\alpha-1)/2}(x)=\begin{cases}\dfrac{\zeta_p(\alpha-1)}{\zeta_p(\alpha)}\phi_{\mathbb{Z}_p}(x)\cdot(1-|px|_p^{\alpha-1}) & p\neq\eta\,,\\[2ex] \dfrac{|x|^{(\alpha-1)/2}}{\zeta_\eta(\alpha)}2\mathrm{K}_{(\alpha-1)/2}(2\pi x) & p=\eta\,,\end{cases}$$

$$\mathcal{K}^\alpha_p(0)=\frac{\zeta_p(\alpha-1)}{\zeta_p(\alpha)},\quad p\geq\eta\,. \qquad (13.3.11)$$

Returning to the function $\tilde\rho^\beta_{\mathbb{A}/\mathbb{Q}^*}(a)$ of (13.1.9), we have for $\mathrm{Re}(\beta)>2$

$$\begin{aligned}\zeta_{\mathbb{A}}(\beta)\cdot\tilde\rho^\beta_{\mathbb{A}/\mathbb{Q}^*}(a)&=\zeta_{\mathbb{A}}(\beta)a^{-\beta/2}\left(\sum_{q\in\mathbb{Q}}\rho^\beta_{\infty\mathbb{A}}(a^{-1}q)-1\right)\\ &=\zeta_{\mathbb{A}}(\beta)a^{-\beta/2}\left(a\cdot\sum_{q\in\mathbb{Q}}\mathcal{F}\rho^\beta_{\infty\mathbb{A}}(a\cdot q)-1\right)\quad\text{by Poisson summation } \mathcal{F}\delta_{V(\mathbb{Q})}=\delta_{V(\mathbb{Q})}\\ &=\zeta_{\mathbb{A}}(\beta)a^{-\beta/2}\left(a\sum_{q\in\mathbb{Q}^*}\frac{a^{(\beta-1)/2}}{\zeta_{\mathbb{A}}(\beta)}\mathrm{K}_{\mathbb{A}}^{(\beta-1)/2}(aq)+a\cdot\frac{\zeta_{\mathbb{A}}(\beta-1)}{\zeta_{\mathbb{A}}(\beta)}-1\right)\\ &\qquad\text{by (13.3.11)}\end{aligned}$$

$$= 2a^{1/2}\sum_{n\geq 1}\mathrm{K}_{\mathbb{A}}^{(\beta-1)/2}(an) + \zeta_{\mathbb{A}}(\beta-1)a^{1-\beta/2} - \zeta_{\mathbb{A}}(\beta)a^{-\beta/2}$$

$$= 2a^{1/2}\sum_{n\geq 1} n^{(\beta-1)/2}\sigma_{1-\beta}(n)2\mathrm{K}_{(\beta-1)/2}(2\pi a n)$$

$$+ \zeta_{\mathbb{A}}(\beta-1)a^{1-\beta/2} - \zeta_{\mathbb{A}}(\beta)a^{-\beta/2}. \qquad (13.3.12)$$

This shows that $\zeta_{\mathbb{A}}(\beta)\cdot\tilde{\rho}^{\beta}_{\mathbb{A}/\mathbb{Q}^*}(a)$ has meromorphic continuation to all $\beta\in\mathbb{C}$, with simple poles only at $\beta=0,1,2$, and residues $1, -(a^{1/2}+a^{-1/2})$, 1, respectively.

The Eisenstein series is defined for $\tau = y + \mathrm{i}x$ in the upper half-plane by

$$\mathrm{E}(\tau,\beta) = \sum_{\gamma\in\mathrm{B}(\mathbb{Z})\backslash\mathrm{PSL}_2(\mathbb{Z})}\mathrm{Im}(\gamma\tau)^{\beta} = \frac{1}{2}\sum_{\substack{m,n\in\mathbb{Z}\\(m,n)=1}}\frac{x^{\beta}}{|m\tau+n|^{2\beta}}, \quad \mathrm{Re}(\beta)>1. \qquad (13.3.13)$$

Comparing it with $\tilde{\rho}^{\beta}_{\mathbb{A}/\mathbb{Q}^*}(a)$ we have

$$\tilde{\rho}^{\beta}_{\mathbb{A}/\mathbb{Q}^*}(a) = \mathrm{E}(\mathrm{i}a,\beta/2) - a^{\beta/2} - a^{-\beta/2}. \qquad (13.3.14)$$

From (13.3.12) we have

$$\zeta_{\mathbb{A}}(\beta)\cdot\mathrm{E}(\mathrm{i}a,\beta/2) = 2a^{1/2}\sum_{n\geq 1}\mathrm{K}_{\mathbb{A}}^{(\beta-1)/2}(an) + \zeta_{\mathbb{A}}(\beta-1)a^{1-\beta/2} + \zeta_{\mathbb{A}}(\beta)a^{\beta/2} \qquad (13.3.15)$$

showing that it has meromorphic continuation to all $\beta\in\mathbb{C}$, with simple poles only at $\beta=0$, 2 (the pole at $\beta=1$ cancels out!), and, moreover, that it is invariant under $\beta\leftrightarrow 2-\beta$. Thus just as summing $\phi_{\mathbb{Z}_{\mathbb{A}}}$ over $\mathbb{Q}\cdot a$ gives the theta function (cf. (13.1.15)), summing $\tilde{\rho}^{\beta}_{\mathbb{A}}$ over '$\mathbb{P}^1(\mathbb{Q})\cdot a$' gives the Eisenstein series.

The Eisenstein series is related to the fundamental projection maps

$$\begin{array}{ccc} & \mathcal{J} = \mathrm{B}(\mathbb{Z})\backslash\mathscr{H} & \\ \Pi_{\mathrm{X}} \swarrow & & \searrow \Pi_{\mathbb{R}^+} \\ \mathrm{X}_0 = \mathrm{SL}_2(\mathbb{Z})\backslash\mathscr{H} & & \mathrm{B}(\mathbb{R})\backslash\mathscr{H} = \mathbb{R}^+ \end{array} \qquad (13.3.16)$$

Here $\mathscr{H} = \mathrm{SL}_2(\mathbb{R})/\mathrm{SL}_2(\mathbb{Z}_\eta)$ is the upper half-plane,

$$\mathrm{B} = \left\{\begin{pmatrix}1 & * \\ 0 & 1\end{pmatrix}\right\},$$

and

$$\mathrm{X}_0 = \mathrm{SL}_2(\mathbb{Z})\backslash\mathscr{H} = \mathrm{SL}_2(\mathbb{Z})\backslash\mathrm{SL}_2(\mathbb{R})/\mathrm{SL}_2(\mathbb{Z}_\eta) = \mathrm{SL}_2(\mathbb{Q})\backslash\mathrm{SL}_2(\mathbb{A})/\prod_{p\geq\eta}\mathrm{SL}_2(\mathbb{Z}_p) \qquad (13.3.17)$$

is the modular curve. The points of $\mathrm{X}_0, \tau \in \mathscr{H} \bmod \mathrm{SL}_2(\mathbb{Z})$, correspond bijectively to (isomorphism classes of) elliptic curves, $E_\tau = \mathbb{C}/\mathbb{Z} + \mathbb{Z}\tau$. Functions on X_0 could be viewed as $\mathrm{SL}_2(\mathbb{A})$-invariant kernels intertwining the $\mathrm{SL}_2(\mathbb{A})$-representations

$$\mathrm{L}_2(\mathrm{SL}_2(\mathbb{Q})\backslash\mathrm{SL}_2(\mathbb{A})) \quad \text{and} \quad \mathrm{L}_2\left(\prod_{p\geq\eta}\mathrm{SL}_2(\mathbb{Z}_p)\backslash\mathrm{SL}_2(\mathbb{A})\right) = \bigotimes_{p\geq\eta}\mathrm{L}_2(\mathrm{SL}_2(\mathbb{Z}_p)\backslash\mathrm{SL}_2(\mathbb{Q}_p)) \tag{13.3.18}$$

associated with the global and the local lattice models (cf. Chapter 12).

The space

$$\mathcal{J} = \mathrm{B}(\mathbb{Z})\backslash\mathscr{H} \xrightarrow{\sim} \mathbb{D}^* = \{q \in \mathbb{C}^* \mid |q| < 1\} \tag{13.3.19}$$

$$\tau \bmod \mathrm{B}(\mathbb{Z}) \mapsto q = \mathrm{e}^{2\pi i r}$$

is the 'moduli space of the q-world', its points $q \in \mathbb{D}^*$ corresponding to the elliptic curve $E_q = \mathbb{C}^*/q^{\mathbb{Z}}$, together with the period $\mathbb{C}^{(1)} \subseteq E_q$. The Eisenstein series is the function on X_0 given by (13.3.13), that is

$$\mathrm{E}(\tau,\beta) = (\Pi_{\mathrm{X}})_*(\Pi_{\mathbb{R}^+})^*(x^\beta)\,. \tag{13.3.20}$$

For a function f on X_0 of rapid decay, $f(y+\mathrm{i}x) = \mathrm{O}(x^{-N})$ as $x \to \infty$ for all N, its constant term is the function on $\mathbb{R}^+$

$$\mathrm{C}(f)(x) = (\Pi_{\mathbb{R}^+})_*(\Pi_{\mathrm{X}})^* f(x) = \int_0^1 f(y+\mathrm{i}x)\mathrm{d}y\,. \tag{13.3.21}$$

Its Mellin transform is given by

$$\begin{aligned}\widehat{\mathrm{C}}(f)(\beta-1) &= \int_{\mathbb{R}^+}\mathrm{C}(f)(x)\cdot x^{\beta-1}\,\mathrm{d}^*x = \int_{\mathcal{J}} f(\tau)\cdot x^\beta\frac{\mathrm{d}y\,\mathrm{d}x}{x^2}\\ &= \int_{\mathrm{X}} f(\tau)\mathrm{E}(\tau,\beta)\mathrm{d}\tau\,,\end{aligned} \tag{13.3.22}$$

where $\mathrm{d}\tau = \mathrm{d}y\,\mathrm{d}x/x^2$ is the $\mathrm{SL}_2(\mathbb{R})$-invariant measure on $\mathscr{H}$, $\tau = y + \mathrm{i}x$.

Arguing as in (13.3.12), we obtain the Fourier expansion'

$$\begin{aligned}&\zeta_{\mathbb{A}}(\beta)\cdot\mathrm{E}(\tau,\beta/2)\\ &\quad= \zeta_{\mathbb{A}}(\beta)\cdot x^{\beta/2} + \zeta_{\mathbb{A}}(\beta-1)\cdot x^{1-\beta/2} + 2x^{1/2}\sum_{n\geq1}\mathrm{K}_{\mathbb{A}}^{(\beta-1)/2}(xn)\cdot\cos(2\pi ny)\,.\end{aligned} \tag{13.3.23}$$

This shows that $\zeta_{\mathbb{A}}(\beta)\cdot\mathrm{E}(\tau,\beta/2)$ has meromorphic continuation to all β, with simple poles only at $\beta = 0, 2$, and is invariant under $\beta \leftrightarrow 2-\beta$. Moreover, we have the eigenfunction equations

$$\Delta_\eta \mathrm{E}(\tau,\beta/2) = \beta/2\cdot(1-\beta/2)\cdot \mathrm{E}(\tau,\beta/2),$$
$$\Delta_\eta = -x^2\left(\frac{\partial^2}{\partial y^2}+\frac{\partial^2}{\partial x^2}\right), \qquad (13.3.24)_\eta$$

$$\mathrm{T}_p\mathrm{E}(\tau,\beta/2) = (p^{\beta/2}+p^{1-\beta/2})\cdot \mathrm{E}(\tau,\beta/2),$$
$$\mathrm{T}_p f(\tau) = f(p\tau) + \sum_{b\in\mathbb{Z}/p\mathbb{Z}} f\left(\frac{\tau+b}{p}\right) \quad p\neq\eta. \qquad (13.3.24)_p$$

Comparing (13.3.22) with (13.3.23), we deduce that $\zeta_{\mathbb{A}}(1+\beta)\cdot\widehat{\mathrm{C}}(f)((\beta-1)/2)$ similarly has a meromorphic continuation, simple poles only at $\beta=\pm1$ with residues $\int_{\mathrm{X}} f(\tau)\mathrm{d}\tau$, and is invariant under $\beta\leftrightarrow-\beta$ ('the Rankin–Selberg method').

From (13.3.22) and (13.3.23) we also see the non-vanishing of $\zeta_{\mathbb{A}}(s)$ on the line $\mathrm{Re}(s)=1$ (equivalent to the prime number theorem $\#\{p<x\}\sim Li(x)$). If $\zeta_{\mathbb{A}}(1+\mathrm{i}t)=0$, then $f(\tau)=\mathrm{E}(\tau,(1+\mathrm{i}t)/2)$ via (13.3.23) decays rapidly and has zero constant term, and so, by (13.3.22)

$$\int_{\mathrm{X}} |f(\tau)|^2\mathrm{d}\tau = \widehat{\mathrm{C}}(f)[-(1+\mathrm{i}t)/2] = 0\,,$$

contradicting $f(\tau)\not\equiv 0$; see Zagier (1981).

One can also formulate the Riemann hypothesis in terms of the geometry of the projections (13.3.16). Consider the closed curve on X_0 given by $\mathcal{C}_x=\Pi_{\mathrm{X}}\Pi^{-1}_{\mathbb{R}^+}(1/x^2)$; it has hyperbolic length x^2. The normalized measure $3/\pi\,\mathrm{d}\tau=\mathrm{d}\tau/2\zeta_{\mathbb{A}}(2)$ induces a probability measure on X_0, and for an open set $U\subseteq\mathrm{X}_0$ let $\mathcal{E}(U)=3/\pi\int_U\mathrm{d}\tau$ denote its expectation. One has

$$\mathrm{length}(\mathcal{C}_x\cap U)=\mathcal{E}(U)\cdot x^2+\mathrm{O}(x^{1+\varepsilon}),\qquad \text{as } x\to\infty,\ \text{all } \varepsilon>0\,. \qquad (13.3.25)$$

The Riemann hypothesis is equivalent to

$$\mathrm{length}(\mathcal{C}_x\cap U)=\mathcal{E}(U)\cdot x^2+\mathrm{O}(x^{1/2+\varepsilon}),\quad \text{as } x\to\infty,\ \text{all } \varepsilon>0\,. \qquad (13.3.26)$$

Letting $f_U(\tau)$ denote the characteristic function of U, we have

$$\mathrm{length}(\mathcal{C}_x\cap U)=\mathrm{C}(f_U)(1/x^2)\cdot x^2\,.$$

Its Mellin transform is given by

$$\int \mathrm{length}(\mathcal{C}_x\cap U)\cdot x^{-s}\mathrm{d}^*x = \frac{1}{2}\,\widehat{\mathrm{C}}(f_U)(s/2-1) = \frac{1}{2}\int_U \mathrm{E}(\tau,s/2)\mathrm{d}\tau$$
$$= \frac{\zeta_{\mathbb{A}}(2-s)}{\zeta_{\mathbb{A}}(s)}\frac{1}{2}\int_U \mathrm{E}(\tau,1-s/2)\mathrm{d}\tau\,. \qquad (13.3.27)$$

This has a simple pole at $s=2$ with residue $\mathcal{E}(U)$, and otherwise is holomorphic in $\mathrm{Re}(s)>\Theta=\sup\{\mathrm{Re}(\alpha)\,|\,\zeta_{\mathbb{A}}(\alpha)=0\}$, in particular in $\mathrm{Re}(s)>1$ from which (13.3.25) follows. Conversely, (13.3.26) implies that (13.3.27) is holomorphic in $\mathrm{Re}(s)>1/2$, except for the simple pole at $s=2$, and this being true for all U implies the Riemann hypothesis: $1/\zeta_{\mathbb{A}}(s)$ is holomorphic in $\mathrm{Re}(s)>1/2$; see Zagier (1981).

For a Schwartz function $f \in \mathcal{S}(\mathbb{R}^+)$, we can form the multiplicative convolution

$$f \circ \theta(a) = \int_{\mathbb{R}^+} \mathrm{d}^* a_0 \, f(a_0)\theta(a_0^{-1}a) = \left(\delta_{V(\mathbb{Q})}, \pi^1_{\mathbb{A}}(a^{-1})\pi^1_{\mathbb{A}}(f)\phi_{\mathbb{Z}_{\mathbb{A}}}\right). \tag{13.3.28}$$

We have from (13.1.18)

$$\begin{aligned} f \circ \theta(a) &= \hat{f}(-1/2)a^{1/2} + \mathrm{O}(a^{-N}) \quad \text{as } a \to \infty \text{ for all } N\,, \\ &= \hat{f}(1/2)a^{-1/2} + \mathrm{O}(a^{N}) \quad \text{as } a \to 0 \text{ for all } N\,. \end{aligned} \tag{13.3.29}$$

Hence if f belongs to the space

$$\mathcal{S}_{00} = \left\{ f \in \mathcal{S}(\mathbb{R}^+) \,\middle|\, \hat{f}(1/2) = \hat{f}(-1/2) = 0 \right\} \tag{13.3.30}$$

then $f \circ \theta$ is a fast decreasing function on $\mathbb{R}^+$; its Mellin transform is

$$(f \circ \theta)\hat{}\,(s) = \hat{f}(s) \cdot \zeta_{\mathbb{A}}(1/2 + s)\,. \tag{13.3.31}$$

We write $a = \mathrm{e}^y$, $\log a = y$, $\mathrm{d}^* a = \mathrm{d}a/a = \mathrm{d}y$, and consider the spaces

$$\mathrm{L}^{\delta}_2(\mathbb{R}^+) \overset{\text{def}}{=} \mathrm{L}_2(\mathbb{R}^+, (1 + (\log a)^2)^{\delta/2}\mathrm{d}^* a) \simeq \mathrm{L}_2(\mathbb{R}, (1 + y^2)^{\delta/2}\mathrm{d}y)\,. \tag{13.3.32}$$

For $\delta > 0$ we have the dense inclusions

$$\mathrm{L}^{\delta}_2(\mathbb{R}^+) \hookrightarrow \mathrm{L}_2(\mathbb{R}^+) \hookrightarrow \mathrm{L}^{-\delta}_2(\mathbb{R}^+)\,. \tag{13.3.33}$$

The inner product of $\mathrm{L}_2(\mathbb{R}^+) = \mathrm{L}^0_2(\mathbb{R}^+)$ sets a duality between $\mathrm{L}^{\delta}_2(\mathbb{R})$ and $\mathrm{L}^{-\delta}_2(\mathbb{R}^+)$:

$$(f_+, f_-)_{\mathrm{L}^0_2} = \int_{\mathbb{R}^+} \mathrm{d}^* a \, f_+(a) f_-(a), \quad f_\pm \in \mathrm{L}^{\pm\delta}_2(\mathbb{R}^+)\,. \tag{13.3.34}$$

Let $\mathscr{H}_\delta$ denote the closure of $\{ f \circ \theta \mid f \in \mathcal{S}_{00} \}$ in $\mathrm{L}^{\delta}_2(\mathbb{R}^+)$, and let $\mathscr{H}^\perp_\delta$ denote its orthogonal complement,

$$\mathrm{L}^{\delta}_2(\mathbb{R}^+) = \mathscr{H}_\delta \oplus \mathscr{H}^\perp_\delta\,. \tag{13.3.35}$$

The dual space of $\mathscr{H}^\perp_\delta$ (with respect to (13.3.34)) is given by

$$(\mathscr{H}^\perp_\delta)^{\vee} = \left\{ g \in \mathrm{L}^{-\delta}_2(\mathbb{R}^+) \,\middle|\, \int_{\mathbb{R}^+} \mathrm{d}^* a \, g(a) f \circ \theta(a) = 0 \text{ for all } f \in \mathcal{S}_{00} \right\}. \tag{13.3.36}$$

Taking the Mellin transform we obtain

$$(\mathscr{H}^\perp_\delta)^{\vee} = \left\{ g \in \mathrm{L}^{-\delta}_2(\mathbb{R}^+) \,\middle|\, \int_{-\mathrm{i}\infty}^{\mathrm{i}\infty} \mathrm{d}s \, \hat{g}(s)\hat{f}(s) \, \zeta_{\mathbb{A}}(1/2 + s) = 0 \text{ for all } f \in \mathcal{S}_{00} \right\}. \tag{13.3.37}$$

We conclude that for $g \in \left(\mathscr{H}^\perp_\delta\right)^{\vee}$ we have as distributions

$$\hat{g}(s) \cdot \zeta_{\mathbb{A}}(1/2 + s) = c_+\delta_{1/2} + c_-\delta_{-1/2} \quad \text{or} \quad g \circ \theta(a) = c_+ a^{1/2} + c_- a^{-1/2} \tag{13.3.38}$$

for some constants c_+ and c_-. But since $g \circ \theta$ is a temperate distribution on $\mathbb{R}^+$ and $a^{\pm 1/2} = \mathrm{e}^{\pm y/2}$ of exponential growth, we conclude that $c_+ = c_- = 0$, and the distribution $\hat{g}(s) \cdot \zeta_{\mathbb{A}}(1/2 + s)$ vanishes identically. We can therefore write

$$\begin{aligned} \hat{g}(s) &= \sum c_\alpha^{(k)} \cdot \partial^k / \partial s^k \quad \delta_{s=\alpha}\,, \\ \text{or } g(a) &= \sum c_\alpha^{(k)} \cdot (\log a)^k a^\alpha\,, \end{aligned} \tag{13.3.39}$$

where the sum is extended over α, k such that $\zeta_{\mathbb{A}}(1/2 + \alpha)$ vanishes at order greater than k, and the $c_\alpha^{(k)}$ are scalars (uniquely determined since the $(\log a)^k a^\alpha$ are linearly independent). Note that $(\log a)^k \cdot a^\alpha = y^k \mathrm{e}^{\alpha y}$ is in the space $\mathrm{L}_2^{-\delta}(\mathbb{R}^+)$ if and only if $\alpha \in \mathrm{i}\mathbb{R}$, $k < (\delta - 1)/2$. We deduce that the set

$$\left\{ (\log a)^k \cdot a^\alpha \,\middle|\, \alpha \in \mathrm{i}\mathbb{R},\ k < \min[(\delta - 1)/2, \operatorname{ord} \zeta_{\mathbb{A}}(1/2 + \alpha)] \right\} \tag{13.3.40}$$

forms a (non-orthogonal) basis for the space $(\mathscr{H}_\delta^\perp)^{\mathrm{v}}$. For $f \in \mathcal{S}(\mathbb{R}^+)$, writing $f \circ$ for the operator of convolution with f, we obtain (cf. Connes 1996)

$$\operatorname{tr}\big(f \circ |\mathscr{H}_\delta^\perp\big) = \operatorname{tr}\big(f \circ |(\mathscr{H}_\delta^\perp)^{\mathrm{v}}\big) = \sum_{\substack{\zeta_{\mathbb{A}}(1/2+\alpha)=0 \\ \alpha \in \mathrm{i}\mathbb{R}}} \hat{f}(\alpha) \cdot \min\{\lceil (\delta + 1)/2 \rceil, \operatorname{ord} \zeta_{\mathbb{A}}(1/2 + \alpha)\}\,. \tag{13.3.41}$$

Note that if

$$f(1) = \frac{1}{2\pi \mathrm{i}} \int_{-\mathrm{i}\infty}^{\mathrm{i}\infty} \mathrm{d}s \hat{f}(s) = \operatorname{tr}(f \circ |\mathrm{L}_2(\mathbb{R}^+))$$

vanishes, then $\operatorname{tr}(f \circ |\mathscr{H}_\delta) = -\operatorname{tr}\big(f \circ |\mathscr{H}_\delta^\perp\big)$ is minus the sum in (13.3.41).

The space $\mathscr{H}_\delta$ is the completion of $\mathcal{S}_{00}$ with respect to the inner product

$$\begin{aligned} (f_1, f_2)_{\mathscr{H}_\delta} &= (f_1 \circ \theta, f_2 \circ \theta)_{\mathrm{L}_2^\delta(\mathbb{R})} \\ &= \int \mathrm{d}^* a \cdot (1 + (\log a)^2)^{\delta/2} \cdot f_1 \circ \theta(a) \cdot \bar{f}_2 \circ \theta(a) \\ &= \left(1 + \frac{\partial^2}{\partial \beta^2}\right)^{\delta/2} \Bigg|_{\beta=1} \int \mathrm{d}^* a \cdot a^{\beta - 1} \cdot f_1 \circ \theta(a) \cdot \bar{f}_2 \circ \theta(a) \\ &= \left(1 + \frac{\partial^2}{\partial \beta^2}\right)^{\delta/2} \Bigg|_{\beta=1} \int \mathrm{d}^* a_1\, f_1(a_1) \cdot \int \mathrm{d}^* a_2\, \bar{f}_2(a_2) \\ &\qquad \cdot |a_1 a_2|^{(\beta-1)/2} \zeta_{\mathbb{A}}(\beta)\, \tilde{\rho}^{\beta}_{\mathbb{A}/\mathbb{Q}^*}(a_2/a_1) \\ &= \left(1 + \frac{\partial^2}{\partial \beta^2}\right)^{\delta/2} \Bigg|_{\beta=1} \int \mathrm{d}^* a_1\, f_1(a_1) \cdot \int \mathrm{d}^* a_2\, \bar{f}_2(a_2) \\ &\qquad \cdot |a_1 a_2|^{(\beta-1)/2} \zeta_{\mathbb{A}}(\beta) \mathrm{E}(\mathrm{i} a_2/a_1, \beta/2). \end{aligned} \tag{13.3.42}$$

Let $\mathcal{E}_\delta$ denote the completion of $\mathcal{S}(\mathbb{R}^+)$ with respect to the inner product $(f_1, f_2)_{\mathcal{E}_\delta}$ given by (the last line of) (13.3.42). We have $[\mathcal{E}_\delta : \mathscr{H}_\delta] = [\mathcal{S}(\mathbb{R}^+) : \mathcal{S}_{00}] = 2$. For $f \in \mathcal{S}_{00}$ the trace of the operator of convolution with f is the same on the space $\mathcal{E}_\delta$ as on the space $\mathscr{H}_\delta$. On the other hand

$$\mathcal{E}_\delta(a_1, a_2) = \left(1 + \frac{\partial^2}{\partial\beta^2}\right)^{\delta/2}\Bigg|_{\beta=1} |a_1 a_2|^{(\beta-1)/2}\zeta_{\mathbb{A}}(\beta)\mathrm{E}(ia_2/a_1, \beta/2) \qquad (13.3.43)$$

is a reproducing kernel for $\mathcal{E}_\delta$, and so we can express (minus) the sum in (13.3.41) also as

$$\begin{aligned}\mathrm{tr}(f \circ |\mathcal{E}_\delta) &= \int \mathrm{d}^*a f(a)\mathcal{E}_\delta(a, 1)\\ &= \left(1 + \frac{\partial^2}{\partial\beta^2}\right)^{\delta/2}\Bigg|_{\beta=1} \int \mathrm{d}^*a\, f(a) a^{(\beta-1)/2} \cdot \zeta_{\mathbb{A}}(\beta) \cdot \mathrm{E}(ia, \beta/2)\,. \qquad (13.3.44)\end{aligned}$$

Note that in the sum (13.3.41) only the zeros on the critical line are counted, while in the explicit sum formula (13.2.4) all the zeros of $\zeta_{\mathbb{A}}(s)$ are counted. Thus the fact that (13.3.44) is equal for δ sufficiently large and for $f \in \mathcal{S}_{00}$ to (minus) the sum in (13.2.4) is equivalent to Riemann's hypothesis.

13.4 THE EISENSTEIN SERIES AND THE INTERTWINING OPERATOR

The Eisenstein series of §13.3 is the $\mathrm{SL}_2(\mathbb{Z}_{\mathbb{A}})$-fixed part of the adelic Eisenstein transform. We would like to recall this transform here and its relation with the intertwining operator. We shall work with the group $\mathrm{G} = \mathrm{PGL}_2$. Let

$$\mathrm{B} = \left\{\begin{pmatrix} a & x \\ 0 & 1\end{pmatrix}\right\}$$

denote the Borel subgroup stabilizing $\infty = (0{:}1)$; $\mathrm{B} = \mathrm{M} \ltimes \mathrm{N}$ is the semi-direct product of the multiplicative group

$$\mathrm{M} = \left\{\begin{pmatrix} a & 0 \\ 0 & 1\end{pmatrix}\right\}$$

and the additive group

$$\mathrm{N} = \left\{\begin{pmatrix} 1 & x \\ 0 & 1\end{pmatrix}\right\}.$$

We denote with a subscript the associated groups of rational points:

$$\mathrm{G}_{\mathbb{Q}_p}, \quad \mathrm{G}_{\mathbb{Z}_p}(= \mathrm{PO}_2,\ p = \eta), \quad \mathrm{M}_{\mathbb{Q}_p},$$

$$\mathrm{M}_{\mathbb{Z}_p}\left(= \left\{\begin{pmatrix} \pm 1 & 0 \\ 0 & 1\end{pmatrix}\right\}, p = \eta\right), \quad \mathrm{N}_{\mathbb{Q}_p}, \quad \mathrm{N}_{\mathbb{Z}_p}(= \{\mathrm{I}\}, p = \eta)\,.$$

The group B is not unimodular: $\mathrm{d}^*a\,\mathrm{d}x\,|a|^{-1}$ (respectively $\mathrm{d}^*a\,\mathrm{d}x$) is a left (respectively right) Haar measure. Let $\mathrm{B}_p^{(1)} = \mathrm{M}_{\mathbb{Z}_p}\mathrm{N}_{\mathbb{Q}_p}$ and $\mathrm{B}_{\mathbb{A}}^{(1)} = \mathrm{M}_{\mathbb{A}}^{(1)}\mathrm{N}_{\mathbb{A}}$ denote the kernels of the modular function

$$\delta\begin{pmatrix} a & x \\ 0 & 1\end{pmatrix} = |a|\,.$$

The local intertwining operator

We begin with the local study. We have the $G_{\mathbb{Q}_p}$-symmetric spaces and the natural projections between them, as given in the diagram

$$
\begin{array}{ccccc}
G_{\mathbb{Q}_p} & \longrightarrow & \Omega_p^{\mathcal{A}} = M_{\mathbb{Z}_p}\backslash G_{\mathbb{Q}_p} & \cdots\cdots\rightarrow & \mathbb{P}_{\mathbb{Q}_p}^{\mathcal{A}} = M_{\mathbb{Q}_p}\backslash G_{\mathbb{Q}_p} \\
(\mathbb{Z}_p)\downarrow & & \downarrow & & \vdots \\
\widetilde{\mathcal{V}}_p = N_{\mathbb{Z}_p}\backslash G_{\mathbb{Q}_p} & \rightarrow & I_p = B_{\mathbb{Z}_p}\backslash G_{\mathbb{Q}_p} & \searrow (\mathbb{P}_{\mathbb{Z}_p}^1) & \vdots (\mathbb{Q}_p) \\
(\mathbb{Q}_p/\mathbb{Z}_p)\ \vdots & & \vdots & \widetilde{X}_p = G_{\mathbb{Z}_p}\backslash G_{\mathbb{Q}_p} & \vdots \\
\mathcal{V}_{\mathbb{Q}_p} = N_{\mathbb{Q}_p}\backslash G_{\mathbb{Q}_p} & \rightarrow & \Omega_p = B_p^{(1)}\backslash G_{\mathbb{Q}_p} & \cdots\cdots\rightarrow & \mathbb{P}_{\mathbb{Q}_p}^1 = B_{\mathbb{Q}_p}\backslash G_{\mathbb{Q}_p}
\end{array}
\tag{13.4.1}
$$

For all the symmetric spaces $Y = H\backslash G_{\mathbb{Q}_p}$, except for $P_{\mathbb{Q}_p}^1 = B_{\mathbb{Q}_p}\backslash G_{\mathbb{Q}_p}$, the group H is unimodular. We then have a $G_{\mathbb{Q}_p}$-invariant measure on Y, and hence a unitary representation of $G_{\mathbb{Q}_p}$ by right translations on the associated L_2 space

$$\pi\colon G_{\mathbb{Q}_p} \to U(L_2(Y)), \qquad \pi(g)f(g_0) = f(g_0 g)\,. \tag{13.4.2}$$

Given two symmetric spaces $Y_1 \to Y_2$, with Y_1 a principal fibration over Y_2, we denote by $\mathrm{pr}(Y_1/Y_2)^*$ the pull-back of functions from Y_2 to Y_1, $\mathrm{pr}(Y_1/Y_2)^* f(H_1 g) = f(H_2 g)$. We denote by $\mathrm{pr}(Y_1/Y_2)_*$ the adjoint operation of integration along the fibers,

$$\mathrm{pr}(Y_1/Y_2)_* f(H_2 g) = \int_{H_1\backslash H_2} dh\, f(H_1 h g)\,.$$

We have the automorphic factor

$$j_p\colon \mathbb{P}_{\mathbb{Q}_p}^1 \times G_{\mathbb{Q}_p}/G_{\mathbb{Z}_p} \to \mathbb{R}^+, \quad j_p(x\!:\!y, g) = \frac{|x,y|_p}{|(x,y)g|_p} \cdot |\det g|_p^{1/2}\,. \tag{13.4.3}$$

It satisfies

$$
\begin{aligned}
j_p(z, g_1 g_2) &= j_p(z, g_1) \cdot j_p(z g_1, g_2)\,, \\
j_p(z, g)^{-1} &= j_p(z g, g^{-1})\,.
\end{aligned}
\tag{13.4.4}
$$

Moreover, with respect to the $G_{\mathbb{Z}_p}$-invariant metric ρ_p on $\mathbb{P}_{\mathbb{Q}_p}^1$ (3.1.2), we have

$$\rho_p(xg, yg) = \rho_p(x, y) \cdot j_p(x, g) \cdot j_p(y, g)\,. \tag{13.4.5}$$

With respect to the $G_{\mathbb{Z}_p}$-invariant beta measure $\tau_p^{1,1}$ on $\mathbb{P}_{\mathbb{Q}_p}^1$, we have

$$\tau_p^{1,1}(zg) = \tau_p^{1,1}(z) \cdot j_p(z, g)^2\,. \tag{13.4.6}$$

The characters $\delta_p^{s/2}\colon \mathrm{B}_{\mathbb{Q}_p} \to \mathbb{R}^+$, $s \in \mathbb{C}/(4\pi\mathrm{i}/\log p)\mathbb{Z}$ ($\mathbb{C}$ for $p = \eta$), give rise to the unramified principal series representations of $\mathrm{G}_{\mathbb{Q}_p}$ via induction: $\mathbb{H}_p^s = \mathrm{Ind}_{\mathrm{B}_{\mathbb{Q}_p}}^{\mathrm{G}_{\mathbb{Q}_p}}(\delta_p^{s/2})$. It acts on the space consisting of functions f on Ω_p satisfying

$$f\left(\begin{pmatrix} a & x \\ 0 & 1 \end{pmatrix} g\right) = |a|_p^{(1+s)/2} \cdot f(g)\,. \tag{13.4.7}$$

Since $\mathrm{G}_{\mathbb{Q}_p} = \mathrm{B}_{\mathbb{Q}_p} \cdot \mathrm{G}_{\mathbb{Z}_p}$, such a function f is determined by its restriction to $\mathrm{G}_{\mathbb{Z}_p}$, which is left $\mathrm{B}_{\mathbb{Z}_p}$-invariant; that is, $f|_{\mathrm{G}_{\mathbb{Z}_p}}$ is a function on $\mathrm{B}_{\mathbb{Z}_p}\backslash\mathrm{G}_{\mathbb{Z}_p} = \mathbb{P}^1_{\mathbb{Q}_p}$, and the Hilbert space structure on $\mathbb{H}_p^s$ is given by identifying it with $\widetilde{\mathscr{H}}_p^{1,1} = \mathrm{L}_2\left(\mathbb{P}^1_{\mathbb{Q}_p}, \tau_p^{1,1}\right)$. We thus have an isomorphism

$$\mathrm{i}_s\colon \mathbb{H}_p^s \xrightarrow{\sim} \widetilde{\mathscr{H}}_p^{1,1}, \quad f \mapsto f|_{\mathrm{G}_{\mathbb{Z}_p}}\,. \tag{13.4.8}$$

Its inverse is given by

$$\mathrm{i}_s^{-1} f(g) = f((0{:}1)g) \cdot \mathrm{j}_{\infty_p}^{1+s}(g)\,, \tag{13.4.9}$$

where

$$\mathrm{j}_{\infty_p}^{1+s}\left(\begin{pmatrix} a & x \\ 0 & 1 \end{pmatrix} k\right) = \mathrm{j}_p\left(0{:}1, \begin{pmatrix} a & x \\ 0 & 1 \end{pmatrix} k\right)^{1+s} = |a|_p^{(1+s)/2} \quad \text{for } k \in \mathrm{G}_{\mathbb{Z}_p}\,. \tag{13.4.10}$$

The group $\mathrm{G}_{\mathbb{Q}_p}$ acts on $\mathbb{H}_p^s$ by bounded invertible operators via right translations, $\pi(g)f(g_0) = f(g_0 g)$, and translating this action to $\widetilde{\mathscr{H}}_p^{1,1}$ we have

$$\pi_p^s(g)f(z) = f(zg) \cdot \mathrm{j}_p(z,g)^{1+s}, \quad f \in \widetilde{\mathscr{H}}_p^{1,1}\,. \tag{13.4.11}$$

It follows from (13.4.6) that the adjoint operator is given by

$$\pi_p^s(g)^* = \pi_p^{-\bar{s}}(g^{-1})\,. \tag{13.4.12}$$

Hence π_p^s is a unitary representation if and only if $s \in \mathrm{i}\mathbb{R}$.

Since $\mathrm{j}_p(z,k) = 1$ for $k \in \mathrm{G}_{\mathbb{Z}_p}$, the restriction of π_p^s to $\mathrm{G}_{\mathbb{Z}_p}$ is independent of s, and we have the associated decomposition of $\widetilde{\mathscr{H}}_p^{1,1}$ into irreducible $\mathrm{G}_{\mathbb{Z}_p}$-representations

$$\widetilde{\mathscr{H}}_p^{1,1} = \bigoplus_{n \geq 0} \widetilde{\mathscr{H}}_p^{1,1}(n)\,. \tag{13.4.13}$$

For $p \neq \eta$ this is just the decomposition of the boundary space of the beta chain ($\alpha = \beta = 1$) on the tree $\widetilde{\mathrm{X}}_p$. For $p = \eta$, $n \geq 1$, $\widetilde{\mathscr{H}}_p^{1,1}(n)$ is two-dimensional spanned by $f_{\pm,n}(x{:}y) = ((x+\mathrm{i}y)/|x+\mathrm{i}y|)^{\pm 2n}$. For $n = 0$, $p \geq \eta$, $\widetilde{\mathscr{H}}_p^{1,1}(0) = \left(\widetilde{\mathscr{H}}_p^{1,1}\right)^{\mathrm{G}_{\mathbb{Z}_p}} = \mathbb{C}\mathbb{1}$ is the trivial representation of $\mathrm{G}_{\mathbb{Z}_p}$.

We define the normalized intertwining operator $\widetilde{\mathrm{R}}_p^s$ on $\widetilde{\mathscr{H}}_p^{1,1}$ by

$$\begin{aligned}\widetilde{\mathrm{R}}_p^s f(y) &= \frac{\zeta_p(1+s)}{\zeta_p(s)} \frac{\zeta_p(1)}{\zeta_p(2)} \int \tau_p^{1,1}(\mathrm{d}x)\rho_p(y,x)^{s-1} f(x) \\ &= \frac{\zeta_p(1+s)}{\zeta_p(s)} \int \frac{\mathrm{d}x}{|1,x|^2} \rho_p(y, 1{:}x)^{s-1} f(1{:}x), \quad \mathrm{Re}(s) > 0\,.\end{aligned} \tag{13.4.14}$$

$\widetilde{\mathrm{R}}_p^s$ has meromorphic continuation to all s, with poles contained in the poles of $\zeta_p(1+s)$. As follows from (13.4.5), it satisfies

$$\widetilde{\mathrm{R}}_p^s \pi_p^s(g) = \pi_p^{-s}(g)\widetilde{\mathrm{R}}_p^s \tag{13.4.15}$$

i.e. $\widetilde{\mathrm{R}}_p^s$ intertwines the representation π_p^s with π_p^{-s}.

$$\widetilde{\mathrm{R}}_p^{-s}\widetilde{\mathrm{R}}_p^s = \mathrm{id}\,; \tag{13.4.16}$$

$$\widetilde{\mathrm{R}}_p^0 = \mathrm{id}, \quad \widetilde{\mathrm{R}}_p^1 = \text{projection onto } \widetilde{\mathscr{H}}_p^{1,1}(0) = \mathbb{C}\mathbb{1}\,; \tag{13.4.17}$$

$$\widetilde{\mathrm{R}}_p^s \mathbb{1} = \mathbb{1} \text{ is 'normalized'}. \tag{13.4.18}$$

In fact since $\widetilde{\mathrm{R}}_p^s$ commutes with the $\mathrm{G}_{\mathbb{Z}_p}$-action, we have with respect to the $\mathrm{G}_{\mathbb{Z}_p}$-decomposition (13.4.13)

$$\widetilde{\mathrm{R}}_p^s = \bigoplus_{n\geq 0} \widetilde{\mathrm{R}}_p^s(n)\cdot \mathrm{id}_{\widetilde{\mathscr{H}}_p^{1,1}(n)}\,. \tag{13.4.19}$$

The scalars $\widetilde{\mathrm{R}}_p^s(n)$ can be computed by the effect of $\widetilde{\mathrm{R}}_p^s$ on the beta basis, the result being

$$\begin{aligned} \widetilde{\mathrm{R}}_p^s(n) &= \begin{cases} 1 & n=0 \\ \dfrac{\zeta_p(1+s)}{\zeta_p(1-s)}p^{-ns} & n\geq 1 \end{cases} \quad p\neq\eta\,, \\ \widetilde{\mathrm{R}}_\eta^s(n) &= \frac{\zeta_\eta(1+s)}{\zeta_\eta(1-s)}\cdot\frac{\zeta_\eta(1-s+2n)}{\zeta_\eta(1+s+2n)}\,. \end{aligned} \tag{13.4.20}$$

This gives the meromorphic continuation of $\widetilde{\mathrm{R}}_p^s$, the properties (13.4.16)–(13.4.18), and, moreover, shows that

$$\begin{aligned} &\text{(i) } \widetilde{\mathrm{R}}_p^s \text{ is unitary} \Longleftrightarrow s\in \mathrm{i}\mathbb{R} \\ &\text{(ii) } \widetilde{\mathrm{R}}_p^s \text{ is self-adjoint} \Longleftrightarrow s\in \mathbb{R} \\ &\text{(iii) } \widetilde{\mathrm{R}}_p^s \text{ is bounded} \Longleftrightarrow \mathrm{Re}(s)\geq 0 \\ &\text{(iv) } \widetilde{\mathrm{R}}_p^s \text{ is Hilbert–Schmidt} \Longleftrightarrow \mathrm{Re}(s) > 1/2 \\ &\text{(v) } \widetilde{\mathrm{R}}_p^s \text{ is trace-class} \Longleftrightarrow \mathrm{Re}(s) > 1 \text{ and } \mathrm{tr}(\widetilde{\mathrm{R}}_p^s) = 0\,, \\ &\text{(vi) } \widetilde{\mathrm{R}}_p^s \text{ is positive definite} \Longleftrightarrow s\in(-1,1), \text{ corresponding to the} \\ &\qquad \text{complementary series representations of } \mathrm{G}_{\mathbb{Q}_p}. \end{aligned} \tag{13.4.21}$$

The operator $\widetilde{\mathrm{R}}_p^s$ is related to the symplectic Fourier transform $\mathcal{F}_V$ (cf. (12.2.6)). Consider the Tate distribution as acting on an arbitrary line in the plane; it gives rise to a map

$$\begin{aligned} &\underline{\mathcal{T}}_p^s\colon \mathcal{S}(V(\mathbb{Q}_p)) \to \mathrm{C}^\infty\left(\mathbb{P}^1_{\mathbb{Q}_p}\right) \\ &\underline{\mathcal{T}}_p^s\varphi(x{:}y) = \int \frac{\mathrm{d}^*a}{\zeta_p(s)}\varphi(ax,ay)\cdot|ax,ay|_p^s, \quad \mathrm{Re}(s)>0\,. \end{aligned} \tag{13.4.22}$$

The map $\underline{\mathcal{I}}_p^s$ has entire analytic continuation to all $s \in \mathbb{C}$; it is onto for $\mathrm{Re}(s) > 0$. At $s = 0$, $\underline{\mathcal{I}}_p^0 \varphi = \varphi(0,0) \cdot \mathbb{1}$ is a rank-1 projection, and $\underline{\mathcal{I}}_p^s$ is normalized in the sense that $\underline{\mathcal{I}}_p^s \Phi_{V(\mathbb{Z}_p)} = \mathbb{1}$. The connection between $\widetilde{\mathrm{R}}_p^s$ and $\mathcal{F}_V$ is expressed in the commutative diagram

$$\begin{array}{ccc} \mathrm{S}(V(\mathbb{Q}_p)) & \xrightarrow[\sim]{\mathcal{F}_V} & \mathrm{S}(V(\mathbb{Q}_p)) \\ \Big\downarrow \underline{\mathcal{I}}_p^{1+s} & & \Big\downarrow \underline{\mathcal{I}}_p^{1-s} \\ \mathrm{C}^\infty\left(\mathbb{P}^1_{\mathbb{Q}_p}\right) & \xrightarrow[\sim]{\widetilde{\mathrm{R}}_p^s} & \mathrm{C}^\infty\left(\mathbb{P}^1_{\mathbb{Q}_p}\right) \end{array} \tag{13.4.23}$$

For the proof of the commutativity of (13.4.23) we can assume by analytic continuation that $\mathrm{Re}(s) \in (-1,1)$. Then

$$\begin{aligned} \underline{\mathcal{I}}_p^{1-s} \mathcal{F}_V \varphi(1{:}z) &= \frac{1}{\zeta_p(1-s)} \int \mathrm{d}^* a \, |a, az|_p^{1-s} \iint \mathrm{d}x \, \mathrm{d}y \, \varphi(x,y) \psi_p(a(y - xz)) \\ &= \frac{|1,z|_p^{1-s}}{\zeta_p(1-s)} \int \mathrm{d}^* a \iint \mathrm{d}x \, \mathrm{d}y \, |a|_p^{1+s} \varphi(ax, ay) \psi_p(y - xz) \\ &= \frac{|1,z|_p^{1-s}}{\zeta_p(1-s)} \int \mathrm{d}^* a \iint \mathrm{d}x \, \mathrm{d}y \, |a|_p^{1+s} \varphi(a, ay) \psi_p(x(y-z)) |x|_p^{-s} \\ &= \frac{|1,z|_p^{1-s}}{\zeta_p(s)} \int \mathrm{d}^* a \int \mathrm{d}y \, |a|_p^{1+s} \varphi(a, ay) |y - z|_p^{s-1} \\ = \frac{\zeta_p(1+s)}{\zeta_p(s)} \int \frac{\mathrm{d}y}{|1,y|^2} & \rho_p(1{:}z, 1{:}y)^{s-1} \, \frac{1}{\zeta_p(1+s)} \int \mathrm{d}^* a \, \varphi(a, ay) \cdot |a, ay|_p^{1+s} \\ &= \widetilde{\mathrm{R}}_p^s \underline{\mathcal{I}}_p^{1+s} \varphi(1{:}z) \, . \end{aligned} \tag{13.4.24}$$

Since conjugation by elements of $\mathrm{M}_{\mathbb{Q}_p}$ stabilizes $\mathrm{B}_p^{(1)}$, we have a well-defined left action of $\mathbb{Q}_p^*/\mathbb{Z}_p^*$ on $\Omega_p = \mathrm{B}_p^{(1)} \backslash \mathrm{G}_{\mathbb{Q}_p}$

$$\pi_\ell(a) f(g) = |a|_p^{1/2} f\left(\begin{pmatrix} a^{-1} & 0 \\ 0 & 1 \end{pmatrix} g \right). \tag{13.4.25}$$

We define the Mellin transform $\mathrm{L}^s f$ for $f \in \mathrm{C}_c(\Omega_p)$ by

$$\mathrm{L}^s f(g) = \int \mathrm{d}^* a \, |a|_p^{(1+s)/2} f\left(\begin{pmatrix} a^{-1} & 0 \\ 0 & 1 \end{pmatrix} g \right); \tag{13.4.26}$$

the inverse transform is given by

$$f(g) = \int_{\mathrm{Re}(s)=\sigma} \mathrm{d}_p(s/2)\, \mathrm{L}^s f(g)\,.$$

Here $s \in \mathbb{C}/(4\pi \mathrm{i}/\log p)\mathbb{Z}$ ($\mathbb{C}$ for $p = \eta$), that is $s/2$ is the usual element of the dual group $(\mathbb{Q}_p^*/\mathbb{Z}_p^*)^{\vee}$. We have $\mathrm{L}^s f \in \mathbb{H}_p^s$, and by Mellin inversion we obtain an isomorphism

$$\mathrm{L}\colon \mathrm{L}_2(\Omega_p) \xrightarrow{\sim} \int_{(\mathbb{Q}_p^*/\mathbb{Z}_p^*)^{\vee}} \mathrm{d}_p(s/2)\, \mathbb{H}_p^s, \quad f \mapsto \mathrm{L}^s f\,. \tag{13.4.27}$$

This gives the decomposition of $\mathrm{L}_2(\Omega_p)$ into the continuous sum of the irreducible $\mathrm{G}_{\mathbb{Q}_p}$-representations $\mathbb{H}_p^s$, $\mathrm{Re}(s) = 0$. Note that Ω_p has multiplicity *two* since $\mathbb{H}_p^s$ and $\mathbb{H}_p^{-s}$ are equivalent $\mathrm{G}_{\mathbb{Q}_p}$-representations.

On the space $\Omega_p^{\phi} = \mathrm{M}_{\mathbb{Z}_p}\backslash \mathrm{G}_{\mathbb{Q}_p}$ we similarly have a well-defined left action of $\mathbb{Q}_p^*/\mathbb{Z}_p^*$, as well as the left action by the Weyl element

$$w = \begin{pmatrix} 0 & -1 \\ 1 & 0 \end{pmatrix},$$

so we obtain an involution on Ω_p^{ϕ}:

$$\mathrm{W} f(g) = f(wg)\,. \tag{13.4.28}$$

We get the following operator acting on functions on Ω_p, commuting with the $\mathrm{G}_{\mathbb{Q}_p}$-action:

$$\mathrm{M}_p f(g) = \mathrm{pr}(\Omega_p^{\phi}/\Omega_p)_* \circ \mathrm{W} \circ \mathrm{pr}(\Omega_p^{\phi}/\Omega_p)^* f(g) = \int \mathrm{d}x\, f\left(w\begin{pmatrix} 1 & x \\ 0 & 1 \end{pmatrix} g\right). \tag{13.4.29}$$

Note that $\mathrm{M}_p f$ is defined for $f \in \mathbb{H}_p^s$, $\mathrm{Re}(s) > 0$, and $\mathrm{M}_p f \in \mathbb{H}_p^{-s}$. In fact, letting $\mathrm{M}_p^{(s)} = \mathrm{i}_{-s}\mathrm{M}_p\mathrm{i}_s^{-1}$ denote the corresponding operator on $\widetilde{\mathscr{H}}_p^{1,1}$, we have

$$\mathrm{M}_p^{(s)} = \frac{\zeta_p(s)}{\zeta_p(1+s)}\widetilde{\mathrm{R}}_p^s\,. \tag{13.4.30}$$

The proof of (13.4.30) is again by straightforward computation

$$\begin{aligned}
\mathrm{M}_p^{(s)} f(1{:}y) &= |1,y|_p^{1-s} \cdot \mathrm{M}_p \mathrm{i}_s^{-1} f\begin{pmatrix} 0 & -1 \\ 1 & y \end{pmatrix} \\
&= |1,y|_p^{1-s} \cdot \int \mathrm{d}x\, \mathrm{i}_s^{-1} f\left(w\begin{pmatrix} 1 & x \\ 0 & 1 \end{pmatrix}\begin{pmatrix} 0 & -1 \\ 1 & y \end{pmatrix}\right) \\
&= |1,y|_p^{1-s} \cdot \int \mathrm{d}x\, \mathrm{i}_s^{-1} f\begin{pmatrix} 1 & y \\ -x & 1-xy \end{pmatrix} \\
&= |1,y|_p^{1-s} \cdot \int \mathrm{d}x\, f(-x{:}1-xy) \cdot |-x, 1-xy|_p^{-(1+s)} \\
&= |1,y|_p^{1-s} \cdot \int \frac{\mathrm{d}x}{|x|_p^2} \cdot f(1{:}y-x) \cdot |x|_p^{1+s} \cdot |1, y-x|_p^{-(1+s)} \\
&= |1,y|_p^{1-s} \cdot \int \mathrm{d}x\, f(1{:}x) \cdot |y-x|_p^{s-1} \cdot |1,x|_p^{-(1+s)} \\
&= \int \frac{\mathrm{d}x}{|1,x|^2} f(1{:}x) \cdot \rho_p(1{:}x, 1{:}y)^{s-1} = \frac{\zeta_p(s)}{\zeta_p(1+s)}\widetilde{\mathrm{R}}_p^s f(1{:}y)\,.
\end{aligned} \tag{13.4.31}$$

The spherical function and Plancherel measure

The spherical function is defined by

$$\mathrm{F}_p^s(g) = (\pi_p^s(g)\mathbb{1}, \mathbb{1}) = \int \tau_p^{1,1}(\mathrm{d}z)\mathrm{j}_p(z,g)^{1+s}. \tag{13.4.32}$$

Note that

$$\mathrm{F}_p^s(g) = \mathrm{F}_p^{-s}(g) = \mathrm{F}_p^s(g^{-1}), \tag{13.4.33}$$

and that F_p^s is left and right $\mathrm{G}_{\mathbb{Z}_p}$-invariant. Since $\mathrm{G}_{\mathbb{Q}_p} = \mathrm{G}_{\mathbb{Z}_p}\mathrm{M}_{\mathbb{Q}_p}^+\mathrm{G}_{\mathbb{Z}_p}$ with

$$\mathrm{M}_{\mathbb{Q}_p}^+ = \left\{\begin{pmatrix} p^n & 0 \\ 0 & 1\end{pmatrix}, n \geq 0\right\}, \quad p \neq \eta, \quad \mathrm{M}_{\mathbb{Q}_\eta}^+ = \left\{\begin{pmatrix} a & 0 \\ 0 & 1\end{pmatrix}, a \in (0,1]\right\}, \tag{13.4.34}$$

F_p^s is determined by its values on $\mathrm{M}_{\mathbb{Q}_p}^+$. For $p \neq \eta$, $n \geq 1$, we have

$$\mathrm{F}_p^s\begin{pmatrix} p^n & 0 \\ 0 & 1\end{pmatrix} = \frac{p^{-n/2}}{(1+p^{-1})}\frac{[p^{(n+1)s/2} - p^{-(n+1)s/2}] - p^{-1}[p^{(n-1)s/2} - p^{-(n-1)s/2}]}{[p^{s/2} - p^{-s/2}]} \stackrel{\text{def}}{=} \mathrm{F}_n(\mathrm{t}_p); \tag{13.4.35$_p$}$$

this is a polynomial of degree n in $\mathrm{t}_p = p^{1/2}(p^{s/2} - p^{-s/2})$. For $p = \eta$

$$\mathrm{F}_\eta^s\begin{pmatrix} a & 0 \\ 0 & 1\end{pmatrix} = |a|^{-(1+s)/2}{}_2\mathrm{F}_1\left(\frac{1+s}{2}, \frac{1}{2}; 1; 1 - a^{-2}\right). \tag{13.4.35$_\eta$}$$

Note that we can realize the representation $\mathbb{H}_p^s$ as the completion of $\mathrm{C}_c(\widetilde{\mathrm{X}}_p)$ with respect to the reproducing kernel $\mathrm{F}_p^s(g_1 \cdot g_2^{-1})$. The Haar measure $\mathrm{d}g$ on $\mathrm{G}_{\mathbb{Q}_p}$, normalized so that $\mathrm{d}g(\mathrm{G}_{\mathbb{Z}_p}) = 1$, induces a measure μ_p on $\mathrm{M}_{\mathbb{Q}_p}^+ \simeq \mathrm{G}_{\mathbb{Z}_p}\backslash\mathrm{G}_{\mathbb{Q}_p}/\mathrm{G}_{\mathbb{Z}_p} = \widetilde{\mathrm{X}}_p/\mathrm{G}_{\mathbb{Z}_p}$. It is given by

$$\mu_p\begin{pmatrix} p^n & 0 \\ 0 & 1\end{pmatrix} = \begin{cases} 1 & n = 0, \\ p^n(1+p^{-1}) & n \geq 1, \end{cases} \quad p \neq \eta, \tag{13.4.36$_p$}$$

$$\mu_\eta\begin{pmatrix} a & 0 \\ 0 & 1\end{pmatrix} = |a^{-1} - a|\mathrm{d}^*a. \tag{13.4.36$_\eta$}$$

The Plancherel measure is the measure $\hat{\mu}_p(s)\mathrm{d}_p(s/2)$ on the dual space

$$\widehat{\mathrm{M}}_{\mathbb{Q}_p}^+ = \left(\mathbb{Q}_p^*/\mathbb{Z}_p^*\right)^V \Big/ s \sim -s \quad (s \text{ and } -s \text{ identified}), \tag{13.4.37}$$

where

$$\hat{\mu}_p(s) = \frac{\zeta_p(1)}{\zeta_p(2)}\left|\frac{\zeta_p(1+s)}{\zeta_p(s)}\right|^2. \tag{13.4.38}$$

We have the isomorphism

$$\mathrm{F}_p\colon \mathrm{L}_2\Big(\mathrm{M}^+_{\mathbb{Q}_p},\mu_p\Big) \xrightarrow{\sim} \mathrm{L}_2\Big(\widehat{\mathrm{M}}^+_{\mathbb{Q}_p},\hat\mu_p(s)\mathrm{d}_p\Big(\frac{s}{2}\Big)\Big) = \mathrm{L}_2\bigg(\mathrm{i}\mathbb{R}\Big/\frac{2\pi\mathrm{i}}{\log p}\mathbb{Z},\hat\mu_p(2\mathrm{it})\mathrm{dt}\cdot\frac{\log p}{2\pi}\bigg)^+ \tag{13.4.39}$$

given by the kernel

$$\mathrm{F}^s_p\begin{pmatrix} a & 0\\ 0 & 1\end{pmatrix},\qquad a\in\mathrm{M}^+_{\mathbb{Q}_p},\, s\in\widehat{\mathrm{M}}^+_{\mathbb{Q}_p}\,;$$

that is, we have inverse isomorphisms

$$f\mapsto \int_{\mathrm{M}^+_{\mathbb{Q}_p}} f(g)\mathrm{F}^s_p(g)\mu_p(\mathrm{d}g) = \int_{\mathrm{G}_{\mathbb{Q}_p}} f(g)\mathrm{F}^s_p(g)\mathrm{d}g = (\pi^s(f)\mathbb{1},\mathbb{1}) \overset{\mathrm{def}}{=} \hat f(s) \tag{13.4.40}$$

$$f(g) = \int_{\mathrm{M}^+_{\mathbb{Q}_p}} \hat f(s)\mathrm{F}^s_p(g)\hat\mu_p(s)\mathrm{d}_p\Big(\frac{s}{2}\Big) \leftarrow\!\shortmid \hat f(s)\,.$$

Convolution on the group $\mathrm{G}_{\mathbb{Q}_p}$ induces a commutative algebra structure on the Hecke algebra $\mathcal{A}_p = \mathrm{C}_c(\mathrm{G}_{\mathbb{Z}_p}\backslash\mathrm{G}_{\mathbb{Q}_p}/\mathrm{G}_{\mathbb{Z}_p}) = \mathrm{C}_c\Big(\mathrm{M}^+_{\mathbb{Q}_p}\Big)$, which corresponds to ordinary multiplication under the transform F_p

$$\mathrm{F}^s_p(f_1 * f_2) = \mathrm{F}^s_p(f_1)\cdot\mathrm{F}^s_p(f_2)\,, \tag{13.4.41}$$

$$\mathrm{F}^s_p(f^*) = \overline{\mathrm{F}^s_p(f)}\,. \tag{13.4.42}$$

Consider next the unitary representation of $\mathrm{G}_{\mathbb{Q}_p}$ on $\mathrm{L}_2(\widetilde{\mathrm{X}}_p)$. The algebra $\mathcal{A}_p$ acts by left convolution on $\mathrm{L}_2(\widetilde{\mathrm{X}}_p)$ generating the commutant of $\mathrm{G}_{\mathbb{Q}_p}$. We have the *constant term*, or *Radon transform*

$$\mathrm{C}_p\colon \mathrm{C}_c(\widetilde{\mathrm{X}}_p) \to \mathrm{C}_c(\Omega_p) \tag{13.4.43}$$

$$\mathrm{C}_p\varphi(g) = \mathrm{pr}(\mathrm{I}_p/\Omega_p)_*\circ\mathrm{pr}(\mathrm{I}_p/\widetilde{\mathrm{X}}_p)^*\varphi(g) = \int \mathrm{d}x\,\varphi\bigg(\begin{pmatrix}1 & x\\ 0 & 1\end{pmatrix}g\bigg)\,.$$

Its formal adjoint is the *Eisenstein integral*

$$\mathrm{E}_p = \mathrm{C}^*_p\colon \mathrm{C}_c(\Omega_p)\to\mathrm{C}_c(\widetilde{\mathrm{X}}_p) \tag{13.4.44}$$

$$\mathrm{E}_p f(g) = \mathrm{pr}(\mathrm{I}_p/\widetilde{\mathrm{X}}_p)_*\circ\mathrm{pr}(\mathrm{I}_p/\Omega_p)^* f(g) = \int_{\mathrm{G}_{\mathbb{Z}_p}} \mathrm{d}k\, f(kg)$$

$$(\mathrm{E}_p f,\varphi)_{\mathrm{L}_2(\widetilde{X}_p)} = (f,\mathrm{C}_p\varphi)_{\mathrm{L}_2(\Omega_p)}\,. \tag{13.4.45}$$

Letting $\mathrm{C}^s_p = \mathrm{L}^s\mathrm{C}_p$ denote the Radon transform followed by the Mellin transform, $\mathrm{C}^s_p\varphi\in\mathbb{H}^s_p$, and we have

$$\begin{aligned}\mathrm{C}^s_p\varphi(g) &= \int \mathrm{d}^*a\,|a|^{-(1+s)/2}\int \mathrm{d}x\,\varphi\bigg(\begin{pmatrix}a & x\\ 0 & 1\end{pmatrix}g\bigg)\\ &= \int_{\widetilde{\mathrm{X}}_p} \mathrm{d}g_0\,\varphi(g_0^{-1})\cdot\mathrm{j}_p(0{:}1,gg_0)^{1+s}\\ &= \pi(\check\varphi)\mathrm{j}^{1+s}_{\infty p}(g),\quad \check\varphi(g)=\varphi(g^{-1})\,.\end{aligned} \tag{13.4.46}$$

From this we obtain

$$\left(C_p^s\varphi_1, C_p^s\varphi_2\right)_{\mathbb{H}_p^s} = \left(\pi(\check{\varphi}_1)j_{\infty p}^{1+s}, \pi(\check{\varphi}_2)j_{\infty p}^{1+s}\right)_{\mathbb{H}_p^s} = \int_{\widetilde{X}_p}\int dg_1\, dg_2\, \varphi_1(g_1)\overline{\varphi_2(g_2)}F_p^s(g_1g_2^{-1}) \tag{13.4.47}$$

and hence

$$(\varphi_1, \varphi_2)_{L_2(\widetilde{X}_p)} = \int_{\widetilde{X}_p} dg\, \varphi_1(g)\overline{\varphi_2(g)} = \int_{\widehat{M}^+_{\mathbb{Q}_p}} d_p\left(\frac{s}{2}\right)\hat{\mu}_p(s)\left(C_p^s\varphi_1, C_p^s\varphi_2\right)_{\mathbb{H}_p^s}. \tag{13.4.48}$$

This gives the continuous decomposition of $L_2(\widetilde{X}_p)$ into $G_{\mathbb{Q}_p}$-irreducible representations

$$L_2(\widetilde{X}_p) \xrightarrow{\sim} \int_{\widehat{M}^+_{\mathbb{Q}_p}} d_p\left(\frac{s}{2}\right)\hat{\mu}_p(s)\mathbb{H}_p^s \tag{13.4.49}$$

$$\varphi \mapsto C_p^s\varphi\,.$$

We have

$$C_p^s(f * \varphi) = F_p^s(f)\cdot C_p^s\varphi, \quad f \in \mathcal{A}_p\,, \tag{13.4.50}$$

that is the action of the Hecke operators $f \in \mathcal{A}_p$ corresponds under the decomposition (13.4.49) to the multiplication by the function $\hat{f}(s) = F_p^s(f)$.

The *p*-adic theory

Let us consider the case $p \neq \eta$ in more detail. We have

$$\mathcal{A}_p = \bigoplus_{n\geq 0} \mathbb{C}\phi_{(n)} \tag{13.4.51}$$

where $\phi_{(n)}$ is the characteristic function of

$$G_{\mathbb{Z}_p}\begin{pmatrix} p^n & 0 \\ 0 & 1 \end{pmatrix}G_{\mathbb{Z}_p}$$

The action of $\phi_{(n)}$ on functions on the $(p+1)$-regular tree $\widetilde{X}_p$ is given by

$$\phi_{(n)} * \varphi(y) = \sum_{\substack{x\in X_p \\ d(x,y)=n}} \varphi(x)\,. \tag{13.4.52}$$

Thus $\phi_{(0)}$ is the identity, and $\phi_{(1)} = T_p$ is the Hecke operator.

We have the recursion relations (coming from the tree structure)

$$T_p^2 = \phi_{(2)} + (p+1), \quad T_p\phi_{(n)} = \phi_{(n+1)} + p\cdot\phi_{(n-1)}\,, \tag{13.4.53}$$

and hence the generating function

$$\sum_{n\geq 0}\phi_{(n)}\cdot z^n = \frac{1-z^2}{1-zT_p+pz^2}\,. \tag{13.4.54}$$

In particular, $\phi_{(n)}$ is a polynomial of degree n in T_p,

$$\phi_{(n)} = \mathrm{F}_n(\mathrm{T}_p)\,, \tag{13.4.55}$$

where F_n are the Mautner–Cartier polynomials $(13.4.35)_p$, and $\mathcal{A}_p = \mathbb{C}[\mathrm{T}_p]$ is a polynomial algebra generated by T_p.

Let the spectral parameter of T_p be denoted by t_p,

$$\mathrm{t}_p = \mathrm{F}_p^s(\mathrm{T}_p) = p^{(1+s)/2} + p^{(1-s)/2}\,; \tag{13.4.56}$$

then we have

$$\mathrm{d}_p\left(\frac{s}{2}\right) = \left(4p - \mathrm{t}_p^2\right)^{-1/2}\frac{\mathrm{dt}_p}{2\pi}, \quad \hat{\mu}_p(s) = (1+p)\frac{4p - \mathrm{t}_p^2}{(1+p)^2 - \mathrm{t}_p^2} \overset{\text{def}}{=} \Box_p(\mathrm{t}_p)\,, \tag{13.4.57}$$

so the Plancherel measure becomes

$$\hat{\mu}_p(s)\mathrm{d}_p\left(\frac{s}{2}\right) = \frac{\left(4p - \mathrm{t}_p^2\right)^{1/2}}{(1+p)^2 - \mathrm{t}_p^2}\frac{\mathrm{dt}_p}{2\pi}(1+p) \quad \text{on } \mathrm{t}_p \in [-2p^{-1/2}, 2p^{1/2}]\,. \tag{13.4.58}$$

The isomorphism (13.4.39) becomes

$$\begin{aligned} \mathrm{L}_2(\mathrm{M}^+_{\mathbb{Q}_p}, \mu_p) &\overset{\sim}{\leftrightarrow} \mathrm{L}_2\left([-2p^{1/2}, 2p^{1/2}], \frac{(4p - \mathrm{t}_p^2)^{1/2}}{(1+p)^2 - \mathrm{t}_p^2}\frac{\mathrm{dt}_p}{2\pi}(1+p)\right) \\ \phi_{(n)} &\leftrightarrow \mathrm{F}_n(\mathrm{t}_p)\,. \end{aligned} \tag{13.4.59}$$

Thus $\mathrm{F}_n(\mathrm{t}_p)$ are the orthogonal polynomials with respect to the measure (13.4.58), and

$$\|\mathrm{F}_n\|^2 = \|\phi_{(n)}\|^2 = \mu_p\begin{pmatrix} p^n & 0 \\ 0 & 1 \end{pmatrix} = \begin{cases} 1 & n = 0\,, \\ p^n(1+p^{-1}) & n \geq 1\,. \end{cases}$$

The Radon transform takes T_p to the operator coming from the left action of $\mathbb{Q}_p^*/\mathbb{Z}_p^*$ on Ω_p,

$$\begin{aligned} \mathrm{C}_p\mathrm{T}_p\varphi(g) &= \mathrm{C}_p\varphi\left(\begin{pmatrix} p^{-1} & 0 \\ 0 & 1 \end{pmatrix}g\right) + p\cdot\mathrm{C}_p\varphi\left(\begin{pmatrix} p & 0 \\ 0 & 1 \end{pmatrix}g\right) \\ &= p^{1/2}(\pi_\ell(p) + \pi_\ell(p)^*)\mathrm{C}_p\varphi(g)\,. \end{aligned} \tag{13.4.60}$$

We can write (13.4.48) in the form

$$\begin{aligned} (\varphi_1, \varphi_2)_{\mathrm{L}_2(\widetilde{\mathrm{X}}_p)} &= \int_{\widehat{M}^+_{\mathbb{Q}_p}} \mathrm{d}_p\left(\frac{s}{2}\right)\hat{\mu}_p(s)(\mathrm{C}_p^s\varphi_1, \mathrm{C}_p^s\varphi_2)_{\mathbb{H}_p^s} \\ &= \int_{(\mathbb{Q}_p^*/\mathbb{Z}_p^*)^\vee} \mathrm{d}_p\left(\frac{s}{2}\right)(\mathrm{L}^s\Lambda_p\mathrm{C}_p\varphi_1, \mathrm{L}^s\Lambda_p\mathrm{C}_p\varphi_2)_{\mathbb{H}_p^s} = (\Lambda_p\mathrm{C}_p\varphi_1, \Lambda_p\mathrm{C}_p\varphi_2)_{\mathrm{L}_2(\Omega_p)} \end{aligned} \tag{13.4.61}$$

so that

$$\Lambda_p \mathrm{C}_p\colon \mathrm{L}_2(\widetilde{\mathrm{X}}_p) \hookrightarrow \mathrm{L}_2(\Omega_p) \quad \text{is an isometry.} \tag{13.4.62}$$

Here Λ_p is the integral operator on $\mathrm{L}_2(\Omega_p)$ coming from the left $\mathbb{Q}_p^*/\mathbb{Z}_p^*$-action

$$\Lambda_p = (p+1)^{1/2} \cdot \frac{\pi_\ell(p) - \pi_\ell(p)^*}{p^{1/2}\pi_\ell(p) - p^{-1/2}\pi_\ell(p)^*}\,; \tag{13.4.63}$$

Λ_p is designed so that

$$\mathrm{L}^s \Lambda_p = (1+p^{-1})^{1/2}\,\frac{\zeta_p(1+s)}{\zeta_p(s)}\,\mathrm{L}^s\,.$$

We see that $(\Lambda_p \mathrm{C}_p)^*(\Lambda_p \mathrm{C}_p)$ is the identity on $\mathrm{L}_2(\widetilde{\mathrm{X}}_p)$, or equivalently

$$\mathrm{id}_{\mathrm{L}_2(\widetilde{\mathrm{X}}_p)} = \mathrm{E}_p \Lambda_p^* \Lambda_p \mathrm{C}_p = \Box_p \mathrm{E}_p \mathrm{C}_p$$

where $\Box_p$ is the integral operator on $\mathrm{L}_2(\widetilde{\mathrm{X}}_p)$ coming from the left $\mathcal{A}_p$-action and given in terms of T_p by

$$\Box_p = (p+1) \cdot \frac{4p - \mathrm{T}_p^2}{(1+p)^2 - \mathrm{T}_p^2}$$

(cf. (13.4.57)). Thus

$$\mathrm{E}_p \mathrm{C}_p = \Box_p^{-1} = \frac{1}{p+1}\,\frac{(1+p)^2 - \mathrm{T}_p^2}{4p - \mathrm{T}_p^2} = 1 + \frac{p-1}{p+1}\sum_{n\geq 1} p^{-n}\phi_{(n)}\,.$$

The Laplacian Δ_p is the operator on $\widetilde{\mathrm{X}}_p$ given by

$$\Delta_p = p + 1 - \mathrm{T}_p \tag{13.4.64}$$

and the associated spectral parameter is

$$\delta_p = \mathrm{F}_p^s(\Delta_p) = p + 1 - \mathrm{t}_p = (1 - p^{1+s/2})(1 - p^{1-s/2}) \in [(p^{1/2}-1)^2, (p^{1/2}+1)^2]\,. \tag{13.4.65}$$

The above discussion can be formulated using Δ_p instead of T_p and δ_p instead of t_p. We also have

$$\Delta_p = \partial_p^* \partial_p, \quad \ell_2(\widetilde{\mathrm{X}}_p) \underset{\partial_p^*}{\overset{\partial_p}{\rightleftarrows}} \ell_2(\widetilde{\mathrm{X}}_p^{(1)}) = \ell_2(\Gamma_0(p)\backslash \mathrm{G}_{\mathbb{Q}_p}) \tag{13.4.66}$$

where the space on the right is the 'tangent space' $\widetilde{\mathrm{X}}_p^{(1)}$, the set of edges of the $(p+1)$-regular tree $\widetilde{\mathrm{X}}_p$, $\Gamma_0(p) = \{g \in \mathrm{G}_{\mathbb{Z}_p} \,|\, g \equiv \begin{pmatrix} * & * \\ 0 & * \end{pmatrix} \bmod p\}$ is the stabilizer of an edge, and ∂_p, ∂_p^* are given by

$$\begin{aligned} \partial_p \varphi(x,y) &= \varphi(x) - \varphi(y)\,, \\ \partial_p^* \varphi(x) &= \sum_{\substack{y \\ (x,y)\in \widetilde{\mathrm{X}}_p^{(1)}}} \varphi(x,y)\,. \end{aligned}$$

Translating Δ_p to Ω_p using C_p shows that it is a product of the 'infinitesimal' generator of the left $p^{\mathbb{Z}}$ multiplicative action and its adjoint, explaining the factorization of δ_p in (13.4.65). We have

$$C_p \Delta_p = D_p D_p^* C_p \tag{13.4.67}$$

with

$$D_p = (1 - p^{1/2}\pi_\ell(p)), \quad L^s D_p = (1 - p^{(1-s)/2}) L^s . \tag{13.4.68}$$

The real theory

For the real prime $p = \eta$, the commutative algebra $C_c(G_{\mathbb{Z}_\eta} \backslash G_{\mathbb{Q}_\eta} / G_{\mathbb{Z}_\eta}) \simeq C_c[1, \infty)$ is not 'algebraic' enough—it is far from being a polynomial algebra. The classical approach is to replace it by the polynomial algebra $\mathcal{A}_\eta = \mathbb{C}[\Delta_\eta] =$ centre $U(\mathrm{sl}_2)$, where Δ_η is the hyperbolic Laplacian given by the action of the Casimir element. The algebra $\mathcal{A}_\eta$ acts by unbounded operators on $L_2(\widetilde{X}_\eta)$ commuting with the $G_{\mathbb{Q}_\eta}$-action. The action of Δ_η is given by

$$\Delta_\eta \varphi \begin{pmatrix} a & x \\ 0 & 1 \end{pmatrix} = -a^2 \left(\frac{\partial^2}{\partial a^2} + \frac{\partial^2}{\partial x^2} \right) \varphi \begin{pmatrix} a & x \\ 0 & 1 \end{pmatrix} . \tag*{(13.4.64)$_\eta$}$$

As the spectral parameter we take, with $s = \mathrm{i}\tau \in \mathrm{i}\mathbb{R}$,

$$\delta_\eta = F_\eta^s(\Delta_\eta) = \left(\frac{1+s}{2} \right) \cdot \left(\frac{1-s}{2} \right) = \frac{1+\tau^2}{4} \in \left[\frac{1}{4}, \infty \right) . \tag*{(13.4.65)$_\eta$}$$

The Plancherel measure is

$$\hat{\mu}_\eta(s) \mathrm{d}_\eta \left(\frac{s}{2} \right) = \left| \frac{\Gamma((1+s)/2)}{\Gamma(s/2)} \right|^2 \frac{\mathrm{d}s}{4\pi \mathrm{i}} = \frac{(\mathrm{e}^{(\pi/2)\tau} - \mathrm{e}^{-(\pi/2)\tau})}{(\mathrm{e}^{(\pi/2)\tau} + \mathrm{e}^{(\pi/2)\tau})} \frac{\tau \mathrm{d}\tau}{8\pi} = \Box_\eta(\delta_\eta) \mathrm{d}_\eta s . \tag*{(13.4.58)$_\eta$}$$

The left $\mathbb{R}^+$-action on Ω_η gives rise to the infinitesimal generator, commuting with the $G_{\mathbb{Q}_\eta}$-action,

$$D_\eta f(g) = \frac{\partial}{\partial a} \bigg|_{a=1} f \left(\begin{pmatrix} a & 0 \\ 0 & 1 \end{pmatrix} g \right) , \tag*{(13.4.68)$_\eta$}$$

and its adjoint

$$D_\eta^* f(g) = \frac{\partial}{\partial a} \bigg|_{a=1} |a| f \left(\begin{pmatrix} a^{-1} & 0 \\ 0 & 1 \end{pmatrix} g \right) ,$$

and we have

$$L^s D_\eta = \left(\frac{1+s}{2} \right) \cdot L^s, \quad L^s D_\eta^* = \left(\frac{1-s}{2} \right) \cdot L^s .$$

The Radon transform takes Δ_η into the product of these operators

$$C_\eta \Delta_\eta = D_\eta D_\eta^* C_\eta . \tag*{(13.4.67)$_\eta$}$$

We have an integral operator Λ_η on $\mathrm{L}_2(\Omega_\eta)$ coming from the left $\mathbb{R}^+$-action such that $\Lambda_\eta \mathrm{C}_\eta$ is an isometry. Λ_η is given by

$$\Lambda_\eta = \frac{\Gamma(\mathrm{D}_\eta)}{\Gamma(\mathrm{D}_\eta - 1/2)} \quad \text{or} \quad \mathrm{L}^s \Lambda_\eta = \left|\frac{\zeta_\eta(1)}{\zeta_\eta(2)}\right|^{1/2} \cdot \frac{\zeta_\eta(1+s)}{\zeta_\eta(s)} \mathrm{L}^s . \tag{13.4.63$_\eta$}$$

We can write

$$\mathrm{id}_{\mathrm{L}_2(\widetilde{\mathrm{X}}_\eta)} = \mathrm{E}_\eta \Lambda_\eta^* \Lambda_\eta \mathrm{C}_\eta = \Box_\eta \mathrm{E}_\eta \mathrm{C}_\eta$$

with $\Box_\eta = \Box_\eta(\Delta_\eta)$ the operator on $\mathrm{L}_2(\widetilde{\mathrm{X}}_\eta)$ given as a function of Δ_η by the function $\Box_\eta$ of $(13.4.58)_\eta$.

Similarly we have

$$\Delta_\eta = \partial_\eta^* \partial_\eta \tag{13.4.66$_\eta$}$$

where ∂_η denotes the ordinary gradient and ∂_η^* the divergent.

The q-theory

We see that the p-adic theory, when formulated using the (bounded) Laplacian Δ_p, is completely analogous to the real theory of the (unbounded) Laplacian Δ_η. There is a q-analogue interpolating between the p-adic and real theories. We replace the Plancherel measure (13.4.38) by its q-analogue

$$\hat{\mu}_q^\beta(s)\mathrm{d}_q\left(\frac{s}{2}\right) = \frac{\zeta_{(q)}(\beta)\cdot\zeta_{(q)}(\beta+1)}{\zeta_{(q)}(2\beta)\cdot\zeta_{(q)}(1)} \cdot \left|\frac{\zeta_{(q)}(\beta+s)}{\zeta_{(q)}(s)}\right|^2 \cdot \mathrm{d}_q\left(\frac{s}{2}\right). \tag{13.4.38$_q$}$$

This is an Askey–Wilson weight function (cf. (10.8.3)) with parameters

$$\{\alpha_1,\alpha_2,\alpha_3,\alpha_4\} = \left\{\frac{\beta}{2}, \frac{\beta}{2}+\frac{1}{2}, \frac{\beta}{2}+\frac{\pi \mathrm{i}}{\log q}, \frac{\beta}{2}+\frac{1}{2}+\frac{\pi\mathrm{i}}{\log q}\right\}. \tag{13.4.69}$$

Considering the elliptic curve $E_q = \mathbb{C}^*/q^{\mathbb{Z}}$, and $[\beta]_{E_q} \in E_q$, we have the natural exact sequence

$$* \to E_q[2] \to E_q \xrightarrow{\{2\}} E_q \to *, \quad \{2\}[\alpha]_{E_q} = [2\alpha]_{E_q}, \tag{13.4.70}$$

and the parameters $\{\alpha_j\}$ are related to β by

$$\{[\alpha_1]_{E_q}, \ldots, [\alpha_4]_{E_q}\} = \{2\}^{-1}[\beta]_{E_q}, \tag{13.4.71}$$

the fiber over $[\beta]_{E_q}$. The associated orthogonal polynomials (10.8.9) and (10.8.11) are the polynomials considered by Rogers, Askey and Ismail, and Macdonald. We shall normalize them and denote them by

$$\begin{aligned} \mathrm{F}_q^\beta(g^n, s) &= \mathrm{F}_n(s) = \mathrm{F}^s(g^n) \\ &:= {}_4\phi_3\left(-n, n+2\beta, \frac{\beta+s}{2}, \frac{\beta-s}{2}; \beta+\frac{1}{2}, \beta+\frac{\pi\mathrm{i}}{\log q}, \beta+\frac{1}{2}+\frac{\pi\mathrm{i}}{\log q}; 1\right). \end{aligned} \tag{13.4.35$_q$}$$

Their norms are given by

$$\mu_q^\beta(g^n) \overset{\text{def}}{=} \| \mathrm{F}_n \|^{-2} = \frac{\zeta_{(q)}(1)}{\zeta_{(q)}(1+n)} \cdot \frac{\zeta_{(q)}(2\beta+n)}{\zeta_{(q)}(2\beta)} \cdot \frac{1-q^{\beta+n}}{1-q^\beta} \cdot q^{-n\beta} \qquad (13.4.36)_q$$

We obtain an isomorphism of Hilbert spaces given by the kernel $\mathrm{F}_q^\beta(g^n, s)$

$$\ell_2(g^{\mathbb{N}}, \mu_q^\beta) \underset{\sim}{\overset{\mathrm{F}_q^\beta}{\longleftrightarrow}} \widehat{\mathscr{H}}^+_{(\alpha_i)} = \mathrm{L}_2\left(\mathrm{i}\mathbb{R}\Big/\frac{2\pi\mathrm{i}}{\log q}\mathbb{Z}, \hat{\mu}_q^\beta(2\mathrm{i}t)\mathrm{d}t \cdot \frac{\log q^{-1}}{2\pi}\right)^+ \qquad (13.4.39)_q$$

$$\phi_{(n)} = \delta_{g^n} \to \|\mathrm{F}_n\|^{-2} \cdot \mathrm{F}_n(\mathrm{i}t)\,.$$

The F_n satisfy the recursion relation

$$(q^{s/2} + q^{-s/2})\mathrm{F}_n = \frac{q^{\beta/2}(1-q^n)}{(1-q^{\beta+n})}\mathrm{F}_{n-1} + \frac{q^{-\beta/2}(1-q^{2\beta+n})}{(1-q^{\beta+n})}\mathrm{F}_{n+1}\,, \qquad (13.4.72)$$

or equivalently

$$\begin{aligned}&\left(1-q^{-[(\beta+s)/2]}\right)\left(1-q^{-[(\beta-s)/2]}\right) \cdot \mathrm{F}_n \\ &\quad = \left(1+q^{-\beta}\right)\mathrm{F}_n - \frac{1-q^n}{1-q^{\beta+n}}\mathrm{F}_{n-1} - \frac{q^{-\beta}(1-q^{2\beta+n})}{1-q^{\beta+n}}\mathrm{F}_{n+1}\,.\end{aligned} \qquad (13.4.73)$$

Introducing the q-Laplacian

$$\Delta_q\varphi(g^n) = (1+q^{-\beta})\varphi(g^n) - \frac{1-q^n}{1-q^{\beta+n}}\varphi(g^{n-1}) - \frac{q^{-\beta}(1-q^{2\beta+n})}{1-q^{\beta+n}}\varphi(g^{n+1})\,, \qquad (13.4.64)_q$$

this can be written as

$$(1-q^{-[(\beta+s)/2]})(1-q^{-[(\beta-s)/2]})\mathrm{F}^s = \Delta_q \mathrm{F}^s\,. \qquad (13.4.65)_q$$

The q-Laplacian $\Delta_q = \partial^*\partial$ is given in terms of the 'gradient' ∂ and its adjoint ∂^*, where

$$\ell_2\left(g^{\mathbb{N}}, \mu_q^\beta\right) \underset{\partial^*}{\overset{\partial}{\rightleftarrows}} \ell_2\left(g^{\mathbb{N}}, v_q^\beta\right), \qquad (13.4.66)_q$$

and the space on the right is the 'tangent' space, with

$$\begin{aligned}v_q^\beta(g^n) &= \frac{1-q^{1+n}}{1-q^{\beta+1+n}}\mu_q^\beta(g^{n+1}) = \frac{\zeta_{(q)}(1)}{\zeta_{(q)}(1+n)}\frac{\zeta_{(q)}(2\beta+n+1)}{\zeta_{(q)}(2\beta)} \cdot \frac{q^{-(n+1)\beta}}{1-q^\beta}\,, \\ \partial\varphi(g^n) &= \varphi(g^n) - \varphi(g^{n+1})\,, \\ \partial^*\varphi(g^n) &= \frac{q^{-\beta}(1-q^{2\beta+n})}{1-q^{\beta+n}}\varphi(g^n) - \frac{1-q^n}{1-q^{\beta+n}}\varphi(g^{n-1})\,.\end{aligned} \qquad (13.4.74)$$

In the p-adic limit Ⓟ:

$$q := p^{-\ell}, \quad \beta := \beta/\ell, \quad s := s/\ell, \quad \ell \to \infty, \quad \text{and} \quad g^n := p^n,$$

$\hat{\mu}_q^\beta$, μ_q^β, v_q^β, Δ_q, ∂^*, ∂, $\mathrm{F}_q^\beta(g^n, \mathrm{s})$ converge to their p-adic counterparts for $\beta = 1$ (and for $\beta = m$ we obtain the theory for the $(p^m + 1)$-regular tree).

In the real limit Ⓝ:

$$q := q^2, \quad \beta := \beta/2, \quad s := s/2, \quad \text{and then } q \to 1, \quad \text{and} \quad n \to \infty$$

(where n is the parameter in $g^n \in g^{\mathbb{N}}$) in such a way that

$$q^n \to a \in (0, 1]\,, \tag{13.4.75}$$

we obtain the real theory for $\beta = 1$ (and a generalization of it for $\beta > 0$).

Note also that if we fix n, and instead vary the parameter $s = i\tau$, defining the compact real limit $Ⓝ_{\text{compact}}$:

$$q := q^2, \quad \beta := \frac{\beta}{2}, \quad \text{and then } q \to 1, \tau \to \infty \text{ in such a way that}$$

$$\frac{1}{2}(q^s + q^{-s}) = \cos(\log q) \cdot \tau \to \cos\theta\,, \tag{13.4.76}$$

our $\mathrm{F}_n = \mathrm{F}_n\big(\frac{1}{2}(q^s + q^{-s})\big)$ converge to $\mathrm{F}_n(\cos\theta)$, the Legendre polynomials (i.e. the Jacobi polynomials with $\alpha = \beta = 2$) which are the zonal spherical functions on $\mathrm{SO}_2\backslash\mathrm{SO}_3/\mathrm{SO}_2$ associated with the *compact* form of SL(2). See Macdonald (1972), and Freund and Zabrodin (1992).

The global theory

Let us now turn to the global theory. The diagram of symmetric spaces is similar to that of (13.4.1):

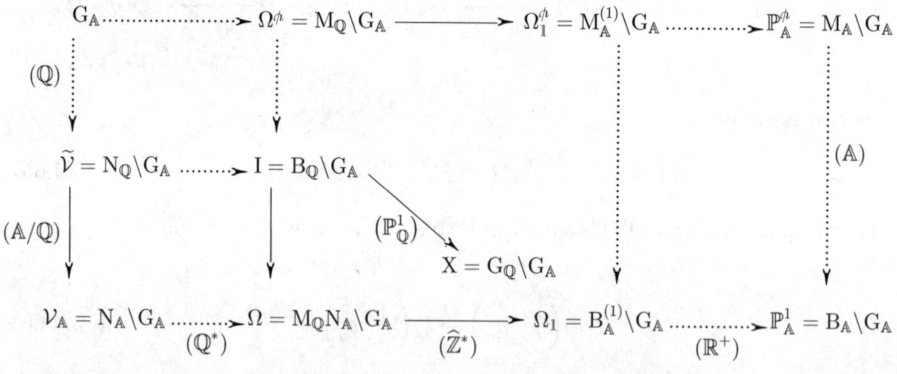

(13.4.77)

In addition we have symmetric spaces $\mathrm{Y}_{\mathbb{A}}$ formed by the (restricted) product of the local symmetric spaces Y_p, so the picture is even richer, cf. Fig. 13.1. All the symmetric spaces except $\mathbb{P}^1_{\mathbb{A}}$ have a $\mathrm{G}_{\mathbb{A}}$-invariant measure. Hence we obtain the unitary representation of $\mathrm{G}_{\mathbb{A}}$ by right translation on the associated L_2-spaces, and we have adjoint maps $\mathrm{pr}(\mathrm{Y}_1/\mathrm{Y}_2)^*$, $\mathrm{pr}(\mathrm{Y}_1/\mathrm{Y}_2)_*$ commuting with $\mathrm{G}_{\mathbb{A}}$-action when $\mathrm{Y}_1 \to \mathrm{Y}_2$.

To explain our notation, note that we have, for example,

$$\mathbb{P}^{\mathcal{A}}_{\mathbb{A}} = \prod_{p \geq \eta} \mathbb{P}^{\mathcal{A}}_{\mathbb{Q}_p} : \mathbb{P}^{\mathcal{A}}_{\mathbb{Z}_p} \tag{13.4.78}$$

with

$$\mathbb{P}^{\pitchfork}_{\mathbb{Q}_p} \cong \left\{(\ell,\ell') \in \mathbb{P}^1_{\mathbb{Q}_p} \times \mathbb{P}^1_{\mathbb{Q}_p} \,\middle|\, \ell \neq \ell'\right\} \quad \text{via } \mathrm{M}_{\mathbb{A}} g \leftrightarrow ((0{:}1)g, (1{:}0)g)\,, \tag{13.4.79}$$

inducing

$$\mathbb{P}^{\pitchfork}_{\mathbb{Z}_p} \cong \left\{(\ell,\ell') \in \mathbb{P}^1_{\mathbb{Q}_p} \times \mathbb{P}^1_{\mathbb{Q}_p} \,\middle|\, \rho_p(\ell,\ell') = 1\right\},$$

so that

$$\mathbb{P}^{\pitchfork}_{\mathbb{A}} \cong \left\{(\ell,\ell') \in \mathbb{P}^1_{\mathbb{A}} \times \mathbb{P}^1_{\mathbb{A}} \,\middle|\, \rho_{\mathbb{A}}(\ell,\ell') \stackrel{\text{def}}{=} \prod_{p\geq\eta} \rho_p(\ell_p,\ell'_p) \neq 0\right\} \tag{13.4.80}$$

is the space of pairs of 'transversal' $\mathbb{A}$-lines in $V(\mathbb{A})$.

The group $\mathbb{Q}^*\backslash\mathbb{A}^*$ (respectively $\mathbb{Z}^*_{\mathbb{A}}\backslash\mathbb{A}^* = \mathbb{Q}^+ \times \mathbb{R}^+$) acts on the left on Ω (respectively $\Omega_{\mathbb{A}} = \mathrm{M}_{\mathbb{Z}_{\mathbb{A}}}\mathrm{N}_{\mathbb{A}}\backslash\mathrm{G}_{\mathbb{A}}$). The Mellin transform gives the $\mathrm{G}_{\mathbb{A}}$-decomposition

$$\mathrm{L}_2(\Omega) = \int_{(\mathbb{Q}^*\backslash\mathbb{A}^*)^{\vee}} \mathrm{d}\chi\, \mathbb{H}^{\chi}_{\mathbb{A}} \quad \left(\text{respectively } \mathrm{L}_2(\Omega_{\mathbb{A}}) = \int_{\left(\mathbb{Z}^*_{\mathbb{A}}\backslash\mathbb{A}^*\right)^{\vee}} \mathrm{d}\chi\, \mathbb{H}^{\chi}_{\mathbb{A}}\right) \tag{13.4.81}$$

$$f \mapsto \mathrm{L}^{\chi} f,\ \mathrm{L}^{\chi} f(g) = \int \mathrm{d}^* a\, \chi(a) |a|^{1/2} f\left(\begin{pmatrix} a^{-1} & 0 \\ 0 & 1 \end{pmatrix} g\right),$$

where $\mathrm{d}\chi$ is the Haar measure dual to the multiplicative measure $\mathrm{d}^* a$, and

$$\begin{aligned} \mathbb{H}^{\chi}_{\mathbb{A}} &= \mathrm{Ind}^{\mathrm{G}_{\mathbb{A}}}_{\mathrm{B}_{\mathbb{A}}}(\chi) = \bigotimes_{p\geq\eta} \mathbb{H}^{\chi_p}_p \\ &= \left\{f\colon \Omega \to \mathbb{C} \,\middle|\, f\left(\begin{pmatrix} a & x \\ 0 & 1 \end{pmatrix} g\right) = |a|^{1/2}\chi(a)\cdot f(g), f|_{\mathrm{G}_{\mathbb{Z}_{\mathbb{A}}}} \in \mathrm{L}_2(\mathrm{G}_{\mathbb{Z}_{\mathbb{A}}})\right\}. \end{aligned} \tag{13.4.82}$$

Note that we can write $\chi \in (\mathbb{Q}^*\backslash\mathbb{A}^*)^V$ (respectively $\chi \in (\mathbb{Z}^*_{\mathbb{A}}\backslash\mathbb{A}^*)^V$) as

$$\chi(a) \doteq \chi_0(a)\cdot |a|^s_{\mathbb{A}}, \quad s \in \mathbb{C},\ \chi_0 \text{ a finite character} \tag{13.4.83}$$

$$\left(\text{respectively } \chi(a) = |a|^{\underline{s}} = \prod_{p\geq\eta} |a_p|^{s_p}_p,\ s = (s_p) \in \mathbb{C} \times \prod_{p\neq\eta} \mathbb{C}\Big/\frac{2\pi\mathrm{i}}{\log p}\mathbb{Z}\right).$$

$\mathrm{H}^{\chi}_{\mathbb{A}}$ is unitary if and only if $s \in \mathrm{i}\mathbb{R}$ (respectively $s_p \in \mathrm{i}\mathbb{R}$ for all $p \geq \eta$), and $\mathbb{H}^{\chi}_{\mathbb{A}}$ is equivalent to $\mathbb{H}^{\chi^{-1}}_{\mathbb{A}}$.

The space $\Omega_1 = \mathrm{B}^{(1)}_{\mathbb{A}}\backslash\mathrm{G}_{\mathbb{A}}$ corresponds to the part of Ω with $\chi_0 = \mathbb{1}$, the trivial character (respectively the part of $\Omega_{\mathbb{A}}$ with $s_p = s \in \mathbb{C}$ for all $p \geq \eta$)

$$\mathrm{L}_2(\Omega_1) = \int_{-\mathrm{i}\infty}^{\mathrm{i}\infty} \frac{\mathrm{d}s}{2\pi\mathrm{i}}\, \mathbb{H}^s_{\mathbb{A}}, \quad \mathbb{H}^s_{\mathbb{A}} = \bigotimes_{p\geq\eta} \mathbb{H}^s_p, \quad f \mapsto \mathrm{L}^s f\,. \tag{13.4.84}$$

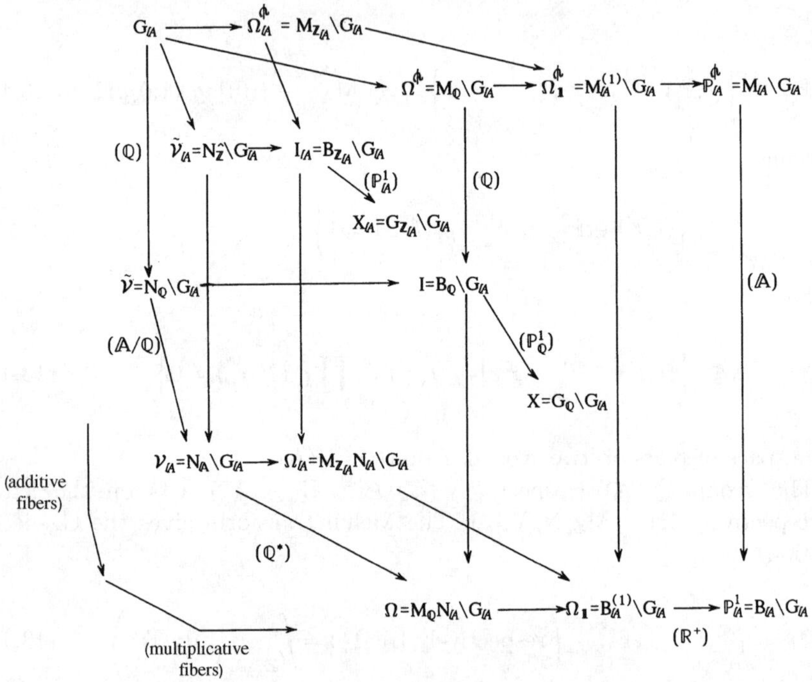

FIG. 13.1 Symmetric spaces for $\mathrm{G}_{\mathbb{A}} = \mathrm{PGL}_2(\mathbb{A})$.

We have the global automorphic factor

$$\mathrm{j}_{\mathbb{A}}\colon \mathbb{P}^1_{\mathbb{A}} \times \mathrm{G}_{\mathbb{A}}/\mathrm{G}_{\mathbb{Z}_{\mathbb{A}}} \to \mathbb{R}^+, \quad \mathrm{j}_{\mathbb{A}}(z,g) = \prod_{p\geq\eta} \mathrm{j}_p(z_p, g_p)\,,$$

satisfying the global analogues of (13.4.4–13.4.6).

We can trivialize the field of Hilbert space $\mathbb{H}^s_{\mathbb{A}}$ via

$$\begin{aligned} \mathbb{H}^s_{\mathbb{A}} \overset{\sim}{\leftrightarrow} \widetilde{\mathscr{H}}^{1,1}_{\mathbb{A}} &= \mathrm{L}_2\left(\mathbb{P}^1_{\mathbb{A}}, \tau^{1,1}_{\mathbb{A}}\right) \qquad (13.4.85)\\ f \mapsto f|_{\mathrm{G}_{\mathbb{Z}_{\mathbb{A}}}} &= \mathrm{i}_s f \\ \mathrm{i}_s^{-1} f(g) = f((0{:}1)g)\mathrm{j}_\infty^{1+s}(g) &\leftarrow\!\shortmid f \end{aligned}$$

and we have the representations $\pi^s_{\mathbb{A}}$ of $\mathrm{G}_{\mathbb{A}}$ on $\widetilde{\mathscr{H}}^{1,1}_{\mathbb{A}}$ (by invertible operators).

Let us turn to the study of the decomposition of $\mathrm{L}_2(\mathrm{X})$. The constant term of a function φ on X is defined by

$$\mathrm{C}\varphi(g) = \mathrm{pr}(\mathrm{I}/\Omega)_* \circ \mathrm{pr}(\mathrm{I}/\mathrm{X})^* \varphi(g) = \int_{\mathbb{Q}\backslash\mathbb{A}} \mathrm{d}x\, \varphi\left(\begin{pmatrix} 1 & x \\ 0 & 1 \end{pmatrix} g\right). \qquad (13.4.86)$$

The most interesting part of $\mathrm{L}_2(\mathrm{X})$ is the space of cusp forms

$$\mathrm{L}_2(\mathrm{X})_{\mathrm{cusp}} = \mathrm{L}_2(\mathrm{X}) \cap \ker \mathrm{C}\,. \qquad (13.4.87)$$

For $f \in C_c^\infty(G_\mathbb{A})$ one shows that $\pi(f)|_{L_2(X)_{\text{cusp}}}$ is Hilbert–Schmidt (since we can subtract from the kernel of $\pi(f)$ its constant term). It follows that $L_2(X)_{\text{cusp}}$ decomposes discretely with finite multiplicity (actually, multiplicity one), so

$$L_2(X)_{\text{cusp}} = \bigoplus_j \mathscr{H}_{\pi_j}, \tag{13.4.88}$$

where π_j are non-equivalent irreducible unitary representations of $G_\mathbb{A}$. We shall focus our attention on the orthogonal complement $L_2(X)_{\text{cusp}}^\perp$.

The Eisenstein series of a function $f \in C_c(\Omega)$ is given by the formal adjoint of C

$$Ef(g) = C^* f(g) = \mathrm{pr}(I/X)_* \circ \mathrm{pr}(I/\Omega)^* f(g) = \sum_{\gamma \in B_\mathbb{Q}\backslash G_\mathbb{Q}} f(\gamma g). \tag{13.4.89}$$

The summation in (13.4.89) is actually a finite sum for f compactly supported, and

$$E\colon C_c(\Omega) \to C_c(X). \tag{13.4.90}$$

E and C are formal adjoints,

$$(\varphi, Ef)_{L_2(X)} = (C\varphi, f)_{L_2(\Omega)}, \quad \varphi \in C_c(X),\ f \in C_c(\Omega), \tag{13.4.91}$$

and it follows that $L_2(X)_{\text{cusp}}^\perp$ is the closure of $E[C_c(\Omega)]$.

Since $\mathbb{P}^1_\mathbb{Q} = \{\infty\} \amalg \mathbb{Q}$, we have Bruhat's decomposition

$$G_\mathbb{Q} = B_\mathbb{Q} \amalg B_\mathbb{Q} w N_\mathbb{Q}, \quad w = \begin{pmatrix} 0 & -1 \\ 1 & 0 \end{pmatrix}, \tag{13.4.92}$$

and hence

$$\begin{aligned} \mathrm{pr}(I/X)^* \circ \mathrm{pr}(I/X)_* f(g) &= \sum_{\gamma \in B_\mathbb{Q}\backslash G_\mathbb{Q}} f(\gamma g) = f(g) + \sum_{q \in \mathbb{Q}} f\left(w\begin{pmatrix} 1 & q \\ 0 & 1 \end{pmatrix} g\right) \\ &= (1 + M_\mathbb{Q}) f(g) \end{aligned} \tag{13.4.93}$$

where $M_\mathbb{Q} f(g) = \mathrm{pr}(\Omega^\pitchfork/I)_* \circ W \circ \mathrm{pr}(\Omega^\pitchfork/I)^* f(g)$.

Putting $\mathrm{pr}(I/\Omega)^* f$ instead of f in the identity (13.4.93), and then acting by $\mathrm{pr}(I/\Omega)_*$, we obtain

$$CE = 1 + M_\mathbb{A}, \tag{13.4.94}$$

where

$$\begin{aligned} M_\mathbb{A} f(g) &= \mathrm{pr}(I/\Omega)_* \circ M_\mathbb{Q} \circ \mathrm{pr}(I/\Omega)^* f(g) = \mathrm{pr}(\Omega^\pitchfork/\Omega)_* \circ W \circ \mathrm{pr}(\Omega^\pitchfork/\Omega)^* f(g) \\ &= \int_\mathbb{A} dx\, f\left(w\begin{pmatrix} 1 & x \\ 0 & 1 \end{pmatrix} g\right). \end{aligned}$$

and $M_\mathbb{A} = \otimes_{p \geq \eta} M_p$ where M_p is the local operator of (13.4.29).

We shall concentrate on the contribution to $L_2(X)_{\text{cusp}}^\perp$ coming from $E[C_c(\Omega_1)]$, corresponding to the trivial character $\chi_0 = \mathbb{1}$; the case of general χ is similar (involving Dirichlet L-functions rather than the zeta function).

Let $f_1, f_2 \in C_c(\Omega_1)$; then from (13.4.94) and Mellin inversion we obtain

$$(\mathrm{E}f_1, \mathrm{E}f_2)_{\mathrm{L}_2(\mathrm{X})} = (\mathrm{CE}f_1, f_2)_{\mathrm{L}_2(\Omega_1)} = (f_1, f_2)_{\mathrm{L}_2(\Omega_1)} + (\mathrm{M}_{\mathbb{A}} f_1, f_2)_{\mathrm{L}_2(\Omega_1)}$$
$$= \int_{-\mathrm{i}\infty}^{\mathrm{i}\infty} \frac{\mathrm{d}s}{4\pi \mathrm{i}} (\mathrm{L}^s f_1, \mathrm{L}^{-\bar{s}} f_2)_{\mathscr{H}^s_{\mathbb{A}}} + \int_{\sigma-\mathrm{i}\infty}^{\sigma+\mathrm{i}\infty} \frac{\mathrm{d}s}{4\pi \mathrm{i}} (\mathrm{M}_{\mathbb{A}} \mathrm{L}^s f_1, \mathrm{L}^{\bar{s}} f_2)_{\mathbb{H}^{-s}_{\mathbb{A}}}, \ \sigma > 1. \tag{13.4.95}$$

Note that $\mathrm{M}_{\mathbb{A}}$ is defined on $\mathbb{H}^s_{\mathbb{A}}$ for $\mathrm{Re}(s) > 1$, and $\mathrm{M}_{\mathbb{A}}\colon \mathbb{H}^s_{\mathbb{A}} \to \mathbb{H}^{-s}_{\mathbb{A}}$.

Denoting by $\mathrm{M}^{(s)}_{\mathbb{A}} = \mathrm{i}_{-s} \mathrm{M}_{\mathbb{A}} \mathrm{i}_s^{-1}$ the associated family of operators on $\mathscr{H}^{1,1}_{\mathbb{A}}$, we have for $\mathrm{Re}(s) > 1$ from (13.4.30),

$$\mathrm{M}^{(s)}_{\mathbb{A}} = \bigotimes_{p \geq \eta} \mathrm{M}^{(s)}_p = \bigotimes_{p \geq \eta} \frac{\zeta_p(s)}{\zeta_p(1+s)} \widetilde{\mathrm{R}}^s_p = \frac{\zeta_{\mathbb{A}}(s)}{\zeta_{\mathbb{A}}(1+s)} \widetilde{\mathrm{R}}^s_{\mathbb{A}} = \frac{\zeta_{\mathbb{A}}(1-s)}{\zeta_{\mathbb{A}}(1+s)} \widetilde{\mathrm{R}}^s_{\mathbb{A}}. \tag{13.4.96}$$

We deduce that $\mathrm{M}^{(s)}_{\mathbb{A}}$ has meromorphic continuation to all s, with a unique pole in $\mathrm{Re}(s) \geq 0$ at $s = 1$; this pole is simple with residue (cf. (13.4.17))

$$\begin{aligned}\operatorname*{Res}_{s=1}(\mathrm{M}_{\mathbb{A}} \mathrm{L}^s f_1, \mathrm{L}^{\bar{s}} f_2)_{\mathbb{H}^{-s}_{\mathbb{A}}} &= \frac{1}{\zeta_{\mathbb{A}}(2)} (\widetilde{\mathrm{R}}^1_{\mathbb{A}} \mathrm{i}_1 \mathrm{L}^1 f_1, \mathrm{i}_1 \mathrm{L}^1 f_2)_{\mathscr{H}^{1,1}_{\mathbb{A}}} \\ &= \frac{1}{\zeta_{\mathbb{A}}(2)} (\mathrm{L}^1 f_1, \mathbb{1})_{\mathscr{H}^{1,1}_{\mathbb{A}}} \cdot (\mathbb{1}, \mathrm{L}^1 f_2)_{\mathscr{H}^{1,1}_{\mathbb{A}}} \\ &= \frac{1}{\zeta_{\mathbb{A}}(2)} \tau^{1,1}_{\mathbb{A}}(\mathrm{L}^1 f_1) \cdot \tau^{1,1}_{\mathbb{A}}(\mathrm{L}^1 \bar{f}_2) = \frac{1}{\zeta_{\mathbb{A}}(2)} (\mathrm{E}f_1, \mathbb{1}) \cdot (\mathbb{1}, \mathrm{E}f_2).\end{aligned} \tag{13.4.97}$$

We can now shift the integral in (13.4.95) to $\sigma = 0$ and write

$$\begin{aligned}(\mathrm{E}f_1, \mathrm{E}f_2)_{\mathrm{L}_2(\mathrm{X})} &= \int_{-\infty}^{\infty} \frac{\mathrm{d}t}{4\pi} \{(\mathrm{L}^{\mathrm{i}t} f_1, \mathrm{L}^{\mathrm{i}t} f_2) + (\mathrm{M}^{(\mathrm{i}t)}_{\mathbb{A}} \mathrm{L}^{\mathrm{i}t} f_1, \mathrm{L}^{-\mathrm{i}t} f_2)\} \\ &\quad + \frac{1}{\zeta_{\mathbb{A}}(2)} (\mathrm{E}f_1, \mathbb{1}) \cdot (\mathbb{1}, \mathrm{E}f_2).\end{aligned} \tag{13.4.98}$$

Note that by (13.4.96) $\mathrm{M}^{(s)}_{\mathbb{A}}$ is unitary for $s \in \mathrm{i}\mathbb{R}$, and satisfies the functional equation

$$\mathrm{M}^{(-s)}_{\mathbb{A}} \mathrm{M}^{(s)}_{\mathbb{A}} = \mathrm{id}_{\mathscr{H}^{1,1}_{\mathbb{A}}}. \tag{13.4.99}$$

We deduce that the contribution of $\mathrm{E}[\mathrm{C}_c(\Omega_1)]$ to $\mathrm{L}_2(\mathrm{X})^{\perp}_{\mathrm{cusp}}$ is

$$\frac{1}{2}(1 + \mathrm{M}_{\mathbb{A}}) \mathrm{L}_2(\Omega_1) \oplus \mathbb{C}\mathbb{1} = \left\{ f = f(s) \in \int_{-\mathrm{i}\infty}^{\mathrm{i}\infty} \frac{\mathrm{d}s}{4\pi \mathrm{i}} \mathbb{H}^s_{\mathbb{A}} \,\middle|\, f(-s) = \mathrm{M}_{\mathbb{A}} f(s) \right\} \oplus \mathbb{C}\mathbb{1}. \tag{13.4.100}$$

That is, it gives the continuous sum of the principal series representatives $\mathbb{H}^s_{\mathbb{A}}$ with multiplicity one, and the one-dimensional trivial representation. Similarly for general Dirichlet character χ_0 we obtain a contribution to $\mathrm{L}_2(\mathrm{X})^{\perp}_{\mathrm{cusp}}$ which is the continuous sum of the principal series representation $\mathbb{H}_{\mathbb{A}}^{\chi_0\|^s}$ with multiplicity one, and a residual contribution when $\chi_0^2 = \mathbb{1}$ given by the one-dimensional representation $\chi_0 \circ \det\colon \mathrm{G}_{\mathbb{A}} \to \pm 1$. In summary:

$$
\begin{aligned}
\mathrm{L}_2(\mathrm{X}) &= \mathrm{L}_2(\mathrm{X})_{\text{cusp}} \oplus \mathrm{L}_2(\mathrm{X})_{\text{cont}} \oplus \mathrm{L}_2(\mathrm{X})_{\text{res}} \\
\mathrm{L}_2(\mathrm{X})_{\text{cont}} &\simeq \frac{1}{2}(1+\mathrm{M}_{\mathbb{A}})\mathrm{L}_2(\Omega) \simeq \int_{(\mathbb{Q}^*\backslash\mathbb{A}^*)^\vee/\chi\sim\chi^{-1}} \mathrm{d}\chi\, \mathbb{H}_{\mathbb{A}}^{\chi} \\
\mathrm{L}_2(\mathrm{X})_{\text{res}} &\simeq \bigoplus_{\chi^2=\mathbb{1}} \mathbb{C}(\chi\circ\det)\,.
\end{aligned}
\tag{13.4.101}
$$

13.5 THE RIESZ POTENTIAL CONNECTION

The Riesz potential R_p^s is the operator acting on functions on $\mathbb{Q}_p$, $p \geq \eta$, given by

$$\mathrm{R}_p^s\varphi = \mathcal{F}|x|_p^{-s}\mathcal{F}^{-1}\varphi\,, \tag{13.5.1}$$

where $\mathcal{F} = \mathcal{F}_p^1$ is the Fourier transform. Alternatively, its action on a function φ is given by convolution with the kernel

$$\mathrm{R}_p^s(x) = \mathcal{F}|x|_p^{-s} = \frac{\zeta_p(1-s)}{\zeta_p(s)}|x|_p^{s-1}, \quad \mathrm{Re}(s) > 0\,, \tag{13.5.2}$$

so

$$\mathrm{R}_p^s\varphi(y) = \frac{\zeta_p(1-s)}{\zeta_p(s)}\int \varphi(x)|y-x|_p^{s-1}\mathrm{d}x\,. \tag{13.5.3}$$

This can be continued meromorphically to all s: for $p \neq \eta$ the continuation is

$$
\begin{aligned}
\mathrm{R}_p^s\varphi(y) = \frac{\zeta_p(1-s)}{\zeta_p(1)}\varphi(y) + \frac{\zeta_p(1-s)}{\zeta_p(s)}\Bigg[&\int_{|x|_p>1}\varphi(y+x)|x|_p^{s-1}\mathrm{d}x \\
&+ \int_{|x|_p\leq 1}(\varphi(y+x)-\varphi(y))|x|_p^{s-1}\mathrm{d}x\Bigg]\,.
\end{aligned}
\tag{13.5.4$_p$}
$$

Similarly for $p = \eta$ and $\mathrm{Re}(s) > -2N$, the continuation is given by

$$
\begin{aligned}
\mathrm{R}_\eta^s\varphi(y) = \frac{\zeta_\eta(1-s)}{\zeta_\eta(s)}\Bigg[&\sum_{0\leq m<N}\frac{1}{(2m)!(s/2+m)}\left(\frac{\partial}{\partial x}\right)^{2m}\varphi(y) + \int_{|x|_\eta>1}\varphi(y+x)|x|_\eta^{s-1}\,\mathrm{d}x \\
&+ \int_{|x|_\eta<1}\left(\varphi(y+x) - \sum_{0\leq m<2N}\varphi^{(m)}(y)\cdot\frac{x^m}{m!}\right)|x|_\eta^{s-1}\mathrm{d}x\Bigg]\,.
\end{aligned}
\tag{13.5.4$_\eta$}
$$

For example, for $\mathrm{Re}(s) > -2$ we have

$$\mathrm{R}_\eta^s\varphi(y) = \frac{\zeta_\eta(1-s)}{\pi\zeta_\eta(2+s)}\varphi(y) + \frac{\zeta_\eta(1-s)}{\zeta_\eta(s)}\left[\int_{|x|_\eta>1}\varphi(y+x)|x|_\eta^{s-1}\mathrm{d}x + \int_{|x|_\eta\leq 1}(\varphi(y+x)-\varphi(y))|x|_\eta^{s-1}\mathrm{d}x\right].$$

Letting $\psi_p^{(x)}(y) = \psi_p(xy)$, where ψ_p is the basic character of $\mathbb{Q}_p$ (12.1.14), we have

$$\mathrm{R}_p^{-s}\psi_p^{(x)}(y) = |x|_p^s \cdot \psi_p^{(x)}(y). \tag{13.5.5}$$

Putting $y = 0$ in this formula, and expanding it using (13.5.4), we obtain

$$\int (1 - \mathrm{Re}\,\psi_p(xz))\left[-\frac{\zeta_p(1+s)}{\zeta_p(-s)}\frac{\mathrm{d}z}{|z|_p^{1+s}}\right] = |x|_p^s. \tag{13.5.6}$$

Note that the measure inside the brackets in (13.5.6) is positive for real $s = \beta \in (0, \lambda_p)$, where

$$\lambda_p = [\bar{\mathbb{Q}}_p{:}\mathbb{Q}_p] = \begin{cases} \infty & p \neq \eta \\ 2 & p = \eta \end{cases} \tag{13.5.7}$$

(and $\lambda_p = 1$ for a complex place of a number field). We deduce from this, using the Levy–Khinchin representation (cf. Berg and Frost 1975), that the function $|x|_p^\beta$ is negative definite for $\beta \in (0, \lambda_p)$. Thus we have a convolution semi-group of probability measures μ_t^β, the symmetric stable semi-group (cf. Feller 1970), given explicitly by

$$\mu_t^\beta = \mathcal{F}\mathrm{e}^{-t|x|_p^\beta} = \sum_{n\geq 1}\frac{(-t)^n}{n!}\frac{\zeta_p(1+n\beta)}{\zeta_p(-n\beta)}\frac{\mathrm{d}x}{|x|_p^{1+n\beta}}. \tag{13.5.8}$$

This semi-group has infinitesimal generator $\mathrm{R}_p^{-\beta}$

$$-\frac{\partial}{\partial t}\bigg|_{t=0}\mu_t^\beta = \mathrm{R}_p^{-\beta}. \tag{13.5.9}$$

For $\beta \in (0,1)$ it is transient with potential kernel R_p^β

$$\int_0^\infty \mu_t^\beta\,\mathrm{d}t = \mathrm{R}_p^\beta. \tag{13.5.10}$$

Consider the path space

$$\Omega_x = \{\mathrm{X}\colon [0,\infty) \to \mathbb{Q}_p \mid \mathrm{X}(0) = x,\ \mathrm{X}\text{ right continuous and having left limits}\}. \tag{13.5.11}$$

We think about such a path $\mathrm{X} \in \Omega_x$ as describing the history of a particle walking in $\mathbb{Q}_p$. By Kolmogorov's theorem there is a unique probability measure P_x^β on Ω_x such that

$$\mathrm{P}_x^\beta[\mathrm{X}(t) \in \mathrm{d}y] = \mu_t^\beta(\mathrm{d}y - x)\,, \tag{13.5.12}$$

and we have the probabilistic interpretation in terms of expectation with respect to P_x^β

$$\mu_t^\beta * \varphi(x) = \mathrm{P}_x^\beta[\varphi(\mathrm{X}(t))]\,, \tag{13.5.13}$$

$$\mathrm{R}_p^\beta \varphi(x) = \mathrm{P}_x^\beta\left[\int_0^\infty \varphi(\mathrm{X}(t))\mathrm{d}t\right]. \tag{13.5.14}$$

Given a compact set $\mathcal{K} \subseteq \mathbb{Q}_p$, and a continuous function $\varphi\colon \mathcal{K} \to \mathbb{C}$, there exists a unique solution h_φ to the 'Dirichlet problem': it has boundary values $\mathrm{h}_\varphi(x) = \varphi(x)$ for $x \in \mathcal{K}$ and is harmonic outside $\mathcal{K}$, that is h_φ is smooth (locally constant for $p \neq \eta$) outside $\mathcal{K}$ and $\mathrm{R}_p^{-\beta}\varphi \equiv 0$ outside $\mathcal{K}$. In terms of the expectation P_x^β it is given by

$$\mathrm{h}_\varphi(x) = \mathrm{P}_x^\beta[\varphi(\mathrm{X}(t_\mathcal{K}))]\,, \tag{13.5.15}$$

where $t_\mathcal{K}(\mathrm{X}) = \inf\{t > 0, \mathrm{X}(t) \in \mathcal{K}\}$ is the hitting time of $\mathcal{K}$.

In the limit $\beta \to \lambda_p$, μ_t^β converges to the 'normal law'

$$\mu_t^{\lambda_p}(x) = \begin{cases} \mathrm{e}^{-t}\phi_{\mathbb{Z}_p}(x) + (1-\mathrm{e}^{-t})p^{-1}\phi_{p^{-1}\mathbb{Z}_p}(x) & p \neq \eta\,, \\ (t/\pi)^{-1/2}\mathrm{e}^{-(\pi^2/t)x^2} & p = \eta\,. \end{cases} \tag{13.5.16}$$

Its infinitesimal generator is the 'Laplacian'

$$-\frac{\partial}{\partial t}\bigg|_{t=0} \mu_t^{\lambda_p}(x) = \begin{cases} \phi_{\mathbb{Z}_p} - p^{-1}\phi_{p^{-1}\mathbb{Z}_p} = \mathcal{F}\phi_{\mathbb{Z}_p^*}\mathcal{F}^{-1} & p \neq \eta\,, \\ -\dfrac{1}{4\pi^2}\dfrac{\partial^2}{\partial x^2} = \mathcal{F}x^2\mathcal{F}^{-1} & p = \eta\,. \end{cases} \tag{13.5.17}$$

Note that for $p \neq \eta$, $\mu_0^{\lambda_p} = \phi_{\mathbb{Z}_p}$, and so $\mu_t^{\lambda_p}$ is a convolution semi-group on $\mathbb{Q}_p/\mathbb{Z}_p$, and not on $\mathbb{Q}_p$.

For $\varphi \in \mathcal{S}(\mathbb{Q}_p)$, $\mathrm{R}_p^s\varphi(y)$ is smooth and is $\mathrm{O}\big(|y|_p^{s-1}\big)$ for $|y|_p \gg 1$, so we can form $\mathrm{R}_p^{s'}\mathrm{R}_p^s\varphi$ for $\mathrm{Re}(s'+s) < 1$. For s and s' such that $0 < \mathrm{Re}(s)$, $\mathrm{Re}(s')$, $\mathrm{Re}(s+s') < 1$, we have from (13.5.1), $\mathrm{R}_p^{s'}\mathrm{R}_p^s\varphi = \mathrm{R}_p^{s+s'}\varphi$, and by analytic continuation we obtain Riesz's reproduction formula

$$\mathrm{R}_p^s\mathrm{R}_p^{s'}\varphi = \mathrm{R}_p^{s+s'}\varphi \qquad \text{for } \mathrm{Re}(s+s') < 1\,. \tag{13.5.18}$$

In particular, R_p^{-s} for $\mathrm{Re}(s) \geq 0$ forms a holomorphic semi-group of operators with infinitesimal generator

$$\mathcal{R}_p = \frac{\partial}{\partial s}\bigg|_{s=0} \mathrm{R}_p^s = \mathcal{F}(-\log|x|_p)\mathcal{F}^{-1}\,. \tag{13.5.19}$$

For a function $f \in \mathrm{C}_c^\infty(\mathbb{R}^+)$, we let $f|_p(x) = f(|x|_p)$ denote the associated function on $\mathbb{Q}_p^*$ obtained by restriction. The connection of the Riesz potentials with the explicit sum is based on the following:

$$\textit{local formula: } \mathcal{R}_p(x^{-1/2}\cdot f)|_p(1) = \mathcal{W}_p(f), \quad p \geq \eta\,. \tag{13.5.20}$$

The proof of the local formula is by direct verification. For $p \neq \eta$ we have

$$\mathcal{R}_p(x^{-1/2}f)|_p(1) = \frac{\partial}{\partial s}\bigg|_{s=0} \mathrm{R}_p^s(x^{-1/2}f)|_p(1)$$

$$= \frac{\partial}{\partial s}\bigg|_{s=0}\left[\frac{\zeta_p(1-s)}{\zeta_p(1)}f(1)\right.$$

$$+ \frac{\zeta_p(1-s)}{\zeta_p(s)}\left(\int_{|y|_p>1}(x^{-1/2}f)|_p(1-y)|y|_p^{s-1}\mathrm{d}y\right.$$

$$\left.\left.+\int_{|y|_p\leq 1}\Big[(x^{-1/2}f)|_p(1-y)-f(1)\Big]|y|_p^{s-1}\mathrm{d}y\right)\right]$$

$$= \zeta_p(1)\log p\left(p^{-1}\cdot f(1)+\int_{|y|_p>1}(x^{-1/2}f)|_p(1-y)\frac{\mathrm{d}y}{|y|_p}\right.$$

$$\left.+\int_{|y|_p\leq 1}\Big[(x^{-1/2}f)|_p(1-y)-f(1)\Big]\frac{\mathrm{d}y}{|y|_p}\right)$$

$$= \zeta_p(1)\log p\left(p^{-1}\cdot f(1)+\int_{|y|_p>1}\frac{f(|y|_p)|y|_p^{-1/2}}{|1-y|_p}\mathrm{d}y\right.$$

$$\left.+\int_{|y|_p<1}\Big[f(|y|_p)|y|_p^{-1/2}-f(1)\Big]\frac{\mathrm{d}y}{|1-y|_p}\right)$$

$$= \log p\cdot\int_{|y|_p\neq 1} f(|y|_p)|y|_p^{1/2}\frac{\mathrm{d}^*y}{|1-y|_p}$$

$$= \log p\cdot\sum_{n\neq 0} f(p^n)p^{-|n|/2} = \mathcal{W}_p(f), \quad \text{cf. (13.2.7)}.$$

Similarly for $p = \eta$ we have

$$\mathcal{R}_\eta(x^{-1/2}f)|_\eta(1) = \frac{\partial}{\partial s}\bigg|_{s=0} \mathrm{R}_\eta^s(x^{-1/2}f)|_\eta(1)$$

$$= \frac{\partial}{\partial s}\bigg|_{s=0}\left[\frac{\zeta_\eta(1-s)}{\pi\zeta_\eta(2+s)}f(1)\right.$$

$$+ \frac{\zeta_\eta(1-s)}{\zeta_\eta(s)}\left(\int_{|y|_\eta>1}(x^{-1/2}f)|_\eta(1-y)|y|_\eta^{s-1}\mathrm{d}y.\right.$$

$$\left.\left.+\int_{|y|_\eta<1}\Big[(x^{-1/2}f)|_\eta(1-y)-f(1)\Big]|y|_\eta^{s-1}\mathrm{d}y\right)\right]$$

$$= (\gamma + \log 2\pi) f(1) + \frac{1}{2}\left(\int_{|y|_\eta > 1} (x^{-1/2} f)|_\eta (1-y) \frac{dy}{|y|_\eta}\right.$$

$$\left. + \int_{|y|_\eta < 1} \left[(x^{-1/2} f)|_\eta (1-y) - f(1) \right] \frac{dy}{|y|_\eta} \right)$$

$$= (\gamma + \log \pi) f(1) + \int_1^2 f(1) \frac{dy}{|y|_\eta} + \frac{1}{2} \int_{|1-y|_\eta > 1} \frac{f(|y|_\eta) |y|_\eta^{-1/2}}{|1-y|_\eta} dy$$

$$+ \frac{1}{2} \int_{|1-y|_\eta < 1} \left[f(|y|_\eta) |y|_\eta^{-1/2} - f(1) \right] \frac{dy}{|1-y|_\eta}$$

$$= (\gamma + \log \pi) f(1) + \int_1^\infty \frac{f(y) y^{-1/2} - f(1) y^{-2}}{1 - y^{-2}} \mathrm{d}^* y + \int_0^1 \frac{f(y) y^{1/2} - f(1) y^2}{1 - y^2} \mathrm{d}^* y$$

$$= \mathcal{W}_\eta(f), \quad \text{cf. (13.2.13)}.$$

Note that we can replace $f(x)$ by $x^{-s/2} f(x)$ without changing the result, so that we also have

$$\left.\frac{\partial}{\partial s}\right|_{s=0} \mathcal{R}_p^s (x^{-(1+s)/2} \cdot f)|_p (1) = \mathcal{W}_p(f). \qquad (13.5.20)'$$

For f with $f(1) = 0$, we have for $p \geq \eta$

$$\begin{aligned} \mathcal{W}_p(f) &= \mathcal{R}_p (x^{-1/2} f)|_p (1) \\ &= \left(\left.\frac{\partial}{\partial s}\right|_{s=0} \frac{\zeta_p(1-s)}{\zeta_p(s)} \right) \cdot \int_{\mathbb{Q}_p} (x^{-1/2} f)|_p (y) \frac{dy}{|1-y|_p} \\ &= \int_{\mathbb{Q}_p} f(|y|_p) |y|_p^{1/2} \frac{\mathrm{d}^0 y}{|1-y|_p}. \end{aligned} \qquad (13.5.20)_0$$

Here $\mathrm{d}^0 y$ is the multiplicative measure, $\mathrm{d}^0 y = n_p \, dy / |y|_p$, normalized by

$$\int_{1 \leq |y|_p < c} \mathrm{d}^0 y \sim \log c \quad \text{as } c \to \infty;$$

the normalizing factor is given by

$$n_p = \left.\frac{\partial}{\partial s}\right|_{s=0} \frac{\zeta_p(1-s)}{\zeta_p(s)} = \begin{cases} \log p / (1 - p^{-1}) & p \neq \eta, \\ \dfrac{1}{2} & p = \eta. \end{cases}$$

Thus $\mathcal{W}_p$ is given by $(13.5.20)_0$ away from 1, and $\mathcal{R}_p$ fixes just the right multiple of δ_1 in the distribution $\mathcal{W}_p$, $p \geq \eta$.

The advantage of writing the Weil distribution in the form (13.5.20) is that it easily translates into a form involving a trace of an operator, indeed in more than one way. The simplest way of doing this is to consider the operator on $\mathrm{L}_2(\mathbb{Q}_p)$ given by $\mathrm{R}_p^s\pi(f|_p)\phi_{\mathbb{Z}_p}$, where $\pi(f|_p)$ is the unitary action of the multiplicative group, and $\phi_{\mathbb{Z}_p}$ is the operator of multiplication by $\phi_{\mathbb{Z}_p}$. We have

$$\begin{aligned}\mathrm{R}_p^s\pi(f|_p)\phi_{\mathbb{Z}_p}\varphi(y) &= \frac{\zeta_p(1-s)}{\zeta_p(s)}\int \mathrm{d}x\,|y-x|_p^{s-1}\int \mathrm{d}^*a\, f|_p(a)|a|_p^{-1/2}\phi_{\mathbb{Z}_p}(a^{-1}x)\varphi(a^{-1}x)\\ &= \frac{\zeta_p(1-s)}{\zeta_p(s)}\int \mathrm{d}^*a\, f|_p(a)|a|_p^{1/2}\int \mathrm{d}x\,|y-ax|_p^{s-1}\phi_{\mathbb{Z}_p}(x)\varphi(x)\,.\end{aligned} \tag{13.5.21}$$

Hence

$$\begin{aligned}\mathrm{tr}(\mathrm{R}_p^s\pi(f|_p)\phi_{\mathbb{Z}_p}) &= \frac{\zeta_p(1-s)}{\zeta_p(s)}\cdot\int \mathrm{d}^*a\, f|_p(a)|a|_p^{1/2}\int \mathrm{d}x\,|x(1-a)|_p^{s-1}\phi_{\mathbb{Z}_p}(x)\\ &= \frac{\zeta_p(1-s)}{\zeta_p(s)}\cdot\int \mathrm{d}a\, f|_p(a)|a|_p^{-1/2}|1-a|_p^{s-1}\int \mathrm{d}^*x\,|x|_p^s\phi_{\mathbb{Z}_p}(x)\\ &= \mathrm{R}_p^s(x^{-1/2}f)|_p(1)\cdot\zeta_p(s)\,.\end{aligned} \tag{13.5.22}$$

Using (13.5.20) we have

$$\frac{1}{\zeta_p(s)}\mathrm{tr}\Big(\mathrm{R}_p^s\pi(f|_p)\phi_{\mathbb{Z}_p}\Big) = f(1) + \mathcal{W}_p(f)\cdot s + \mathrm{O}(s^2), \qquad \text{as } s\to 0,\ p\geq\eta\,. \tag{13.5.23}$$

Here (and elsewhere) we mean that the trace on the left-hand side exists for $\mathrm{Re}(s)$ large, and has meromorphic continuation satisfying the right-hand side.

The problem with this approach is that for $p\neq\eta$ we have

$$\mathrm{R}_p^s\phi_{\mathbb{Z}_p}(y) = \begin{cases}\dfrac{\zeta_p(1-s)}{\zeta_p(1)} & |y|_p\leq 1\,,\\[2mm] \dfrac{\zeta_p(1-s)}{\zeta_p(s)}|y|_p^{s-1} & |y|_p > 1\,,\end{cases} \tag{13.5.24}$$

so that the operator $\otimes_p\mathrm{R}_p^s$ is not defined on $\mathbb{A}$. The 'good' operator is convolution with the Tate distribution

$$\begin{aligned}&\tau_p^s = \frac{\zeta_p(1)}{\zeta_p(1-s)}\mathrm{R}_p^s, \quad \tau_p^s\varphi(y) = \int \tau_p^s(\mathrm{d}x)\varphi(y+x)\,;\\ &\tau_{\mathbb{A}}^s = \bigotimes_p \tau_p^s \text{ is well-defined on } \mathbb{A}\,,\end{aligned} \tag{13.5.25}$$

$$\tau_{\mathbb{A}}^s\varphi(y) = \int \tau_{\mathbb{A}}^s(\mathrm{d}x)\varphi(y+x)\,. \tag{13.5.26}$$

But replacing R_p^s with τ_p^s in (13.5.23), $\mathcal{W}_p(f)$ on the right-hand side is replaced by $\mathcal{W}_p(f) - (\log p)/(p-1)\cdot f(1)$.

$\mathcal{W}_p$ via the cut-off

We see that $\mathcal{W}_p(f)$ appears as a secondary term in a normalization of the trace of the operator $\pi(f|_p)$. Another way of writing this is as follows (cf. Connes 1998). Let

$$\mathrm{B}_c(x) = \begin{cases} 1 & |x|_p \leq c\,, \\ 0 & |x|_p > c\,, \end{cases} \tag{13.5.27}$$

and view B_c as a projection on $\mathrm{L}_2(\mathbb{Q}_p)$ via multiplication (cf. §10.6). Let $\widehat{\mathrm{B}}_c = \mathcal{F}\mathrm{B}_c\mathcal{F}^{-1}$ denote the dual projection given by convolution with $\mathcal{F}\mathrm{B}_c$, and let $\pi^0(f|_p)\varphi(x) = \int \mathrm{d}^0 a\; f(|a|_p)|a|_p^{1/2}\varphi(ax)$ denote the integrated action π_p with respect to the normalized measure $\mathrm{d}^0 a = n_p\,\mathrm{d}a/|a|_p$ of $(13.5.20)_0$. The operator $\widehat{\mathrm{B}}_c\mathrm{B}_c\pi^0(f|_p)$ is given by

$$\widehat{\mathrm{B}}_c\mathrm{B}_c\pi^0(f|_p)\varphi(y) \\ = \int \mathrm{d}x\,\varphi(x)\int \mathrm{d}^0 a\, f|_p(a)|a|_p^{-1/2}\mathrm{B}_c(a^{-1}x)\int \mathrm{d}z\,\mathrm{B}_c(z)\psi_p((y-a^{-1}x)z)\,. \tag{13.5.28}$$

Its trace is therefore

$$\int \mathrm{d}x \int \mathrm{d}^0 a\; f|_p(a)|a|_p^{-1/2}\mathrm{B}_c(a^{-1}x)\int \mathrm{d}z\,\mathrm{B}_c(z)\psi_p((1-a^{-1})xz)\,,$$

$$x := ax,\, z := z/x$$

$$\begin{aligned} &= \int \mathrm{d}x\,\mathrm{B}_c(x)\int \frac{\mathrm{d}z}{|x|}\mathrm{B}_c\Big(\frac{z}{x}\Big)\int \mathrm{d}^0 a\; f|_p(a)|a|_p^{1/2}\psi_p((a-1)z) \\ &= \int_{|z|_p\leq c^2} \mathrm{d}z \int_{|z|_p/c\leq |x|_p\leq c} \mathrm{d}^0 x \int \mathrm{d}a\; f_p|(a)|a|_p^{-1/2}\psi_p((a-1)z) \\ &= \int_{|z|_p\leq c^2} \mathrm{d}z(2\log c - \log|z|_p)\mathcal{F}(f|_p(x)|x|_p^{-1/2})(z)\bar{\psi}_p(z) \end{aligned}$$

and writing the integral over $|z|_p \leq c^2$ as the integral over all z minus the integral over $|z|_p > c^2$, then this becomes by the local formula (13.5.20)

$$\begin{aligned} \mathrm{tr}(\widehat{\mathrm{B}}_c\mathrm{B}_c\pi^0(f|_p)) &= (2\log c)\mathcal{F}^{-1}\mathcal{F}(f|_p(x)|x|_p^{-1/2})(1) \\ &\quad + \mathcal{F}^{-1}(-\log|z|_p)\mathcal{F}(f|_p(x)|x|_p^{-1/2})(1) \\ &\quad - \int_{|z|_p>c^2} \mathrm{d}z(2\log c - \log|z|_p)\mathcal{F}(f|_p(x)|x|_p^{-1/2})(z)\bar{\psi}_p(z) \\ &= (2\log c)f(1) + \mathcal{W}_p(f) + \mathrm{O}(c^{-N}) \text{ as } c\to\infty, \text{for all } N \end{aligned} \tag{13.5.29}$$

(where for $p = \eta$ one can also replace $\widehat{\mathrm{B}}_c\mathrm{B}_c$ by the projection onto the subspace $\mathscr{H}^1_\eta\langle c\rangle$, cf. (10.6.32)).

$\mathcal{W}_p$ via the pure basis

Another way of normalizing the trace of the operator $\pi(f|_p)$ is by using the heat kernel of the number operator associated with the pure basis. For $p \neq \eta$ the basis is (cf. (10.6.9)) $\varphi_{p,0} = \phi_{\mathbb{Z}_p}$, $\varphi_{p,n,\pm} = (1 \pm \mathcal{F})\phi_{p^{-n}\mathbb{Z}_p^*}$, $n \geq 1$, and the heat kernel is given by

$$\mathrm{e}^{-t\widetilde{\mathbb{N}}_p}\varphi_{p,n,\pm} = p^{-tn} \cdot \varphi_{p,n,\pm}, \tag{13.5.30}$$

$$\begin{aligned}\mathrm{e}^{-t\widetilde{\mathbb{N}}_p}(y,x) &= \sum_{n,\pm} \|\varphi_{p,n,\pm}\|^{-2}\, \varphi_{p,n,\pm}(y), \varphi_{p,n,\pm}(x) p^{-tn} \\ &= \phi_{\mathbb{Z}_p}(y)\phi_{\mathbb{Z}_p}(x) + \zeta_p(1) \sum_{n\geq 1} p^{-(1+t)n} \Big[\phi_{p^{-n}\mathbb{Z}_p^*}(x)\phi_{p^{-n}\mathbb{Z}_p^*}(y) \\ &\qquad + \mathcal{F}\phi_{p^{-n}\mathbb{Z}_p^*}(y)\mathcal{F}\phi_{p^{-n}\mathbb{Z}_p^*}(x)\Big].\end{aligned}$$

We have

$$\mathrm{e}^{-t\widetilde{\mathbb{N}}_p}\pi(f|_p)\varphi(y) = \int \mathrm{d}x\, \varphi(x) \int \mathrm{d}^*a\, f|_p(a)|a|_p^{1/2} \mathrm{e}^{-t\widetilde{\mathbb{N}}_p}(y, ax). \tag{13.5.31}$$

Hence

$$\begin{aligned}\mathrm{tr}(\mathrm{e}^{-t\widetilde{\mathbb{N}}_p}\pi(f|_p)) &= \int \mathrm{d}^*a\, f|_p(a)|a|_p^{1/2} \int \mathrm{d}x\ \mathrm{e}^{-t\widetilde{\mathbb{N}}_p}(x, ax) \\ &= f(1) \cdot \int \mathrm{d}x\ \mathrm{e}^{-t\widetilde{\mathbb{N}}_p}(x,x) + \sum_{n\neq 0} f(p^n)p^{n/2} \int \mathrm{d}x\, \phi_{\mathbb{Z}_p}(x) \cdot \phi_{\mathbb{Z}_p}(p^{-n}x) \\ &= f(1) \cdot \frac{1+p^{-t}}{1-p^{-t}} + \sum_{n\neq 0} f(p^n)p^{-|n|/2}.\end{aligned} \tag{13.5.32}$$

This can be written as

$$\begin{aligned}\frac{1}{\zeta_p(t,t)}\mathrm{tr}(\mathrm{e}^{-t\widetilde{\mathbb{N}}_p}\pi(f|_p)) &= f(1) + \frac{1-p^{-t}}{1+p^{-t}} \sum_{n\neq 0} f(p^n)p^{-|n|/2} \\ &= f(1) + \frac{t}{2}\mathcal{W}_p(f) + \mathrm{O}(t^2), \quad \text{as } t \to 0,\ p \neq \eta.\end{aligned} \tag{13.5.33}$$

Something similar holds for $p = \eta$, where the pure basis is given in terms of the (even) Hermite polynomials (cf. (10.6.12)). The heat kernel is given in terms of the Mehler–Fock kernel

$$z^{\mathbb{N}^1_\eta}\varphi_{\eta,n} = z^{2n} \cdot \varphi_{\eta,n}, \tag{13.5.34}$$

$$\begin{aligned}z^{\mathbb{N}^1_\eta}(y,x) &= \sum_{n\geq 0} \|\varphi_{\eta,n}\|^{-2}\varphi_{\eta,n}(y)\varphi_{\eta,n}(x)z^{2n} \\ &= [2(1-z^2)]^{-1/2}\mathrm{e}^{-\pi(x^2+y^2)(1+z^2)/(1-z^2)}[\mathrm{e}^{4\pi xyz/(1-z^2)} + \mathrm{e}^{-4\pi xyz/(1-z^2)}].\end{aligned}$$

We obtain

$$\operatorname{tr}(z^{\mathbb{N}^1_\eta}\pi(f|_\eta)) = \int \mathrm{d}^*a\, f(|a|)|a|^{1/2} \int \mathrm{d}x\, z^{\mathbb{N}^1_\eta}(x, ax)$$

$$= \int \mathrm{d}^*a\, f(|a|)|a|^{1/2} \frac{1}{2}\left\{ \left[\frac{(1+a^2)(1+z^2)}{2} + 2az\right]^{-1/2} + \left[\frac{(1+a^2)(1+z^2)}{2} - 2za\right]^{-1/2} \right\}$$

$$= \frac{1}{2}\int_{|1-a|>1} \mathrm{d}^*a\, f(a)|a|^{1/2}|1-a|^{-1} + \frac{1}{2}\int_{|1-a|<1} \mathrm{d}^*a\, [f(a)|a|^{1/2} - f(1)|a|]\, |1-a|^{-1} + f(1)\frac{1}{2}\int_0^2 \mathrm{d}a \left[\frac{(1+a^2)(1+z^2)}{2} - 2az\right]^{-1/2} + \mathrm{O}(z-1), \text{ as } z \to 1,$$

$$= \frac{1}{2}\int_{|1-a|>1} \mathrm{d}^*a\, f(a)|a|^{1/2}|1-a|^{-1} + \frac{1}{2}\int_{|1-a|<1} \mathrm{d}^*a [f(a)|a|^{1/2} - f(1)|a|]\, |1-a|^{-1} + f(1)\cdot[\log 2 - \log(1-z)] + \mathrm{O}(z-1)$$

$$= f(1)[\log(1-z)^{-1} - \log\pi - \gamma] + \mathcal{W}_\eta(f) + \mathrm{O}(z-1), \text{cf. } (13.2.13)\,. \tag{13.5.35}$$

Setting $z = \mathrm{e}^{-u}$, $f_n = \|\varphi_{\eta,n}\|^{-2}(\pi(f|_\eta)\varphi_{\eta,n}, \varphi_{\eta,n})$, we have

$$\theta_f(u) = \operatorname{tr}(\mathrm{e}^{-u\mathbb{N}^1_\eta}\pi(f|_\eta)) = \sum_{n\geq 0} f_n \mathrm{e}^{-2nu}$$

$$= -f(1)\log u + [\mathcal{W}_\eta(f) - (\gamma + \log\pi)f(1)] + \mathrm{O}(u) \text{ as } u \to 0\,.$$

Thus for $r > 0$

$$\operatorname{tr}\left(\left(r + \mathbb{N}^1_\eta\right)^{-t} \pi\left(f|_\eta\right)\right) = \sum_{n\geq 0} f_n(r+2n)^{-t} = \frac{1}{\Gamma(t)}\int_0^\infty \theta_f(u)\mathrm{e}^{-ru}u^t\,\frac{\mathrm{d}u}{u}$$

$$= \frac{f(1)}{t} + [\mathcal{W}_\eta(f) - f(1)\log\pi] + \mathrm{O}(t) \text{ as } t \to 0, \text{independent of } r > 0\,. \tag{13.5.36}$$

We can take $r = 1$, and set $\widetilde{\mathbb{N}}_\eta = \frac{1}{2}\log\left(1 + \mathbb{N}^1_\eta\right)$, so that

$$\mathrm{e}^{-t\widetilde{\mathbb{N}}_\eta}\varphi_{\eta,n} = (2n+1)^{-t/2}\varphi_{\eta,n}, \qquad \text{for } n \geq 0\,,$$

and

$$\operatorname{tr}\left(\mathrm{e}^{-t\widetilde{\mathbb{N}}_\eta}\pi(f|_\eta)\right) = \frac{f(1)}{t/2} + [\mathcal{W}_\eta(f) - f(1)\log\pi] + \mathrm{O}(t) \quad \text{as } t \to 0\,.$$

This can be written as in (13.5.33),

$$\frac{2}{\zeta_\eta(t,t)} \operatorname{tr}\left(\mathrm{e}^{-t\widetilde{\mathbb{N}}_\eta}\pi(f|_\eta)\right) = f(1) + \frac{t}{2}[\mathcal{W}_\eta(f) - f(1)\log\pi] + \mathrm{O}(t^2) \quad \text{as } t \to 0\,. \tag{13.5.37}$$

$\mathcal{W}_p$ via the intertwining operator

For $p \geq \eta$, the Riesz potential satisfies

$$\mathrm{R}_p^s \rho_{\infty_p}^{1+s} = \frac{\zeta_p(1-s)}{\zeta_p(1+s)} \rho_{\infty_p}^{1-s}\,. \tag{13.5.38}$$

This can be verified directly, and is equivalent to

$$\mathcal{F}\rho_{\infty_p}^{1+s} = \frac{\zeta_p(1-s)}{\zeta_p(1+s)} |x|_p^s \mathcal{F}\rho_{\infty_p}^{1-s}\,,$$

which follows from (13.3.11). This suggests considering the operator acting on functions on $\mathbb{P}^1(\mathbb{Q}_p)$ given by

$$\begin{aligned}
\widetilde{\mathrm{R}}_p^s\, \varphi(1{:}y) &= \left(\zeta_p(1-s)\rho_{\infty_p}^{1-s}\right)^{-1} \circ \mathrm{R}_p^s \circ \zeta_p(1+s)\rho_{\infty_p}^{1+s}\, \varphi(1{:}y) \\
&= \frac{\zeta_p(1+s)}{\zeta_p(s)} \int \mathrm{d}x\, \rho_{\infty_p}^{s-1}(1{:}y)|y-x|_p^{s-1}\, \rho_{\infty p}^{1+s}(1{:}x)\varphi(1{:}x) \\
&= \frac{\zeta_p(1+s)}{\zeta_p(s)} \int \frac{\mathrm{d}x}{|1,x|_p^2} \rho_p(1{:}y,1{:}x)^{s-1}\varphi(1{:}x)\,,
\end{aligned} \tag{13.5.39}$$

that is the normalized intertwining operator of §13.4.

We can translate the local formula (13.5.20) into a formula involving the trace of $\widetilde{\mathrm{R}}_p^s$. Restricting the representation π_p^s of $\mathrm{PGL}_2(\mathbb{Q}_p)$ to $\mathrm{M}(\mathbb{Q}_p) = \mathbb{Q}_p^*$, we have

$$\begin{aligned}
&\widetilde{\mathrm{R}}_p^s \pi_p^s(f|_p)\varphi(y) \\
&= \frac{\zeta_p(1+s)}{\zeta_p(s)} \frac{\zeta_p(1)}{\zeta_p(2)} \int \tau_p^{1,1}(\mathrm{d}x)\rho_p(y,x)^{s-1} \int \mathrm{d}^*a\, f|_p(a)\varphi\left(1{:}x\begin{pmatrix} a & 0 \\ 0 & 1\end{pmatrix}\right) \cdot \\
&\qquad\qquad \mathrm{j}_p\left(1{:}x, \begin{pmatrix} a & 0 \\ 0 & 1\end{pmatrix}\right)^{1+s} \\
&= \frac{\zeta_p(1+s)}{\zeta_p(s)} \int \frac{\mathrm{d}x}{|1,x|^2} \varphi(1{:}x) \int \mathrm{d}^*a\, f|_p(a)|a|_p^{(1-s)/2} \left(\frac{|y-ax|_p}{|1,x|_p \cdot |1,y|_p}\right)^{s-1}.
\end{aligned} \tag{13.5.40}$$

Thus its trace is given by

$$\mathrm{tr}\Big(\widetilde{\mathrm{R}}_p^s\pi_p^s(f|_p)\Big) = \frac{\zeta_p(1+s)}{\zeta_p(s)}\int\!\mathrm{d}x\,\frac{|x|_p^{s-1}}{|1,x|_p^{2s}}\int\!\mathrm{d}^*a\, f|_p(a)|a|_p^{(1-s)/2}|1-a|_p^{s-1}$$

$$= \frac{\zeta_p(s)^2}{\zeta_p(2s)}\,\frac{\zeta_p(1+s)}{\zeta_p(s)}\int\!\mathrm{d}a\, f|_p(a)\ |a|_p^{-(1+s)/2}|1-a|_p^{s-1}\,. \qquad (13.5.41)$$

Setting

$$\check{\mathrm{M}}_p^s = \frac{\zeta_p(1-s)}{\zeta_p(1+s)}\cdot\widetilde{\mathrm{R}}_p^s = \frac{\zeta_p(1-s)}{\zeta_p(s)}\cdot\mathrm{M}_p^{(s)},\ \text{cf. } (13.4.30)\,, \qquad (13.5.42)$$

we obtain from (13.5.41) and (13.5.20)′,

$$\frac{1}{\zeta_p(s,s)}\mathrm{tr}(\check{\mathrm{M}}_p^s\pi_p^s(f|_p)) = f(1) + \mathcal{W}_p(f)\cdot s + \mathrm{O}(s^2), \quad \text{as } s\to 0,\ p\geq\eta\,. \quad (13.5.43)$$

$\mathcal{W}_p$ via the symmetric beta chain, $\alpha=\beta=1$

The operator $\check{\mathrm{M}}_p^s$ is not normalized. For $p\geq\eta$ we have (13.4.19, 20)

$$\check{\mathrm{M}}_p^s = \bigoplus_{n\geq 0}\check{\mathrm{M}}_p^s(n)\cdot\mathrm{id}_{\widetilde{\mathscr{H}}_p^{1,1}(n)} \qquad (13.5.44)$$

$$\check{\mathrm{M}}_p^s(n) = \begin{cases}\zeta_p(1-s)/\zeta_p(1+s) & n=0\\ p^{-ns} & n\geq 1\end{cases}\quad p\neq\eta\,,$$

$$\check{\mathrm{M}}_\eta^s(n) = \frac{\zeta_\eta(1-s+2n)}{\zeta_\eta(1+s+2n)},\qquad n\geq 0\,.$$

Thus, $\check{\mathrm{M}}_p^s$ is closely related to the heat kernel of the number operator $\mathbb{N}_p^{1,1}$ of the symmetric beta chain. For $p\neq\eta$,

$$\mathrm{e}^{-t\mathbb{N}_p^{1,1}} = \bigoplus_{n\geq 0}p^{-nt}\cdot\mathrm{id}_{\widetilde{\mathscr{H}}^{1,1}{}_p(n)} = \check{\mathrm{M}}_p^t + \left(1-\frac{\zeta_p(1-t)}{\zeta_p(1+t)}\right)\mathbb{1} \qquad (13.5.45)$$

with $\mathbb{1}$ denoting the projection onto the constants $\widetilde{\mathscr{H}}_p^{1,1}(0)$. We get

$$\frac{1}{\zeta_p(t,t)}\mathrm{tr}\Big(\mathrm{e}^{-t\mathbb{N}_p^{1,1}}\pi_p^t(f|_p)\Big) = \frac{1}{\zeta_p(t,t)}\mathrm{tr}\Big(\check{\mathrm{M}}_p^t\pi_p^t(f|_p)\Big) + \frac{[1-\zeta_p(1-t)/\zeta_p(1+t)]}{\zeta_p(t,t)}(\pi_p^t(f|_p)\mathbb{1},\mathbb{1})\,. \qquad (13.5.46)$$

Hence we can write (13.5.43) also in the form

$$\frac{1}{\zeta_p(t,t)}\mathrm{tr}\Big(\mathrm{e}^{-t\mathbb{N}_p^{1,1}}\pi^t(f|_p)\Big) = f(1) + \mathcal{W}_p(f)\cdot t + \mathrm{O}(t^2), \qquad \text{as } t\to 0\,. \qquad (13.5.47)$$

Something similar holds also at $p=\eta$. The Jacobi base at $\alpha=\beta=1$ is given by the Chebyshev polynomials

$$\varphi_{\eta,n}(x{:}y) = \mathrm{Re}\left(\frac{x+\mathrm{i}y}{|x+\mathrm{i}y|}\right)^{2n} = \sum_k\binom{2n}{2k}(-1)^k\frac{x^{2(n-k)}y^{2k}}{(x^2+y^2)^n}\,. \qquad (13.5.48)$$

$\check{\mathrm{M}}_\eta^t$ multiplies $\varphi_{\eta,n}$ by the eigenvalue

$$\begin{aligned}
\frac{\zeta_\eta(1-t+2n)}{\zeta_\eta(1+t+2n)} &= \pi^t \cdot \frac{\Gamma((1-t)/2+n)}{\Gamma((1+t)/2+n)} \\
&= \pi^t \cdot \frac{[(1-t)/2+n]^{-t/2+n}}{[(1+t)/2+n]^{t/2+n}} \cdot \mathrm{e}^t \left(\frac{1+1/[6n+3(1-t)]+\mathrm{O}(1/n^2)}{1+1/[6n+3(1+t)]+\mathrm{O}(1/n^2)} \right) \\
&= \frac{\pi^t}{n^t} \mathrm{e}^{(-t/2+n)\log[1+(1-t)/2n]-(t/2+n)\log[1+(1+t)/2n]+t} \left[1+\mathrm{O}\left(\frac{1}{n^2}\right)\right] \\
&= \frac{\pi^t}{n^t} \mathrm{e}^{\mathrm{O}(1/n^2)} \left[1+\mathrm{O}\left(\frac{1}{n^2}\right)\right] = \frac{\pi^t}{n^t}\left[1+\mathrm{O}\left(\frac{1}{n^2}\right)\right]. \qquad (13.5.49)
\end{aligned}$$

Define the number operator $\mathrm{N}_\eta^{1,1}$ by

$$\mathrm{N}_\eta^{1,1}\varphi_{\eta,n} = \begin{cases} (\log n)\cdot\varphi_{\eta,n} & n \geq 1\,, \\ 0 & n = 0\,. \end{cases} \qquad (13.5.50)$$

We have, with $f_n^t = \|\varphi_{\eta,n}\|^{-2}(\pi^t(f|_\eta)\varphi_{\eta,n},\varphi_{\eta,n})$,

$$\begin{aligned}
\pi^t \cdot \mathrm{tr}(\mathrm{e}^{-t\mathrm{N}_\eta^{1,1}}\pi^t(f|_\eta)) &= \sum_{n\geq 0}\left(\frac{\pi}{n}\right)^t \cdot f_n^t \\
&= \mathrm{tr}(\check{\mathrm{M}}_\eta^t \pi^t(f|_\eta)) - \sum_{n \geq 0}\left[\frac{\zeta_\eta(1-t+2n)}{\zeta_n(1+t+2n)} - \left(\frac{\pi}{n}\right)^t\right]\cdot f_n^t\,. \qquad (13.5.51)
\end{aligned}$$

Note that the last sum in (13.5.51) converges for $\mathrm{Re}(t) > -1$, and vanishes at $t = 0$. Hence we can rewrite (13.5.43) for $p = \eta$ also in the form

$$\frac{\pi^t}{\zeta_\eta(t,t)}\mathrm{tr}\left(\mathrm{e}^{-t\mathrm{N}_\eta^{1,1}}\pi^t(f|_\eta)\right) = f(1) + \mathcal{W}_\eta(f)\cdot t + \mathrm{O}(t^2), \qquad \text{as } t \to 0\,. \qquad (13.5.52)$$

Global formulations

Note that by using $\pi_p(f|_p)$ in the above formulae, we are automatically projecting onto the subspace of $\mathbb{Z}_p^*$-invariant functions. Note also that for a given $f \in \mathrm{C}_c^\infty(\mathbb{R}^+)$, $f|_p(x_p) = f(1)\cdot\phi_{\mathbb{Z}_p^*}(x_p)$ for all but finitely many p. We have for $p \neq \eta$

$$\begin{aligned}
\mathrm{R}_p^s\phi_{\mathbb{Z}_p^*}(y) &= \left\{ \begin{array}{ll} \dfrac{\zeta_p(1-s)}{\zeta_p(s)\zeta_p(1)}|y|_p^{s-1} & |y|_p > 1 \\ \zeta_p(1-s)\cdot(1-2p^{-1}+p^{-1-s}) & |y|_p = 1 \\ \dfrac{\zeta_p(1-s)}{\zeta_p(s)\zeta_p(1)} & |y|_p < 1 \end{array}\right\} \\
&= \frac{\zeta_p(1-s)}{\zeta_p(s)\zeta_p(1)}\rho_{\infty_p}^{1-s}(y) + p^{-s}\phi_{\mathbb{Z}_p^*}(y). \qquad (13.5.53)
\end{aligned}$$

The derivative of this with respect to s, at $s = 0$, gives

$$\mathcal{R}_p\phi_{\mathbb{Z}_p^*}(y) = \log p\cdot(\rho_{\infty_p}(y) - \phi_{\mathbb{Z}_p^*}(y))\,. \qquad (13.5.54)$$

In particular we obtain

$$\mathrm{R}_p^s \phi_{\mathbb{Z}_p^*}(1) = 1 + \mathrm{O}(s^2), \qquad \text{as } s \to 0\,, \tag{13.5.55}$$

$$\mathcal{R}_p \phi_{\mathbb{Z}_p^*}(1) = 0\,. \tag{13.5.55$'$}$$

Thus for the operator

$$\mathcal{R}_{\mathbb{A}} = \sum_{p \geq \eta} \mathcal{R}_p = \sum_{p \geq \eta} \mathcal{F}(-\log |x_p|_p)\mathcal{F}^{-1} = \mathcal{F}(-\log |x|_{\mathbb{A}})\mathcal{F}^{-1}\,, \tag{13.5.56}$$

$\mathcal{R}_{\mathbb{A}} \tilde{f}(a)$ is well defined for $\tilde{f} \in \mathcal{S}(\mathbb{A}^*)$, and for $a \in \mathbb{A}^*$.

To our function $f \in \mathrm{C}_c^\infty(\mathbb{R}^+)$, we associate the function $\tilde{f} = f \otimes \phi_{\hat{\mathbb{Z}}^*}$ on $\mathbb{A}^*$,

$$\tilde{f} = \bigotimes_{p \geq \eta} f_p \quad \text{with} \quad f_p = \begin{cases} \phi_{\mathbb{Z}_p^*} & p \neq \eta\,, \\ f & p = \eta\,, \end{cases}$$

so that

$$\tilde{f}(x) = \begin{cases} f(x_\eta) & \text{if} |x_p| = 1 \text{ for all } p \neq \eta, \text{ and } x_\eta > 0\,, \\ 0 & \text{otherwise}\,. \end{cases} \tag{13.5.57}$$

We let $\tilde{f}_q(x) = \tilde{f}(qx)$ for $q \in \mathbb{Q}^*, x \in \mathbb{A}^*$.

We identify $\mathbb{Q}_p^* \hookrightarrow \mathbb{A}^*$ by $x_p \mapsto x_{(p)} = (1 \ldots 1, x_p, 1 \ldots)$. With these conventions we have for all $p \geq \eta$:

$$f|_p(x_p) = \sum_{q \in \mathbb{Q}^*} \tilde{f}_q(x_{(p)})\,. \tag{13.5.58}$$

We can now paste together the local formulae (13.5.20) into the

$$\textit{global formula}: \quad \hat{f}(1/2) + \hat{f}(-1/2) - \sum_{\zeta_{\mathbb{A}}(1/2+s)=0} \hat{f}(s) = \sum_{q \in \mathbb{Q}^*} \mathcal{R}_{\mathbb{A}}(x^{-1/2} f)_q^{\sim}(1)\,. \tag{13.5.59}$$

Indeed, the left-hand side is equal by the explicit sum formula (13.2.4) to

$$\begin{aligned} \sum_{p \geq \eta} \mathcal{W}_p(f) &= \sum_{p \geq \eta} \mathcal{R}_p(x^{-1/2} f)|_p(1) \text{ by the local formula}\,, \\ &= \sum_{q \in \mathbb{Q}^*} \sum_{p \geq \eta} \mathcal{R}_p(x^{-1/2} f)_q^{\sim}(1) \text{ using (13.5.58)}\,. \end{aligned}$$

Note that the operator $\mathcal{R}_{\mathbb{A}}$ commutes with $\mathbb{Q}^*$-translations, so we can write the global formula also in the form

$$\hat{f}(1/2) + \hat{f}(-1/2) - \sum_{\zeta_{\mathbb{A}}(1/2+s)=0} \hat{f}(s) = \sum_{q \in \mathbb{Q}^*} \mathcal{R}_{\mathbb{A}}(x^{-1/2} f)^{\sim}(q)\,. \tag{13.5.59$'$}$$

There are many ways of writing the global formula. For example, note that by (13.5.53), we have for $p \neq \eta$

$$\frac{\zeta_p(1,1)}{\zeta_p(1+s, 1-s)} \mathrm{R}_p^s \phi_{\mathbb{Z}_p^*}(1) = 1 - \frac{p^{-2}(1-p^{-s})^2}{(1-p^{-1})^2} \tag{13.5.60}$$

so that if we set

$$\underline{\mathcal{R}}^s_{\mathbb{A}}\tilde{f}(y) = \bigotimes_{p\geq\eta} \frac{\zeta_p(1,1)}{\zeta_p(1+s,1-s)} \mathrm{R}^s_p|y|_p^{-1/2}\tilde{f}(y)\,, \tag{13.5.61}$$

it is well defined for $\tilde{f} \in \mathcal{S}(\mathbb{A}^*)$ and $y \in \mathbb{A}^*$, and analytic in s for $\mathrm{Re}(s) > -1/2$. Noting that the normalizing factor satisfies $\zeta_p(1,1)/\zeta_p(1+s,1-s) = 1 + \mathrm{O}(s^2)$, we have

$$\begin{aligned}
\sum_{p\geq\eta} \mathcal{W}_p(f) &= \sum_{p\geq\eta} \frac{\partial}{\partial s}\bigg|_{s=0} \frac{\zeta_p(1,1)}{\zeta_p(1+s,1-s)}.\mathrm{R}^s_p|x|_p^{-1/2} f|_p(1) \\
&= \sum_{q\in\mathbb{Q}^*}\sum_{p\geq\eta} \frac{\partial}{\partial s}\bigg|_{s=0} \frac{\zeta_p(1,1)}{\zeta_p(1+s,1-s)}.\mathrm{R}^s_p|x|_p^{-1/2} \tilde{f}_q(1) \\
&= \sum_{q\in\mathbb{Q}^*} \frac{\partial}{\partial s}\bigg|_{s=0} \underline{\mathcal{R}}^s_{\mathbb{A}}\tilde{f}_q(1)
\end{aligned}$$

so that we can rewrite the global formula in the form

$$\hat{f}(1/2) + \hat{f}(-1/2) - \sum_{\zeta_{\mathbb{A}}(1/2+\alpha)=0} \hat{f}(\alpha) = \sum_{q\in\mathbb{Q}^*} \frac{\partial}{\partial s}\bigg|_{s=0} \underline{\mathcal{R}}^s_{\mathbb{A}}\tilde{f}(q)\,. \tag{13.5.62}$$

We can use the local translations of the Weil distribution into traces to obtain such a global translation. Using the pure basis formulation, (13.5.33) and (13.5.37), and putting

$$\varepsilon_p = \begin{cases} 1 & p \neq \eta\,, \\ 2\pi^{t/2} & p = \eta\,, \end{cases}$$

we have

$$\begin{aligned}
\sum_{p\geq\eta} \mathcal{W}_p(f) &= \sum_{p\geq\eta} \frac{\partial}{\partial t}\bigg|_{t=0} \frac{2\varepsilon_p}{\zeta_p(t,t)} \mathrm{tr}(\mathrm{e}^{-t\widetilde{\mathbb{N}}_p}\pi(f|_p)) \\
&= \sum_{q\in\mathbb{Q}^*}\sum_{p\geq\eta} \frac{\partial}{\partial t}\bigg|_{t=0} \frac{2\varepsilon_p}{\zeta_p(t,t)} \mathrm{tr}(\mathrm{e}^{-t\widetilde{\mathbb{N}}_p}\pi(f_p(qx_p))) \\
&\qquad\qquad \cdot \left[\prod_{p'\neq p} \frac{\varepsilon_{p'}}{\zeta_{p'}(t,t)} \mathrm{tr}(\mathrm{e}^{-t\widetilde{\mathbb{N}}_{p'}}\pi(f_{p'}(qx_{p'})))\right]_{t=0} \\
&= \sum_{q\in\mathbb{Q}^*} \frac{\partial}{\partial t}\bigg|_{t=0} \frac{4\pi^{t/2}}{\zeta_{\mathbb{A}}(t,t)} \mathrm{tr}\left(\mathrm{e}^{-t\sum_{p\geq\eta}\widetilde{\mathbb{N}}_p}\pi(\tilde{f}_q)\right).
\end{aligned} \tag{13.5.63}$$

Defining the global number operator by

$$\widetilde{\mathbb{N}}_{\mathbb{A}} = \sum_{p\geq\eta} \widetilde{\mathbb{N}}_p - \frac{1}{2}\log\pi\,, \tag{13.5.64}$$

and noting that

$$\frac{1}{\zeta_{\mathbb{A}}(t,t)} = -t/2[1 + \mathrm{O}(t^2)] \text{ as } t \to 0\,, \tag{13.5.65}$$

we can rewrite the global formula in the form

$$\hat{f}(1/2) + \hat{f}(-1/2) - \sum_{\zeta(1/2+\alpha)=0} \hat{f}(\alpha) = -2\sum_{q\in\mathbb{Q}^*} \operatorname*{CT}_{t=0} \operatorname{tr}(\mathrm{e}^{-t\widetilde{\mathbb{N}_{\mathbb{A}}}}\pi(\tilde{f}_q))\,, \tag{13.5.66}$$

$\mathrm{CT}_{t=0}$ denoting the constant term in the Taylor series about $t = 0$.

We can similarly use the operator

$$\mathbb{N}_{\mathbb{A}}^{1,1} = \sum_{p\geq\eta} \mathbb{N}_p^{1,1} - \log\pi \tag{13.5.67}$$

to glue together the local formulae (13.5.47) and (13.5.52) in the form

$$\sum_{p\geq\eta} \mathcal{W}_p(f) = \sum_{q\in\mathbb{Q}^*} \frac{\partial}{\partial t}\bigg|_{t=0} \frac{1}{\zeta_{\mathbb{A}}(t,t)} \operatorname{tr}\left(\mathrm{e}^{-t\mathbb{N}_{\mathbb{A}}^{1,1}}\pi^t(\tilde{f}_q)\right)\,, \tag{13.5.68}$$

so that using (13.5.65) we can rewrite the global formula as

$$\hat{f}(1/2) + \hat{f}(-1/2) - \sum_{\zeta_{\mathbb{A}}(1/2+\alpha)=0} \hat{f}(\alpha) = -1/2\sum_{q\in\mathbb{Q}^*} \operatorname*{CT}_{t=0} \operatorname{tr}\left(\mathrm{e}^{-t\mathbb{N}_{\mathbb{A}}^{1,1}}\pi^t(\tilde{f}_q)\right). \tag{13.5.69}$$

Note how the minus sign appears in (13.5.69) and (13.5.66) from (13.5.65). Putting $f \circ f^\circ$ for f, with $\hat{f}(\pm 1/2) = 0$, the Riemann hypothesis is equivalent via (13.5.66) to the positivity

$$\sum_{q\in\mathbb{Q}^*} \operatorname*{CT}_{t=0} \operatorname{tr}\left(\mathrm{e}^{-t\widetilde{\mathbb{N}_{\mathbb{A}}}}\pi(\tilde{f})\pi(\tilde{f})^*\pi(q)\right) \geq 0\,, \tag{13.5.70}$$

and similarly via (13.5.69) to the positivity

$$\sum_{q\in\mathbb{Q}^*} \operatorname*{CT}_{t=0} \operatorname{tr}\left(\mathrm{e}^{-t\mathbb{N}_{\mathbb{A}}^{1,1}}\pi^t(\tilde{f})\pi^t(\tilde{f})^*\pi^t(q)\right) \geq 0\,. \tag{13.5.71}$$

We see that expressing the Weil distribution using an *additive* convolution as in (13.5.20) opens up a whole spectrum of possibilities for expressing the Weil distribution using *traces*, all of which are normalizations of the trace of the action of the multiplicative group $\mathbb{R}^+ = \mathbb{A}^*/\mathbb{A}^{(1)}$ on the ('non-commutative') space $\mathbb{A}/\mathbb{A}^{(1)}$, or $\mathbb{P}^1(\mathbb{A})/\mathbb{A}^{(1)}$ (cf. Haran 1991).

Alain Connes has recently found an alternative way to rewrite our local formula, thus 'explaining' the positivity of the Weil distribution. Recall the $(2c^2+1)$-dimensional space $\mathscr{H}^1_\eta\langle c\rangle$ of (10.6.32). Let $\mathscr{H}^1_\eta\langle c\rangle_0 = \{\varphi \in \mathscr{H}^1_\eta\langle c\rangle \mid \varphi(0) = 0 = \mathcal{F}\varphi(0)\}$, and for $\varphi \in \mathscr{H}^1_\eta\langle c\rangle_0$ consider the fast decreasing function on $\mathbb{R}^+$

$$\mathrm{E}\varphi(a) = 2a^{1/2}\sum_{n\geq 1}\varphi(a\cdot n)\,. \tag{13.5.72}$$

We thus have a $(2c^2-1)$-dimensional subspace

$$\mathrm{E}\Big[\mathscr{H}^1_\eta\langle c\rangle_0\Big] \subseteq \mathrm{L}_2(\mathbb{R}^+, \mathrm{d}^*a)\,, \tag{13.5.73}$$

and let $\mathrm{E}^1_{\langle c\rangle}$ denote the associated projection.

For $\varphi \in \mathscr{H}^1_\eta\langle c\rangle$, supp $\varphi \subseteq [-c, c]$, and hence supp $\mathrm{E}\varphi \subseteq (0, c]$. Also, the Fourier transform $\mathcal{F}\varphi$ is essentially concentrated in $[-c, c]$; hence $\mathrm{E}\varphi(a) = \mathrm{E}\mathcal{F}\varphi(a^{-1})$ is essentially concentrated in $[c^{-1}, \infty)$. In summary, $\mathrm{E}\varphi$ is essentially supported in $[c^{-1}, c]$. Let $\mathbb{1}_{[c^{-1},c]}$ denote the characteristic function of $[c^{-1}, c]$, and view $\mathbb{1}_{[c^{-1},c]}$ as a projection on $\mathrm{L}_2(\mathbb{R}^+, \mathrm{d}^*a)$. Then essentially $\mathbb{1}_{[c^{-1},c]} \geq \mathrm{E}^1_{\langle c\rangle}$, and $\mathbb{1}_{[c^{-1},c]} - \mathrm{E}^1_{\langle c\rangle}$ is essentially positive. Connes conjectures that for $f \in \mathrm{C}^\infty_c(\mathbb{R}^+)$, with $\hat{f}(\pm 1/2) = 0$,

$$\operatorname{tr}\Big((\mathbb{1}_{[c^{-1},c]} - \mathrm{E}^1_{\langle c\rangle})\pi(f)\pi(f)^*\Big) \underset{c\to\infty}{\longrightarrow} -\sum_{p\geq\eta} \mathcal{W}_p(f\circ f^\circ)\,. \tag{13.5.74}$$

Since the operator in (13.5.74) is essentially positive, the limit $c \to \infty$ of its trace is positive, and we obtain Weil's positivity and hence the Riemann hypothesis. Since

$$\operatorname{tr}(\mathbb{1}_{[c^{-1},c]}\pi(f)) = 2(\log c)\cdot f(1)\,, \tag{13.5.75}$$

(13.5.74) is equivalent to the statement
Connes trace formula:

$$\operatorname{tr}(\mathrm{E}^1_{\langle c\rangle}\pi(f)) = 2(\log c)\cdot f(1) + \sum_{p\geq\eta} \mathcal{W}_p(f) + \circ(1), \text{ as } c\to\infty\,. \tag{13.5.76}$$

Since the cut-off basis $\varphi^1_{\eta,c,n}$ approaches the pure ($\beta = 1$, even Hermite) basis $\varphi^1_{\eta,n}$ as $c \to \infty$ (cf. (10.6.57)–(10.6.61)), one can replace the projection $\mathrm{E}^1_{\langle c\rangle}$ in (13.5.76) by the projection $\mathrm{P}^1_{\langle c\rangle}$ onto the $(2c^2-1)$-dimensional subspace of $\mathrm{L}_2(\mathbb{R}^+, \mathrm{d}^*a)$ spanned by the $\mathrm{E}\varphi(a)$, with φ varying in

$$\mathscr{H}^1_{\eta(c),0} \stackrel{\text{def}}{=} \Big\{\varphi \in \mathscr{H}^1_{\eta(c)} = \bigoplus_{0\leq n\leq 2c^2+1} \mathbb{C}\varphi^1_{\eta,n} \,\big|\, \varphi(0) = 0 = \mathcal{F}\varphi(0)\Big\}\,. \tag{13.5.77}$$

Note that the Mellin transform of $\mathrm{E}\varphi(a)$ for φ in $\mathscr{H}^1_{\eta(c),0}$ is equal to $\zeta_{\mathbb{A}}(1/2+\mathrm{i}t)\cdot\hat{g}(t)$, where $\hat{g}(t) = \tau_\eta^{1/2+\mathrm{i}t}(\varphi)$ is a polynomial in t of degree less than or equal to $2c^2+1$. Moreover, the condtions $\varphi(0) = 0 = \mathcal{F}\varphi(0)$ translate into the conditions $\hat{g}(\mathrm{i}/2) = 0 = \hat{g}(-\mathrm{i}/2)$, so that the polynomial $\hat{g}(t)$ is divisible by $(t-\mathrm{i}/2)(t+\mathrm{i}/2) = t^2+1/4$. We are thus led to consider the *global orthogonal polynomials* with respect to the measure

$$\hat{\tau}^1_{\mathbb{Q},0}(\mathrm{d}t) = \left|\left(t^2+\frac{1}{4}\right)\cdot\zeta_{\mathbb{A}}\left(\frac{1}{2}+\mathrm{i}t\right)\right|^2\frac{\mathrm{d}t}{4\pi}\,. \tag{13.5.78}$$

Note that the fast decay of the real component $\left|\zeta_\eta(1/2+\mathrm{i}t)\right|^2$ as $t \to \pm\infty$ makes the polynomials integrable with respect to $\hat{\tau}^1_{\mathbb{Q},0}(\mathrm{d}t)$, and the moment problem is determined: the polynomials are dense in $\widehat{\mathscr{H}}^1_{\mathbb{Q},0} = \mathrm{L}_2\Big(\mathbb{R}, \hat{\tau}^1_{\mathbb{Q},0}(\mathrm{d}t)\Big)$. The projection

$\widehat{\mathrm{P}}^1_{\langle c\rangle}$ onto the $(2c^2-1)$-dimensional subspace of $\widehat{\mathscr{H}}^1_{\mathbb{Q},0}$ spanned by the polynomials of degree less than or equal to $2(c^2-1)$ is given by the Christoffel–Darboux formula in terms of these global polynomials. The Riemann hypothesis can be formulated in terms of these polynomials by replacing $\mathrm{E}^1_{\langle c\rangle}$ in (13.5.76) by $\mathrm{P}^1_{\langle c\rangle}$ and passing to the analytic realization. It reads

$$\mathrm{tr}\left(\widehat{\mathrm{P}}^1_{\langle c\rangle}\hat{f}\right) = 2(\log c)f(1) + \sum_{p\geq\eta} \mathcal{W}_p(f) + \mathrm{o}(1) \text{ as } c\to\infty\,, \tag{13.5.79}$$

where $\hat{f}$ is the operator on $\widehat{\mathscr{H}}^1_{\mathbb{Q},0}$ given by multiplication by the Mellin transform $\hat{f}(it)$ of f.

We can approximate the non-commutative space $\left(\mathbb{A}, dx = \tau^1_{\mathbb{A}}\right)/\mathbb{A}^{(1)}$ by the ordinary spaces $\left(\mathbb{A}^*, \tau^\beta_{\mathbb{A}}\right)/\mathbb{A}^{(1)} = (\mathbb{R}^+, a^\beta\, d^*a/\zeta_{\mathbb{A}}(\beta))$; that is, replace the additive Haar measure $dx = \tau^1_{\mathbb{A}}$ on the adeles by the Tate measure $\tau^\beta_{\mathbb{A}}$, which for $\beta>1$ is supported on the ideles. Thus for $\beta>1$, the characteristic function B_c of the set $\{a\in\mathbb{A}^* \mid |a|\leq c\}$ gives a well-defined projection on $\mathscr{H}^\beta_{\mathbb{A}} = \mathrm{L}_2\left(\mathbb{A}^*/\mathbb{Z}^*_{\mathbb{A}}, \tau^\beta_{\mathbb{A}}\right)$. Using the Fourier–Bessel operator $\mathcal{F}^\beta_{\mathbb{A}}$ we obtain the dual projection $\widehat{\mathrm{B}}^\beta_c = \mathcal{F}^\beta_{\mathbb{A}}\mathrm{B}_c\mathcal{F}^\beta_{\mathbb{A}}$. Note that B_c (respectively $\mathcal{F}^\beta_{\mathbb{A}}$) induces a self-adjoint projection (respectively unitary) operator on the $\mathbb{Q}^+$-quotient

$$\mathscr{H}^\beta_{\mathbb{Q}} = \mathrm{L}_2\left(\mathbb{A}^*/\mathbb{Z}^*_{\mathbb{A}}\mathbb{Q}^*, \tau^\beta_{\mathbb{A}}\right) \cong \mathrm{L}_2\left(\mathbb{R}^+, \frac{a^\beta d^*a}{\zeta_{\mathbb{A}}(\beta)}\right). \tag{13.5.80}$$

Using the lifting $\tilde{f}$ of f (13.5.57) we have the operator $\pi^\beta(\tilde{f})$ acting on $\mathscr{H}^\beta_{\mathbb{A}}$

$$\pi^\beta(\tilde{f})\varphi(x) = \int_{\mathbb{A}^*} d^*a\, \tilde{f}(a)|a|^{-\beta/2}\varphi(a^{-1}x)\,.$$

The operator $\mathrm{B}_c\widehat{\mathrm{B}}^\beta_c\pi^\beta(\tilde{f})$ acting on $\mathscr{H}^\beta_{\mathbb{A}}$ is given by a kernel $\widetilde{\mathrm{K}}^{\beta,c}_f(y,x)$:

$$\begin{aligned}\mathrm{B}_c\widehat{\mathrm{B}}^\beta_c\pi^\beta(\tilde{f})\varphi(y) &= \mathrm{B}_c(y)\int_{\mathbb{A}^*}\frac{d^*z}{\zeta_{\mathbb{A}}(\beta)}|z|^\beta\mathcal{F}^\beta_{\mathbb{A}}(yz)\mathrm{B}_c(z)\int_{\mathbb{A}^*}\frac{d^*x}{\zeta_{\mathbb{A}}(\beta)}|x|^\beta\mathcal{F}^\beta_{\mathbb{A}}(zx)\pi^\beta(\tilde{f})\varphi(x)\\ &= \int_{\mathbb{A}^*}\frac{d^*x}{\zeta_{\mathbb{A}}(\beta)}|x|^\beta\varphi(x)\widetilde{\mathrm{K}}^{\beta,c}_f(y,x)\,,\end{aligned} \tag{13.5.81}$$

with

$$\widetilde{\mathrm{K}}^{\beta,c}_f(y,x) = \mathrm{B}_c(y)\int\frac{d^*z}{\zeta_{\mathbb{A}}(\beta)}|z|^\beta\mathcal{F}^\beta_{\mathbb{A}}(yz)\mathrm{B}_c(z)\int d^*a\,\tilde{f}(a)|a|^{\beta/2}\mathcal{F}^\beta_{\mathbb{A}}(zxa)\,. \tag{13.5.82}$$

$\mathrm{B}_c\widehat{\mathrm{B}}^\beta_c\pi^\beta(\tilde{f})$ commutes with the $\mathbb{Q}^+$-action, $\widetilde{\mathrm{K}}^{\beta,c}_f(y,x) = \widetilde{\mathrm{K}}^{\beta,c}_f(y\cdot q, x\cdot q)$ for $q\in\mathbb{Q}^*$, and we obtain a well-defined operator $\mathrm{B}_c\mathcal{F}^\beta_{\mathbb{Q}}\mathrm{B}_c\mathcal{F}^\beta_{\mathbb{Q}}\pi^\beta(f)$ on the $\mathbb{Q}^+$-quotient space $\mathscr{H}^\beta_{\mathbb{Q}}$; it is given by the kernel

$$\mathrm{K}^{\beta,c}_f(y,x) = \sum_{q\in\mathbb{Q}^+}\widetilde{\mathrm{K}}^{\beta,c}_f(y, x\cdot q)\,. \tag{13.5.83}$$

The trace of the operator $\mathrm{B}_c\mathcal{F}^{\beta}_{\mathbb{Q}}\mathrm{B}_c\mathcal{F}^{\beta}_{\mathbb{Q}}\pi^{\beta}(f)$ is given by

$$\begin{aligned}
\mathrm{tr}\Big(\mathrm{B}_c\mathcal{F}^{\beta}_{\mathbb{Q}}\mathrm{B}_c\mathcal{F}^{\beta}_{\mathbb{Q}}\pi^{\beta}(f)\Big) &= \int_{\mathbb{R}^+}\frac{\mathrm{d}^*x}{\zeta_{\mathbb{A}}(\beta)}|x|^{\beta}\mathrm{K}^{\beta,c}_f(x,x)\\
&= \sum_{q\in\mathbb{Q}^+}\int_{\mathbb{R}^+}\frac{\mathrm{d}^*x}{\zeta_{\mathbb{A}}(\beta)}|x|^{\beta}\mathrm{B}_c(x)\int_{\mathbb{A}^*}\frac{\mathrm{d}^*z}{\zeta_{\mathbb{A}}(\beta)}|z|^{\beta}\mathcal{F}^{\beta}_{\mathbb{A}}(xz)\mathrm{B}_c(z)\int_{\mathbb{A}^*}\mathrm{d}^*a\,\tilde{f}(a)|a|^{\beta/2}\mathcal{F}^{\beta}_{\mathbb{A}}(zxqa)\\
&\qquad\qquad (a := a\cdot q^{-1}, z := z/x)\\
&= \sum_{q\in\mathbb{Q}^+}\int_{\mathbb{A}^*}\frac{\mathrm{d}^*z}{\zeta_{\mathbb{A}}(\beta)}|z|^{\beta}\mathcal{F}^{\beta}_{\mathbb{A}}(z)\int_{\mathbb{R}^+}\mathrm{d}^*x\,\mathrm{B}_c(x)\mathrm{B}_c\Big(\frac{z}{x}\Big)\cdot\mathcal{F}^{\beta}_{\mathbb{A}}\Big(|\cdot|^{-\beta/2}\cdot\tilde{f}_q\Big)(z)\\
&= \sum_{q\in\mathbb{Q}^+}\int_{\mathbb{A}^*}\frac{\mathrm{d}^*z}{\zeta_{\mathbb{A}}(\beta)}|z|^{\beta}\mathcal{F}^{\beta}_{\mathbb{A}}(z)\cdot\Big[\mathrm{B}_{c^2}(z)\cdot\log\frac{c^2}{|z|}\Big]\cdot\mathcal{F}^{\beta}_{\mathbb{A}}\Big(|\cdot|^{-\beta/2}\tilde{f}_q\Big)(z)\\
&= \sum_{q\in\mathbb{Q}^+}\mathcal{F}^{\beta}_{\mathbb{A}}\Big[\mathrm{B}_{c^2}(z)\log\frac{c^2}{|z|}\Big]\mathcal{F}^{\beta}_{\mathbb{A}}\Big(|\cdot|^{-\beta/2}\tilde{f}_q\Big)(1)\\
&= \sum_{q\in\mathbb{Q}^+}\mathcal{F}^{\beta}_{\mathbb{A}}\Big[\log\frac{c^2}{|z|}\Big]\mathcal{F}^{\beta}_{\mathbb{A}}|\,|^{-\beta/2}\tilde{f}_q(1)+\mathcal{R}^{\beta}_c\\
&= 2(\log c)f(1)+\sum_{q\in\mathbb{Q}^+}\sum_{p\ge\eta}\mathcal{F}^{\beta}_p\Big[\log|z|_p^{-1}\Big]\mathcal{F}^{\beta}_p|\cdot|^{-\beta/2}\tilde{f}_q(1)+\mathcal{R}^{\beta}_c\\
&\overset{\mathrm{def}}{=} 2(\log c)f(1)+\mathcal{W}^{\beta}(f)+\mathcal{R}^{\beta}_c,
\end{aligned}\tag{13.5.84}$$

where the 'remainder' $\mathcal{R}^{\beta}_c$ is given by

$$\mathcal{R}^{\beta}_c = \sum_{q\in\mathbb{Q}^+}\mathcal{F}^{\beta}_{\mathbb{A}}\Big[(1-\mathrm{B}_{c^2})(z)\cdot\log\frac{c^2}{|z|}\Big]\mathcal{F}^{\beta}_{\mathbb{A}}|\cdot|^{-\beta/2}\tilde{f}(q)\,. \tag{13.5.85}$$

For a function $\mathcal{R}=\mathcal{R}(\beta,c)$ of our parameters β and c let us write $\mathcal{R}=o_{\beta,c}$ if for every $\varepsilon>0$ there exists β_ε and c_ε such that $|\mathcal{R}(\beta,c)|<\varepsilon$ for all $\beta\in(1,\beta_\varepsilon), c\in(c_\varepsilon,\infty)$; if R is also independent of β (respectively c) we write $\mathcal{R}=o_c$ (respectively o_β).

Let $\mathrm{E}^{\beta}_{\langle c\rangle}$ denote the projection onto the $(2c^2-1)$-dimensional subspace of $\mathrm{L}_2(\mathbb{R}^+,\mathrm{d}^*a)$ spanned by $a^{(\beta-1)/2}\cdot\mathrm{E}\varphi(a)$, with φ in

$$\mathscr{H}^{\beta}_{\eta(c),0}\overset{\mathrm{def}}{=}\Big\{\varphi\in\bigoplus_{n\le 2c^2}\mathbb{C}\varphi^{\beta}_{\eta,c,n}\;\Big|\;\mathcal{F}\,\mathcal{F}^{\beta}_{\eta}\varphi(0)=0=\mathcal{F}\varphi(0)\Big\}\,. \tag{13.5.86}$$

The Riemann hypothesis follows, via (13.5.76), from the following equalities:

$$\begin{aligned}
\mathrm{tr}\Big(\mathrm{E}^{1}_{\langle c\rangle}\pi^{1}(f)\Big) &= \mathrm{tr}\Big(\mathrm{E}^{\beta}_{\langle c\rangle}\pi^{\beta}(f)\Big)+o_{\beta,c} &(13.5.87)\\
&= \mathrm{tr}\Big(\mathrm{B}_c\mathcal{F}^{\beta}_{\mathbb{Q}}\mathrm{B}_c\mathcal{F}^{\beta}_{\mathbb{Q}}\pi^{\beta}(f)\Big)+o_{\beta,c} &(13.5.88)\\
&= 2(\log c)f(1)+\mathcal{W}^{\beta}(f)+o_{\beta,c} &(13.5.89)\\
&= 2(\log c)f(1)+\sum_{p\ge\eta}\mathcal{W}_p(f)+o_c\,. &(13.5.90)
\end{aligned}$$

Equation (13.5.87) follows from the continuity in β, c of the eigenfunctions $\varphi^{\beta}_{\eta,c,n}$; (13.5.89) follows from the above calculation (13.5.84) and the estimate for the remainder (13.5.85), $\mathcal{R}^{\beta}_{c} = o_{\beta,c}$; and (13.5.90) follows since $\mathcal{W}^{\beta}(f) = \mathcal{W}^{1}(f) + o_{\beta}$, and $\mathcal{W}^{1}(f) = \sum_{p\geq\eta} \mathcal{W}_p(f)$ by our global formula (13.5.59). Since the difference between the left-hand side of (13.5.87) and (13.5.90) is independent of β, we can replace $\mathrm{o}_{\beta,c}$ by o_c, obtaining Connes trace formula (13.5.76).

The equality (13.5.87 = 88) follows in the function field case since $E^{\beta}_{\langle c\rangle} \subseteq \mathrm{B}_c$, $\mathcal{F}^{\beta}_{\mathbb{Q}}\mathrm{B}_c\mathcal{F}^{\beta}_{\mathbb{Q}}$, and using the trace calculation (13.5.84), and the explicit sum, with f the characteristic function of the units, we obtain

$$\mathrm{tr}\left(\mathrm{B}_c\mathcal{F}^{\beta}_{\mathbb{Q}}\mathrm{B}_c\mathcal{F}^{\beta}_{\mathbb{Q}}\right) = 2(n-g)+1 = \dim E^{\beta}_{\langle c\rangle} \text{ for } c = q^n \text{ large}.$$

In the number field case we have to localize further in the (non-compact) dual $\mathrm{i}\mathbb{R}$ of the (non-discrete) multiplicative group $\mathbb{R}^{+} = \mathbb{Q}^{*}_{\eta}/\pm 1$.

Typically for the Riemann hypothesis, closing the gap (13.5.87 = 88) in the above proof could be a formidable task. However, as the way to a solution often yields more fruit than the proof itself, we can already see clear evidence of close connections between the Riemann hypothesis and the uncertainty principle, with its singular expressions over the real numbers.

The final proof of the Riemann hypothesis will probably require a more additive and global construction of the finite-dimensional spaces of 'band- and time-limited' functions on $\mathbb{A}/\mathbb{Q}^{*}$.

Bibliography

RECOMMENDED READING

Chapter 1	See Weil (1939). For Arakelov theory see Faltings (1992), Soulé *et al.* (1992), and Arakelov (1974).
Chapter 2	For the global theory and associated factors, see Araki and Woods (1968), Takesaki (1970), Connes (1994), and Bost and Connes (1995).
Chapter 4	For Markov chains, see Kemeny *et al.* (1966) and Doob (1959). For trees see Serre (1980) and Cartier (1972).
Chapter 5	For the Farey tree, see Huxley (1996).
Chapter 6	See Askey (1980). For a physicist's approach see Freund and Zabrodin (1992). For classical elliptic functions see Lang (1973) and Weil (1976).
Chapter 7	See Andrews and Askey (1985) and Nikirov *et al.* (1991).
Chapter 8	For the Jacobi and Laguerre polynomials, see Szegö (1959). For the finite approximations see Beckner (1975), Mueller and Weissler (1982) and Trotter (1958).
Chapter 9	See Porat (1999).
Chapter 10	For the q-Fourier–Bessel transform, see Koornwinder and Swarttouw (1992). For Bessel functions, see Watson (1922). For the pure gamma basis, $\beta = 1$, see Bump and Sz-Ng (1986). For the prolate spheroidal wave functions, see Flammer (1957), Robin (1959), Landau and Pollak (1962), Slepian and Pollak (1961) and Slepian (1964*a*). For the Askey–Wilson polynomials, see Askey and Wilson (1985).
Chapter 11	See Drinfeld (1987, 1988), Jimbo (1985), Manin (1988), and the books by Lusztig (1993) and Magid (1995). For $\mathrm{SU}_q(1,1)$, see Masuda *et al.* (1990). For analogous $q \to 0$ basis, see Kashiwara (1990, 1991).
Chapter 12	See Folland (1989), Hörmander (1979), Haran (1993b), Howe (1979, 1988), Weil (1964) and Bargmann (1961). For §12.7, see Auslander (1977), Cartier (1966), Igusa (1974) and Mumford (1983).
Chapter 13	For §13.1, see Tate (1967). For §13.2, see Weil (1952, 1972). For §13.3, see Zagier (1981). For §13.4, see Duflo and Labesse (1971), Faddeev (1967), Gelfand *et al.* (1969), Gelbart and Jacquet (1979), Helgason (1980, 1981), Jacquet and Langlands (1970), Kubota (1973), Lang (1975), Selberg (1956) and Onn (2001). For §13.5, see Berg and Frost (1975), Feller (1970), Haran (1990, 1991, 1993*a*), and the recent papers by Connes (1998) and Haran (1999).

Andrews, G. E. and Askey, R. (1985). Classical orthogonal polynomials. In *Polynômes orthogonaux et applications* (ed. C. Brezinski *et al.*), Lecture Notes in Mathematics, **1171**, pp. 36–62. Springer-Verlag, Berlin.

Arakelov, S. J. (1974). Intersection theory of divisors on an arithmetic surface. *Matematiceskii USSR Invstija*, **8**, 1167–80.

Araki, H. and Woods, E. J. (1986). A classification of factors. *Publications of the Research Institute of Mathematical Science, Kyoto University*, **4**, 51–130.

Artin, E. (1965). *Collected papers*. Addison-Wesley, Reading, MA.

Askey, R. (1980). Ramanujan's extensions of the gamma and beta functions. *American Mathematics Monthly*, **87**, 346–59.

Askey, R. and Wilson, J. (1985). Some basic hypergeometrical orthogonal polynomials that generalize Jacobi polynomials. *Memoirs of the American Mathematical Society*, 319.

Atiyah, M. F. and Bott, R. (1976). A Lefchetz fixed point formula for elliptic complexes. *Annals of Mathematics*, **86**, 374–407.

Auslander, L. (1977). *Lecture notes on nil-theta functions*, AMS Regional Conference Series in Mathematics, **34**.

Bargmann, V. (1961). On a Hilbert space of analytic functions and an associated integral transform. *Communications in Pure and Applied Mathematics*, **14**, 187–214.

Bastiaans, M. J. (1980). Gabor's expansion of a signal into Gaussian elementary signals. *Proceedings of the IEEE*, **68**(4).

Baxter, R. J. (1982). *Exactly solvable models in statistical mechanics.* Academic Press, London.

Beckner, W. (1975). Inequalities in Fourier analysis. *Annals of Mathematics*, (2), **102**, 159–82.

Berg, C. and Frost, G. (1975). *Potential theory on locally compact abelian groups.* Springer-Verlag, Berlin.

Berry, M. (1986). *Riemann's zeta function: a model of quantum chaos.* Lecture notes in Physics, **263**. Springer-Verlag, Heidelberg.

Blakadar, B. (1977). The regular representation of restricted direct product groups. *Journal of Functional Analysis*, **25**, 267–74.

Bost, J. B. and Connes, A. (1995). Hecke algebras, type III factors, and phase transitions with spontaneous symmetry breaking in number theory. *Selecta Mathematica New Series 1*, (**3**), 411–57.

Brateli, O. and Robinson, D. W. (1979). *Operator algebras and quantum statistical mechanics.* Springer-Verlag, Berlin.

Bump, D. and Sz-Ng, E. K. (1986). On Riemann's zeta function. *Mathematische Zeitschrift*, **192**, 195–204.

Cartier, P. (1966). Quantum mechanical commutation relations and theta functions. *Proceedings of the Symposium on Pure Mathematics*, **9**, AMS, 361–83.

Cartier, P. (1972). Fonctions harmoniques sur un arbre. *Symposia Mathematica*, **9**, 203–7.

Chari, V. and Pressley, A. N. (1994). *A guide to quantum groups.* Cambridge University Press.

Chui, C. K. (1992). *An introduction to wavelets.* Academic Press, New York.

Connes, A. (1994). *Non-commutative geometry.* Academic Press, San Diego and London.

Connes, A. (1996). Formule de trace en géométrie non-commutative et hypothèse de Riemann, *Comptes Rendus Académie de l' Science de Paris*, **323**, 1231–6.

Connes, A. (1998). Trace formula in noncommutative geometry and the zeros of the Riemann zeta function. Preprint, Schrödinger Institute, Vienna.

Daubechies, I. (1988). Time-frequency localization operators: a geometric phase space approach. *IEEE Transactions on Information Theory*, **34**, 605–12.

Daubechies, I. (1992). Ten lectures on wavelets. CBMS-NSF Regional Conference Series in Applied Mathematics, **61**.

Daubechies, I. and Grossmann, A. (1988). Frames of entire functions in the Bargmann space. *Communications in Pure and Applied Mathematics*, **41**, 151–64.

Daubechies, I., Jaffard, S., and Journé, J. L. (1991). A simple Wilson orthonormal basis with exponential decay. *SIAM Journal of Mathematical Analysis*, **22**, 554–72.

de Branges, L. (1968). *Hilbert spaces of entire functions*, Prentice Hall Series in Modern Analysis. Prentice Hall, New Jersey.

de Branges, L. (1972). Gauss spaces of entire functions. *Journal of Mathematical Analysis and Applications*, **37**, 1–41.

Deninger, C. (1992). Local L-factors of motives and regularized determinants. *Inventiones Mathematical*, **107**, 135–50.

Deninger, C. (1993). Evidence for a cohomological approach to analytic number theory. Preprint.

Deshouillers, J. M. and Iwaniec, H. (1982). Kloosterman sums and Fourier coefficients of cusp forms. *Inventiones Mathematic'ae*, **70**, 219–88.

Donoghue, W. F. (1969). *Distributions and Fourier transform.* Academic Press, New York.

Doob, J. L. (1959). Discrete potential theory and boundaries. *Journal of Mathematics and Mechanics*, **8**, 433–58.

Drinfeld, V. G. (1987). Quantum groups. *Proceedings of the ICM*, pp. 798–820. AMS, Providence, RI.

Drinfeld, V. G. (1988). Quantum groups. *Journal of Soviet Mathematics*, **41**, 18–49.

Drinfeld, V. G. (1989). Quasi-Hopf algebras and Knizhnik–Zamolodchikov equations. In *Problems of Modern Quantum Field Theory* (ed. A. V. Klimyk *et al.*). Springer-Verlag, Berlin.

Drinfeld, V. G. (1990). On almost cocommutative Hopf algebras and quasi-Hopf algebras. *Leningrad Mathematics Journal*, **1**, 321–342, 1419–57.

Drinfeld, V. G. (1991). On quasitriangular quasi-Hopf algebras and a certain group closely connected with Gal($\bar{\mathbb{Q}}/\mathbb{Q}$). *Leningrad Mathematics Journal*, **2**, 829–60.

Drinfeld, V. G. (1992). On the structure of the quasitriangular quasi-Hopf algebras. *Functional Analysis and Applications*, **26**, 63–5.

Duflo, M. and Labesse, J. P. (1971). Sur la formula des traces de Selberg. *Annales Scientifiques de l' École Normale Supèrieure*, **4**, 193–284.

Edwards, H. (1974). *Riemann's zeta function.* Academic Press, New York.

Eichler, M. and Zagier, D. (1985). The theory of Jacobi forms. *Progress in Mathematics*, **55**, Birkhäuser, Boston.

Erdélyi, A., Magnus, W., Oberhettinger, F. and Tricomi, F. G. (1953). *Higher transcendental functions, Bateman Manuscript Project.* McGraw-Hill, New York.

Exton, H. (1983). *q-hypergeometric functions and applications.* John Wiley, New York.

Faddeev, L. D. (1967). Expansion in eigenfunctions of the Laplace operator on the fundamental domain of a discrete group on the Lobacevskii plane. *Transactions of the Moscow Mathematical Society*, **17**, 357–86.

Faltings, G. (1992). *Lectures on the arithmetic Riemann–Roch theorem*, Annals of Mathematical Studies, **127**. Princeton University Press.

Feller, W. (1970). *An introduction to probability theory and its application.* John Wiley, New York.

Feller, W. (1991). Diffusion processes in genetics. *Proceedings of the 2nd Berkeley Symposium on Mathematical Statistics and Probability*, pp. 227–46.

Flammer, C. (1957). *Spheroidal wave functions.* Stanford University Press.

Folland, G. B. (1989). *Harmonic analysis in phase space.* Princeton University Press.

Ford, L. (1951). *Automorphic functions.* Chelsea, New York.

Freund, P. G. O. and Zabrodin, A. V. (1992). Macdonald polynomials from Sklyanin algebras: a conceptual basis for the p-adic quantum group connection. *Communications in Mathematical Physics*, **147**, 277–94.

Gelbart, S. (1976). *Weil's representation and the spectrum of the Metaplectic group*, Lecture Notes in Mathematics, **530**. Springer-Verlag, Berlin.

Gelbart, S. and Jacquet, H. (1979). Forms on GL(2) from the analytic point of view. *Proceedings of the Symposia of Pure Mathematics*, **33**, 213–15.

Gelfand, I. M., Graev, M., and Piateski-Shapiro, I. I. (1969). *Automorphic functions and representation theory.* W. B. Saunders, Philadelphia.

Goldfeld, D. (1989). Explicit formulae as trace formulae. In *Number theory, trace formulas and discrete groups* (ed. K. E., Aubert, E. Bombieri, and D. Goldfeld), pp. 281–8, Academic Press.

Goldfeld, D. (1994). *A spectral interpretation of Weil's explicit formula*, Lecture Notes in Mathematics, **1593**, pp. 135–52, Springer-Verlag, Berlin.

Guillemin, V. (1977). Lectures on spectral theory of elliptic operators. *Duke Mathematical Journal*, **44**(3), 485–517.

Guillemin, V. and Sternberg, S. (1977). *Geometric asymptotics*, AMS Mathematical Surveys, **14**.

Haran, S. (1990). Riesz potentials and explicit sums in arithmetic. *Inventiones Mathematical*, **101**, 697–703.

Haran, S. (1991). Index theory, potential theory, and the Riemann hypothesis. In *L-functions and arithmetic*, Durham, 1990, *LMS Lecture Notes*, **153**, pp. 257–70. Academic Press, London.

Haran, S. (1993*a*). Analytic potential theory over the p−adics. *Annales de l' Institut Fourier*, **43**(4), 905–44.

Haran, S. (1993*b*). Quantization and symbolic calculus over the p−adic numbers. *Annales de l' Institut Fourier*, **43**(4), 997–1053.

Haran, S. (1993*c*). On Riemann's zeta function and the mysteries of the prime at infinity. Preprint.

Haran, S. (1999). On Riemann's zeta function. *Proceedings of the Conference on Dynamical, Spectral, and Arithmetic Zeta Functions, Contemporary Mathematics* (ed. M. van Frankenhuysen and L. Lapidus). AMS, Providence, RI.

Hardy, G. H. (1940). *Ramanujan.* Chelsea, New York.

Hardy, G. H. (1966, 1979). *Collected papers.* Oxford University Press.

Hardy, G. H. and Ramanujan, S. (1918). Asymptotic formula in combinatory analysis. *Proceedings of the London Mathematical Society* (2), **17**, 75–115.

Hejhal, D. (1976, 1983). *The Selberg trace formula for* $\mathrm{PSL}_2(\mathbb{R})$, Lecture Notes in Mathematics, **548**, p. 1001. Springer, New York.

Helgason, S. (1980). *The Radon transform.* Birkäuser, Boston.

Helgason, S. (1981). *Topics in harmonic analysis on homogeneous spaces.* Birkhäuser, Boston.

Hermite, C. (1905–1917). *Oeuvres.* Gauthier-Villars, Paris.

Hörmander, L. (1979). The Weyl calculus of pseudodifferential operators. *Communications in Pure and Applied Mathematics*, **32**, 359–443.

Howe, R. (1979). Theta series and invariant theory. *Proceedings of the Symposia on Pure Mathematics*, **33**, AMS, pp. 275–85.

Howe, R. (1988). The oscillator semigroup. *Proceedings of the Symposia on Pure Mathematics*, **48**, AMS, pp. 61–132.

Huxley, M. N. (1996). *Area, lattice points and exponential sums*, LMS Monographs, New Series **13**. Clarendon Press, Oxford.

Igusa, J. I. (1974). *Theta functions.* Springer, New York.

Igusa, J. I. (1978). *Lectures on forms of higher degree.* Tata Institute, Bombay.

Jackson, F. H. (1910). On q-definite integrals. *Quarterly Journal of Pure and Applied Mathematics*, **41**, 193–203.

Jacobi, C. G. J. (1969). *Collected works.* Chelsea, New York.

Jacquet, H. and Langlands, R. P. (1970). *Automorphic forms on GL(2)*, Lecture Notes in Mathematics, **114**. Springer-Verlag, Berlin.

Janssen, A. J. E. M. (1988). The Zak transform: a signal transform for sampled time-continuous signals. *Philips Journal of Research*, **43**, 23–69

Jimbo, M. (1985). A q-difference analog of U(g) and the Yang–Baxter equation. *Letters in Mathematical Physics*, **10**, 63–9.

Kac, M. (1947). Random walk and Brownian motion. *American Mathematics Monthly*, **54**, 369–91.

Kac, M. (1959). *Statistical independence in probability, analysis, and number theory.* Carus Mathematical Monographs **12**, John Wiley, New Jersey.

Kashiwara, M. (1990). Crystallizing the q-analogue of universal enveloping algebras. *Communications in Mathematical Physics*, **133**, 249–60.

Kashiwara, M. (1991). On crystal bases of the q-analogue of universal enveloping algebras. *Duke Mathematical Journal*, **63**, 465–516.

Katz, N. and Sarnak, P. (1999). Zeros of zeta functions, their spacing and spectral nature. *Bulletin of the AMS*, **36**, 1–26.

Kemeny, J. G., Snell J. L., and Knapp, A. W. (1966). *Denumerable Markov chains.* Van Nostrand, New York.

Koelink, H. T. and Koornwinder, T. H. (1989). The Clebsch–Gordon coefficients for the quantum group SU(2) and q-Hahn polynomials, *Nederlands Akademie Wetenschap Proceedings Series A*, **92**, 443–56.

Koornwinder T. H. (1989). Representations of the twisted SU(2) quantum group and some q-hypergeometric orthogonal polynomials, *Nederlands Akademie Wetenschap Proceedings Ser. A*, **92**, 97–117.

Koornwinder, T. H. (1990). Orthogonal polynomials in connection with quantum groups. *Orthogonal Polynomials: Theory and Practice* (ed. P. Nevai), NATO ASI Series C, **294**, 257–292.

Koornwinder, T. H. and Swarttouw, R. F. (1992). On q-analogues of the Fourier and Hankel transforms. *Transactions of the American Mathematical Society*, **333**, 445–61.

Kubota, T. (1973). *Elementary theory of Eisenstein series.* Halsted Press, New York.

Kuznetsov, N. V. (1981). Petersson's conjecture for cusp forms of weight zero and Linnik's conjecture, sums of Kloosterman sums. *Mathematiceskii USSR Sbornik*, **39**, 299–342.

Laguerre, E. (1972). *Collected works.* Chelsea, New York.

Landau, H. J. and Pollak, H. O. (1962). Prolate spheroidal wave functions, Fourier analysis, and uncertainty II, III. *Bell Systems Technical Journal*, **40** (1961), 65–84, 1295–336.

Lang, S. (1973). *Elliptic functions.* Addison-Wesley, Reading, MA.

Lang, S. (1975). $SL(2,\mathbb{R})$. Addison-Wesley, Reading, MA.

Lang, S. (1990). *Cyclotomic fields* (2nd edn). Springer-Verlag, Berlin.

Laumon, G. (1987). Transformation de Fourier, constantes d'équations fonctionelles et conjecture de Weil. *Publication de IHES*, **65**, 131–210.

Lusztig, G. (1993). *Introduction to quantum groups.* Birkhäuser, Boston.

Maass, H. (1964). *Lectures on modular functions of one complex variable.* Tata Institute, Bombay.

Macdonald, I. G. (1972). Affine root systems and Dedekind's eta function. *Inventiones Mathematical*, **15**, 91–143.

Magid, S. (1995). *Foundations of quantum group theory.* Cambridge University Press.

Manin, Yu. I. (1988). Quantum groups and non-commutative geometry. Technical Report, Centre de Recherches Mathematique, Montreal.

Manin, Yu. I. (1991). Three-dimensional hyperbolic geometry as ∞-adic Arakelov geometry. *Inventiones Mathematical*, **104**, 223–44.

Masuda, T., Mimachin, K., Nakagami, Y., Noumi, M., and Kono, K. (1988). Representation of quantum groups and q-analogue orthogonal polynomials, *Comptes Rendus de l' Acadèmie de Sciences de Paris, Serie I, Mathematique*, **307**, 559–64.

Masuda, T., Mimachi, K., Nakagami, Y., Noumi, M., Saburi, Y., and Ueno K. (1990). Unitary representations of the quantum group $SU_q(1,1)$, I & II, *Letters in Mathematical Physics*, **19**, 187–194, 195–204.

Masuda, T., Mimachi, K., Nakagami, Y., Noumi, M., and Ueno, K. (1991). Representations of the quantum group $SU_q(2)$ and the little q-Jacobi polynomials, *Journal of Functional Analysis*, **99**, 357–86.

Moak, D. S. (1981). The q-analogue of the Laguerre polynomials. *Journal of Mathematical Analysis and Applications*, **81**, 20–47.

Montgomery, H. (1973). The pair correlation of zeros of the zeta function. *Analytic number theory*, AMS, Providence, RI.

Mueller, C. E. and Weissler, F. B. (1982). Hypercontractivity for the heat semigroup for ultraspherical polynomials and on the n-sphere. *Journal of Functional Analysis*, **48**, 252–83.

Mumford, D. (1983–1991). *Tata Lectures on Theta.* Birkhäuser, Boston.

Nikirov, A. F., Suslov, S. K., and Uvarov, V. B. (1991). *Classical orthogonal polynomials of a discrete variable.* Springer-Verlag, Berlin.

Onn, Y. (2001). Thesis. Technion—Israel Institute of Technology.

Polya, G. (1974). *Collected papers.* MIT Press, Cambridge, MA.

Porat, H. (1999). Thesis. Technion—Israel Institute of Technology.

Porat, O. (1994). Thesis. Technion—Israel Institute of Technology.

Ramanujan, S. (1962). *Collected papers.* Chelsea, New York.

Riemann, B. (1953). *Collected works.* Dover, New York.

Robin, L. (1959). *Fonctions sphériques de Legendre et fonctions sphéroidales*, Collection Technique et Scientifique du CNET. Gauthiers-Villars, Paris.

Ruskai, M. B., Beylkin, G., Coifman, R., Daubechies, I., Mallat, S., Meyer, Y., and Raphael, L. (1992). *Wavelets and their application.* Jones and Baetlett, Boston.

Selberg, A. (1956). Harmonic analysis and discontinuous groups. *Journal of the Indian Mathematical Society*, **20**, 47–87.

Selberg, A. (1989). *Collected papers.* Springer-Verlag, Heidelberg.

Serre, J. P. (1977). *Linear representations of finite groups*, GTM **42**, Springer-Verlag, New York.

Serre, J. P. (1980). Trees. Springer-Verlag, Berlin.

Sklyanin, E. K. (1982). Some algebraic structures connected with the Yang–Baxter equation. *Functional Analysis and Applications*, **16**, 263–70; **17**, 273–84.

Slepian, D. (1964*a*). Prolate spheroidal wave functions, Fourier analysis and uncertainty IV: Extensions to many dimensions; generalized prolate spheroidal functions. *Bell Systems Technical Journal*, 3009–57.

Slepian, D. (1964*b*). Some asymptotic expansions for prolate spheroidal wave functions, Journal of Mathermatical Physics, **44**(2), 99–140.

Slepian, D. (1977). Prolate spheroidal wave functions, Fouries analysis, and uncertainty V: the discrete case. *Bell Systems Technical Journal*, **57**(5), 1371–430.

Slepian, D. and Pollak, H. O. (1961). Prolate spheroidal wave functions, Fourier analysis and uncertainty. *Bell Systems Technical Journal*, **40**, 43–63.

Soibelman, Ya. S. and Vaksman, L. L. (1992). On some problems in the theory of quantum groups, representation theory and dynamical systems. *Advances in Soviet Mathematics*, **9**, 3–55, AMS.

Soulé, C., Abramovich, D., Burnol, J. F., and Kramer, J. (1992). *Lectures on Arakelov geometry*, Cambridge Studies in Advanced Mathematics, **33**. Cambridge University Press.

Stanton, D. (1984). Orthogonal polynomials and Chevalley groups. In *Special functions: Group theoretical aspects and applications* (ed. R. Askey, T. H. Koornwrinder, and W. Schenpp). Reidel, Dordrecht.

Stieltjes, T. J. (1889). Recherches sur les fractions continues. *Annales de Faculté de Science, Toulouse*, **3**, H1–17; **8**, J1–122; **9**, A1–47.

Szegö, G. (1959). *Orthogonal polynomials*, AMS Colloquium Publication **XXIII**.

Taibleson, M. (1975). *Fourier analysis on local fields.* Princeton University Press.

Takesaki, M. (1970). *Tomita's theory of modular Hilbert algebras and its applications.* Lecture Notes in Mathematics **128**. Springer-Verlag, Berlin.

Tate, J. (1967). Fourier analysis in number fields and Hecke's zeta function. In *Algebraic Number Theory* (ed. J. W. S. Cassels and A. Fröhlich). Academic Press, London.

Tchebychef, P. L. (1961). *Collected works.* Chelsea, New York.

Terras, A. (1985). *Harmonic analysis on symmetric spaces and applications.* Springer-Verlag, Berlin.

Titchmarsh, E. C. (1951). *The theory of Riemann's zeta function.* Oxford University Press.

Trotter, H. F. (1958). Approximation of semi-groups of operators. *Pacific Journal of Mathematics*, **8**, 887–919.

Ueno, K. (1990). Spectral analysis for the Casimir operator on the quantum group $SU_q(1,1)$. *Proceedings of the Japan Academy*, **66**, Series A, 42–4.

Vaksman, L. L. (1989). q-analogues of Clebsch–Gordan coefficients and the algebra of functions on the quantum group SO(2). *Soviet Mathematics Doklady*, **39**, 467–70.

Vaksman, L. L. and Korogodskii, L. I. (1991). Spherical functions on the quantum group SU(1,1) and the q-analogue of the Mehler–Fock formula. *Functional Analysis and Applications*, 48–49.

Vilenkin, N. J. (1968). *Special functions and the theory of group representations*, trans. Mathematical Monographs **22**. AMS, Providence, RI.

Watson, G. N. (1922). *Theory of Bessel functions.* Cambridge University Press.

Watson, G. N. (1934). Notes on generating functions of polynomials. Jacobi polynomials. *Journal of the London Mathematical Society*, **9**, 22–8.

Weil, A. (1939). Sur l'analogie entre les corps de nombres algébriques et les corps de fonctions algébriques. Revue Scientifique, **77**, 104–6.

Weil, A. (1952). Sur les fromules explicites de la théorie des nombres premiers. *Proceedings of the Royal Physiographical Society Lund*, **21**, 252–65.

Weil, A. (1964). Sur certains groupes d'opérateurs unitaires. *Acta Mathematica*, **111**, 143–211.

Weil, A. (1966). Fonction zeta et distributon. Séminaire Bourbaki, **312**.

Weil, A. (1972). Sur les formules explicites de la théorie des nombres. *Isvestiya matematiceskii Nauk*, **36**, 3–18.

Weil, A. (1974). Basic number theory. Springer, New York.

Weil, A. (1976). Elliptic functions according to Eisenstein and Kronecker. Springer, New York.

Weyl, H. (1950). *The theory of groups and quantum mechanics.* Dover, New York.

Zagier, D. (1981). Eisenstein series and the Riemann zeta function. In *Automorphic forms, representation theory and arithmetic*, Bombay Colloquium. Springer-Verlag, Berlin.

Index

addition identity 43 (6.3.31)
adic limit 37 (6.2.8)
annihilation operator 81 (8.3.37), 151 (12.4.15), 155 (12.5.3), 165 (12.6.29)
antipode 135 (11.3.5)
Arakelov 4–5
Askey–Wilson basis, polynomials 127 (10.8.9), 200 (13.4.35)$_\eta$
automorphic factor 189 (13.4.3)
automorphic forms 207 (13.4.101)
  on the Heisenberg group 167–72 (12.7)

Bargmann–Fock model 150
basic basis, q- 105 (10.4.2)
  η- 107 (10.4.2)$_\eta$
  p- 107
basic chain 44 (7.1.1)
Bessel function, J- 101 (10.2.4), 102 (10.2.7)
  K- 181 (13.3.1)
Bessel semi-group 182 (13.3.11)
beta chain, p-adic 28 (4.4.3), 29 (Fig. 4.3)
  p-adic symmetric 24 (4.3.2), 26 (Fig. 7.2)
  q- 49 (7.2.7), 50 (Fig. 7.3)
  q-symmetric 48 (7.2.3), 48 (Fig. 7.2)
  η, real 65 (8.1.2), 66 (Fig. 8.1)
beta distribution, beta measure 11 (3.2.1)
beta function 36 (6.1.16)
  higher dimensional 87 (9.2.6)
beta integral 11 (3.2.3)
beta sum, q- 37 (6.3.1), 38 (6.3.2)
  symmetric 38 (6.3.6)
binom, q- 39 (6.3.8)
  coefficients 39 (6.3.10), 87 (9.2.5)
  operator 40 (6.3.18)
Braid group 140 (11.4.11–12)
  relation 140 (11.4.9)

Cayley transform 14 (3.3.3), 70 (8.2.15)
chain
  basic 44 (7.1.1)
  beta, p-adic 29 (Fig. 4.3)
    p-adic symmetric 26 (Fig. 4.2)
      q- 49 (7.2.7), 50 (Fig. 7.3)
      q-symmetric 47 (7.2.1), 48 (Fig 7.2)
    higher dimensional p-adic 91 (9.2.32)
      q 87 (9.2.2)
      real 91 (9.2.37)
  direct product 18 (4.1.7)
  q-gamma 45 (7.1.8)
  p-gamma 22 (4.2.3), 23 (Fig. 4.1)
  non-stop 18 (4.1.9)
  semi-direct product 18 (4.1.8)
  universal covering 19 (4.1.11)
Cebotarev density theorem vi (0.4)
central limit theorem 7, 81
characteristic polynomial 114 (10.6.17)

Chebyshev polynomial 217 (13.5.48)
Chu–Vandermonde 40 (6.3.15)
  dual 40 (6.3.15)′
q-commutativity 40 (6.3.17)
compact form, real limit 202 (13.4.76)
complex structure 150 (12.4.5)
Connes trace formula 222 (13.5.76)
constant term 195 (13.4.43)
continued fraction 32–3 (5.3)
creation operator 81 (8.3.37), 151 (12.4.14), 155 (12.5.3), 165 (12.6.29)
cusp forms 204 (13.4.87)
cut-off basis, *see*: prolate spheroidal wave functions

Dirichlet problem 209 (13.5.15)
dual orthogonality 139 (11.3.22)
duality identity 40 (6.3.16)

Eisenstein integral 195 (13.4.44)
Eisenstein series 183 (13.3.13), 184 (13.3.23), 205 (13.4.89)
elliptic curves 36–7 (6.2), 125, 171, 184
  isogeny of 36 (6.2.3), 124–5, 200 (13.4.70)
ergodic action 9, 16
Euler's constant 179 (13.2.12)
explicit sums 176–7 (13.2.4)
  see also: local and global formula
q-exponential 41 (6.3.24)

q-factorial 39 (6.3.10)
Farey tree 33 (Fig. 5.1)
flag space 39 (6.3.12)
functional equation of theta and zeta 175

gamma chain, q- 45 (7.1.8)
  p- 22 (4.2.3), 23 (Fig. 4.1)
gamma distribution 7 (2.2.5)
q-gamma function 35 (6.1.11)
gamma integral 7 (2.2.1)
q-gamma sum 38 (6.3.5)
Gaussian 7 (2.2.2)
generating function
  for p-adic Laguerre basis 24 (4.2.19)
  for real Laguerre basis 75 (8.2.42–3)
  for Hermite basis 82 (8.3.41)
global formula 219 (13.5.59), 220 (13.5.62)
  via pure basis 221 (13.5.66)
  via beta chain 221 (13.5.69)
global orthogonal polynomials 222 (13.5.78)
graph 17
Grassmannian 39 (6.3.12)
Green kernel 19 (4.1.17)
  see also: Martin kernel

Hahn basis, p-adic 30 (4.4.10)
  q- 58 (7.3.21)
  real 69 (8.2.11)
  continuous symmetric polynomials 123 (10.7.6)
    q- 127 (10.8.12)
Hankel inversion 101 (10.2.2)
harmonic function 19 (4.1.16), 209 (13.5.15)
harmonic measure 20 (4.1.21)
heat kernel
  for p-adic Laguerre basis 24 (4.2.20)
  for real Laguerre basis 76 (8.2.44)
  for Hermite basis 82 (8.3.42)
Hecke operator 185 $(13.3.24)_p$, 196 (13.4.52)
q-Heisenberg algebra 108 (10.4.14–17)
Heisenberg relations 81 (8.3.38), 151 (12.4.16)

Heisenberg relations up the ladder
  for Askey–Wilson basis 126 (10.8.8)
  for basic basis 108 (10.4.16), (10.4.22)
  for elementary symmetric functions 81 (8.3.32)
  for q-Hahn basis 54 (7.3.12)
  for real Hahn basis 69 (8.2.9)
  for multidimensional q-Hahn basis 89 (9.2.17)
  for multidimensional real Hahn basis 93 (9.2.46)
  for analytic Jacobi basis 123 (10.7.8)
  for q-Jacobi basis 59 (7.3.28)
  for real Jacobi basis 70 (8.2.18)
  for multidimensional q-Jacobi basis 90 (9.2.27)
  for multidimensional real Jacobi basis 94 (9.2.54)
  for q-Laguerre basis 62 (7.3.41), 63 (7.3.49)
  for real Laguerre basis 73 (8.2.29)
  for pure gamma basis 115 (10.6.20)
  for weight basis 132 (11.2.12)
Hermite polynomials 82 (8.3.39), 155 (12.5.4), 156, 214 (13.5.34), 222 (13.5.77)
highest weight module 131 (11.2.7)
highest weight vector 131 (11.2.1–2)
hyperbolic Laplacian 185 $(13.3.24)_\eta$, 198 (13.4.64), 199 $(13.4.64)_\eta$, 201 $(13.4.64)_q$
hypercontractivity 83 (8.3.45)
hypergeometric functions, q- 43 (6.3.32)

intertwining operator 190 (13.4.14)
isogeny, *see*: elliptic curves

Jacobi basis, p-adic 30 (4.4.9)
  p-adic symmetric 123 (10.7.4)
  q- 59 (7.3.29)
  real 71 (8.2.20)
Jacobi group 171 (12.7.35)
Jacobi polynomials 71 (8.2.20)

Laguerre basis, p-adic 23 (4.2.11)
  q- 63 (7.3.50)
  real 73 (8.2.31)
Laguerre polynomials 73 (8.2.31)
Laplacian 209 (13.5.17)
  see also: hyperbolic Laplacian
local formula 209 (13.5.20), 211 $(13.5.20)'$, 212 (13.5.23)
  via beta chain 217 (13.5.47), 218 (13.5.52)
  via cut-off 213 (13.5.29)
  via intertwining operator 217 (13.5.43)
  via pure basis 214 (13.5.33)
localization operator 149 (12.3.33)
logarithmic Sobolev inequality 83 (8.3.44)

Markov chain 17 (4.1.1)
  see: chain
Martin kernel 19 (4.1.18)
  for beta chain, p-adic symmetric 25 (4.3.6)
    q- 52 (7.2.15)
    real 67 (8.1.9)
    multidimensional, q- 88 (9.2.11)
    multidimensional, real 92 (9.2.43)
  for gamma chain, p-adic 22 (4.2.6)
    q- 47 (7.1.14)
  for real units 79 (8.3.14)
Martin metric 19 (4.1.19)
Mautner–Cartier polynomials 197 (13.4.55), 194 $(13.4.35)_p$
Mehler–Fock kernel 82 (8.3.42), 214 (13.5.34)
Meixner–Pollaczeck polynomials 113 (10.6.10)
  q- 128 (10.8.15)
metaplectic group 171 (12.7.34)
metric on $\mathbb{Q}_p^{\oplus N}$ 4 (1.3.1)
  on $\mathbb{P}^1(\mathbb{Q}_p)$ 10 (3.1.2)
q-Möbius sum 39 (6.3.9)
modular curve 171 (12.7.36), 183 (13.3.17)

normal law 209 (13.5.16)
number operator 21 (4.1.29)
q-numbers 34 (6.1.4)
  symmetric 34 (6.1.1)

operator *q*-binom 40 (6.3.18)
orthogonal group 4 (1.3.3), 14 (3.3.4)

path space 208 (13.5.11)
Plancherel measure, $p \geq \eta$ 194 (13.4.38)
  p-adic 197 $(13.4.58)_p$
  q- 200 $(13.4.38)_q$
  real 199 $(13.4.58)_\eta$
Poisson summation formula 162 (12.6.10)
potential kernel 208 (13.5.10)
prime number theorem vi (0.5)
prolate spheroidal wave functions 116
  asymptotic behaviour 121 (10.6.65), 122 (10.6.67)
  q-difference equation 119 (10.6.46)
  differential equation 119 (10.6.52)
  integral equation 118 (10.6.38)
pure basis, gamma, *q*- 113 (10.6.6), 127 (10.8.15)
    p-adic 113 (10.6.8)
    real 113 (10.6.10)
  beta, *q*- 127 (10.8.12)
    p-adic 123 (10.7.2)
    real 123 (10.7.6)

quantum Yang–Baxter equation 139 (11.4.8)

R matrix 139 (11.4.3)
Radon transform 195 (13.4.43)
Ramanujan 42 (6.3.27), 181 (13.3.5)
random walk 17 (4.1.3)
  over a chain 17 (4.1.4)
Rankin–Selberg method 185
real limit 37 (6.2.9)
Riemann hypothesis (equivalent formulations) 33 (5.3.7), 177 (13.2.5–6), 180 (13.2.20–21), 185 (13.3.26), 188 (13.3.44), 221 (13.5.70–1), 222 (13.5.74), (13.5.76), 223 (13.5.79), 224 (13.5.87=88).
Riesz, potential 207 (13.5.1)
  kernel 207 (13.5.2)
  reproduction formula 209 (13.5.18)

Schrödinger model 144 (12.1.17)
spherical function 194 (13.4.32), (13.4.35)
standard metric 4 (1.3.1)
Stirling's formula 7 (2.1.7)
Stone–von Neumann theorem 143, 146
symmetric stable semi-group 208 (13.5.8)
symplectic Fourier transform 146 (12.2.6)

Tate distribution 7 (2.2.4),
  q- 98 (10.1.1)
theta function 36 (6.2.4), 42 (6.3.30), 43 (6.3.31), 169 (12.7.18–20), 170 (12.7.24–5), 175 (13.1.15)
timed chain 20
Toeplitz algebra 134 (11.2.27)
Toeplitz operator 11 (3.1.8)
twisted convolution 145 (12.2.3)
twisted multiplication 146 (12.2.8)

uncertainty principle 148 (12.3.22)
unit shift 20 (4.1.22)
unitary group 5
universal covering 19 (4.1.11)
universal R matrix 139 (11.4.3)

vacuum, *q*- 98 (10.1.2)

wave packet expansion 149 (12.3.27)
Weber's first formula 101 (10.2.2)
  second formula 107 $(10.4.11)_\eta$
Weil–Berezin model 164 (12.6.24–5)
Weil distribution 177 (13.2.4)
  beta 178 (13.2.8–9)
  gamma 178 (13.2.10)
  see also: local and global formulae
Weil representation 171 (12.7.34)
Weyl quantization 146 (12.2.7)
Wigner transform 148 (12.3.14)

Yang–Baxter equation 139 (11.4.8)

zeta function 6 (2.1.1), 175
  p- 166 (12.6.38)

Beta chains

	p (symmetric)		q	η
finite approximations:	$X^{\square}_{p(N)}$	$(X_{p(N)})$	$X^{\square}_{q(N)}$	$X_{\eta(N)}$
measure:	$\tau^{(\alpha)\beta}_{p(N)}$	$\left(\tau^{\alpha,\beta}_{p(N)}\right)$	$\tau^{(\alpha)\beta}_{q(N)}$	$\tau^{\alpha,\beta}_{\eta(N)}$
	25 (4.3.5)		51 (7.2.10)	65 (8.1.4)
Hilbert space:	$\mathscr{H}^{(\alpha)\beta}_{p(N)}$	$\left(\mathscr{H}^{\alpha,\beta}_{p(N)}\right)$	$\mathscr{H}^{(\alpha)\beta}_{q(N)}$	$\mathscr{H}^{\alpha,\beta}_{\eta(N)}$
	30 (4.4.8)		52 (7.3.1)	68 (8.2.1)
Hahn basis:	$\varphi^{(\alpha)\beta}_{p(N),m}$	$\left(\varphi^{\alpha,\beta}_{p(N),\pm,m}\right)$	$\varphi^{(\alpha)\beta}_{q(N),m}$	$\varphi^{\alpha,\beta}_{\eta(N),\,m}$
	30 (4.4.10)	(113 (10.6.9))	58 (7.3.22)	69 (8.2.11)
boundary:	$\partial X^{\square}_{p}$	(∂X_p)	$\partial X^{\square}_{q}$	∂X_η
harmonic measure:	$\tau^{(\alpha)\beta}_{p}$	$\left(\tau^{\alpha,\beta}_{p}\right)$	$\tau^{(\alpha)\beta}_{q}$	$\tau^{\alpha,\beta}_{\eta}$
	29	(11 (3.2.1))	52 (7.2.16)	11 (3.2.1)
Hilbert space:	$\mathscr{H}^{(\alpha)\beta}_{p}$	$\left(\mathscr{H}^{\alpha,\beta}_{p}\right)$	$\mathscr{H}^{(\alpha)\beta}_{q}$	$\mathscr{H}^{\alpha,\beta}_{\eta}$
	30 (4.4.7)	(12 (3.2.10))	52 (7.3.1)	12 (3.2.10)
Jacobi basis:	$\varphi^{(\alpha)\beta}_{p,m}$	$\left(\varphi^{\alpha,\beta}_{p,\pm,m}\right)$	$\varphi^{(\alpha)\beta}_{q,m}$	$\varphi^{\alpha,\beta}_{\eta,m}$
	30 (4.4.9)	(123 (10.7.4))	59 (7.3.29)	71 (8.2.20)

The higher dimensional analogues of the above are all in §9.2.

The symmetric boundary spaces have analytic realizations:

	p	q	η
measure:	$\hat{\tau}^{\alpha,\beta}_{p}$		$\hat{\tau}^{\alpha,\beta}_{\eta}$
	110 (10.5.2)		
Hilbert space:	$\widehat{\mathscr{H}}^{\alpha,\beta}_{p}$	12 (3.2.10)	$\widehat{\mathscr{H}}^{\alpha,\beta}_{\eta}$
Pure beta basis:	$\hat{f}^{\alpha,\beta}_{p,\pm,m}$	$\hat{f}^{\alpha,\beta}_{q,n}$	$\hat{f}^{\alpha,\beta}_{\eta,m}$
	123 (10.7.2)	127 (10.8.12)	123 (10.7.6)

Gamma chains

	p	q	η
finite approximations:	$X^o_{p(N)}$	$X^o_{q(N)}$	
measure:	$\tau^\beta_{p(N)}$	$\tau^\beta_{q(N)}$	
	22 (4.2.5)	44 (7.1.4)	
Hilbert space:	$\mathscr{H}^\beta_{p(N)}$	$\mathscr{H}^\beta_{q(N)}$	
	23 (4.2.12), 116 (10.6.25)	61 (7.3.38)	
basis:	$\varphi^\beta_{p(N),m}$	$\varphi^\beta_{q(N),m}$	
	24 (4.2.13)	62 (7.3.43)	
boundary:	∂X^o_p	∂X^o_q	∂X^o_η
measure:	$\tau^\beta_{\mathbb{Z}_p}$	$\tau^\beta_{\mathbb{Z}_q}$	$\tau^\beta_{\mathbb{Z}_\eta}$
	7 (2.2.5)	47 (7.1.15)	7 (2.2.5)
Hilbert space:	$\mathscr{H}^\beta_{\mathbb{Z}_p}$	$\mathscr{H}^\beta_{\mathbb{Z}_q}$	$\mathscr{H}^\beta_{\mathbb{Z}_\eta}$
	8 (2.2.8)	62 (7.3.46)	8 (2.2.8)
Laguerre basis:	$\varphi^\beta_{\mathbb{Z}_p,m}$	$\varphi^\beta_{\mathbb{Z}_q,m}$	$\varphi^\beta_{\mathbb{Z}_\eta,m}$
	23 (4.2.11)	63 (7.3.51)	73 (8.2.31)
		Tate measure (analytic)	
	$\tau^\beta_p \quad \left(\hat{\tau}^\beta_p\right)$	$\tau^\beta_q \quad \left(\hat{\tau}^\beta_q\right)$	$\tau^\beta_\eta \quad \left(\hat{\tau}^\beta_\eta\right)$
	7 (2.2.4)	98 (10.1.1)	7 (2.2.4)
Hilbert space:	$\mathscr{H}^\beta_p \quad \left(\widehat{\mathscr{H}}^\beta_p\right)$	$\mathscr{H}^\beta_q \quad \left(\widehat{\mathscr{H}}^\beta_q\right)$	$\mathscr{H}^\beta_\eta \quad \left(\widehat{\mathscr{H}}^\beta_\eta\right)$
	8 (2.2.8)	98 (10.1.6) (99 (10.1.12))	8 (2.2.8)

	p	q	η
Pure gamma basis:	$\varphi^{\beta}_{p,\pm,m}$ $\left(\hat{f}^{\beta}_{p,\pm,m}\right)$	$\left(\hat{f}^{\beta}_{q,\pm,m}\right)$ $\left(\hat{f}^{\beta}_{q,m}\right)$	$\varphi^{\beta}_{\eta,m}$ $\left(\hat{f}^{\beta}_{\eta,m}\right)$
	113 (10.6.9) (113 (10.6.8))	(113 (10.6.6)) (128 (10.8.15))	114 (10.6.12) (113 (10.6.10))
related spaces and bases:			
Laguerre basis:	$\mathscr{H}^{\beta}_{\mathbb{Z}_p}, \varphi^{\beta}_{\mathbb{Z}_p,m}$	$\phi_{\mathbb{Z}_q} \cdot \mathscr{H}^{\beta}_{\mathbb{Z}_q}, \phi_{\mathbb{Z}_q} \cdot \varphi^{\beta}_{\mathbb{Z}_q,m}$	$\phi_{\mathbb{Z}_\eta} \cdot \mathscr{H}^{\beta}_{\mathbb{Z}_\eta}, \phi_{\mathbb{Z}_\eta} \cdot \varphi^{\beta}_{\mathbb{Z}_\eta,m}$
basic basis:	$\mathscr{H}^{\beta}_{\mathbb{Q}_p/\mathbb{Z}_p}, \phi_{p,m}$	$\mathscr{H}^{\beta}_{\mathbb{Q}_q/\mathbb{Z}_q}$, $\phi_{(q),m}$	$\mathscr{H}^{\beta}_{\mathbb{Q}_\eta/\mathbb{Z}_\eta}$, $\phi_{\eta,m}$
		105 (10.4.4) 105 (10.4.2)	107 $(10.4.4)_\eta$ 107 (10.4.9)
weight basis:	$\mathcal{V}^{\beta}_{p}, v^{\beta}_{p,m}$	$\mathcal{V}^{\beta}_{q}, v^{\beta}_{q,m}$	$\mathcal{V}^{\beta}_{\eta}, v^{\beta}_{\eta,m}$
		131 (11.2.7)	
cut-off basis:	$\mathscr{H}^{\beta}_{p(N)}$, $\varphi^{\beta}_{p(N),m}$	$\mathscr{H}^{\beta}_{q,c}$, $\varphi^{\beta}_{q,c,m}$	$\mathscr{H}^{\beta}_{\eta,c}$, $\varphi^{\beta}_{\eta,c,m}$
	116 (10.6.25), 23 (4.2.11)	118 (10.6.34), 118 (10.6.38)	116 (10.6.28), 117 (10.6.31)

Models for the Heisenberg group representation

	p-adic	η, *real*	$\mathbb{A}$, *adelic*
Schrödinger:	$L_2(\mathbb{Q}_p)$	$L_2(\mathbb{Q}_\eta)$	$L_2(\mathbb{A})$
		144 (12.1.17)	
local lattice:	$L_2(V(\mathbb{Q}_p))_{\phi_p}$	$L_2(V(\mathbb{Q}_\eta))_{\phi_\eta}$	$L_2(V(\mathbb{A}))_{\phi_{\mathbb{A}}}$
	150 (12.4.3)	150 (12.4.8)	
global lattice:	$\mathscr{H}^{(p)}_{V(\mathbb{Q})}$	$\mathscr{H}^{(\eta)}_{V(\mathbb{Q})}$	$L_2(V(\mathbb{A}))_{\delta_{V(\mathbb{Q})}}$
	165 (12.6.30–4)	164 (12.6.24–6)	161 (12.6.2–3)